The Anatomy of the Human Embryo

A Scanning Electron-Microscopic Atlas

Dedicated to the memory of
the great anatomists and embryologists
Wilhelm His (1831–1904)
and
Erich Blechschmidt (1904–1992)

Gerd Steding

The Anatomy of the Human Embryo

A Scanning Electron-Microscopic Atlas

818 figures, 2009

KARGER

Basel · Freiburg · Paris · London · New York · Bangalore ·
Bangkok · Shanghai · Singapore · Tokyo · Sydney

Dr. med. Gerd Steding
Former Professor of Anatomy and
Head of the Department of Embryology
Centre of Anatomy
University of Göttingen
Germany

Library of Congress Cataloging-in-Publication Data

Steding, Gerd.
The anatomy of the human embryo : a scanning electron-microscopic atlas / Gerd Steding.
p. ; cm.
Includes bibliographical references.
ISBN 978-3-8055-8361-9 (hard cover : alk. paper)
1. Embryology, Human--Atlases. 2. Scanning electron microscopy--Atlases. I. Title.
[DNLM: 1. Embryo, Mammalian--anatomy & histology--Atlases. 2. Microscopy, Electron, Scanning--Atlases. QS 617 S812t 2009]
QM602.S738 2009
612.6'40222--dc22
2008026596

www.karger.com
Printed in Switzerland on acid-free and non-aging paper (ISO 9706)
by Reinhardt Druck, Basel
ISBN 978–3–8055–8361–9

Contents

Introduction

In this atlas, the forms of the major human organs during the early weeks of prenatal development are depicted. Illustrations of the external form of human embryos have been published in great numbers and are to be found in numerous textbooks dealing with human embryology. However, we predominantly find a description of organ development with pictures of two-dimensional histological sections or with more or less schematic drawings often made according to historical originals. It is not rare to find in textbooks, even today, pictures of animal embryos as material for human development. Since it has been shown that results for animal embryos cannot always be simply and uncritically applied to human development, in this atlas of human embryology, only original images of the major organs of human embryos during the first few weeks of prenatal development are shown.

For technical reasons, modern developmental biology is almost exclusively concerned with animal embryos. However, it would be somewhat irresponsible to expect that the use of these new procedures could replace the investigation of human embryos with the traditional tools for the description of form. Sometimes, however, it seems that nowadays only molecular biological techniques and methods of developmental biology are considered to be a suitable approach to understanding developmental processes, whereas the classical morphological methods and procedures are increasingly seen as historical and, thereby, as hardly capable of producing good results. Regrettably, with this approach, it is often ignored that reports on the biochemical processes taking place during development can hardly provide a hint as to the ways and means in which they influence form. The fact that the steps taking place at the molecular level during the development of a leg of a mouse or chicken are to a large extent similar, however, offers no explanation as to how the final results of these processes are so different in form. To establish under what circumstances and conditions the arm, for example, of a human being is given its characteristic form, first precise knowledge of the course of development is necessary. We need to know which steps are involved, from the development of the anlage to the formation of the 'finished' human form.

Only when the course of each individual step in these external changes in form has been understood, can there be a chance of also estimating the molecular incidents in the stages of the development of form.

Therefore, even nowadays, the classical observation, description and analysis of form, its development and change cannot be ignored altogether, neither within the framework of developmental biological nor embryological medical investigations. In this atlas, all that is depicted are those changes in position and form of the organs, which are visible during embryonic development. It is obvious that, together with these visible changes, a succession of processes in submicroscopic molecular dimensions, which are not directly visible, take place.

There has been much conjecture as to the relationship, so to speak, between macroscopic and submicroscopic processes and much has been alleged or even postulated without there having been any success in disproving any one of the hypotheses. Therefore, I restrict myself solely to illustrating phenomena of human organ development which are directly visible and I will keep my views on theories and hypotheses to myself. The reader's concept of human organ development will, on the basis of our specimens, hopefully become more vivid and plastic.

Just as in the anatomical dissection course the organs concealed under the skin are exposed and become visible by dissection, layer for layer, the organs of an embryo can also be exposed and, thereby, made visible in their position and spatial form. As in an anatomical atlas the position and form of human organs are shown, in this atlas the changes in position and form of some of the human organs during prenatal development are seen in photographs of the specimens.

For the selection of the photographs, I made sure that the developmental stages were as close together as possible to make the changes in shapes of the organs clear and comprehensible. To demonstrate the relationships in form and position more reliably than the standard aspects are able to do, images of the same organs photographed from different directions are shown. This cannot, of course, be equally successful for every organ, simply because the small size of some objects sometimes makes unexpectedly high demands on the patience and dexterity of the author.

It would hardly be surprising if some of the photographs shown here might appear peculiar and strange to the experienced observer of histological and schematic illustrations in textbooks. This could be due to the fact that these specimens also do not appear to be as simple and familiar as the two-dimensional sections and repeatedly drawn schemata so often seen. As in everyday life, also in prenatal development, it is often the inconspicuous events which are most important. Temporally close stages may sometimes appear hardly distinguishable from each other and only upon more exact inspection can a characteristic developmental step be recognized.

Even if it is customary to speak of 'the' human embryo, it must be remembered that, already in his prenatal life, the human being is a unique and unmistakable individual. As far as possible, I have also tried to keep this aspect of the individual development of form in mind. Since the growth of the embryo is precisely in the early weeks a fundamental factor in the development of form, the pictures of all stages of an organ are, in general, shown at the same magnification.

The introductory text to each chapter hardly offers experts anything new. It cannot and should not replace a textbook, but rather explains, only briefly, some of the most significant developmental steps documented in the illustrations. To instruct or teach the readers is not the aim of this atlas, rather it simply offers them an opportunity to develop a more precise conception of the anatomy of the human embryo through the power of their own observations.

In this context, I must accept the fact that the choice of photographs and their number cannot satisfy every reader in equal measure and that unnecessary or even absent aspects may be criticized. However, I have endeavoured to give not only human embryologists but also readers on the periphery of human embryology and interested physicians and also laymen an opportunity to make their own observations.

As expected from an atlas, the pictures of the specimens are what is most important. Some of these pictures offer an unusual aspect and often also only one small detail from a region. In order to make orientation in the pictures easier for those who are strangers to human embryology, some explanatory terms have been added in small frames with deliberately simple and small line drawings. These should not distract the reader from looking at the photographs and cannot replace the photographs. In a few large format photographs, abbreviations refer to important terms.

To avoid burying the photographs beneath nomenclature, only those illustrations in a chapter have been labelled whose items are not self-explanatory from the previous figure. The nomenclature is oriented on the anatomy of adults as far as possible and on everyday speech and only when unavoidable have technical terms been employed.

Materials and Procedures

The embryos shown here were collected and dissected over a period of more than 25 years. I am very grateful to many colleagues, among them friends in many parts of the world, for their willingness to provide me with their material.

Human beings are not sacrificed especially for the procurement of cadavers for anatomical dissection courses and, for the same reasons, human embryos are neither killed nor bred for the sake of investigation in the field of human embryology. All of the embryos, without exception, were obtained from legal or medically indicated abortions and were all exclusively employed for the investigations described here. For these reasons, the early stages of human development are not depicted since only embryos of at least 4 weeks of age are available.

Following the employment of various techniques, the embryos were fixed in glutaraldehyde or paraformaldehyde, partly postfixed in Bouin's solution, then dehydrated in an ascending alcohol series and dried with CO_2 using the critical point method. The embryos were mounted on a specimen support and were cool-sputtered with gold or gold-palladium to a thickness of 30 nm.

When the surface layers of a specimen had been photographically documented with the scanning electron microscope (Zeiss Novascan or DSM 960 Zeiss), the deeper layers were exposed with the help of specially made microinstruments, the specimens resputtered and re-photographed with the scanning electron microscope. Dependent on the region under investigation, up to ten layers were exposed and photographed for each embryo.

Age Determination

It is usually not possible to determine the exact developmental age of an embryo because the time point of fertilization of the egg remains unknown. However, estimation of age based on measurements of the greatest length of the embryo or the crown-rump length was attempted. Since, however, embryos shrink by up to 20% as a result of fixation and the following dehydration dependent upon the techniques employed and, moreover, the extent of shrinkage is dependent on the length of time spent in the various solutions and, of course, just as for adults, there are shorter and taller human embryos, the evaluation of the length cannot be a highly reliable technique for the evaluation of the age.

Streeter's classification of embryos[1] using his term, developmental horizons, was modified by O'Rahilly and Müller[2] who used the term, developmental stages. The disadvantage of this system for the classification into stages is that for the first 8 embryonic weeks only 23 stages are suggested.

Due to the low number of stages in this system, the single stages are not characterized by typical developmental states of all organs, only reached at this stage, but include quite a lot of different maturation states of the organs. Since not all organ systems develop concurrently, but rather, for example, the heart can be somewhat further developed than the stomach, the use of developmental stages seems impractical.

Therefore, only approximate age estimations are given in weeks, estimated from an assumed time point of fertilization of the egg. The classification of the embryo according to developmental weeks is made on the basis of a comparison of the devel-

---

1 Streeter GL: Developmental Horizons in Human Embryos. Carnegie Inst Wash Publ Contrib Embryol 1942;30:211–245; 1945;31:27–63; 1948;32:133–203; 1949;33:149–167; 1951;34:165–196.

2 O'Rahilly R, Müller F: Developmental Stages in Human Embryos, Including a Revision of Streeter's 'Horizons' and a Survey of the Carnegie Collection. Washington, Carnegie Institution of Washington, 1987, Publ No 637.

opmental state of as many embryonic organs as possible with embryos of tolerably exact developmental ages described in the literature.

Since, in obstetrics, for the determination of the age of the embryo the 'first day of the last period' is taken to be the beginning of the pregnancy, the gynaecological age is approximately 2 weeks more than the embryological age given here.

Acknowledgements

I am very grateful to all the co-workers of my old department for their work in collecting and preparing the material for this atlas.

In particular, Hans-Georg Sydow deserves respect and thanks for his untiring and conscientious efforts in collecting and preparing the samples for the scanning electron microscopy and his unique dexterity in the difficult job of making the microinstruments for the dissection. I am also very grateful for his invaluable help in the solution of difficult technical problems with the scanning electron microscope. Furthermore, in completing the scanning of the negatives, his work proved to be excellent and of inestimable value.

I would also like to thank Kirsten Falk-Stietenroth for her photographic work in the darkroom which she carried out with unsurpassed conscientiousness, great commitment and unique aesthetic competence.

I would like to express my gratitude to Anja Aue for taking on the building up of the archives of the specimens and the photographs with great thoroughness, thus making an important contribution to the maintenance of order.

I am also grateful to Cyrilla Maelicke for translating the text.

And, finally, I would like to thank Dr. Jörg Männer who as acting head of my old department generously made it possible for me to complete this atlas in my old familiar setting after my retirement. Also following the restructuring of the departments, I was able to continue my work calmly and without interruption in the Centre of Anatomy, for which I expressly thank Dr. Viebahn.

Finally, I am highly indebted to the head of the publishing house and all members of staff for their meticulous reproduction of the substantial number of illustrations and the outstanding design of the atlas.

1

External Aspects

1.
External Aspects

1.1 The External Form of the Embryo

According to the regulations applied here to obtain the specimens, human embryos of the first 3 weeks could not be obtained. Furthermore, the modern techniques of abortion usually do not provide the conditions necessary for obtaining uninjured specimens.

Gaps are consequently inevitable and, therefore, the stages presented here are barely sufficient to give a detailed outlook of the development of the external appearance of human embryos.

Week 4

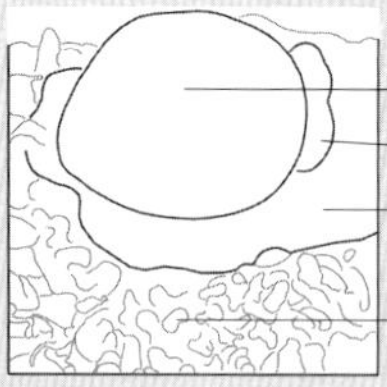

1.
External Aspects

1.1
The External Form of the Embryo

1.1.01–1.1.13
Development of the external form of the embryo. Lateral views.

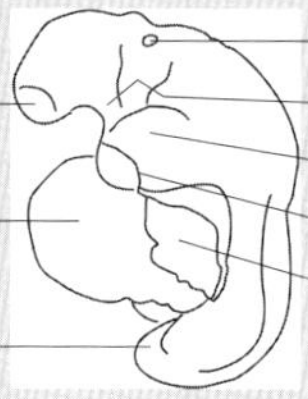

Week 5

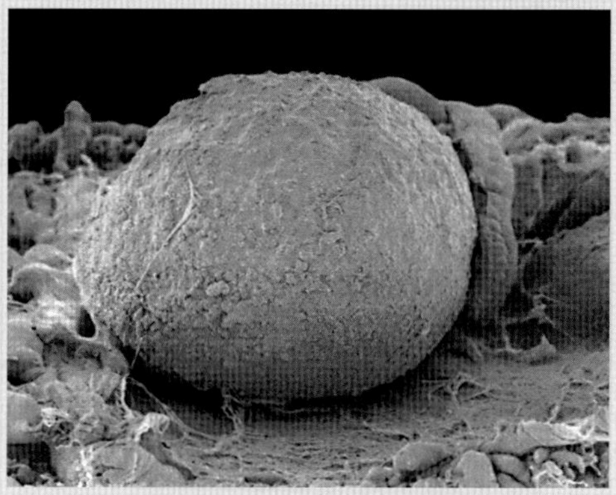

1.1.01 Embryo at the beginning of week 4. The amnion has been removed.

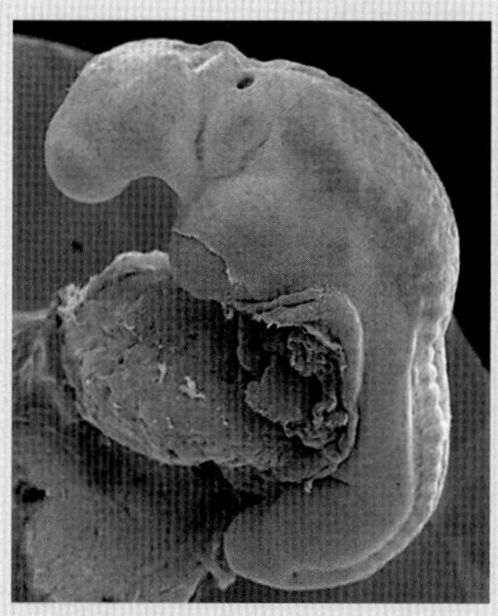

1.1.02 Embryo at the end of week 4. The umbilical vesicle is partly collapsed.

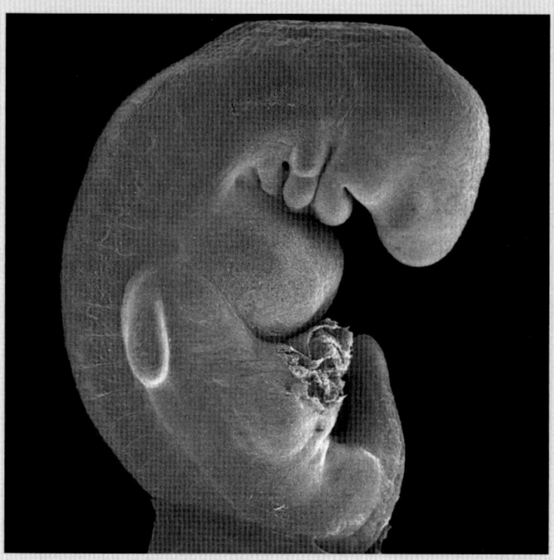

1.1.03 Embryo at week 5.

Week 5

1.1.01–1.1.13
Development of the external form of the embryo. Lateral views.

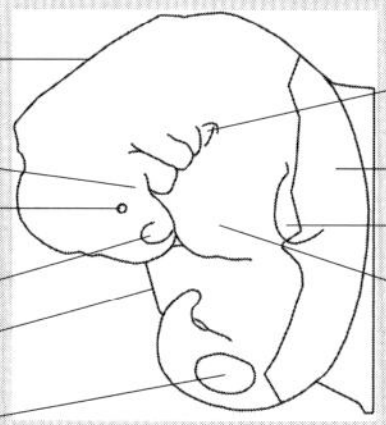

1.1.04

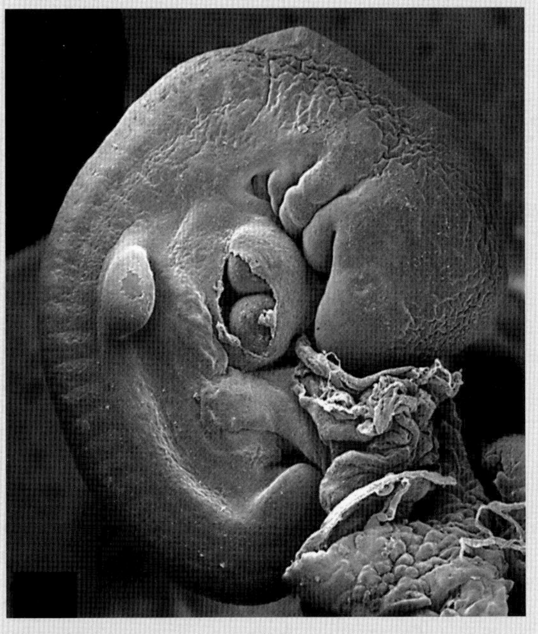

Embryos at week 5.

1.1.05

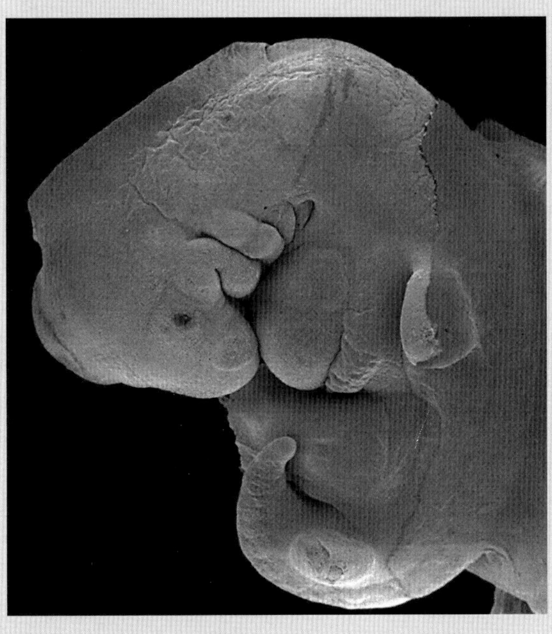

Week 5

1.
External Aspects

1.1
The External Form of the Embryo

1.1.01–1.1.13
Development of the external form of the embryo. Lateral views.

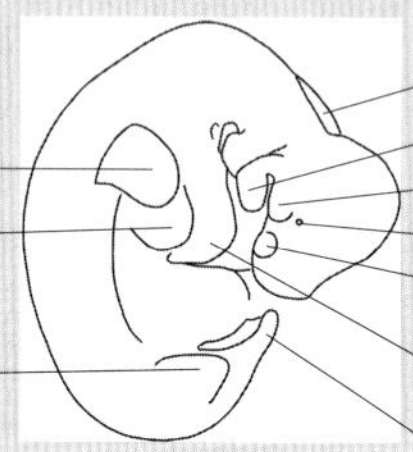

Week 6

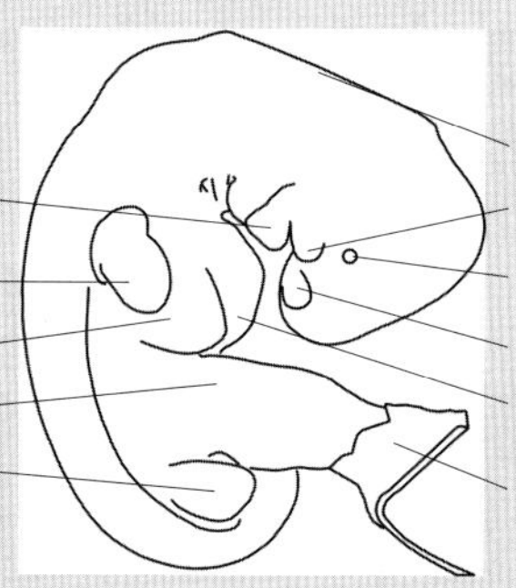

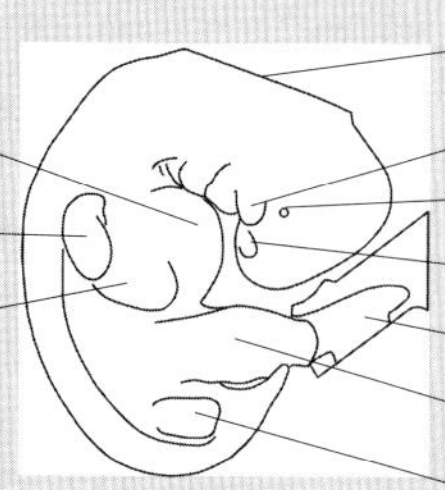

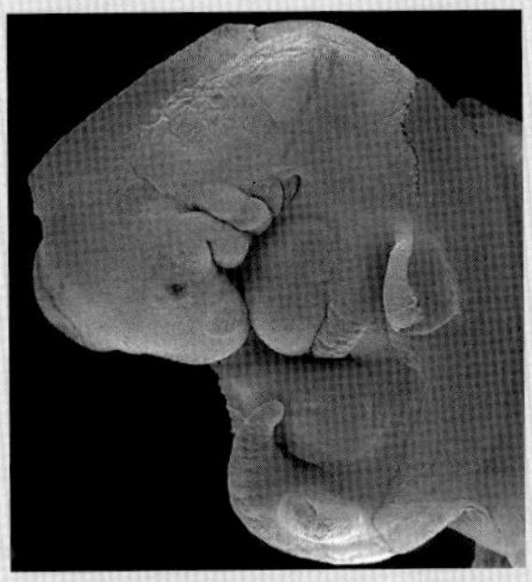

1.1.05a Same embryo as in figure 1.1.05.

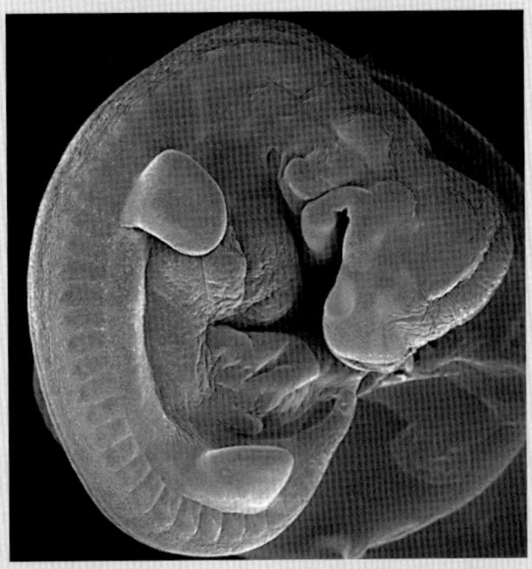

1.1.06 Embryo at the end of week 5.

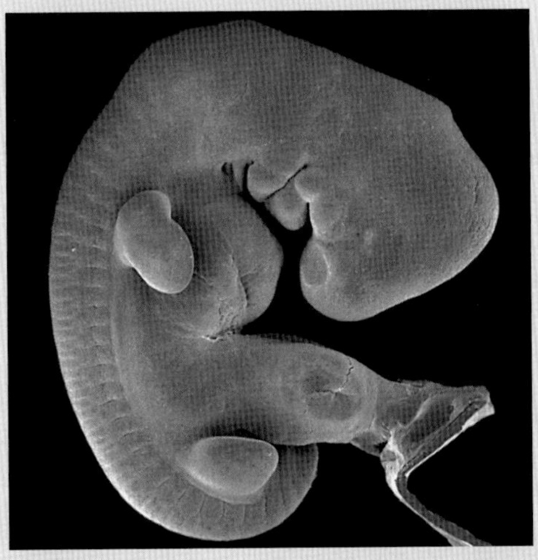

1.1.07 Embryos at week 6.

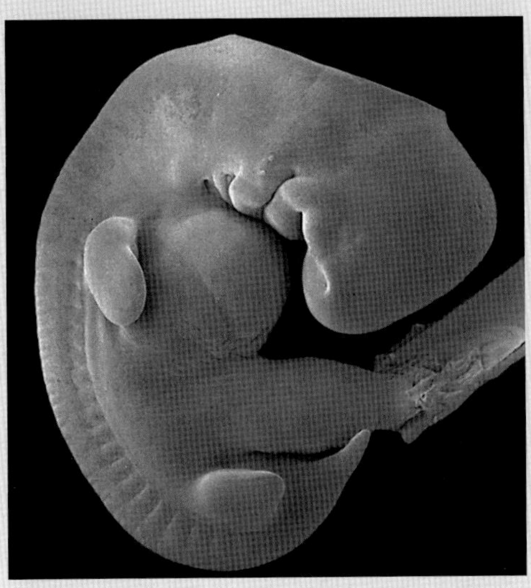

1.1.08

1.1.01–1.1.13

Development of the external form of the embryo. Lateral views.

Week 6

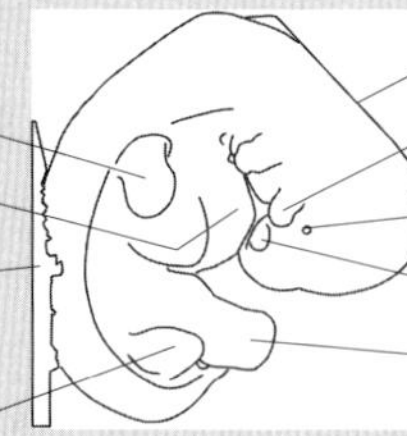

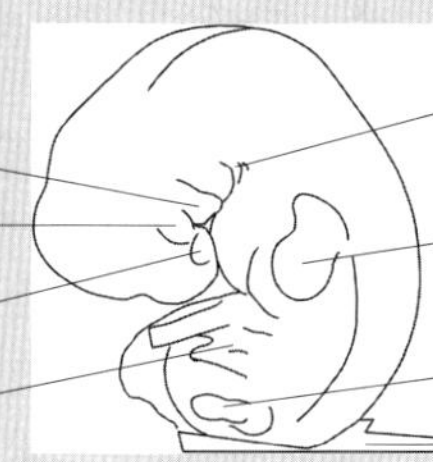

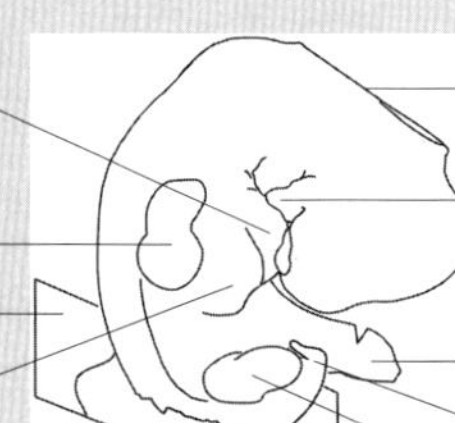

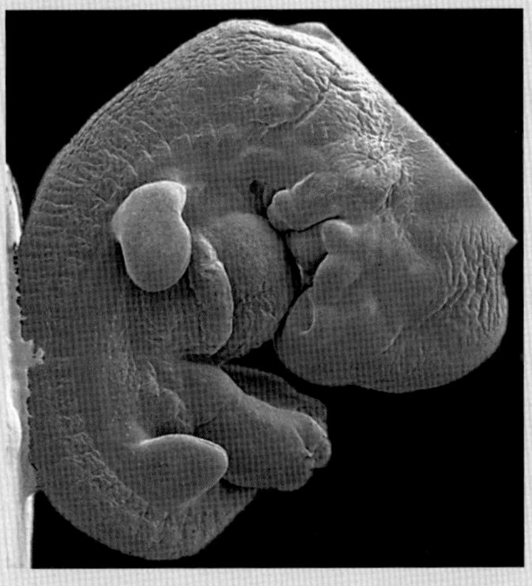

1.1.09 Embryo at week 6.

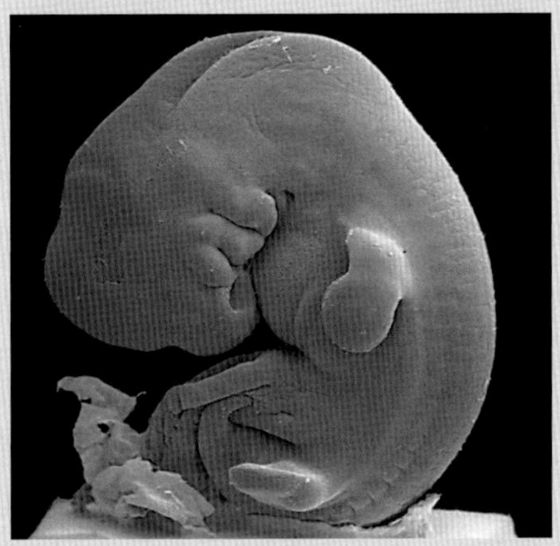

1.1.10 Embryos at the end of week 6.

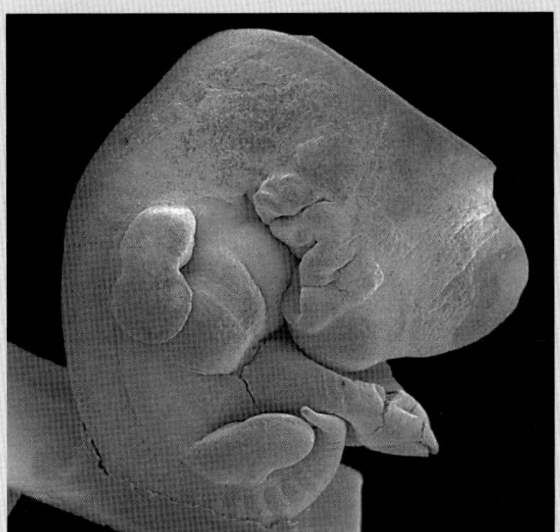

1.1.11

1.1.01–1.1.13

Development of the external form of the embryo. Lateral views.

Week 6

Week 7

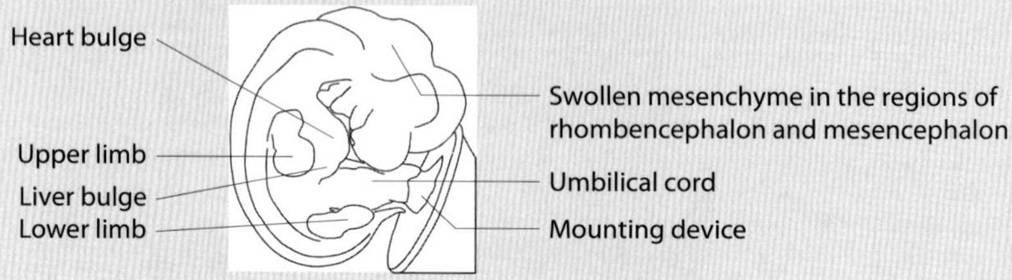

Week 8

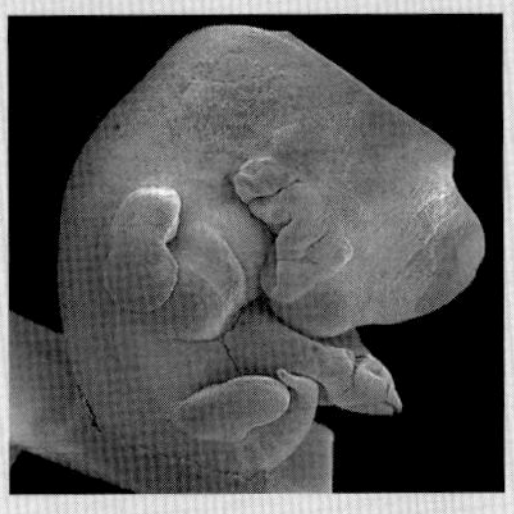

1.1.11a Same embryo as in figure 1.1.11.

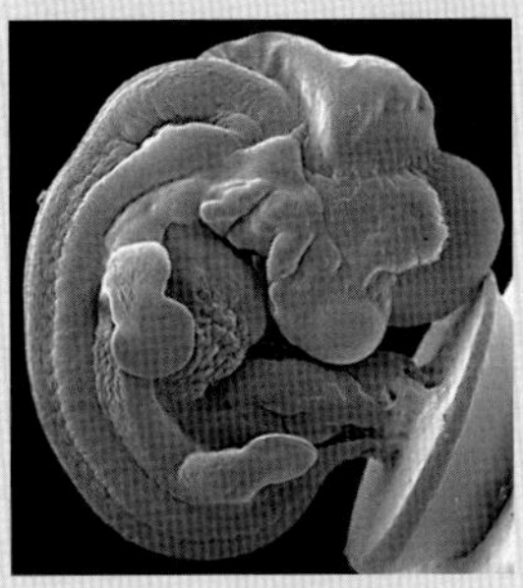

1.1.12 Embryo at the end of week 7.

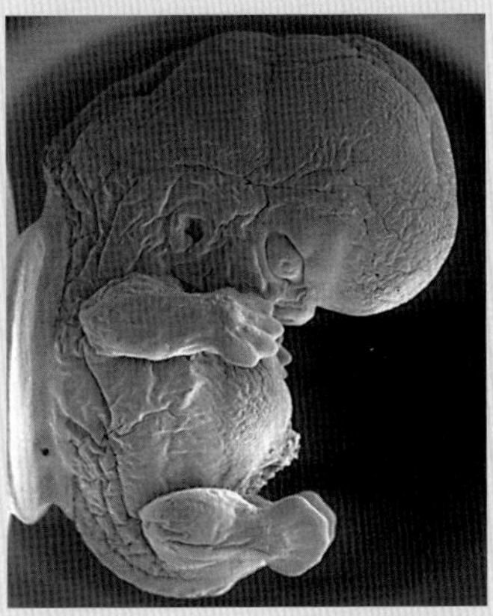

1.1.13 Embryo at week 8.

1.
External Aspects

1.2
The Development of the Face

From a morphological point of view the adult face can be seen as a system of manifold high reliefs and deepenings. In the young embryo the foldings of the face are more simple: there are neither nose nor chin, neither eyes nor cheeks (fig. 1.2.01). In week 4, we only find the deepening of the future mouth bordered by the prominences of the first pharyngeal arches and the prominence of the lower side of the brain (fig. 1.2.02, 1.2.18).

After the maxillary eminences have been formed (fig. 1.2.05, 1.2.06), the foldings of the nasal placodes develop (fig. 1.2.05–1.2.09). The nasal placode deepens to form the nasal pit which is bordered by the foldings of the medial and the lateral nasal prominences (fig. 1.2.09–1.2.12). Now we find a clear-cut upper lip moulded by the right and the left maxillary eminences and the medial nasal prominences (fig. 1.2.11, 1.2.12). It should be mentioned that the medial and the lateral nasal prominences and the maxillary eminence are in contact with each other from the very outset and do not fuse (fig. 1.2.07–1.2.14).

Due to the growth of the brain the eyes change their position primarily at the lateral side of the head (fig. 1.2.19–1.2.21) and become ventrally transposed (fig. 1.2.11–1.2.17). The integration of the eyes into the front of the face has now accomplished the formation of the supraorbital (eyebrows) and infraorbital (maxillary) eminences (fig. 1.2.15–1.2.17, 1.2.28, 1.2.29).

Meanwhile the entire face has changed its originally horizontal oval shape into a longitudinal oval by integration of the forehead and considerable longitudinal growth (fig. 1.2.01, 1.2.17). The forward growth of the external nose and also of the mandible (pharyngeal arch I) completes the transformation of the embryonic face into a face of the adult type (fig. 1.2.33–1.2.37).

1.2.01–1.2.17
Developmental stages of the face. Ventral view.

Week 4

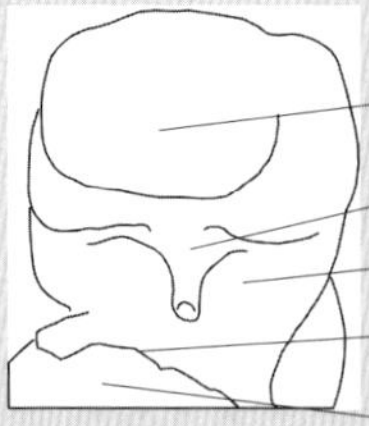

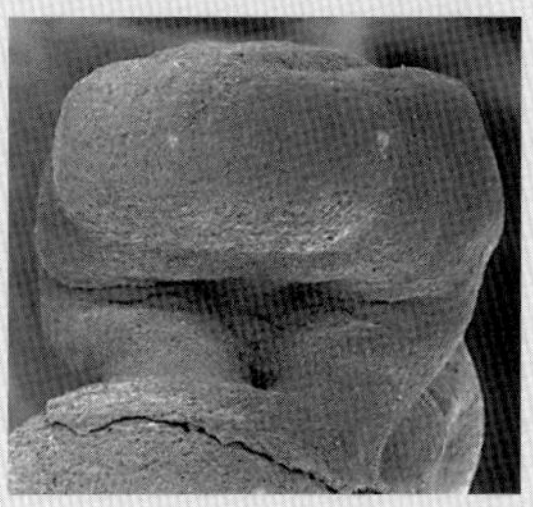

1.2.01

Embryos at week 4.

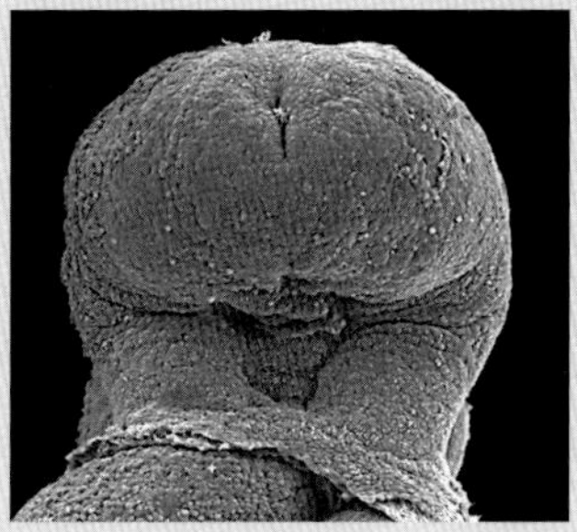

1.2.02

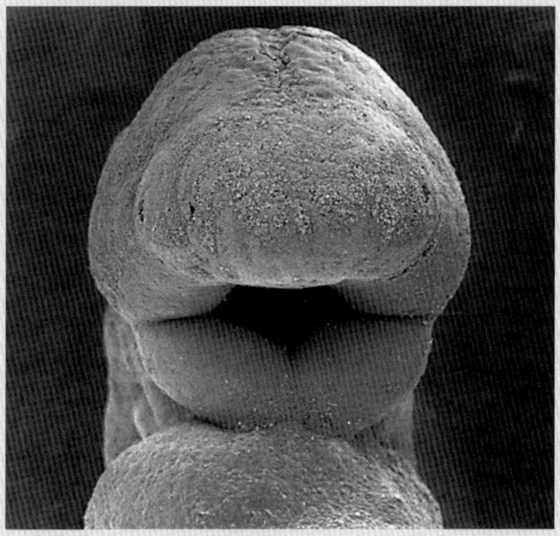

1.2.03

Week 4

Week 5

1.2.01–1.2.17

Developmental stages of the face. Ventral view.

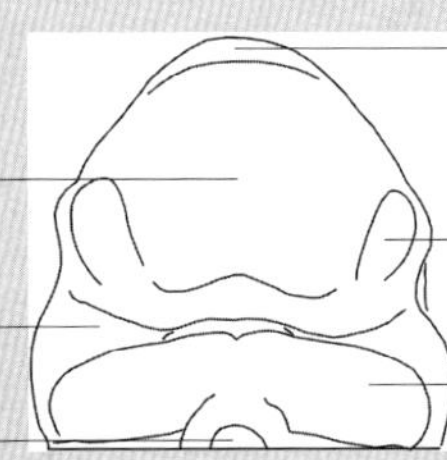

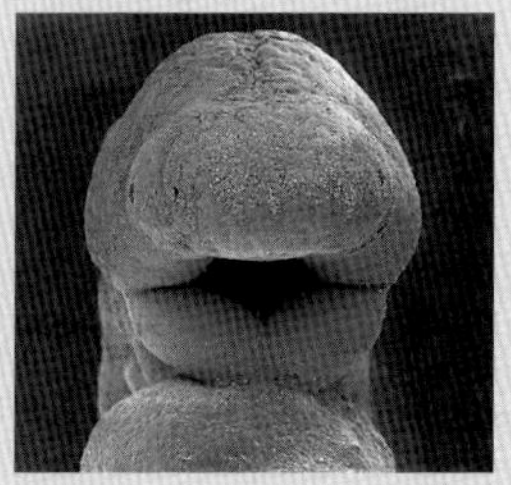

1.2.03a

Same embryo as in figure 1.2.03.

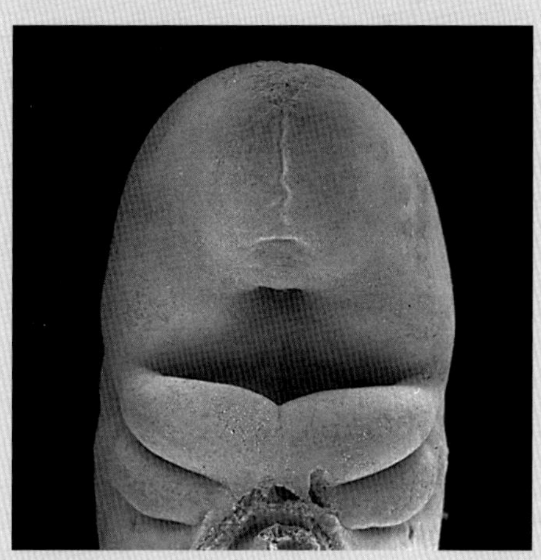

1.2.04

Embryos at week 5.

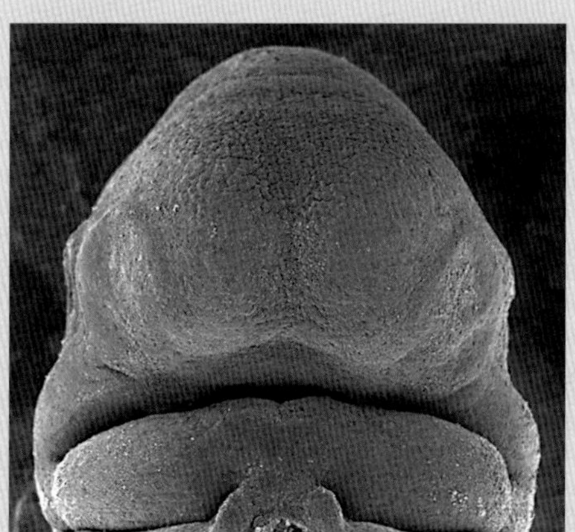

1.2.05

Week 5

1.2.01–1.2.17
Developmental stages of the face. Ventral view.

1.2.06

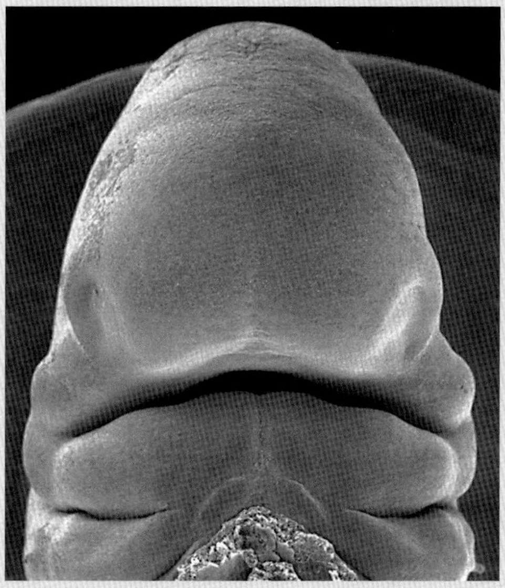

Embryos at week 5.

1.2.07

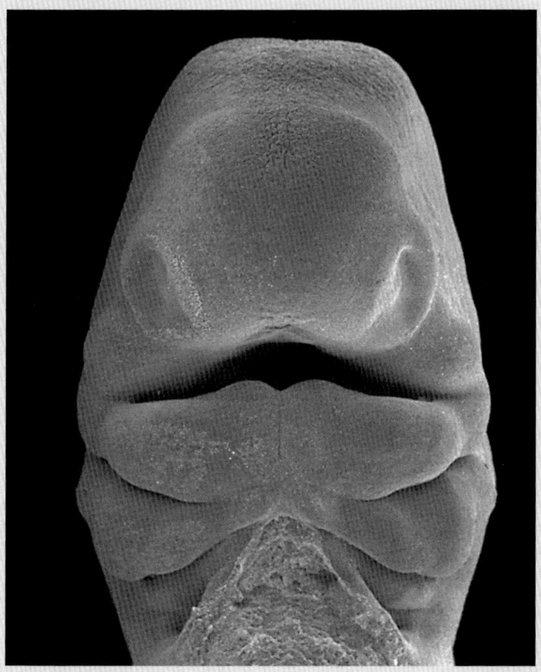

Week 5

Week 5 / 6

1.2.01–1.2.17
Developmental stages of the face. Ventral view.

Week 6

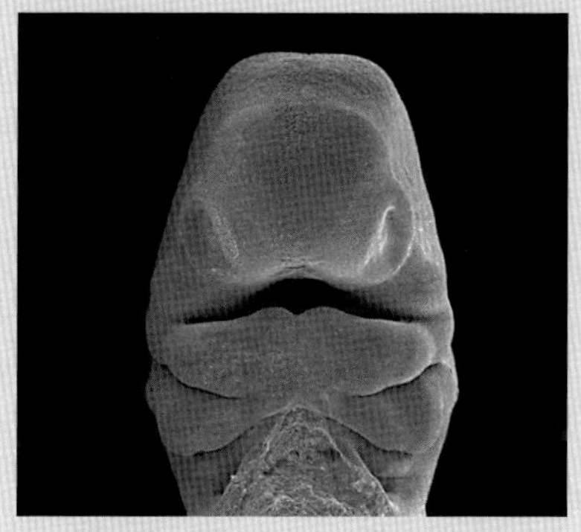

1.2.07a Same embryo as in figure 1.2.07.

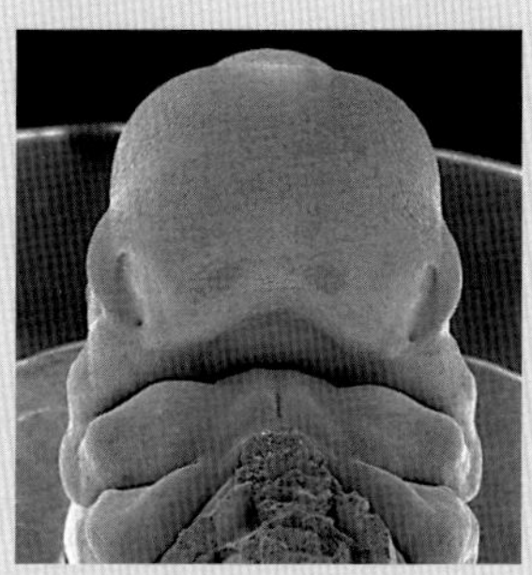

1.2.08 Embryo at the end of week 5/beginning of week 6.

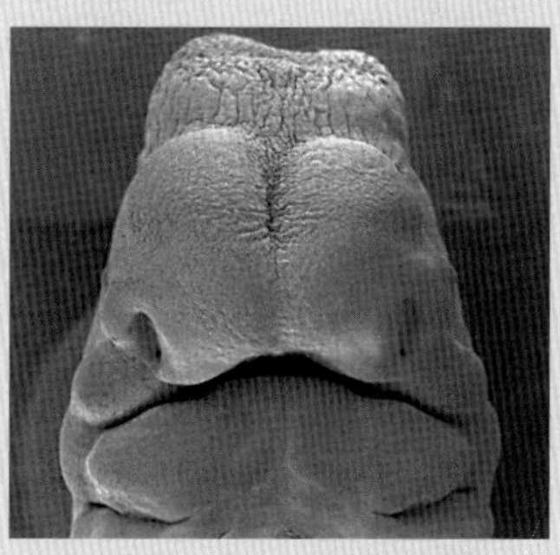

1.2.09 Embryos at week 6.

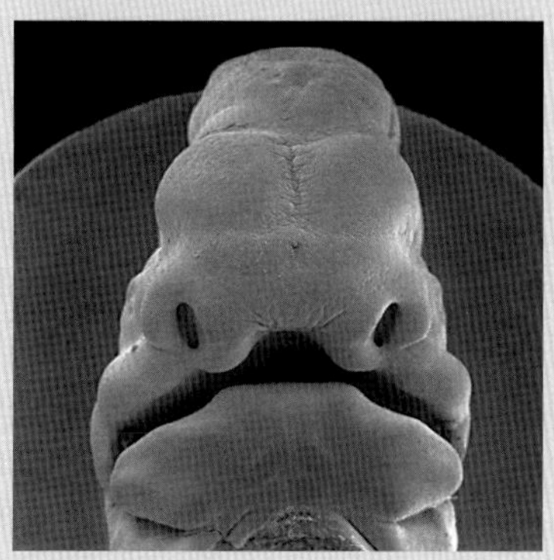

1.2.10

1.2.01–1.2.17
Developmental stages of the face. Ventral view.

Week 6

Week 7

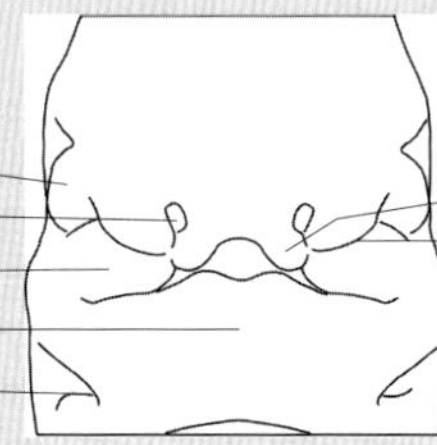

1.2.11

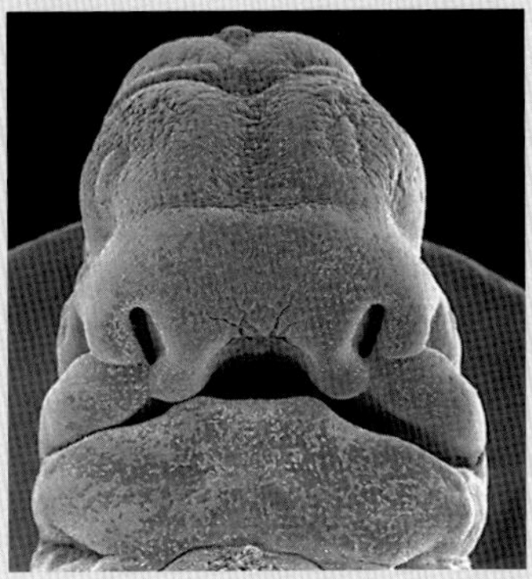

Embryos at week 6.

1.2.12

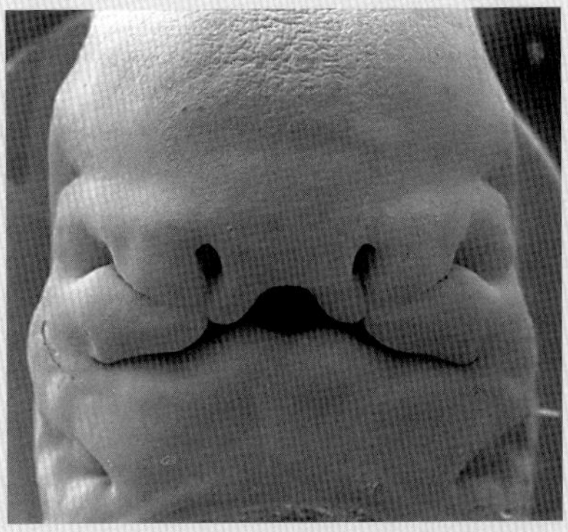

Embryos at week 7.

1.2.13

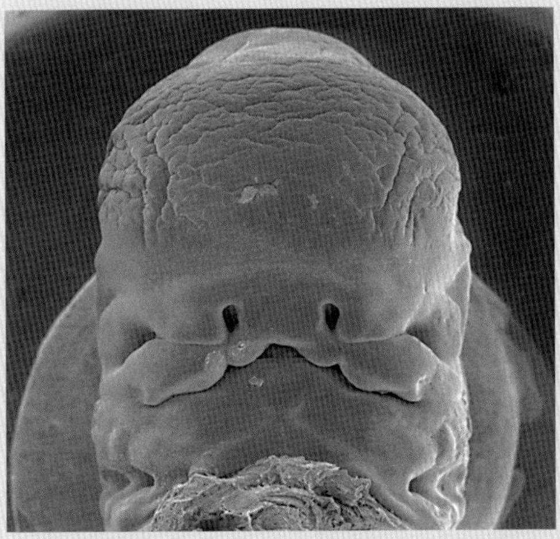

1.2.14

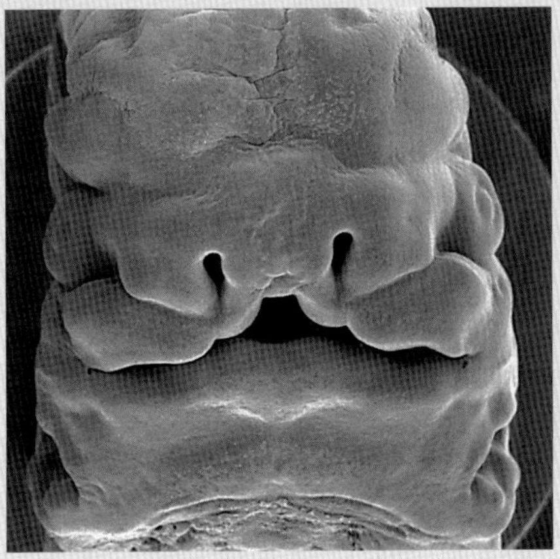

Week 7

1.2.01–1.2.17
Developmental stages of the face. Ventral view.

Week 8

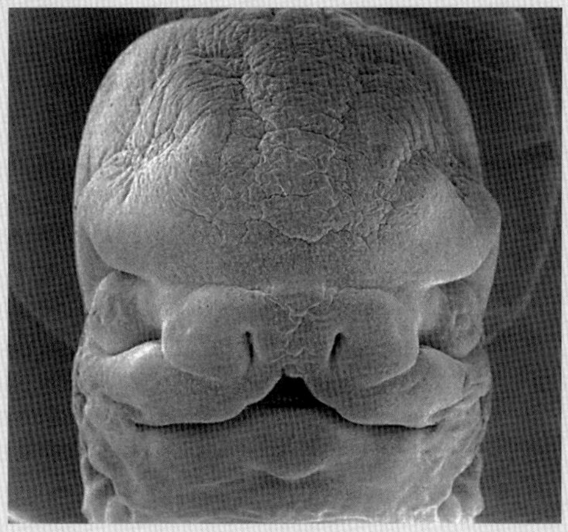

1.2.15 Embryo at week 7.

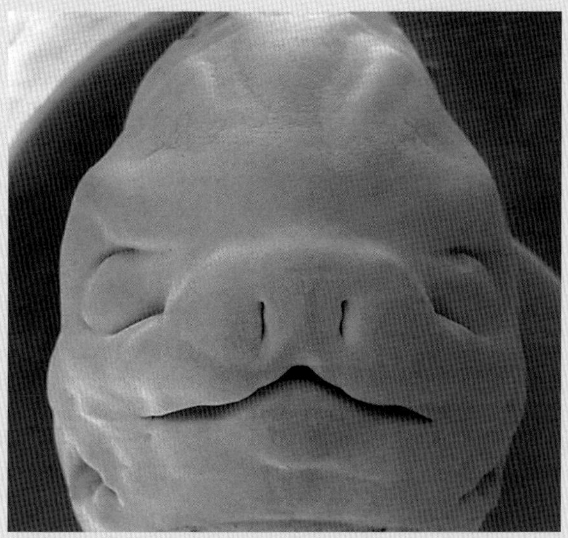

1.2.16 Embryos at week 8.

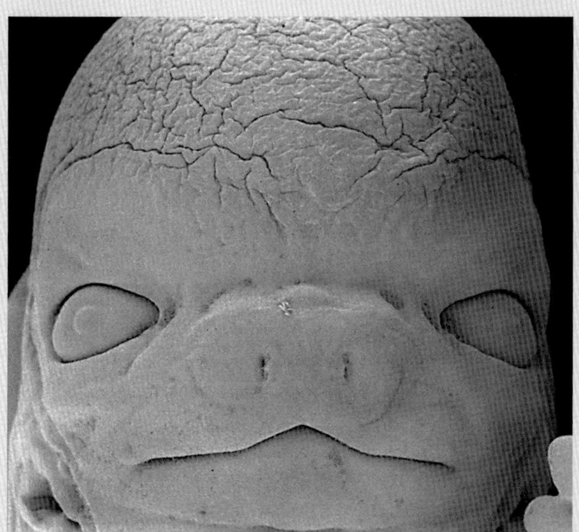

1.2.17

1.2.18–1.2.29
Developmental stages of the face. Ventral-left view.

30:1

Week 4

Week 5

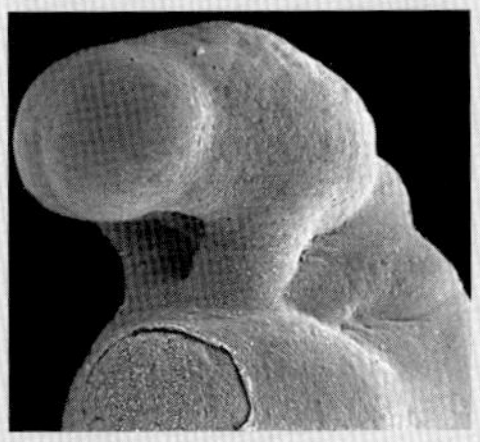

1.2.18 Embryo at week 4.

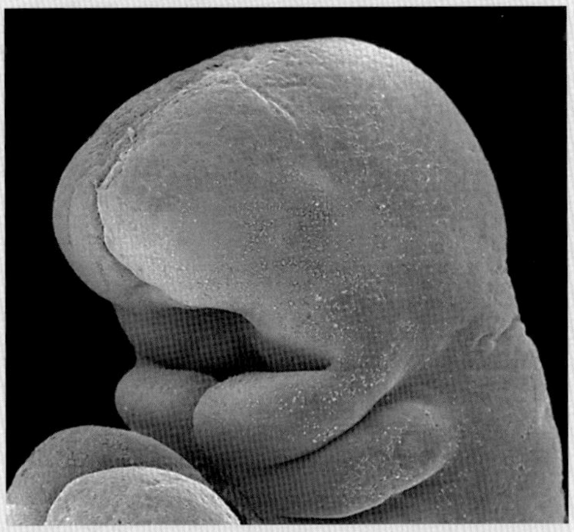

1.2.19 Embryos at week 5.

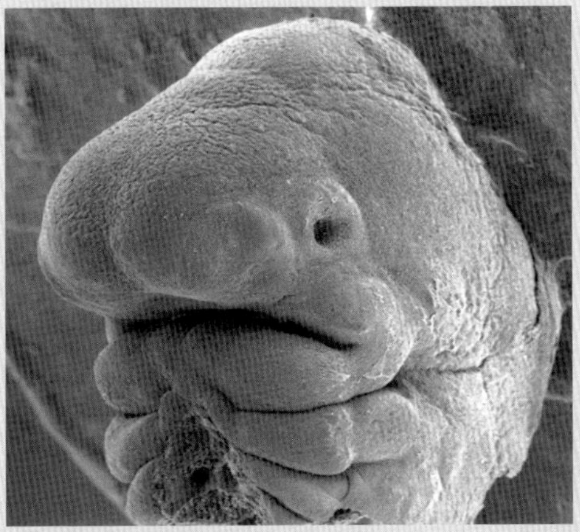

1.2.20

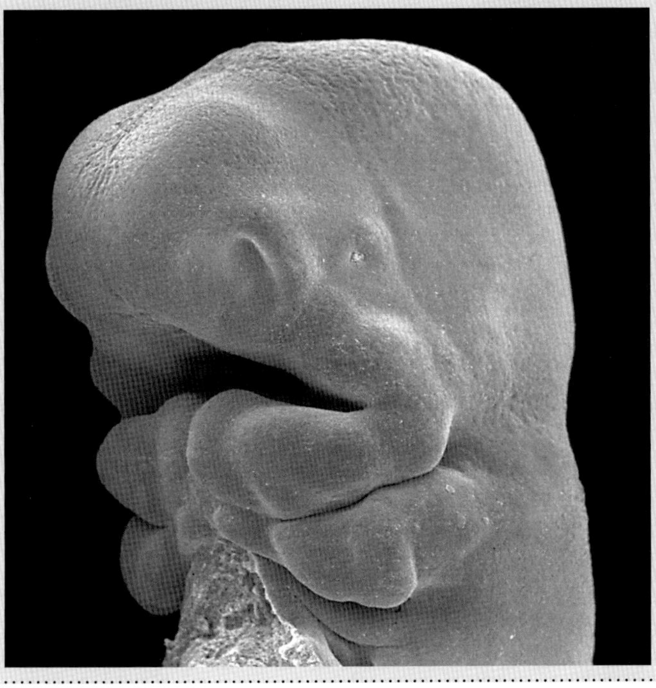

1.2.21

Week 5

Week 6

1.2.18–1.2.29
Developmental stages of the face. Ventral-left view.

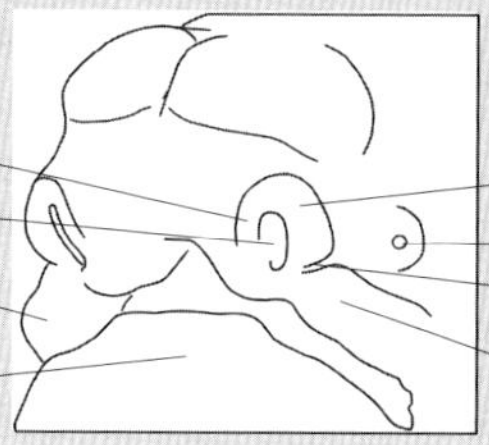

Week 7

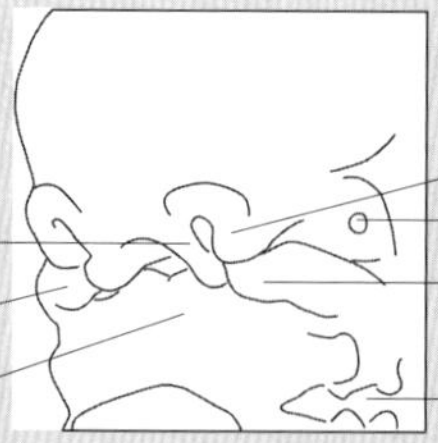

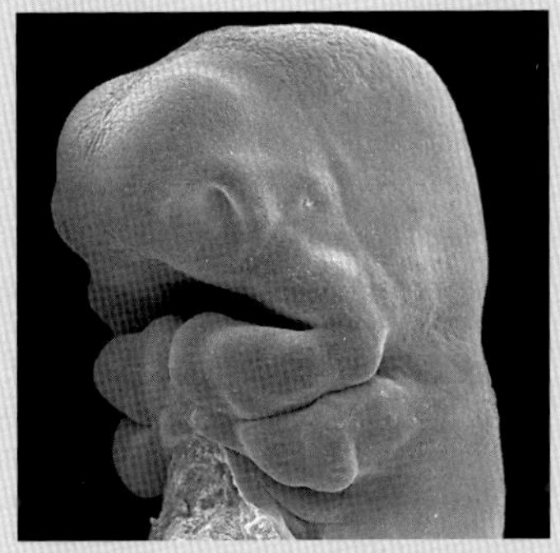

1.2.21a Same embryo as in figure 1.2.21.

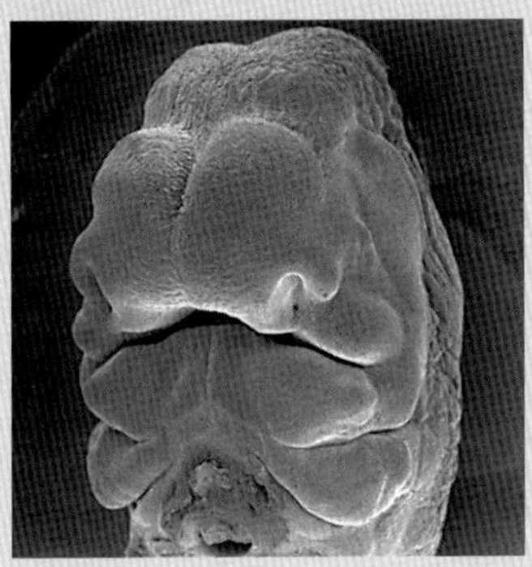

1.2.22 Embryos at week 6.

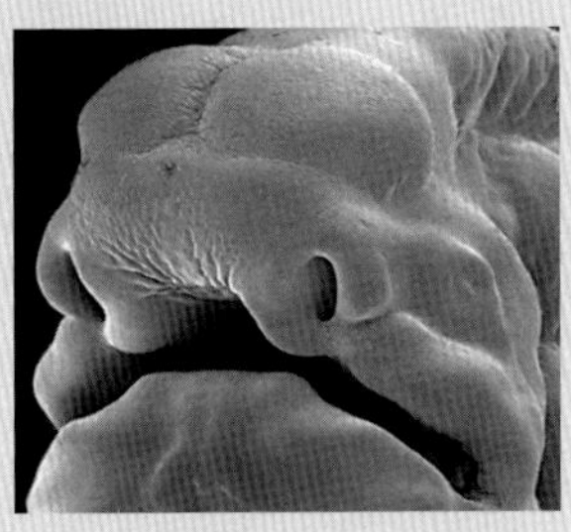

1.2.23

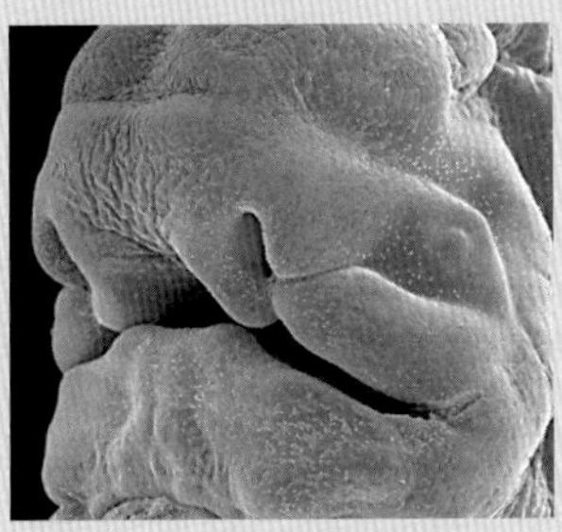

1.2.24

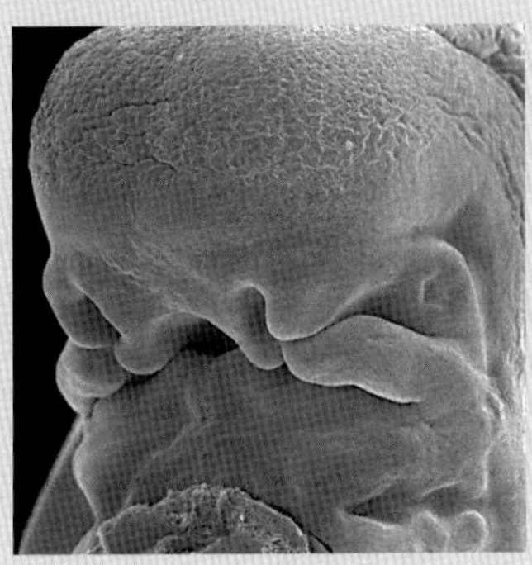

1.2.25 Embryo at week 7.

Week 7

1.2.18–1.2.29
Developmental stages of the face. Ventral-left view.

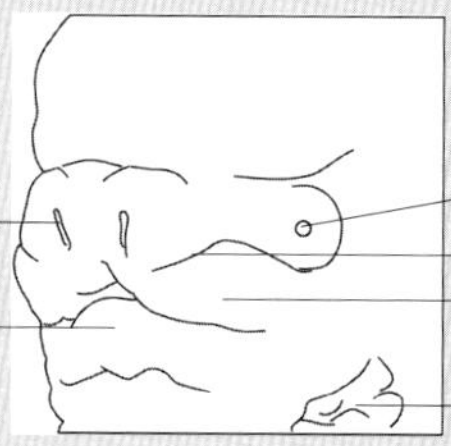

Week 8

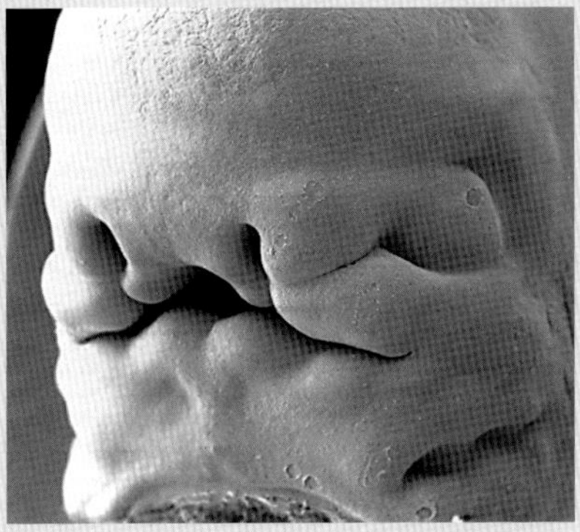

1.2.26 Embryos at week 7.

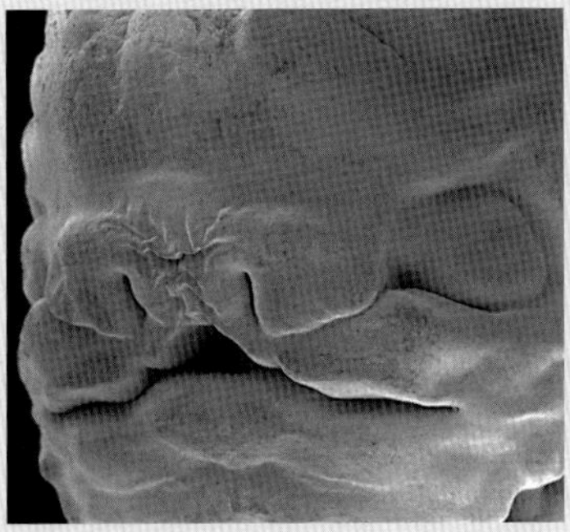

1.2.27

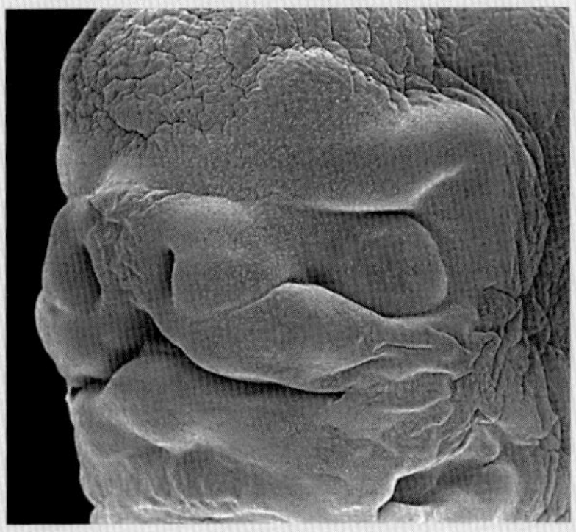

1.2.28

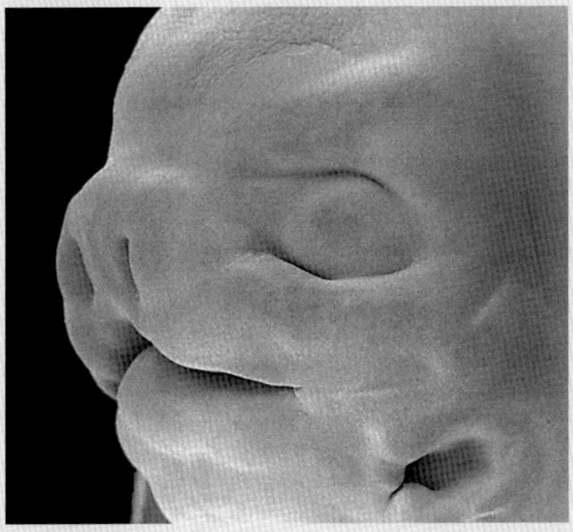

1.2.29 Embryo at week 8.

Week 4 / 5

1.2.30–1.2.37
Developmental stages of the face. Left view.

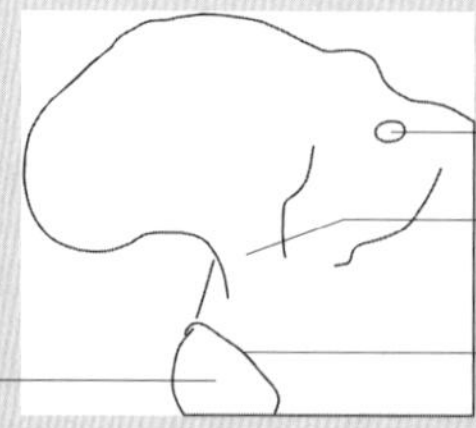

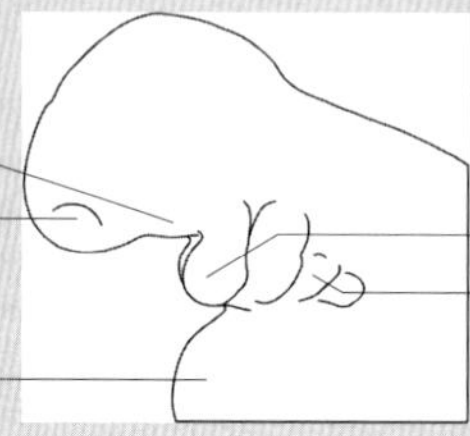

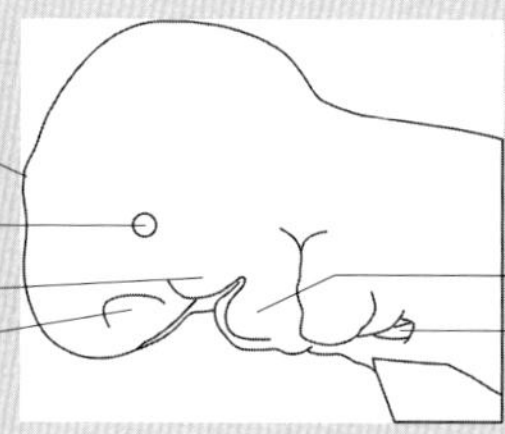

1.2.30

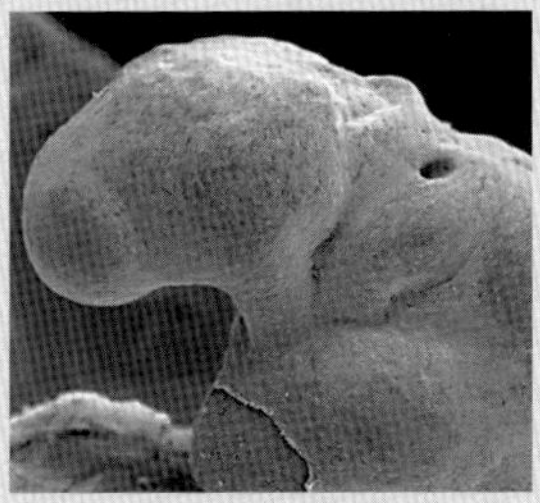

Embryos at the end of week 4 (1.2.30) and at the end of week 5.

1.2.31

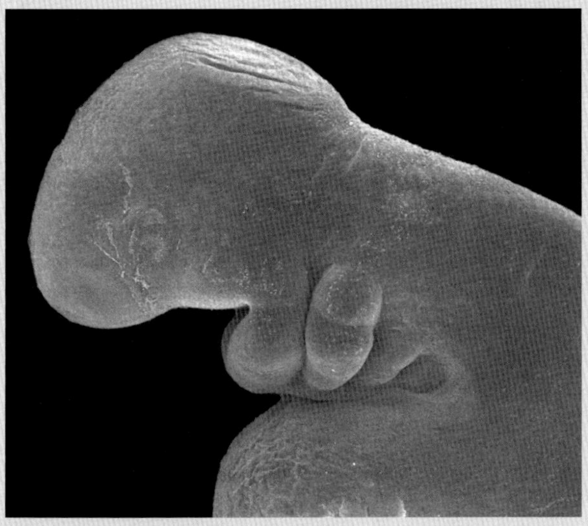

1.2.32

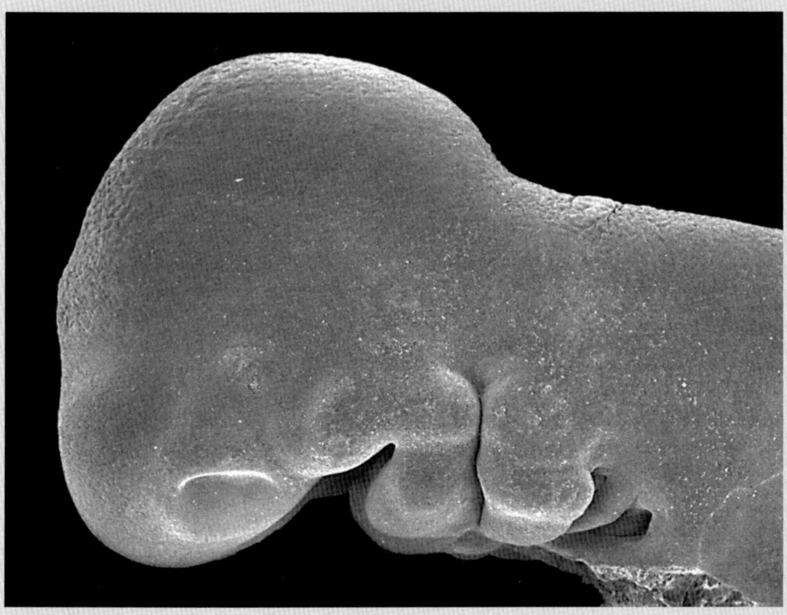

Week 5

Week 6

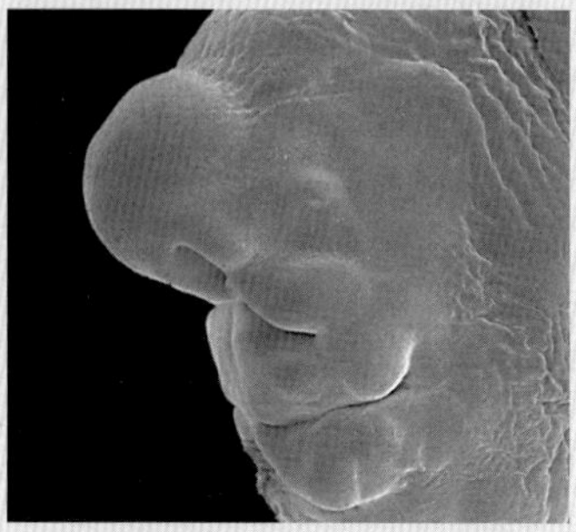

1.2.30–1.2.37
Developmental stages of the face. Left view.

Week 7

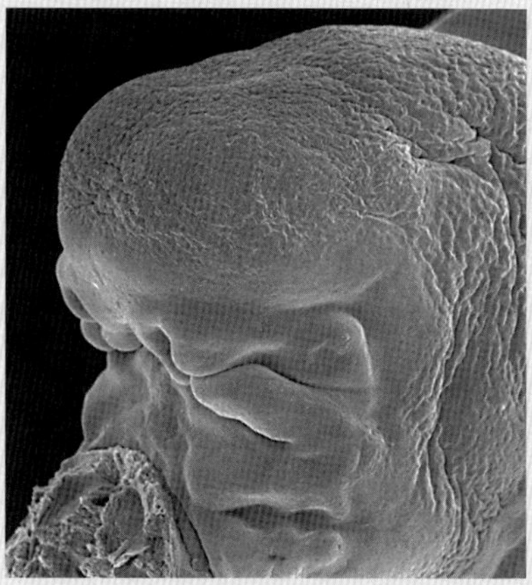

Week 8

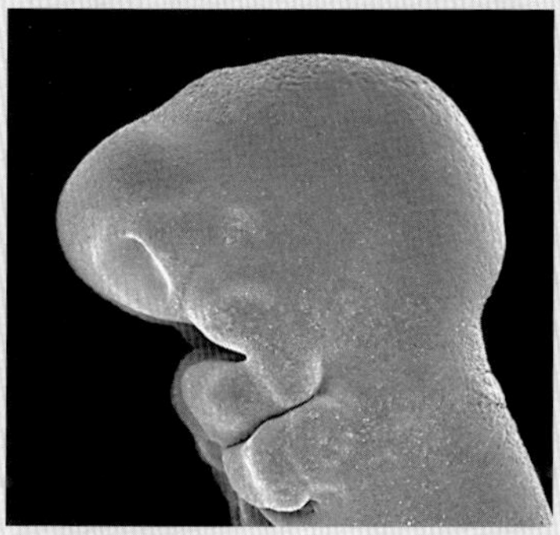

1.2.32a

Same embryo as in figure 1.2.32. In contrast to figure 1.2.32 the head is oriented such that the lower jaw lies in a nearly horizontal position.

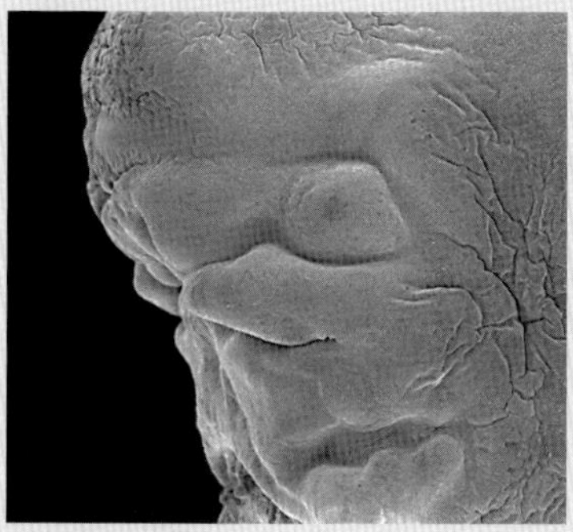

1.2.33 / 34

Embryos at week 6.

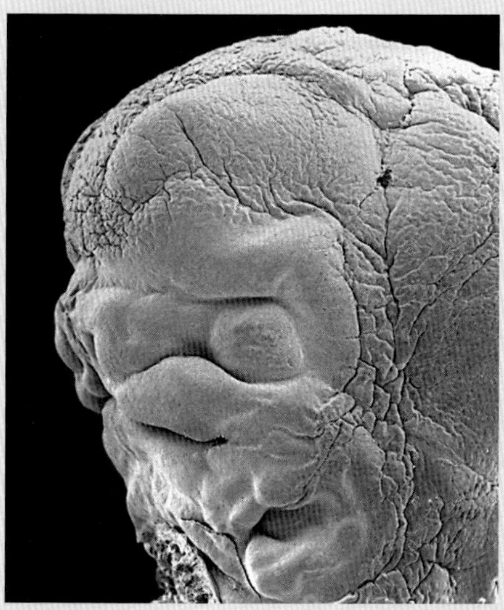

1.2.35 / 36

Embryos at week 7.

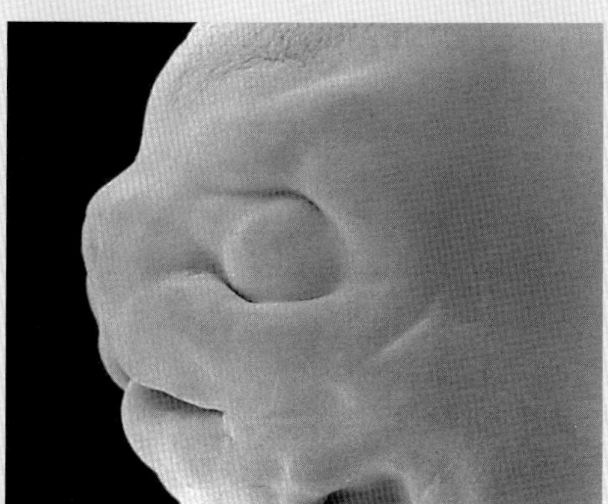

1.2.37

Embryo at week 8.

1.3 The Development of the Pharyngeal Arches

The foldings in the embryonic cervical region met with unmerited publicity since these organs were assumed to serve as proof of the repetition of the phylogeny during embryonic development of a recent individual. Unfortunately, at first glance, these cervical foldings of young human embryos were thought to resemble those of fish. Because in the middle of the 19th century, when this assumption was introduced by Haeckel, three-dimensional stereoscopic inspection was not yet possible, the remarkable differences between the anlagen of the branchial arches of the fish and the human pharyngeal arches were not clearly discernible or, for philosophical reasons, were possibly neglected.

Already in 1874, the great embryologist Wilhelm His recognised that the illustrations of Haeckel, by which the alleged identities of the anlagen of the gills in fish and the pharyngeal arches in human embryos were demonstrated, were falsifications. This view of recapitulation of evolution during recent embryonic life is repeated in text books, even today.

More detailed imaging with the stereomicroscope or with the scanning electron microscope permits recognition of the great differences in position, form, and structure of the cervical foldings of vertebrate embryos from different species. Whereas the endodermal branchial arches of fish develop into gills, the endodermal human pharyngeal arches give rise to the thymus, the parathyroid glands, the ultimopharyngeal body and a series of lymphoid tissues.

The cervical arches are built up of ectodermal, endodermal and mesenchymal tissues. The external ectodermal foldings are bordered by ectodermal valleys; the grooves (fig. 1.3.01–1.3.13) and the internal endodermal foldings are pouches of the cervical foregut, the pharynx (fig. 2.1.02–2.1.04). In each of the mesenchymal bulks between ectoderm and endoderm, an artery and cartilaginous tissue arise, and a cerebral nerve grows into them (fig. 6.1.58, 6.1.59).

Initially, just caudal to the mouth, the first arch develops, which will give rise to the mandible. Shortly afterwards, the second one, called the hyoid arch, arises and then the third and the fourth one develop (fig. 1.3.01–1.3.08). In week 6, the second arch grows in a caudal direction over the third and the fourth arches which become submerged (fig. 1.3.18–1.3.22) forming a sort of cave, the cervical sinus. Due to ectodermal growth this cave is filled up and as a consequence the caudal rim of the second arch disappears (fig. 1.3.31–1.3.35, 1.3.46–1.3.48).

In week 7, the first groove and the bordering bulges of the first and the second arches are present, forming the external auditory duct and the foldings of the external ear (fig. 1.3.36–1.3.61). Figures 1.3.40/1.3.41, 1.3.42/1.3.45, 1.3.46/1.3.47, 1.3.48/1.3.51, 1.3.52/1.3.53, 1.3.54/1.3.55, 1.3.56/1.3.57, 1.3.58/1.3.59, and 1.3.60/1.3.61 show the individual features of the external ear of embryos at the same developmental stage.

Week 4

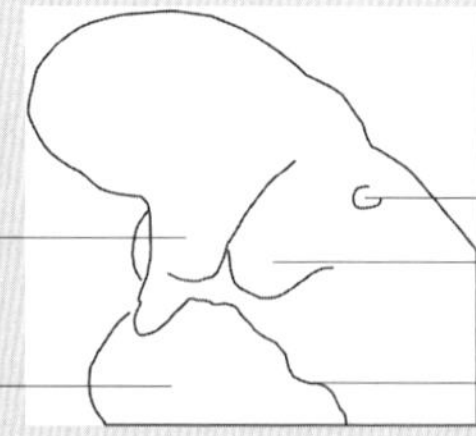

1.3.01–1.3.22
Developmental stages of the external form of the pharyngeal arches. Lateral views.

1.3.01

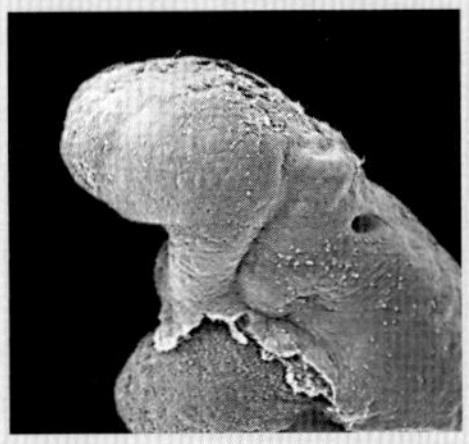

1.3.02

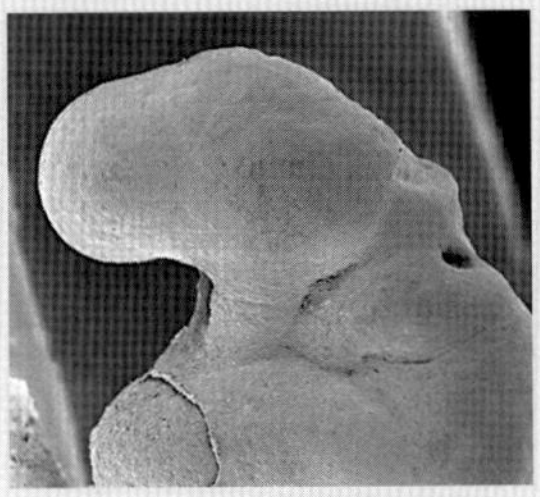

1.3.03

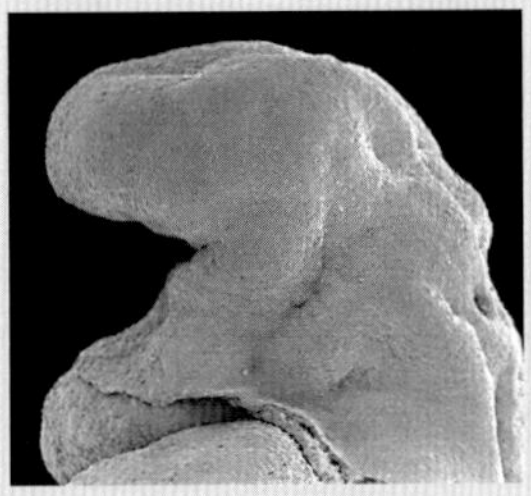

1.3.04

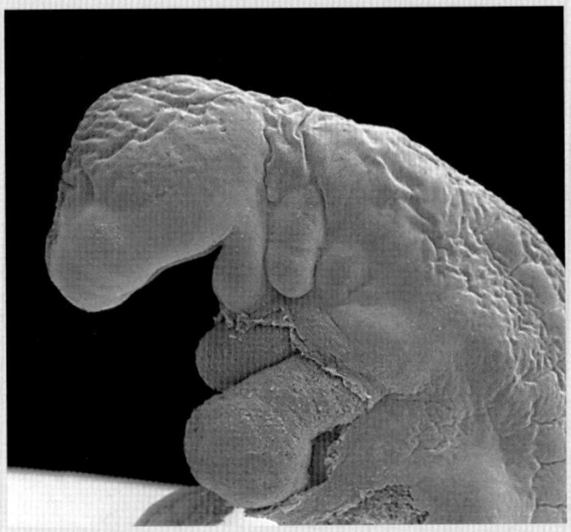

Head and pharyngeal arches. Left view. Embryos at week 4.

Week 5

1.
External Aspects

1.3
The Development of the Pharyngeal Arches

1.3.01–1.3.22
Developmental stages of the external form of the pharyngeal arches. Lateral views.

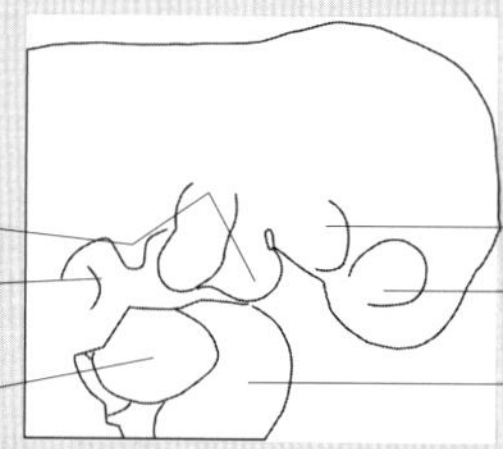

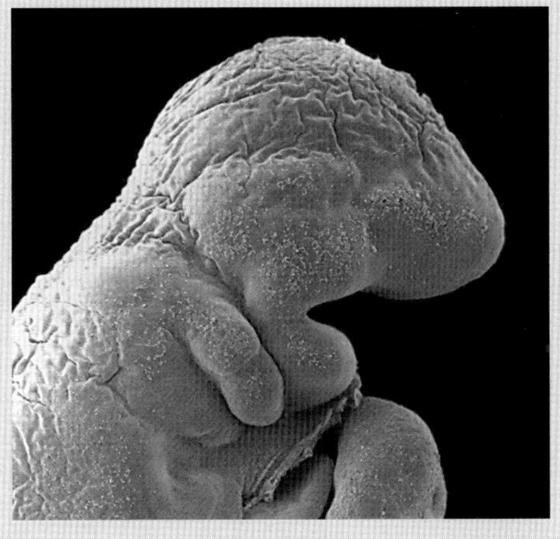

1.3.05

Head and pharyngeal arches. Right view. Embryos up to the middle of week 5.

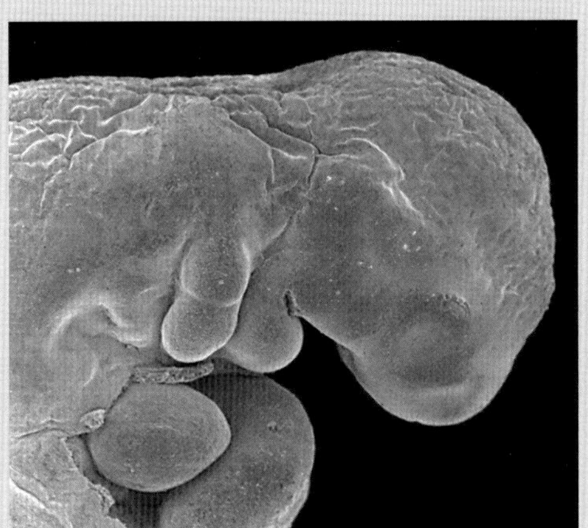

1.3.06

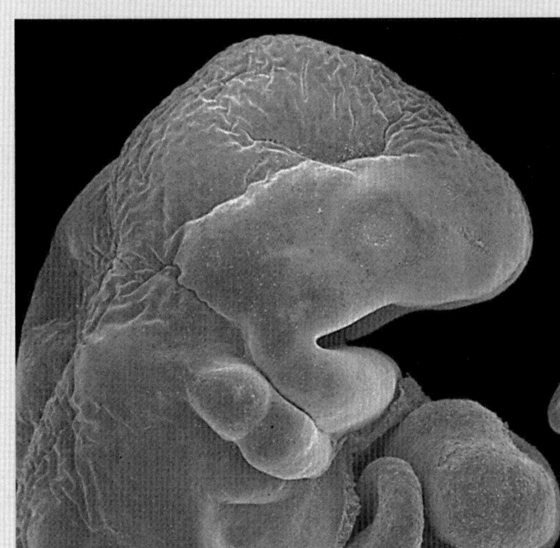

1.3.07

Week 5

1.3.01–1.3.22

Developmental stages of the external form of the pharyngeal arches. Lateral views.

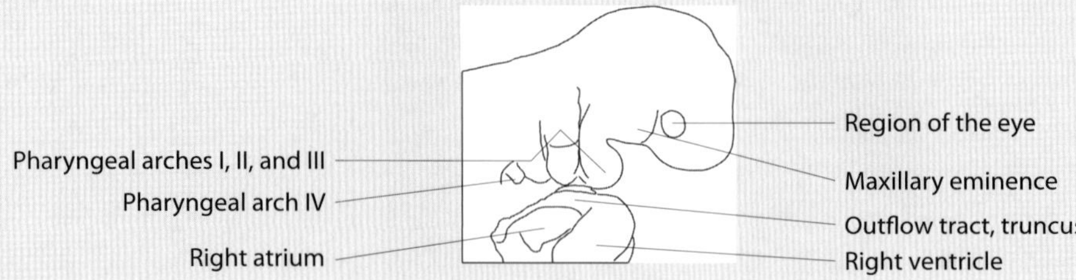

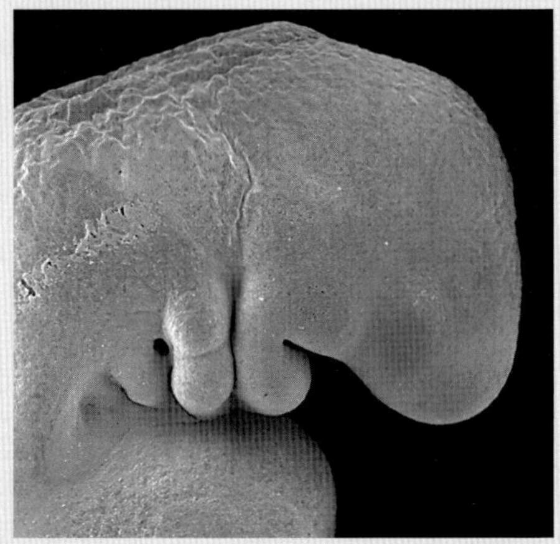

1.3.08

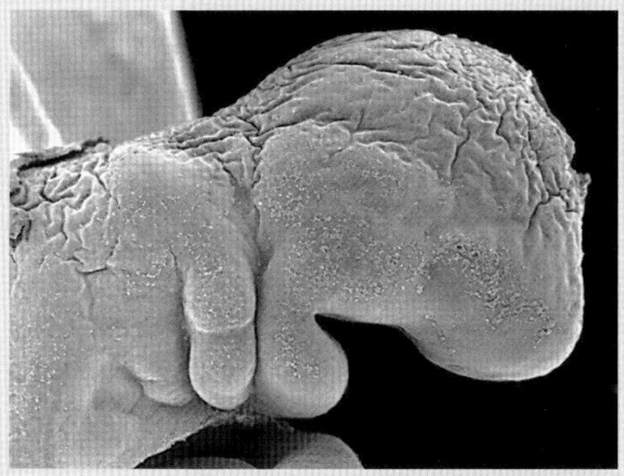

Head and pharyngeal arches. Right view. Embryos up to the middle of week 5.

1.3.09 / 10

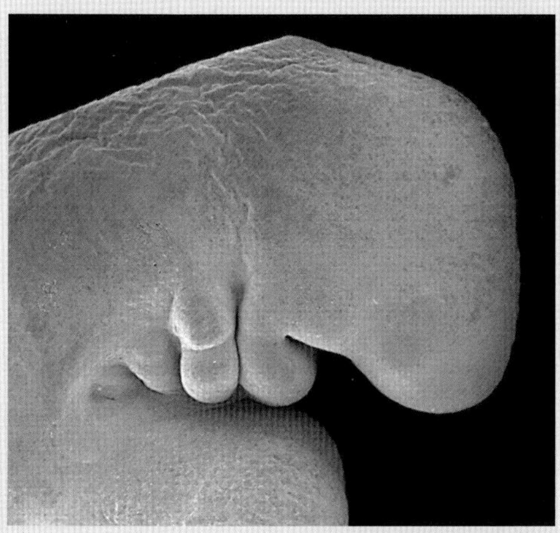

1.3.11

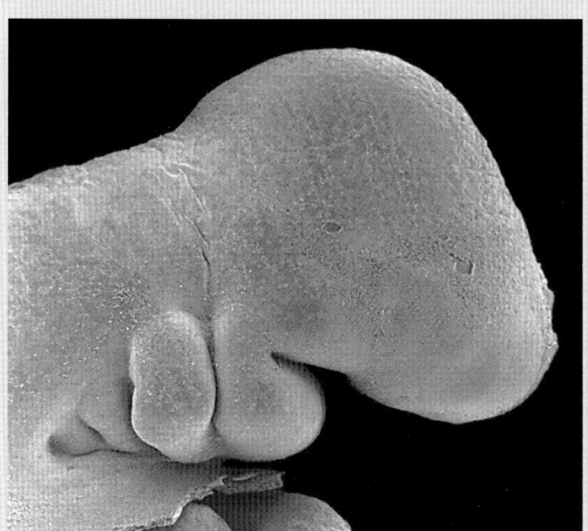

Head and pharyngeal arches. Right view. Embryos in the middle of week 5.

Week 5

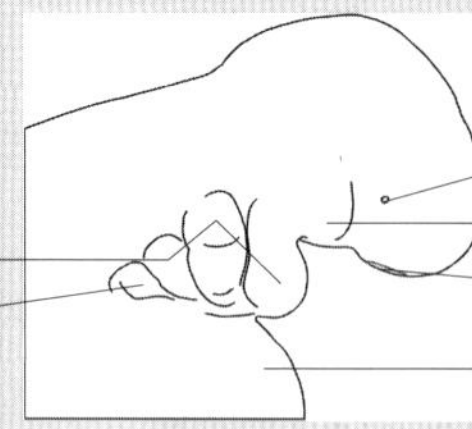

1.3.01–1.3.22
Developmental stages of the external form of the pharyngeal arches. Lateral views.

Week 6

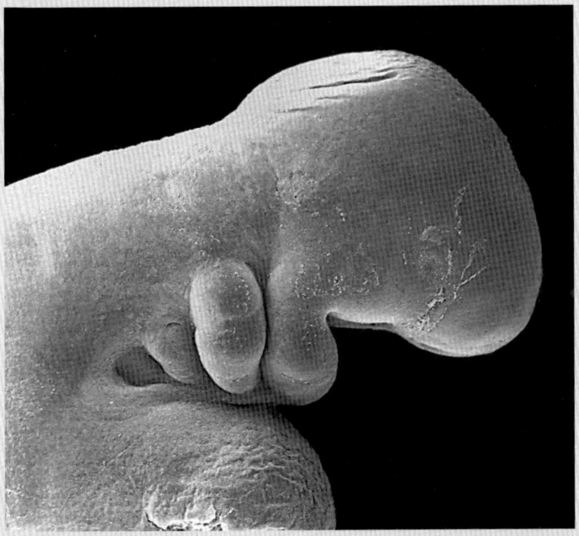

1.3.12 Head and pharyngeal arches. Right view. Embryos in the middle of week 5.

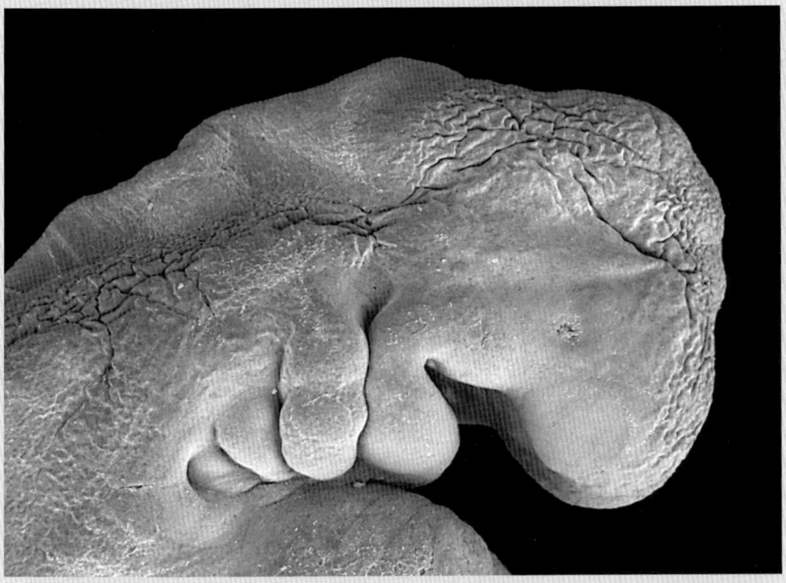

1.3.13

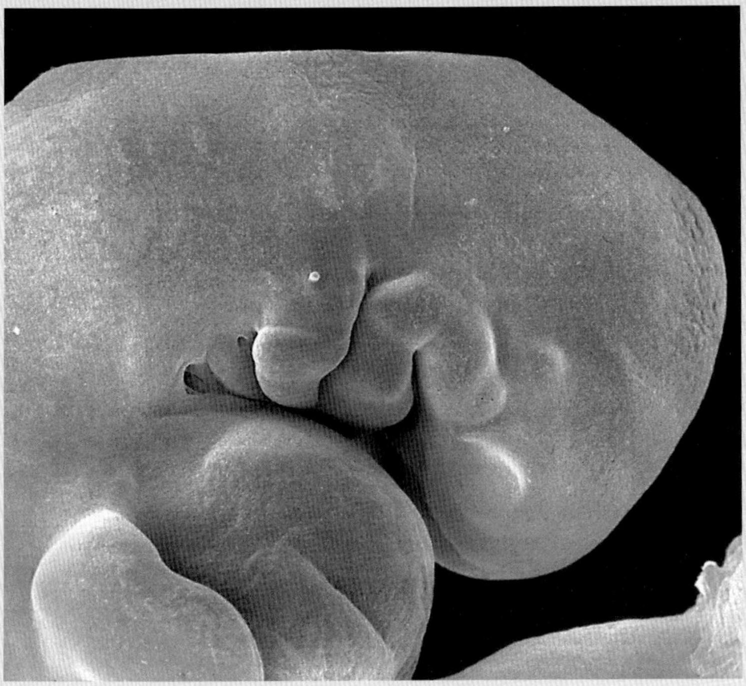

1.3.14 Head and pharyngeal arches. Right view. Embryo early in week 6.

1.3.01–1.3.22
Developmental stages of the external form of the pharyngeal arches. Lateral views.

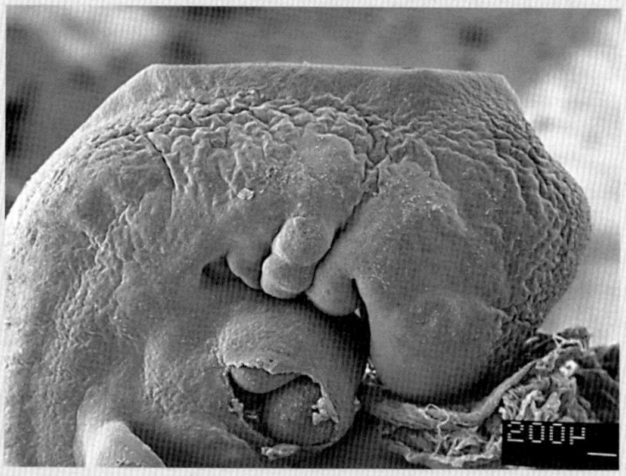

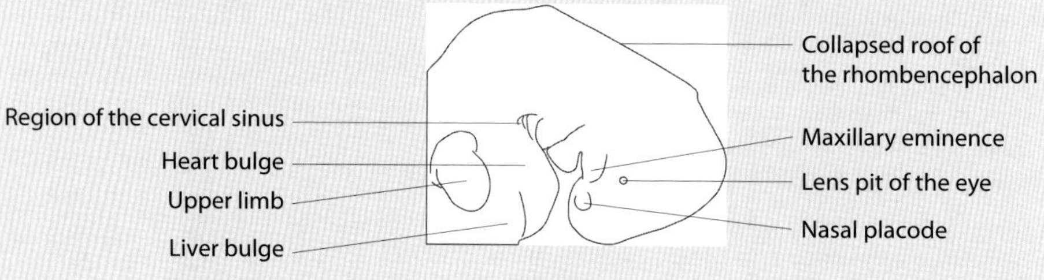

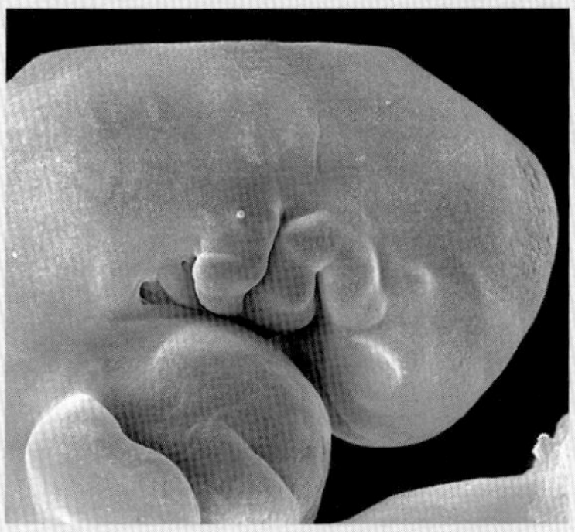

1.3.14a

Same embryo as in figure 1.3.14.

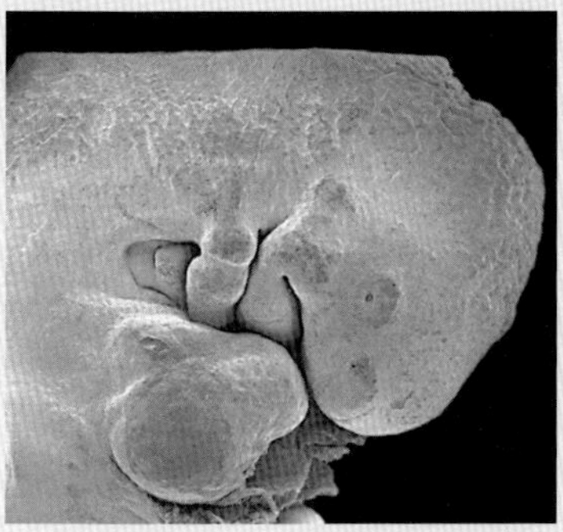

1.3.15

Head and pharyngeal arches. Right view. Embryos at week 6.

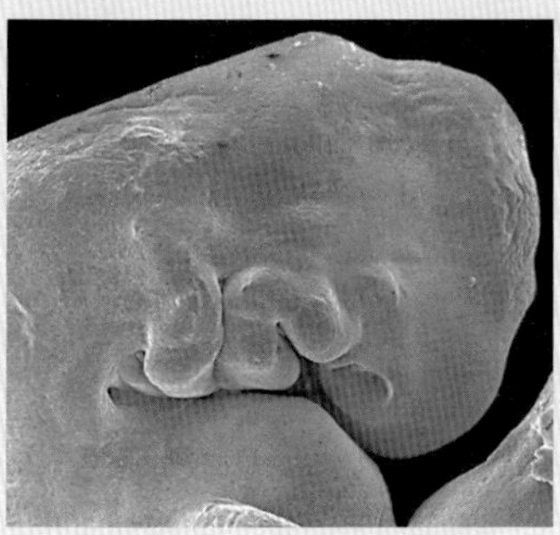

1.3.16 / 17

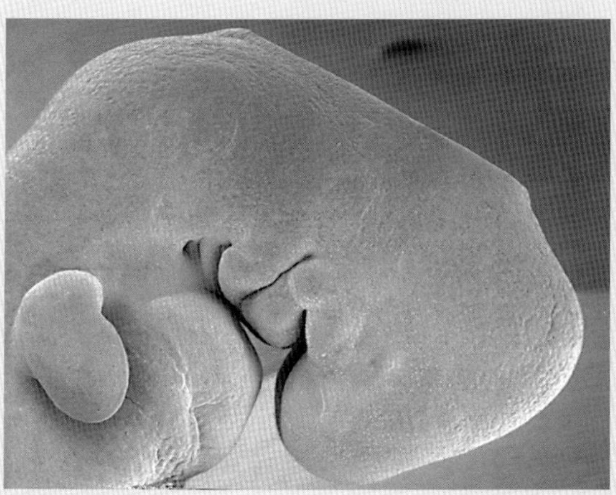

1.3.18

15:1

Week 6

1.3.01–1.3.22

Developmental stages of the external form of the pharyngeal arches. Lateral views.

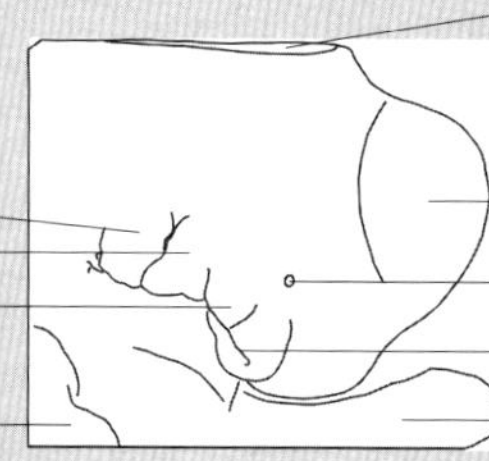

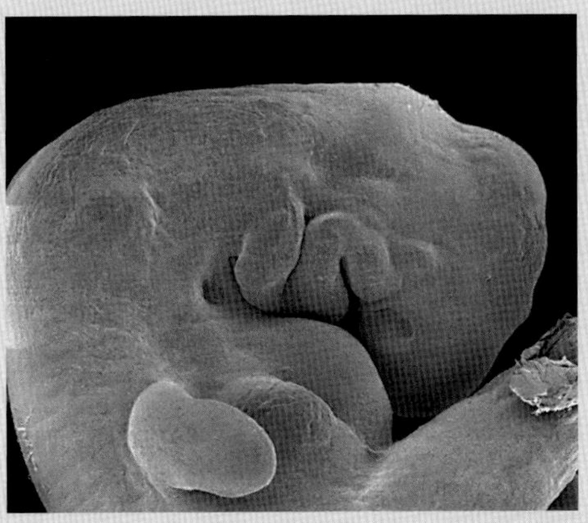

1.3.19

Head and pharyngeal arches. Right view. Embryos in the middle of week 6.

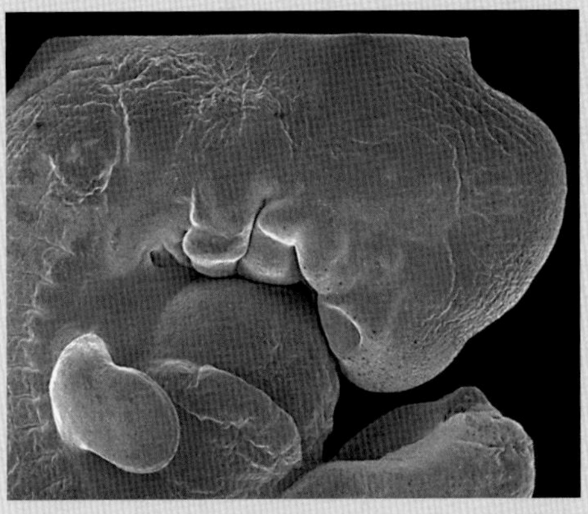

1.3.20

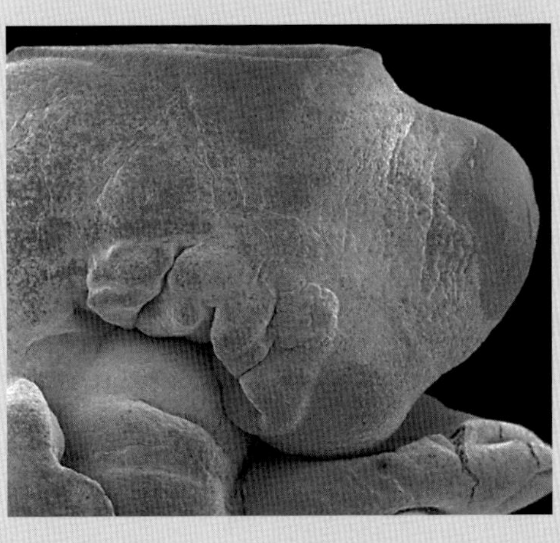

1.3.21

Head and pharyngeal arches. Right view. Embryos late in week 6.

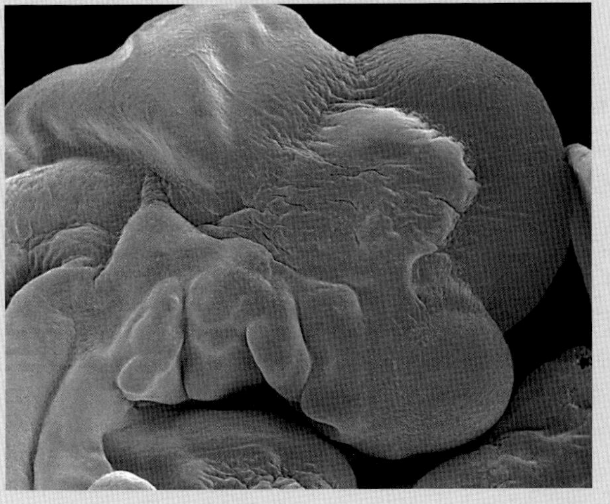

1.3.22

1.3.23–1.3.35
Pharyngeal arches in ventral view.

Week 4

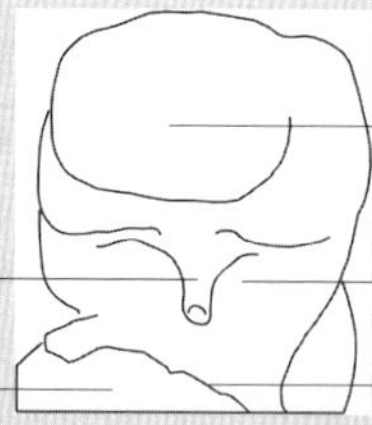

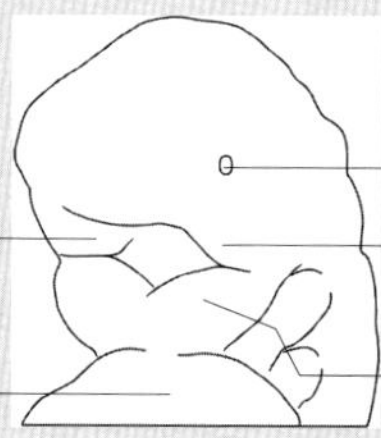

Week 5

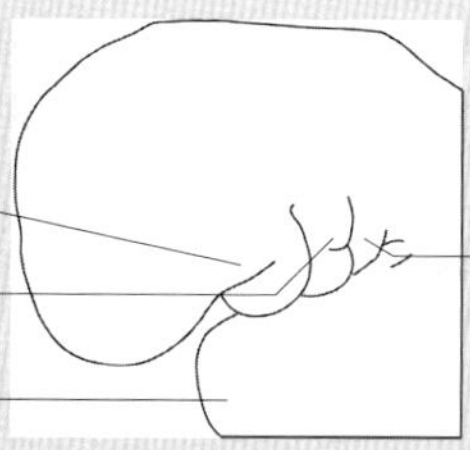

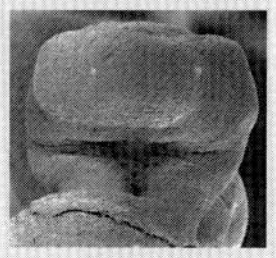

1.3.23

Head and pharyngeal arches in ventral view. Embryos at week 4.

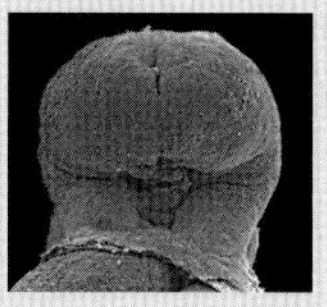

1.3.24

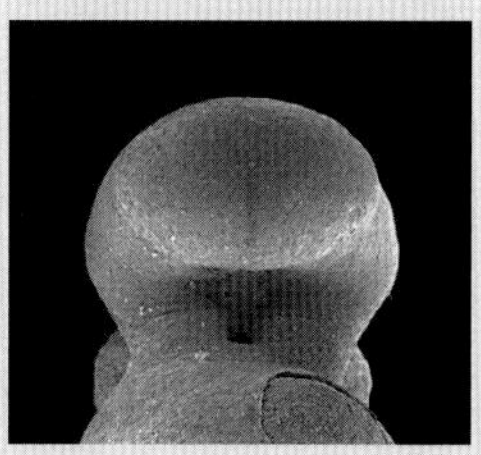

1.3.25

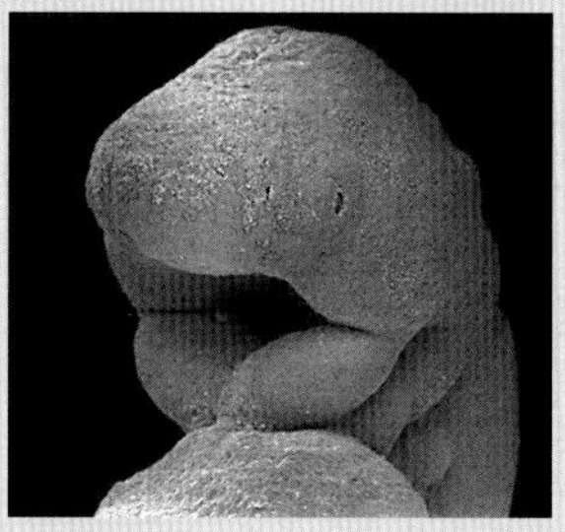

1.3.26

Head and pharyngeal arches in ventral-left view. Embryos late in week 4 and early in week 5.

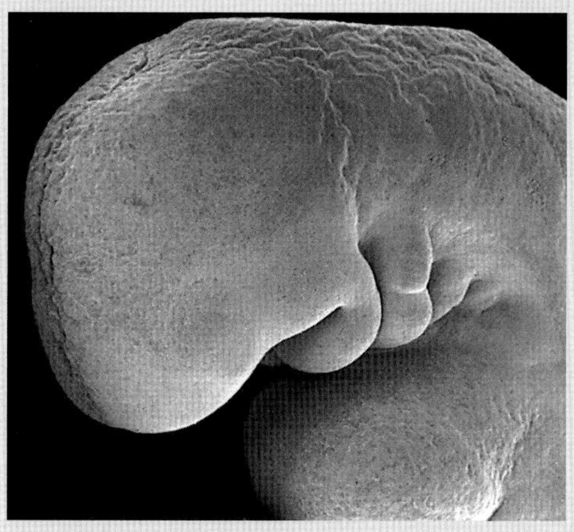

1.3.27

Week 5

1.3.23–1.3.35
Pharyngeal arches in ventral view.

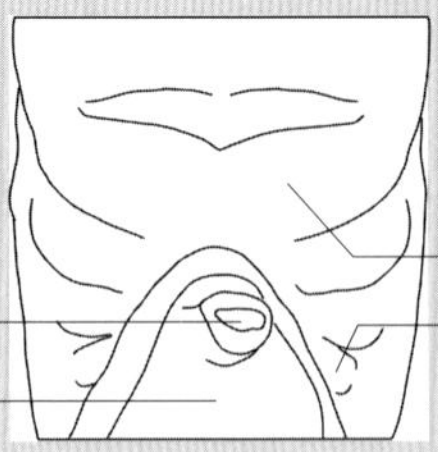

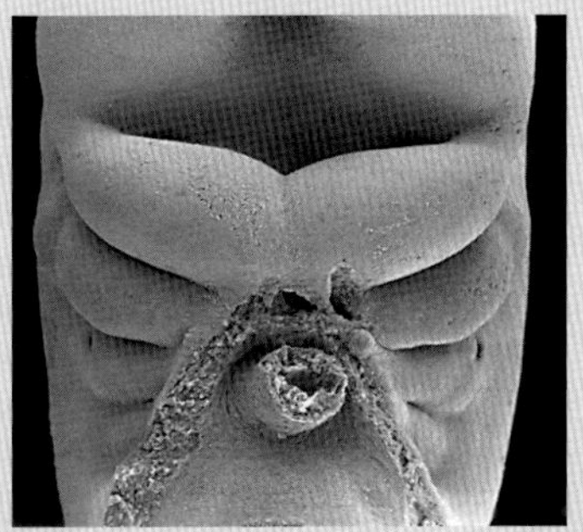

1.3.28 Pharyngeal arches in ventral view. Embryos at week 5.

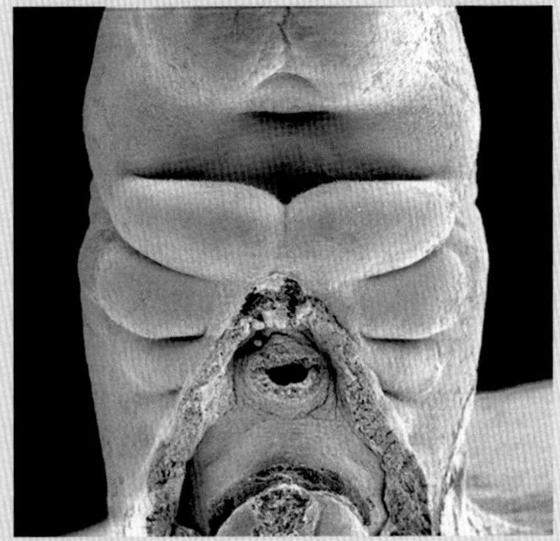

1.3.29

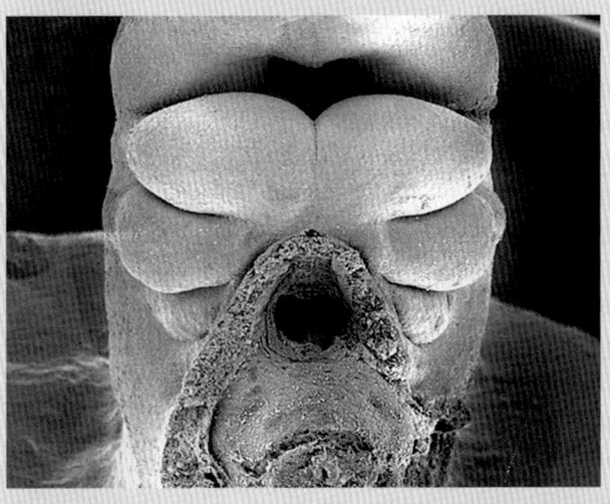

1.3.30

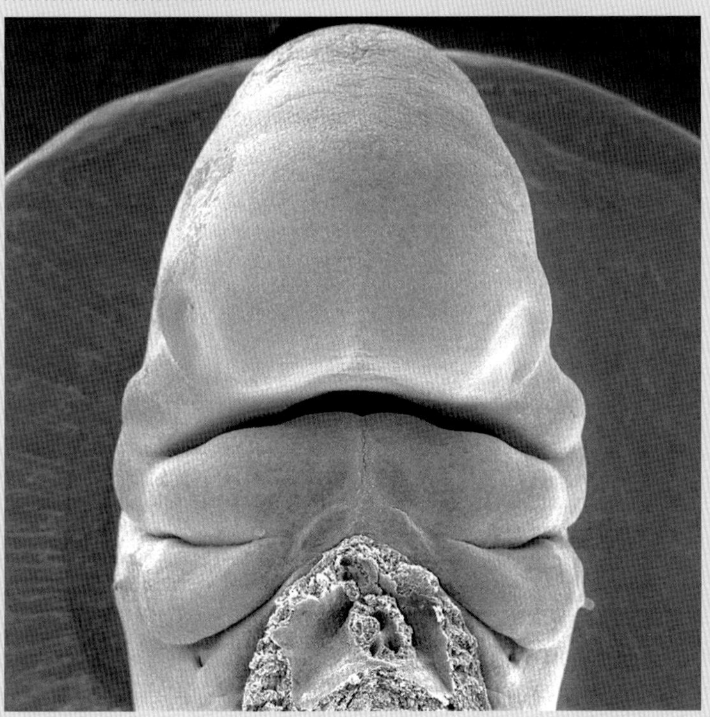

1.3.31 Head and pharyngeal arches in ventral view. Embryo at week 5.

Week 5

Week 6

1.3.23–1.3.35
Pharyngeal arches in ventral view.

Week 7

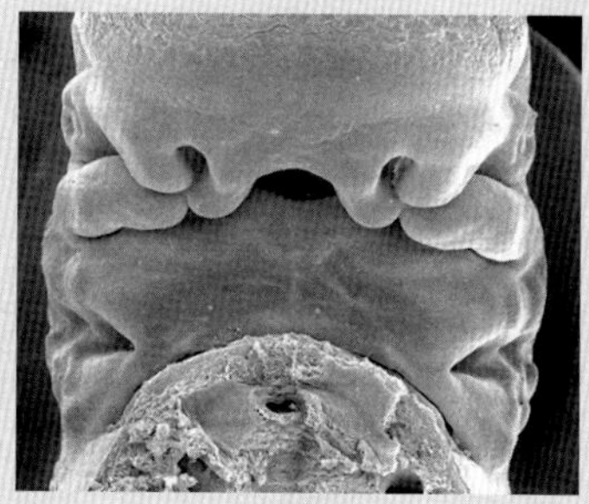

Week 8

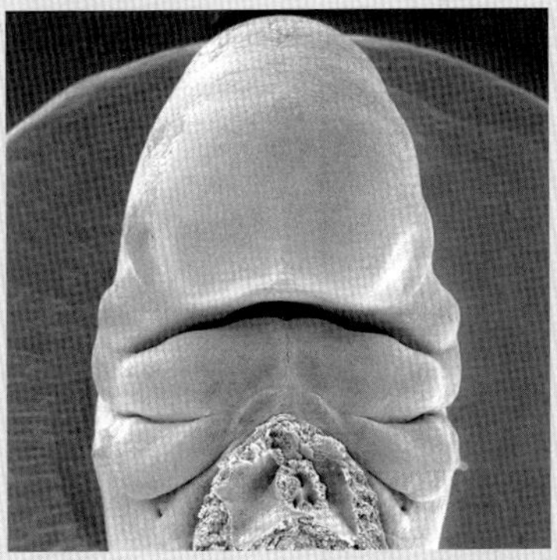

1.3.31a

Head and pharyngeal arches in ventral view. Embryo at week 5.

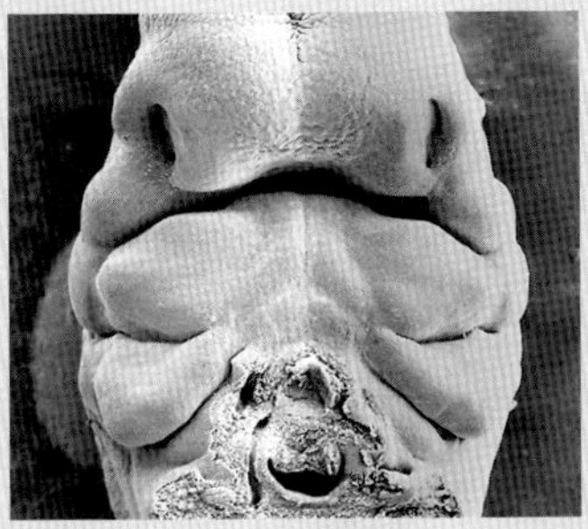

1.3.32

Face and pharyngeal arches in ventral view. Embryo at the beginning or in the middle of week 6.

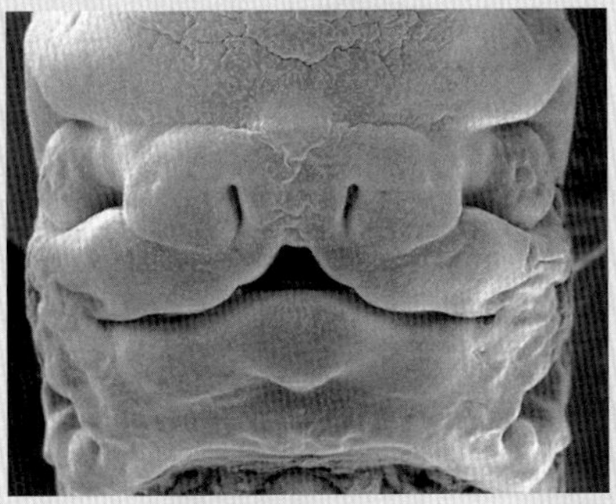

1.3.33 / 34

Face and pharyngeal arches in ventral view. Embryos early and late in week 7.

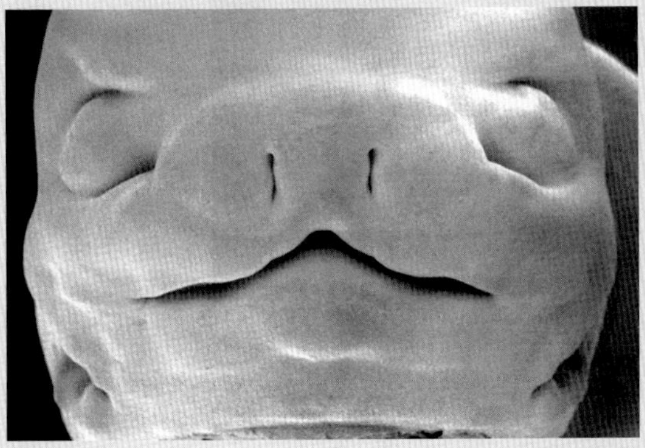

1.3.35

Face and pharyngeal arches in ventral view. Embryo at week 8.

1.3.36–1.3.61
Developmental stages of the external ear.

Week 4

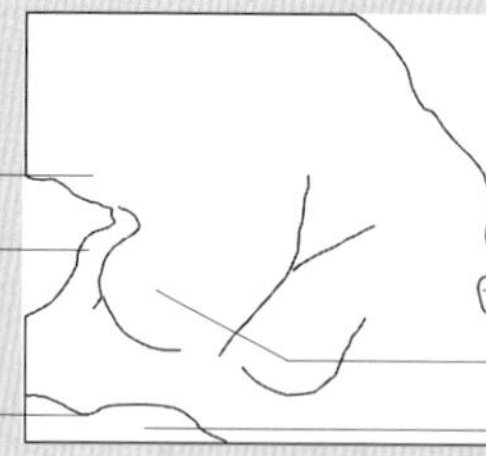

Week 5

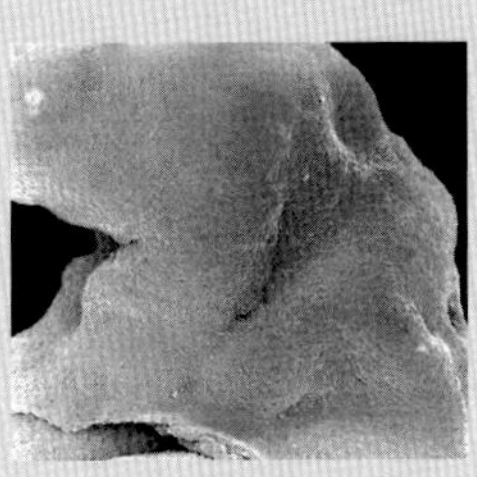

1.3.36 Pharyngeal arches, left view. Embryos at week 4.

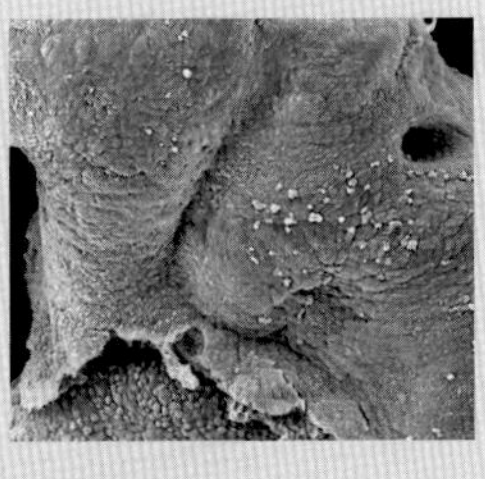

1.3.37

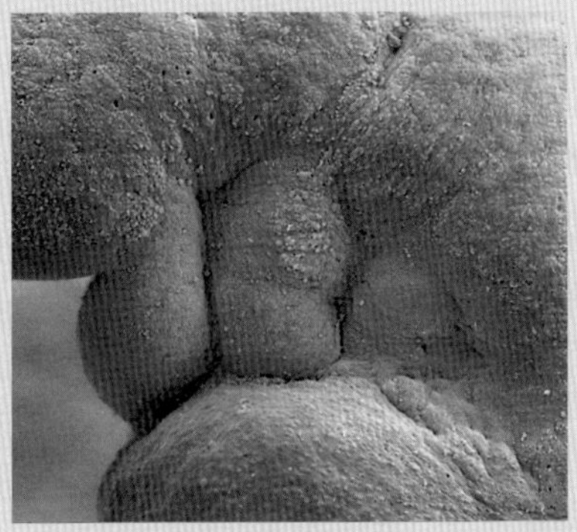

1.3.38 Pharyngeal arches, left view. Embryos late in week 5.

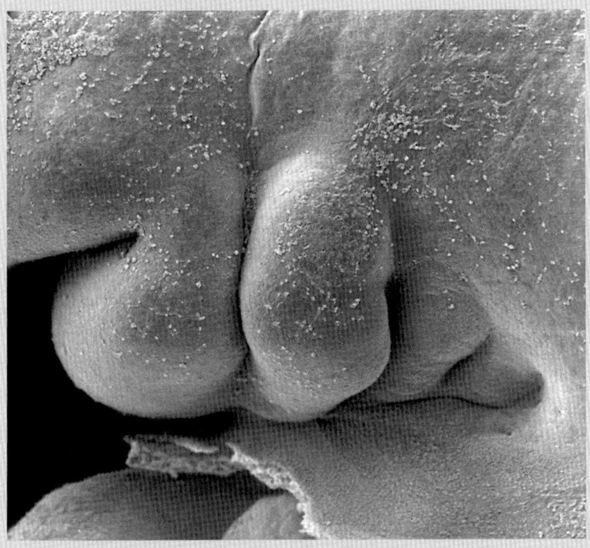

1.3.39

Week 5

1.3.36–1.3.61
Developmental stages of the external ear.

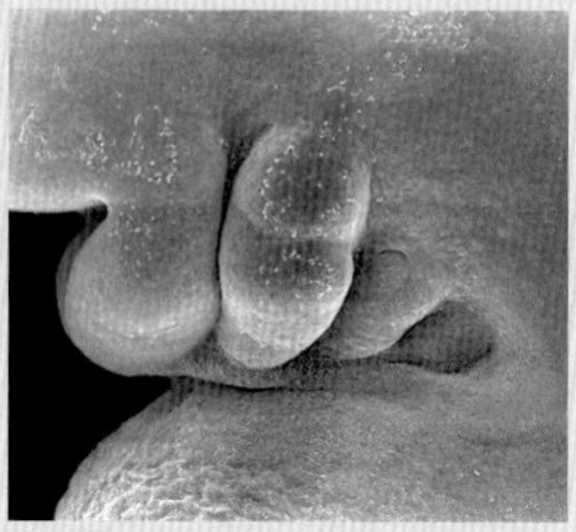

Week 6

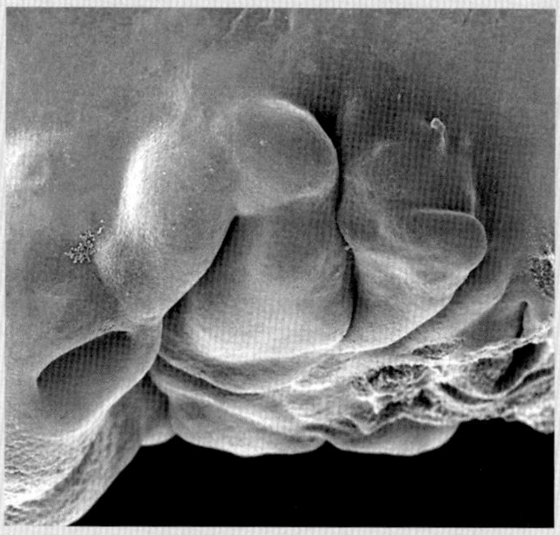

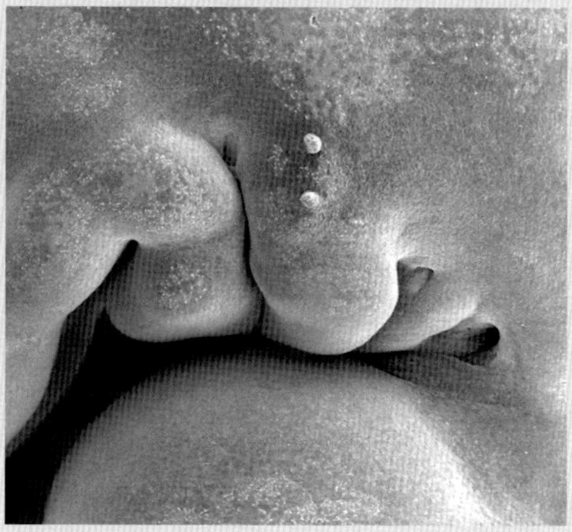

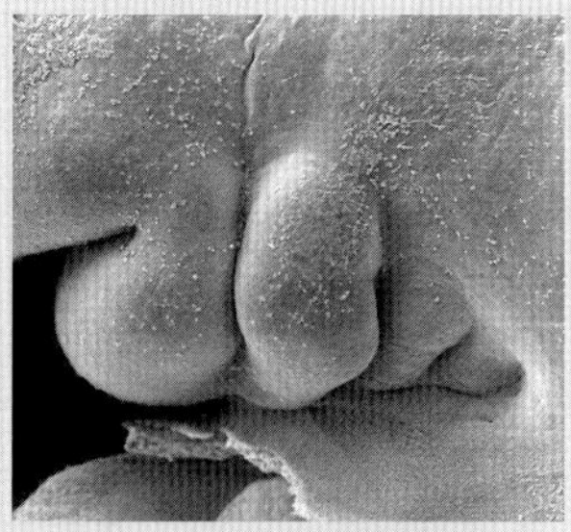

1.3.39a

Same embryo as in figure 1.3.39.

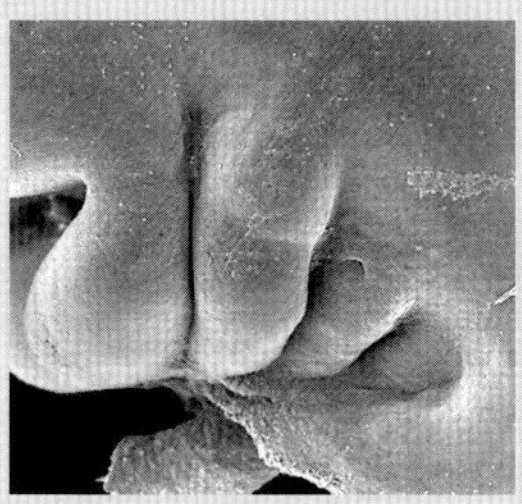

1.3.40 / 41

Pharyngeal arches, left view. Embryos at week 5.

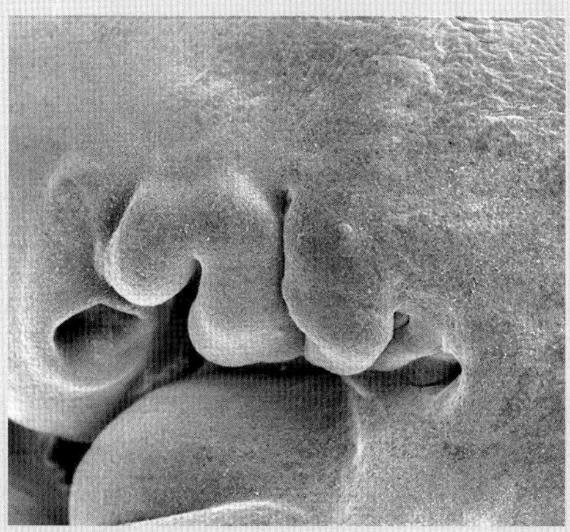

1.3.42 / 43

Pharyngeal arches, left view. Embryos at week 6.

1.3.42–44
Ectodermal 'warts' on arch II are not rare. Their origin and fate are unknown.

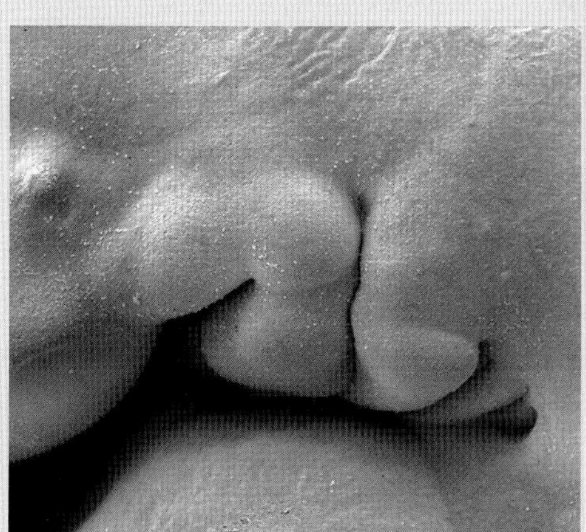

1.3.44 / 45

1.3.36–1.3.61
Developmental stages of the external ear.

Week 6

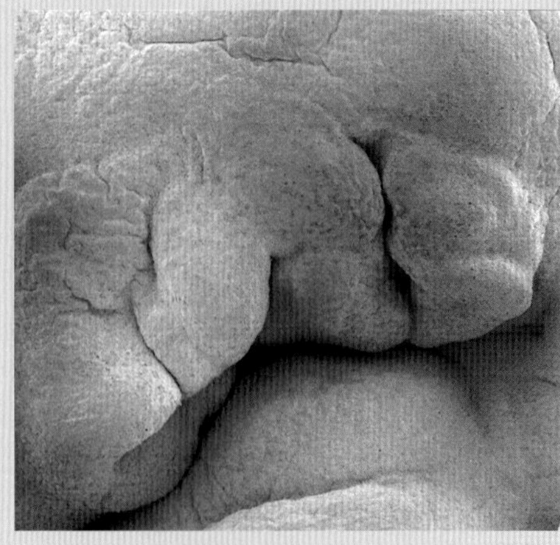

Week 7

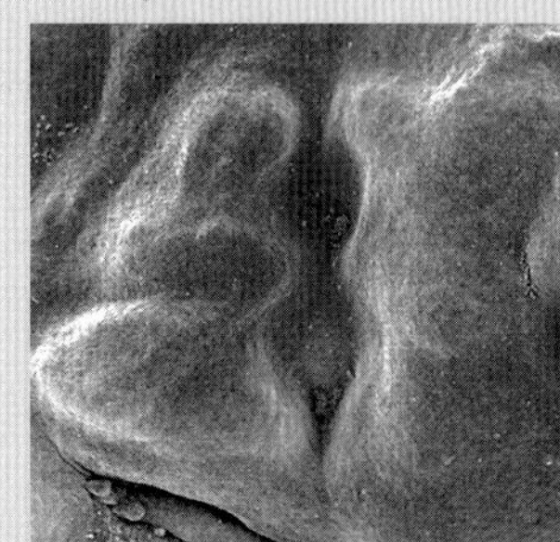

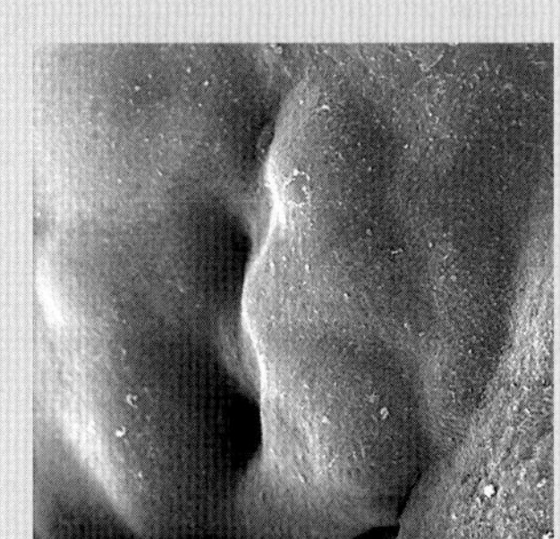

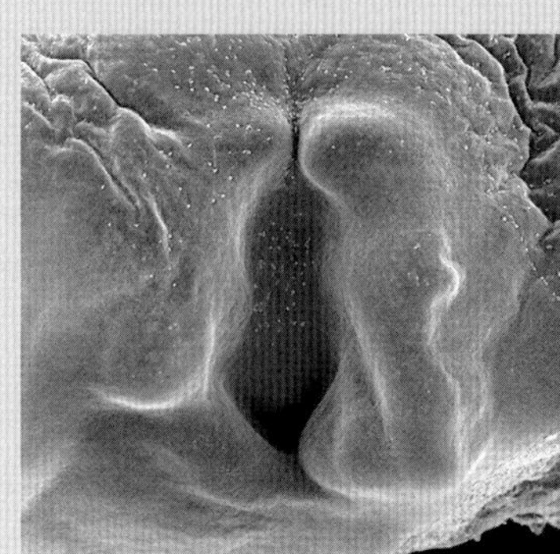

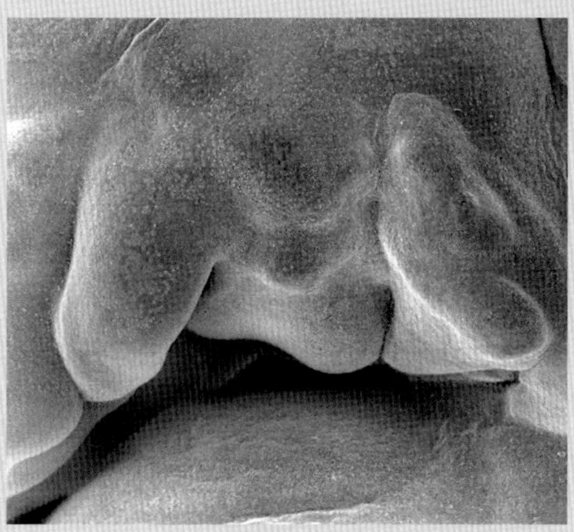

1.3.46 / 47

Pharyngeal arches, left view. Embryos at week 6.

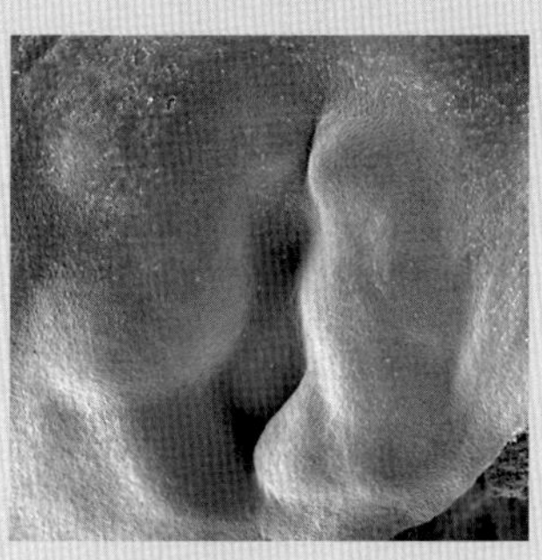

1.3.48 / 49

External ear, left view. Embryos early in week 7.

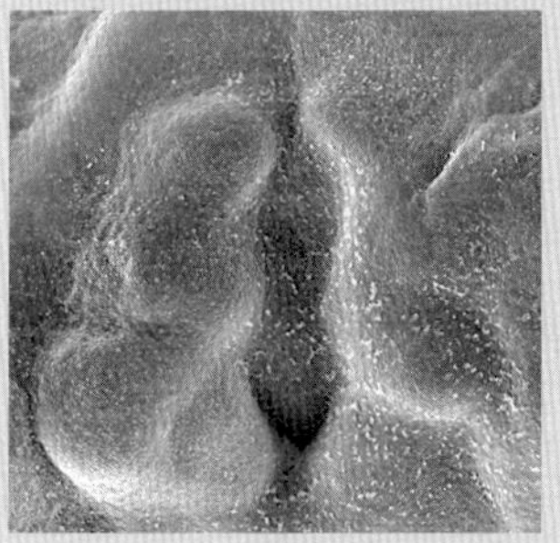

1.3.50 / 51

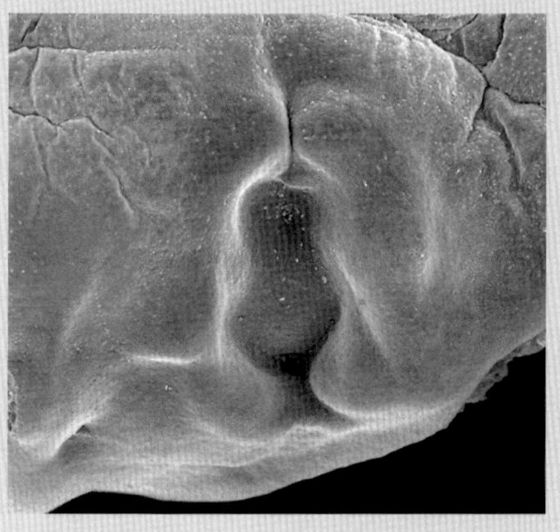

1.3.52 / 53

External ear, left view. Embryos in the middle of week 7.

1.3.36–1.3.61
Developmental stages of the external ear.

Week 8

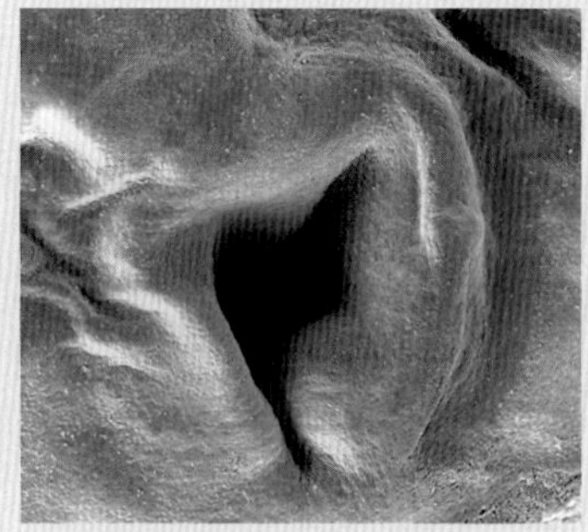

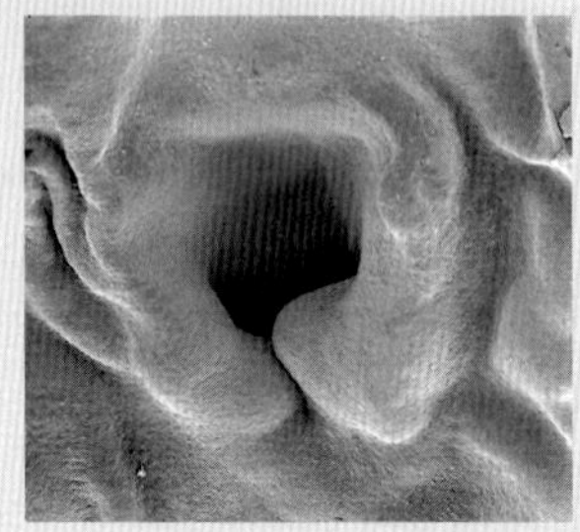

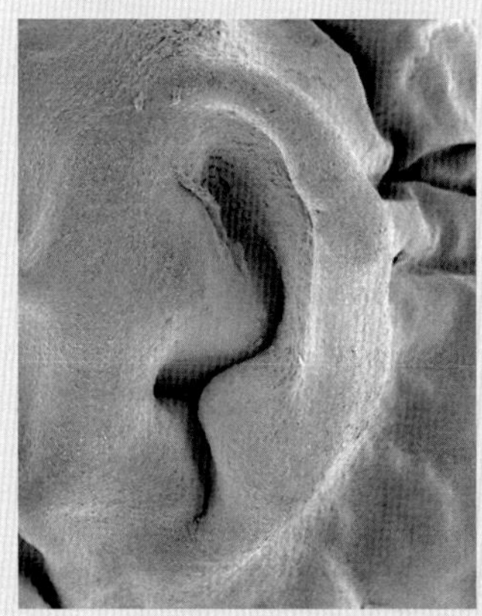

Week 9

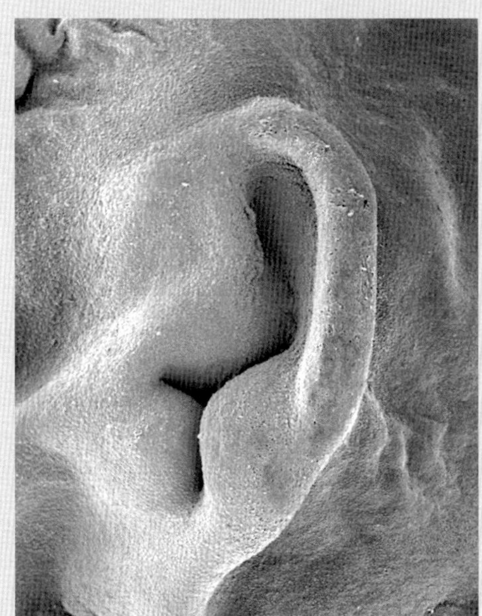

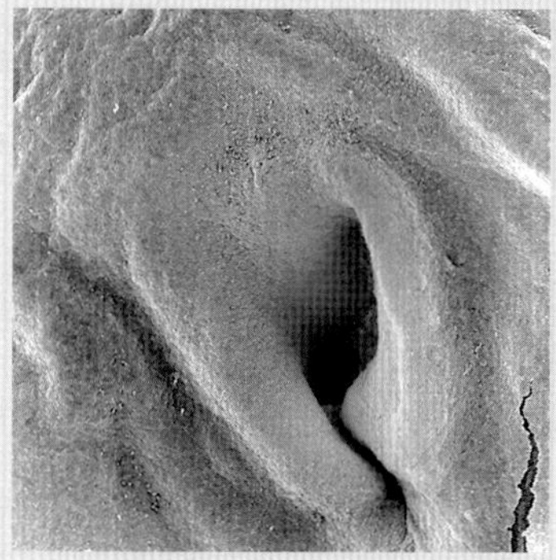

1.3.54 / 55

External ear, left view. Embryos at the beginning of week 8.

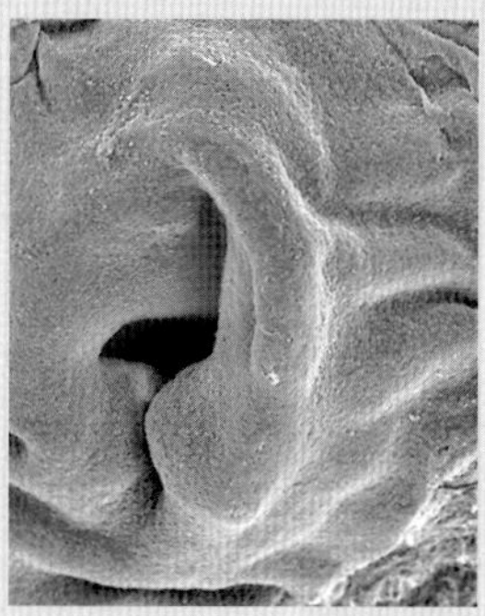

1.3.56 / 57

External ear, left view. Embryos in the middle of week 8.

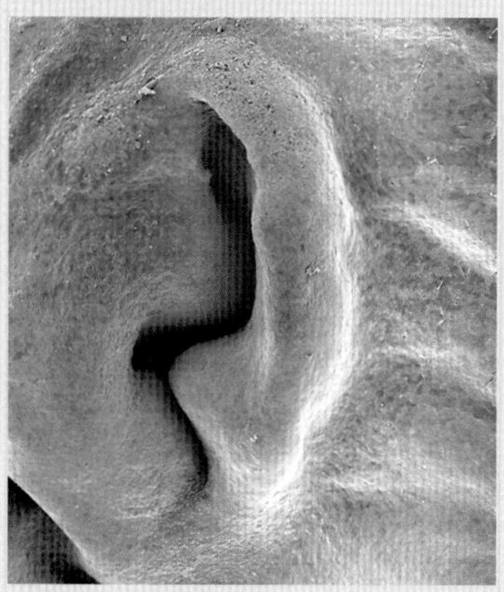

1.3.58 / 59

External ear, left view. Embryos at the end of week 8.

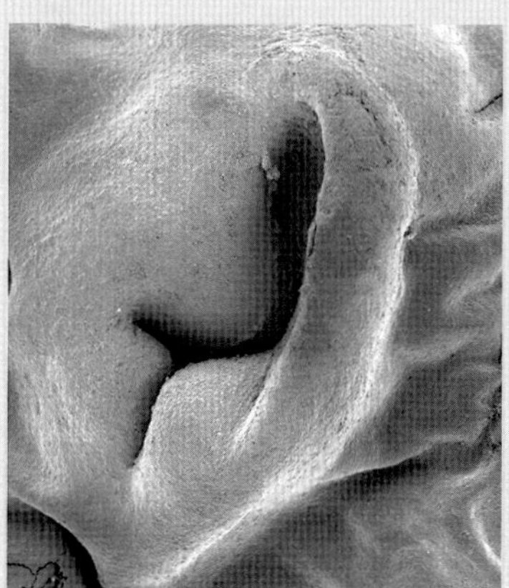

1.3.60 / 61

External ear, left view. Embryos at week 9.

1.
External Aspects

1.4

The Development of the Upper Limb

The first anlage of the upper limb becomes visible at the beginning of week 4 as a slight ectodermal fold bulging off from the body wall (fig. 1.4.01–1.4.03, 1.1.03 and 1.1.04). During development, the upper limb achieves its typical shape due to its dorsal/ventral and cranial/caudal and proximodistal differential growth. The cranial regions of the bud increase more than the caudal regions, and as a consequence the arm moves downward (fig. 1.4.04–1.4.07, 1.4.29–1.4.33). The dorsal surface enlarges more than the ventral surface (fig. 1.4.34–1.4.37) and as a consequence the arm performs a ventrad movement (adduction) in the upper and, a bit later, in the lower arm, and the hand (fig. 1.4.38, 1.4.39).

The distal portion of the bud becomes broader than the proximal region and will form the hand (fig. 1.4.07–1.4.12).

The above-mentioned movements of the entire limb are a consequence of local differential ectodermal growth. Additionally to these movements, the adjacent regions of the shoulder girdle, the upper arm, the forearm, and the hand become kinked in their transition regions (upper arm vs. shoulder girdle: fig. 1.4.34 and 1.4.35; forearm vs. lower arm: fig. 1.4.10 and 1.4.13; hand vs. forearm: fig. 1.4.13 and 1.4.14).

These developmental movements are the main causes for the localization of the joints of the shoulder (glenohumeral), the elbow and the wrist. Even the extent and directions of excursions in these joints of the adult are determined by the range of embryonic movement patterns.

Other consequences of these early ectodermal growth conditions are the local differentiations of the mesenchyme into cartilage, bones, and muscles, as well as the patterning of the vascular system.

1.
External Aspects

1.4
The Development of the Upper Limb

1.4.01–1.4.40
Developmental stages of the upper limb.

1.4.04–1.4.11
Left upper limb. Lateral view.

Week 4

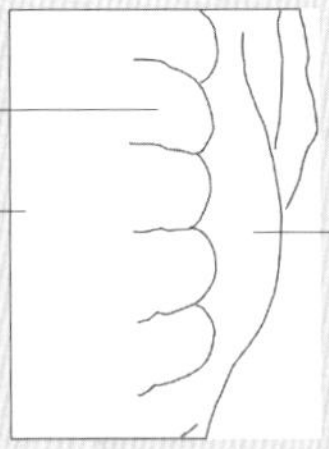

Week 5

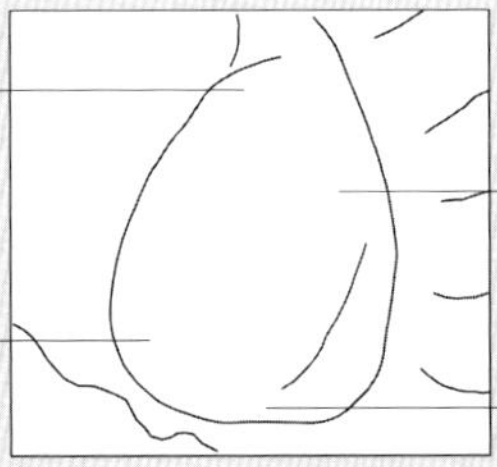

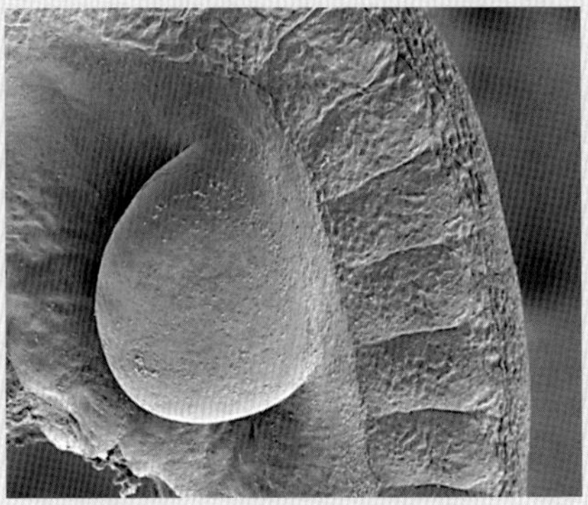

1.4.01

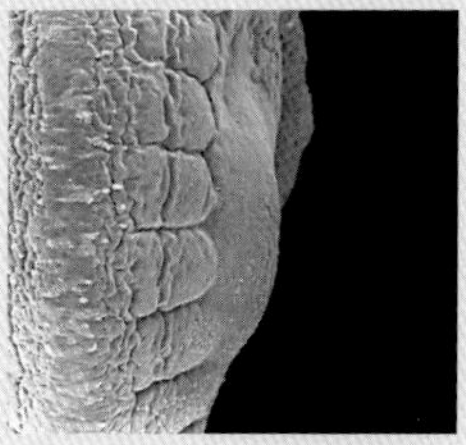

Right upper limb bud. Lateral-dorsal view. Embryos early, in the middle of and late in week 4.

1.4.02

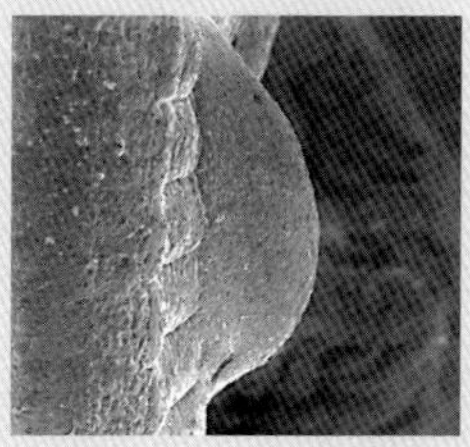

1.4.03

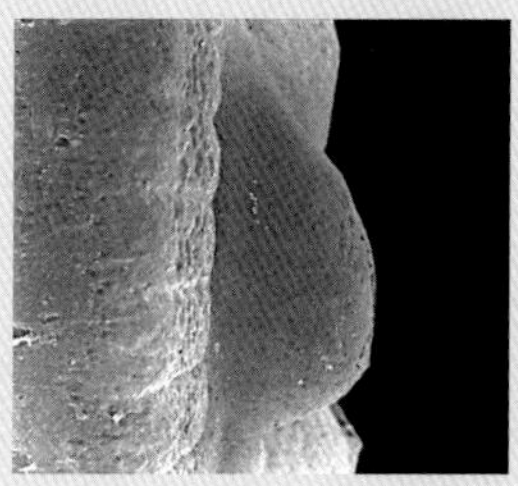

1.4.04

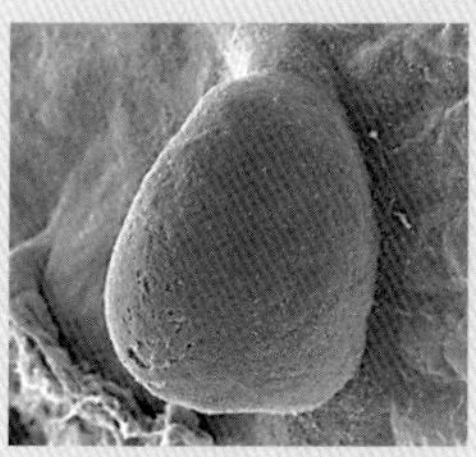

Embryos at week 5.

1.4.05 / 06

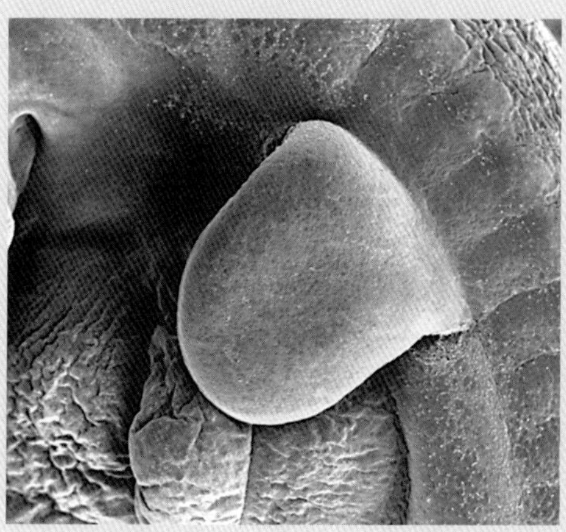

1.4.01–1.4.40
Developmental stages of the upper limb.

1.4.04–1.4.11
Left upper limb. Lateral view.

Week 6

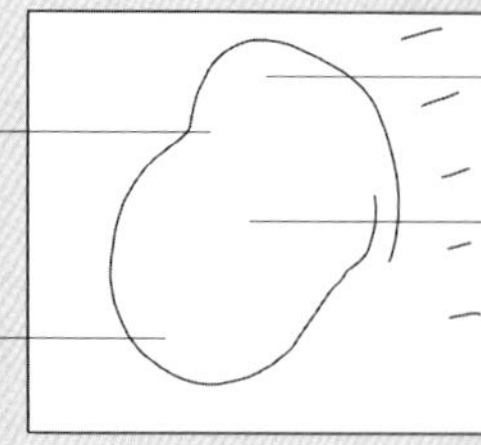

Week 7

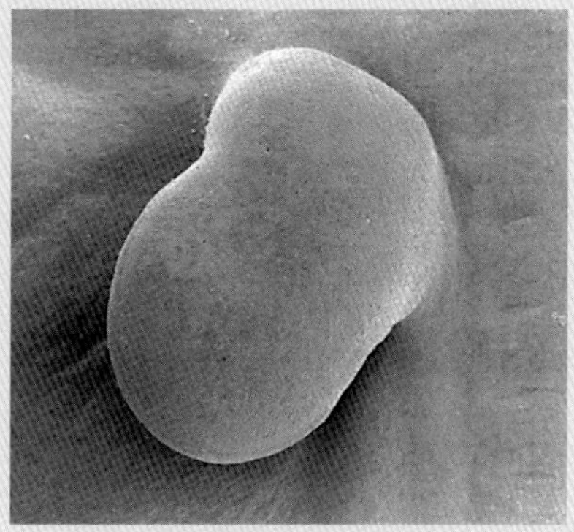

1.4.07 Embryos at week 6.

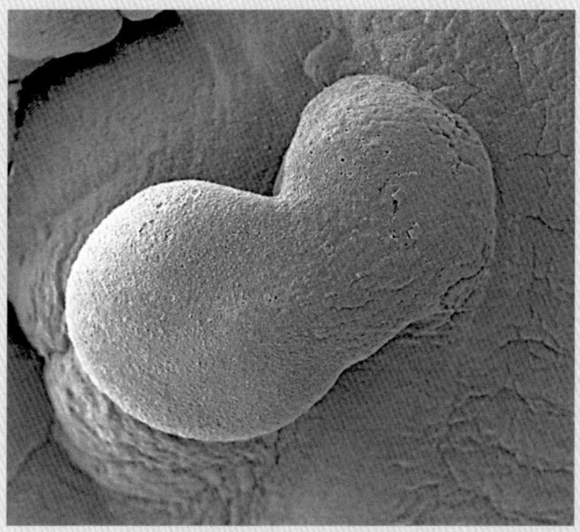

1.4.08

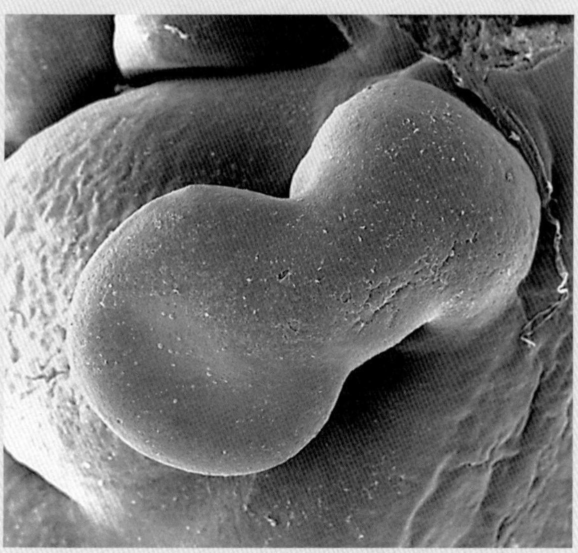

1.4.09 Embryo at week 7.

Week 7

1.4.01–1.4.40
Developmental stages of the upper limb.

1.4.04–1.4.11
Left upper limb. Lateral view.

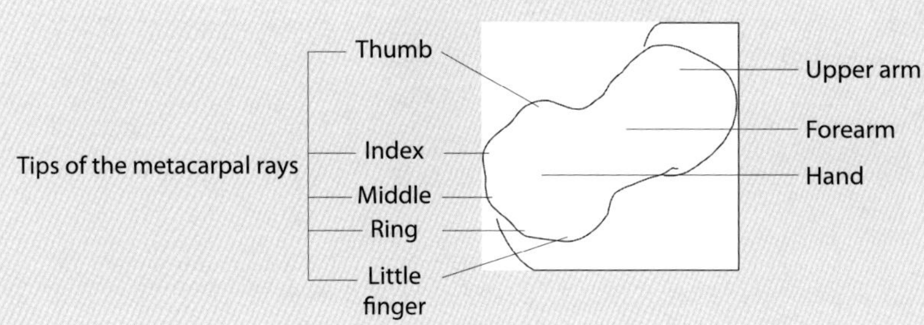

1.4.10

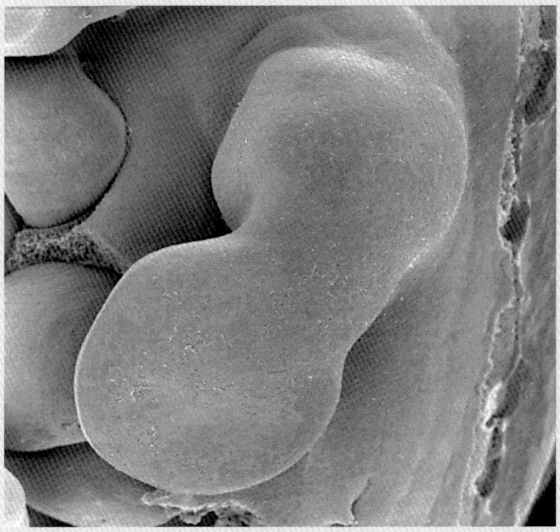

Embryos at week 7.

1.4.11

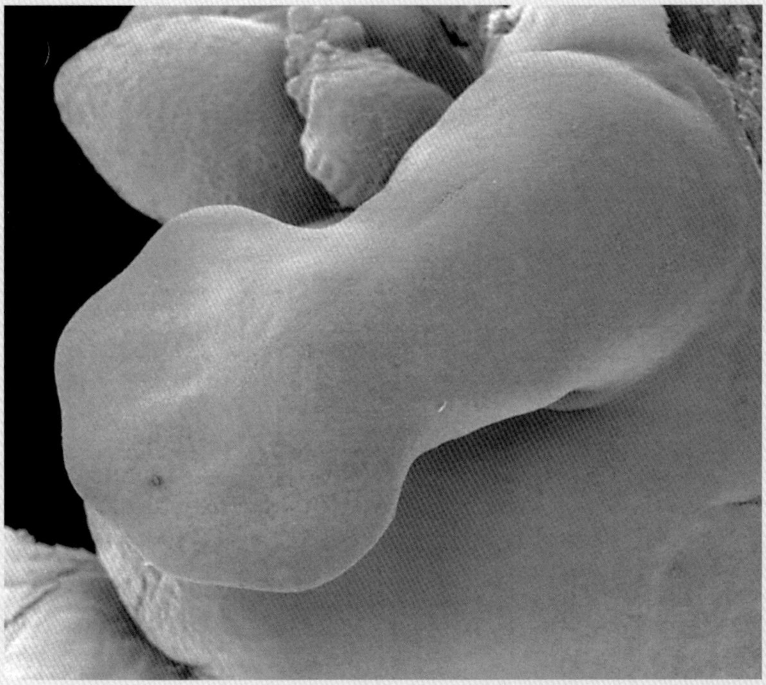

1.4.01–1.4.40
Developmental stages of the upper limb.

1.4.11a–1.4.13
Left upper limb. Lateral view.

Week 7

Week 8

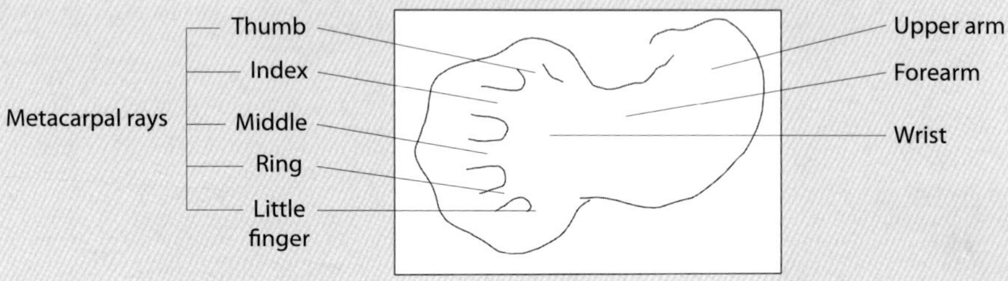

Week 9

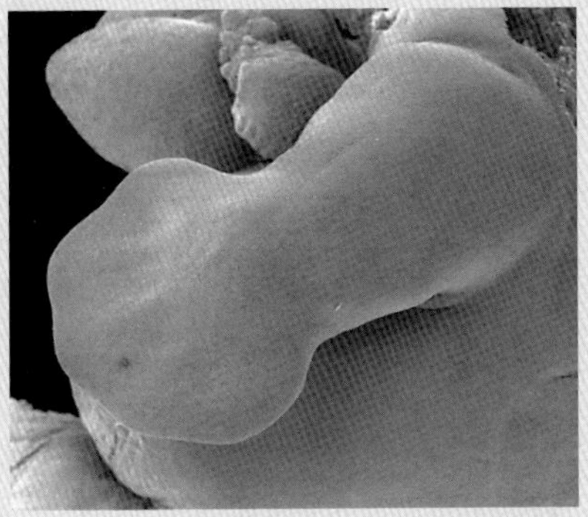

1.4.11a Same embryo as in figure 1.4.11.

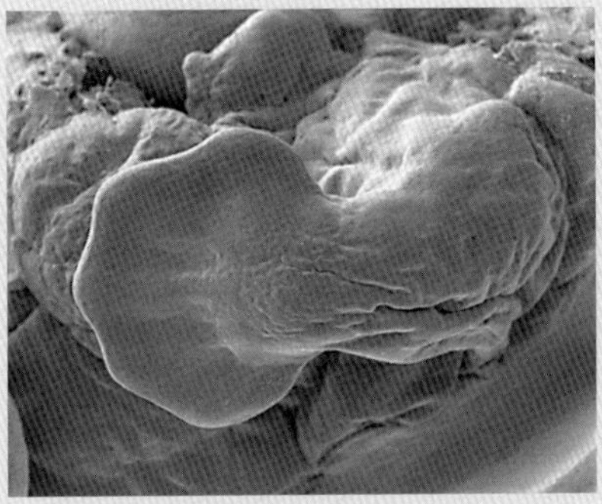

1.4.12 Embryo early in week 8.

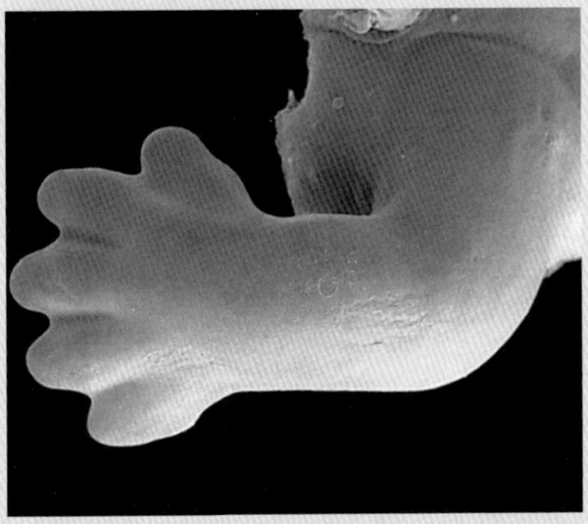

1.4.13 Embryo at the end of week 8.

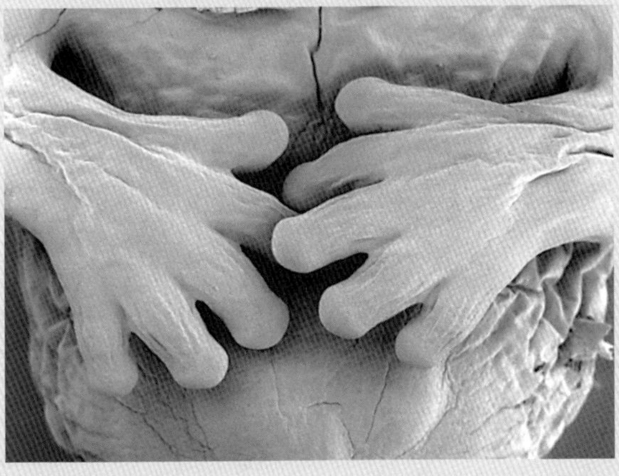

1.4.14 Embryo at week 9. Ventral view.

Week 8

1.4.01–1.4.40
Developmental stages of the upper limb.

1.4.15–1.4.19
Developmental stages of the left hand and the fingers. Volar view.

Week 8 / 9

Week 9

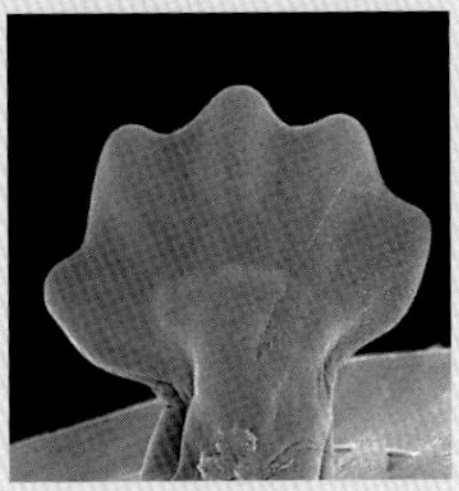

1.4.15 Embryo in the middle of week 8.

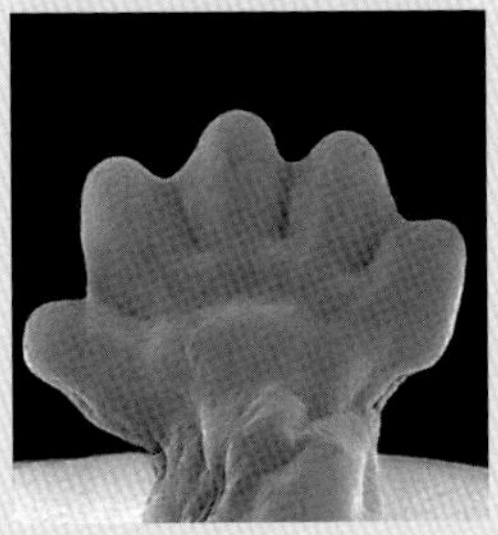

1.4.16 Embryo late in week 8.

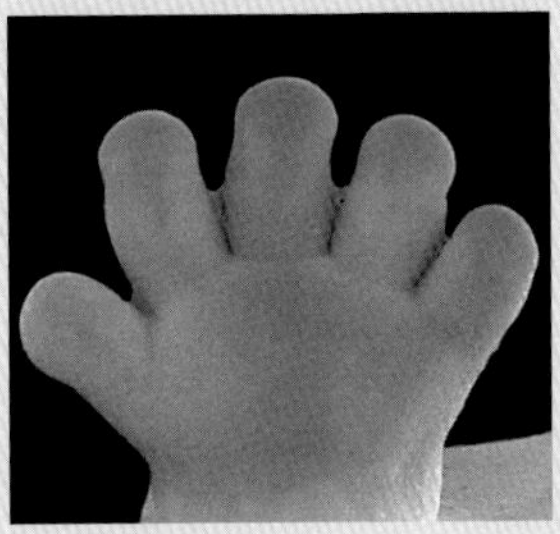

1.4.17 Embryo late in week 8/ early in week 9.

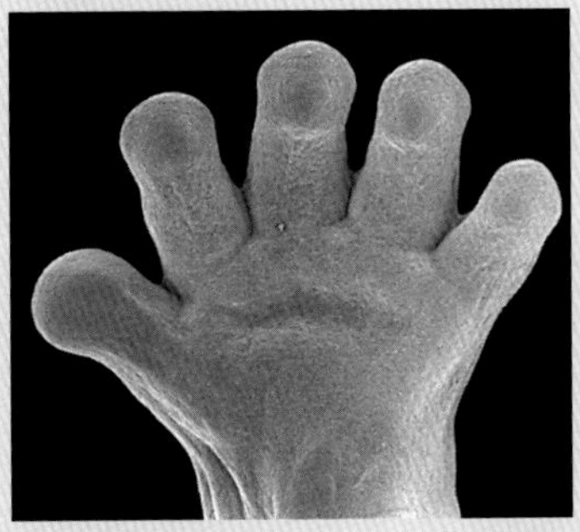

1.4.18 Embryo at week 9.

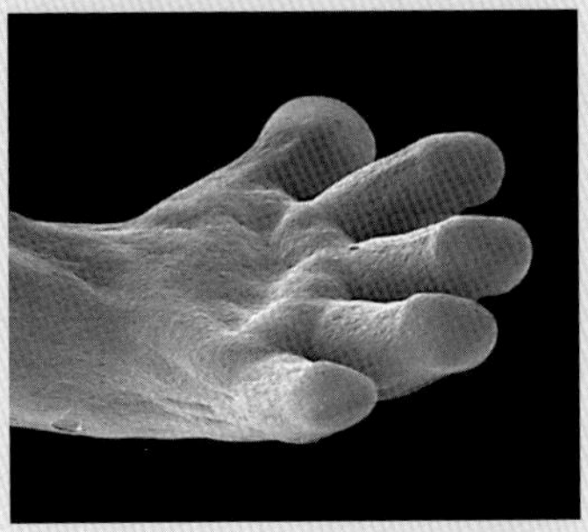

1.4.19 Same embryo as in figure 1.4.18.

Week 4 / 5

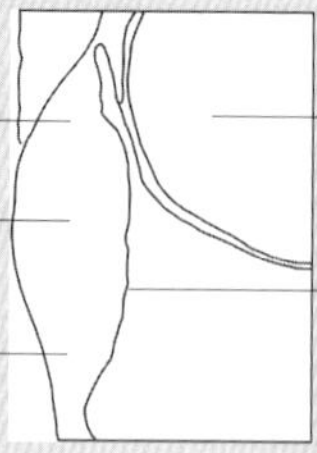

1.
External Aspects

1.4
The Development of the Upper Limb

1.4.01–1.4.40
Developmental stages of the upper limb.

1.4.20–1.4.28
Developmental stages of the upper limb. Ventral view.

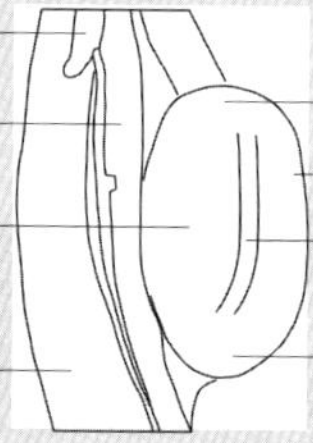

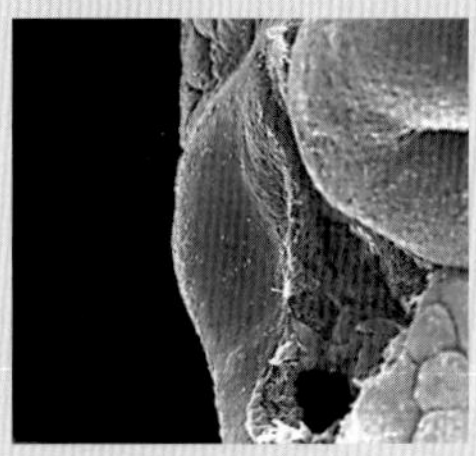

1.4.20

Right arm. Embryos at week 4 and early in week 5.

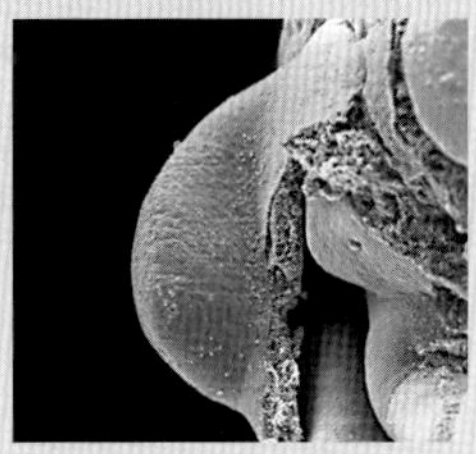

1.4.21

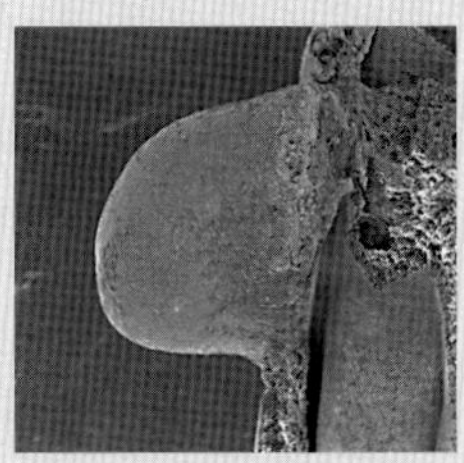

1.4.22

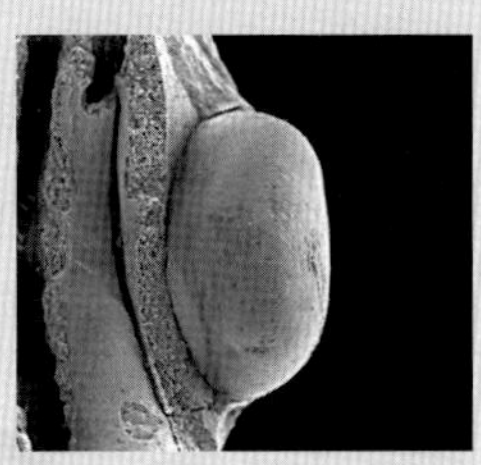

1.4.23

Left arm. Embryos in the middle and at the end of week 5.

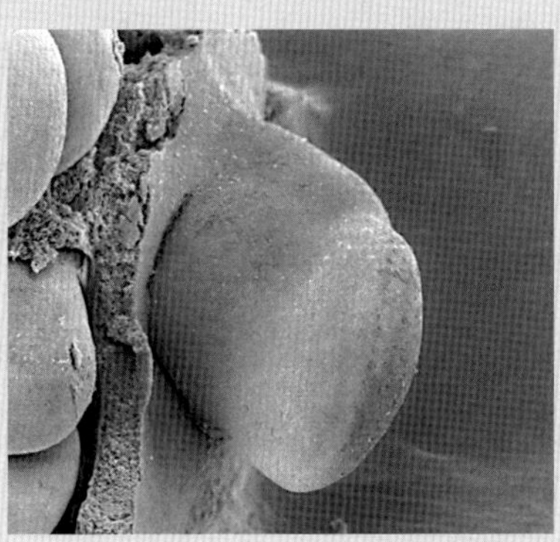

1.4.24

Week 6

1.4.01–1.4.40
Developmental stages of the upper limb.

1.4.20–1.4.28
Developmental stages of the upper limb. Ventral view.

Week 8

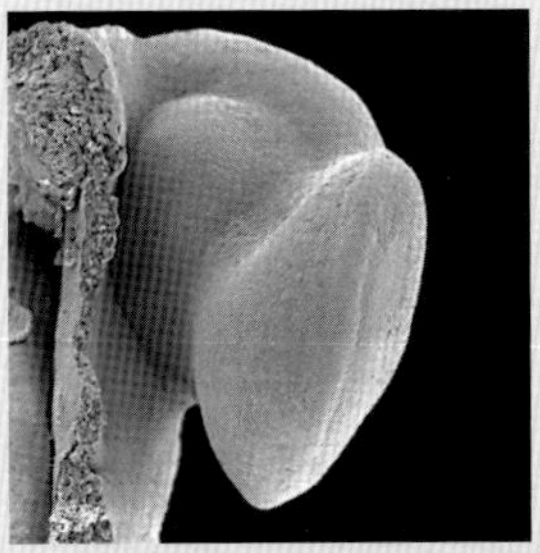

1.4.25

Left arm. Embryos at week 6. In the stage shown in 1.4.27, the palm has started its downward movement (pronation).

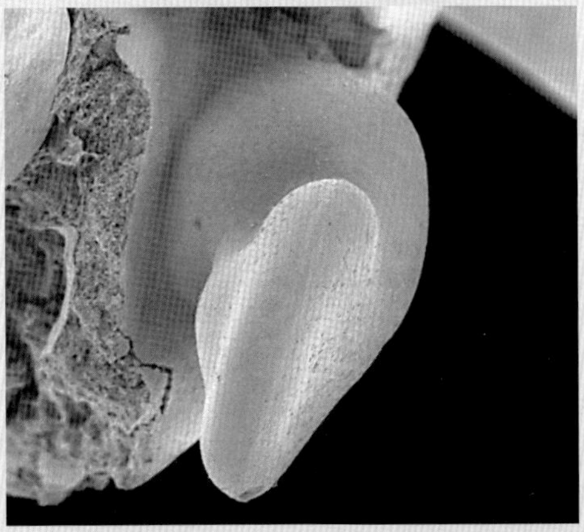

1.4.26

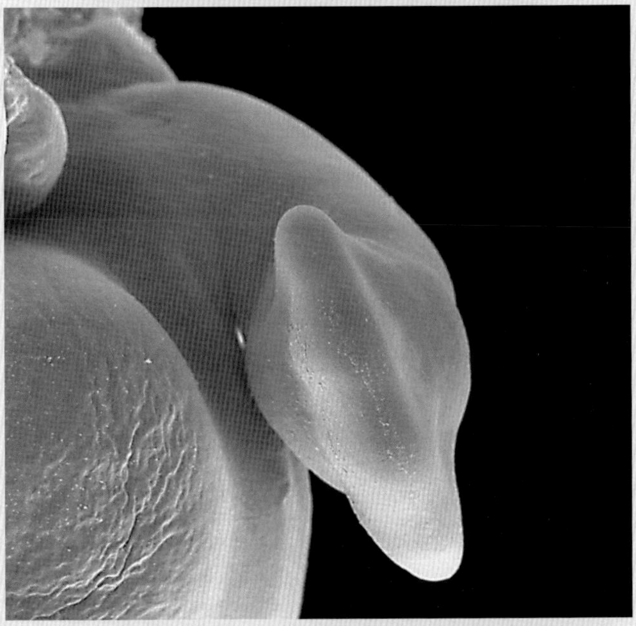

1.4.27

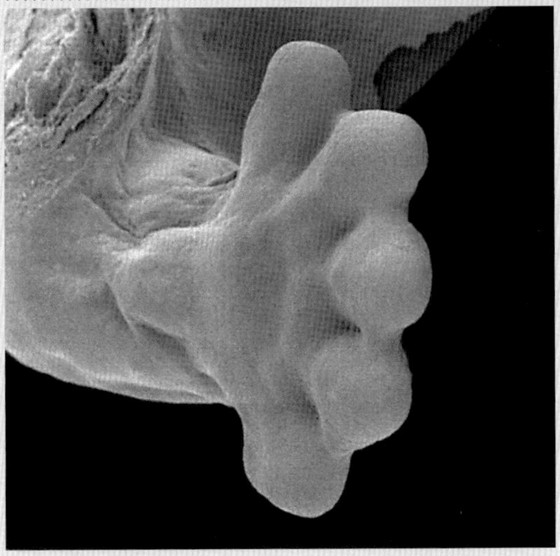

1.4.28

Left arm. Embryo at week 8.

Week 5

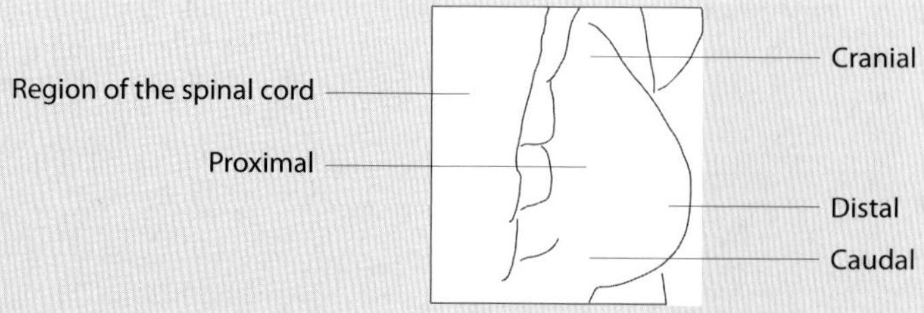

1.4.01–1.4.40
Developmental stages of the upper limb.

1.4.29–1.4.33
Developmental stages of the upper limb. Dorsal view.

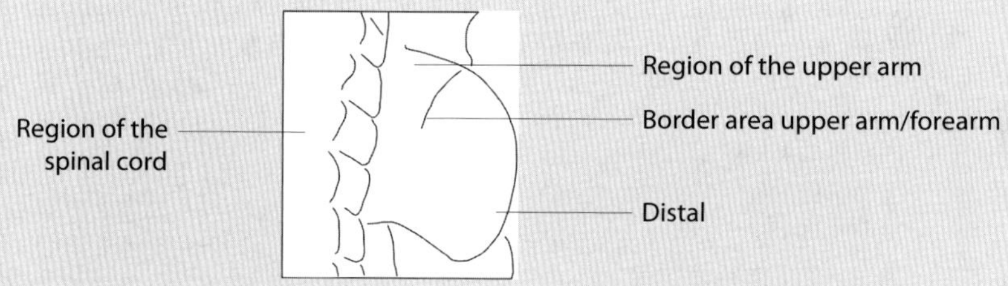

Week 6 / 7

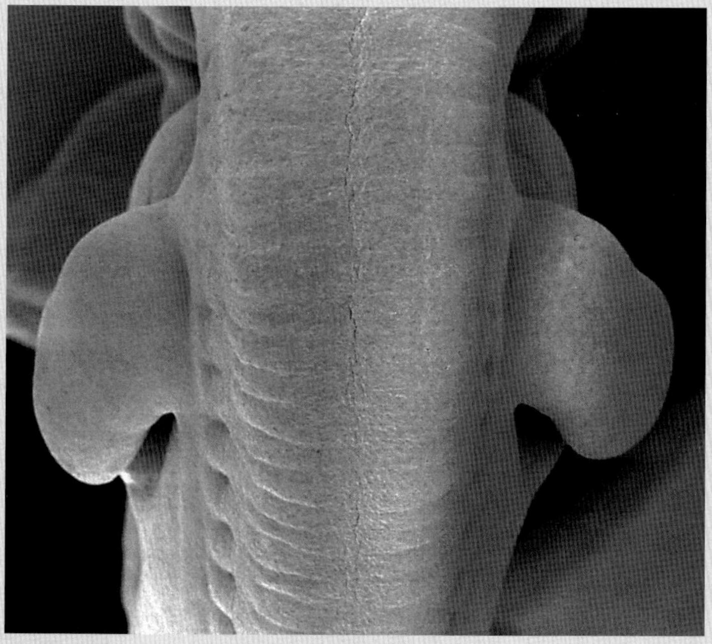

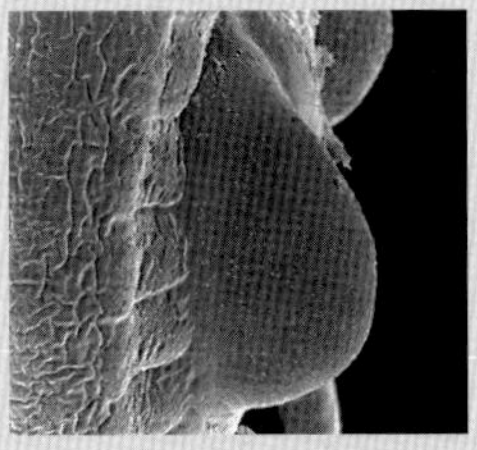

1.4.29 Embryos at week 5.

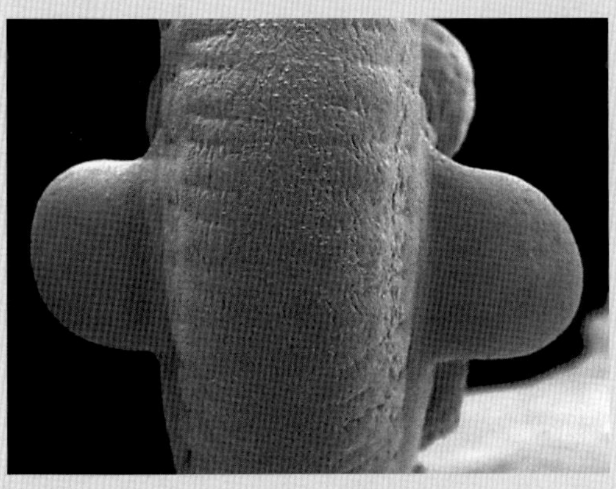

1.4.30

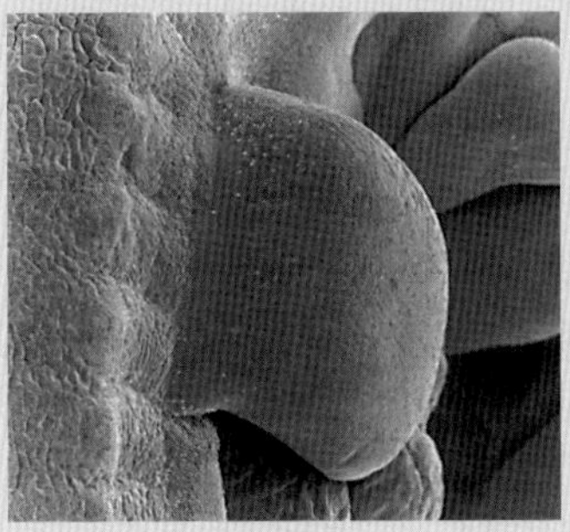

1.4.31

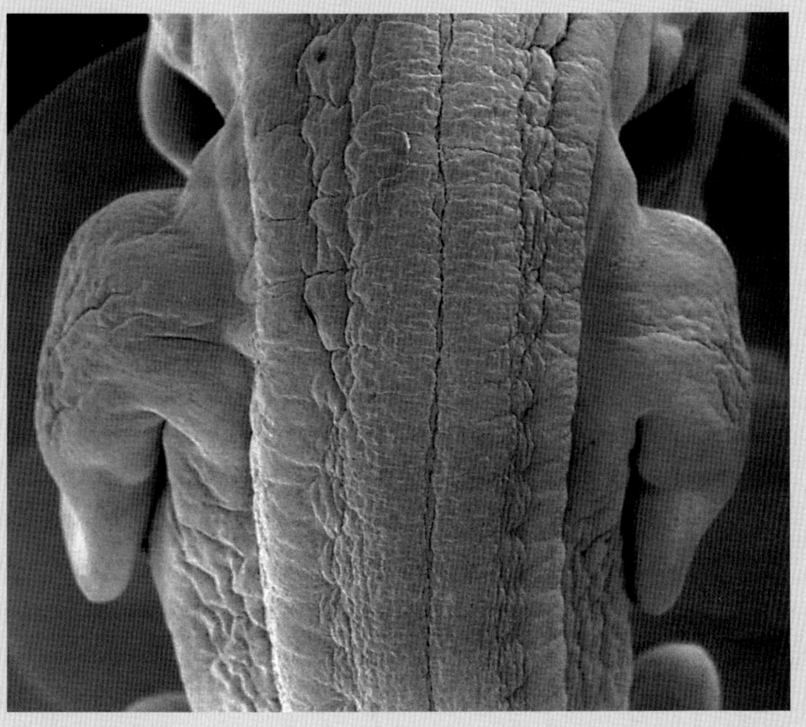

1.4.32 / 33 Embryos at weeks 6 and 7.

Week 5

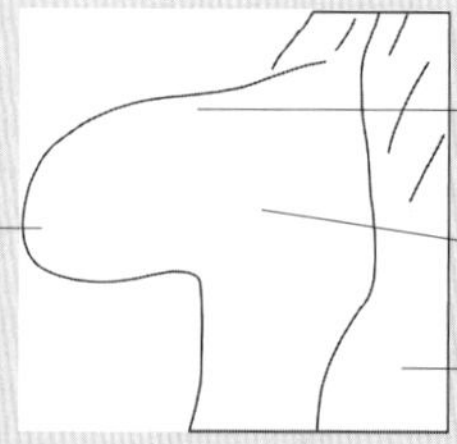

Week 6

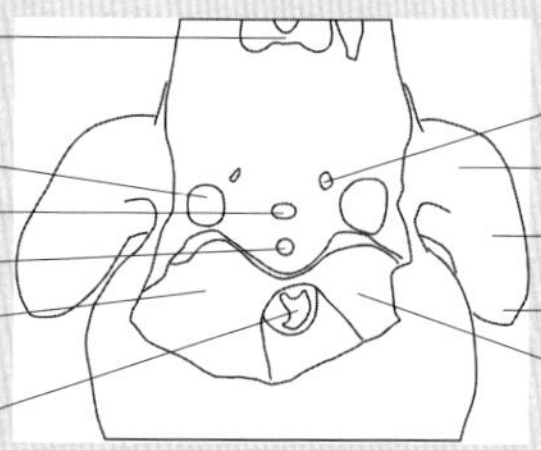

1.4.01–1.4.40
Developmental stages of the upper limb.

1.4.34–1.4.40
Developmental stages of the upper limb. Cranial view.

1.4.34 Embryos at week 5.

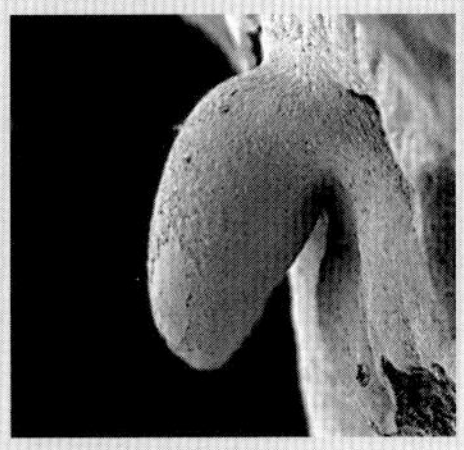

1.4.35

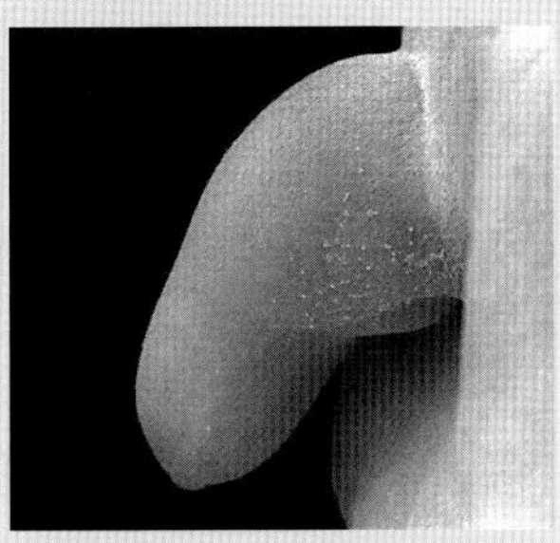

1.4.36

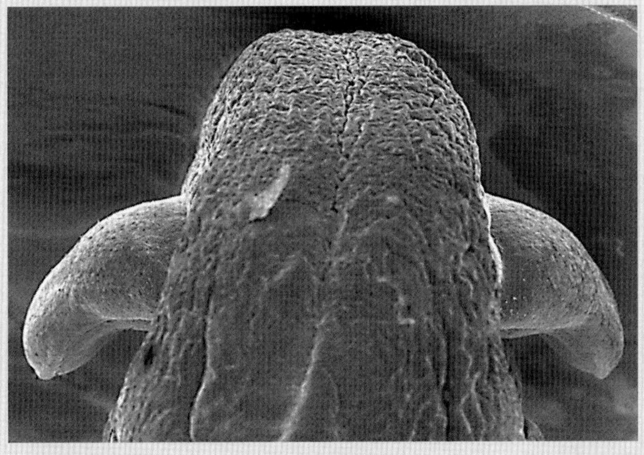

1.4.37 Embryos at week 6.

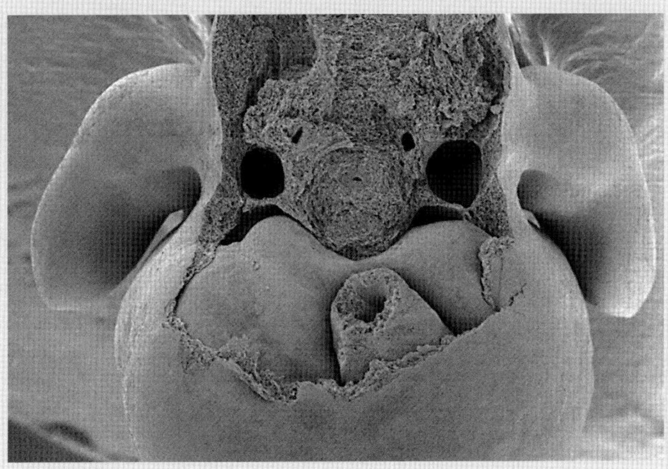

1.4.38

30:1

Week 7

1.4.01–1.4.40
Developmental stages of the upper limb.

1.4.34–1.4.40
Developmental stages of the upper limb. Cranial view.

15:1

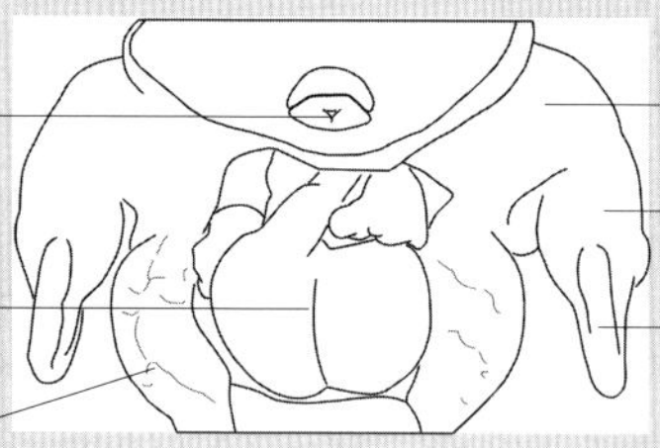

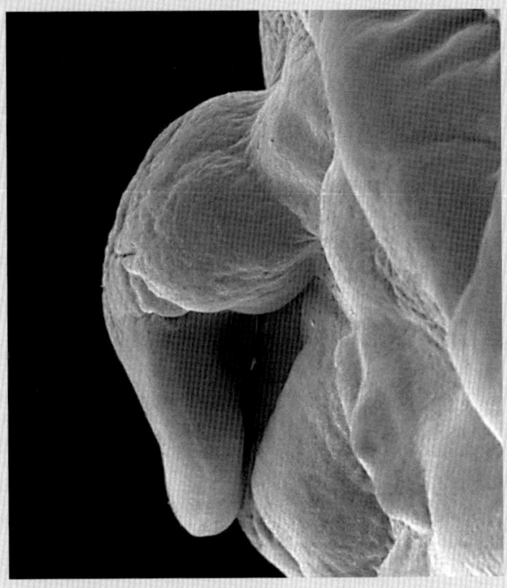

1.4.39 Embryo at week 7.

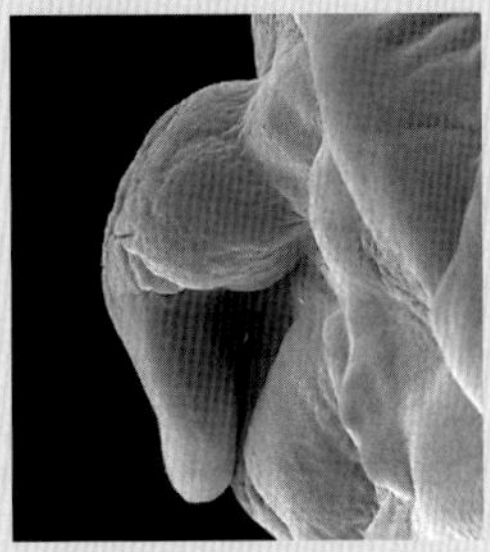

1.4.39a Embryo at week 7.

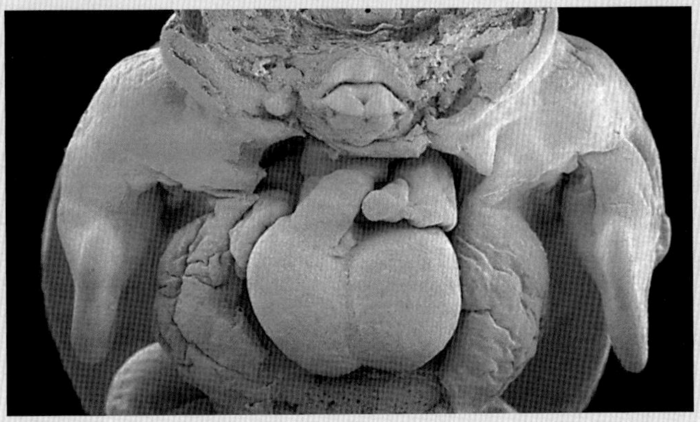

1.4.40 Embryo at the end of week 7. The thorax and the pericardial cavity have been opened.

1.
External Aspects

1.5

The Development of the Lower Limb

The first anlage of the lower limb becomes visible at the beginning of week 4, shortly after that of the upper limb, as a slight ectodermal fold bulging off from the body wall (fig. 1.5.01–1.5.03 and 1.1.03–1.1.05). Beginning with the first appearance of the lower limb bud, its form is inconspicuously different from that of the arm (see fig. 1.5.01 and 1.4.01/1.4.02, 1.5.04/1.5.05 and 1.4.04/1.4.05, and 1.5.23 and 1.4.21: whereas the tip of the arm is more rounded, the lower limb is more tapering).

During development, the limb achieves its typical shape due to its dorsal/ventral and cranial/caudal and proximodistal differential growth. The dorsal surface enlarges more than the ventral surface (fig. 1.5.17a–1.5.19) and as a consequence the leg performs a ventrad movement in the future thigh region and, shortly afterwards, in the lower thigh and foot region (fig. 1.5.26a–1.5.30). The cranial regions of the bud increase more than the caudal regions, and as a consequence the leg moves downward (fig. 1.5.14–1.5.19, 1.5.23–1.5.26). The distal portion of the bud becomes broader than the adjoining region and will form the foot (fig. 1.5.09).

The movements of the entire limb noted above are a consequence of local differential ectodermal growth. Additionally to these movements, the adjacent regions of the hip, the thigh, the lower thigh, and the foot become kinked in their transition regions (thigh vs. hip: fig. 1.5.25 and 1.5.26; lower thigh vs. upper thigh: fig. 1.5.20–1.5.22; foot vs. lower thigh: fig. 1.5.22 and 1.5.26).

These developmental movements are the main reasons for the localization of the joints of the hip, the knee and the foot. Even the extent and directions of excursions in these joints of the adult are laid down by the range of embryonic movement patterns.

Other consequences of these early ectodermal growth conditions are the local differentiations of the mesenchyme into cartilage, bones, and muscles, as well as the patterning of the vascular system.

1.
External Aspects

1.5
The Development of
the Lower Limb

1.5.01–1.5.30
Developmental stages of the lower limb.

1.5.01–1.5.09
Lateral-left view of the left lower limb.

Week 4 / 5

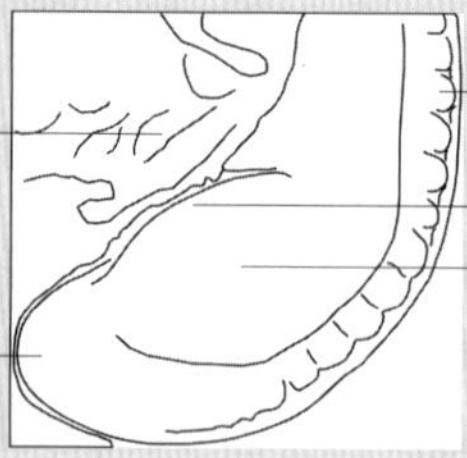

Week 5

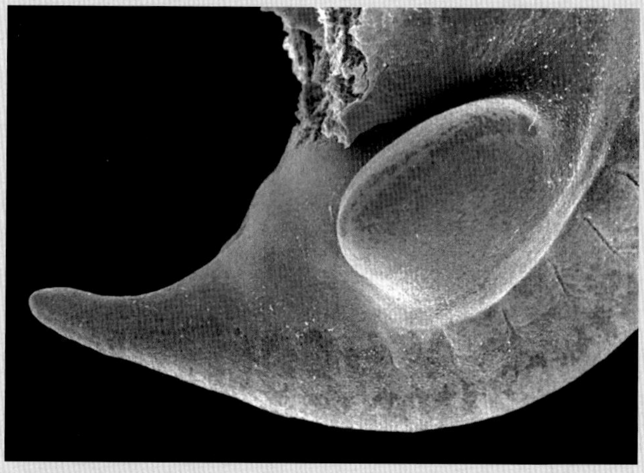

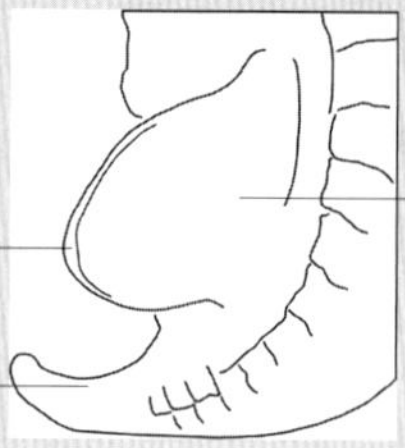

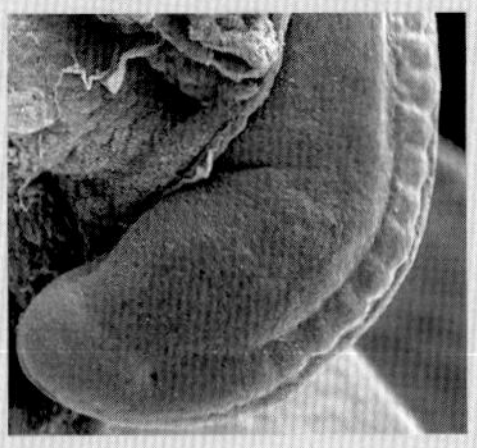

1.5.01

Embryos at the end of week 4 and early in week 5.

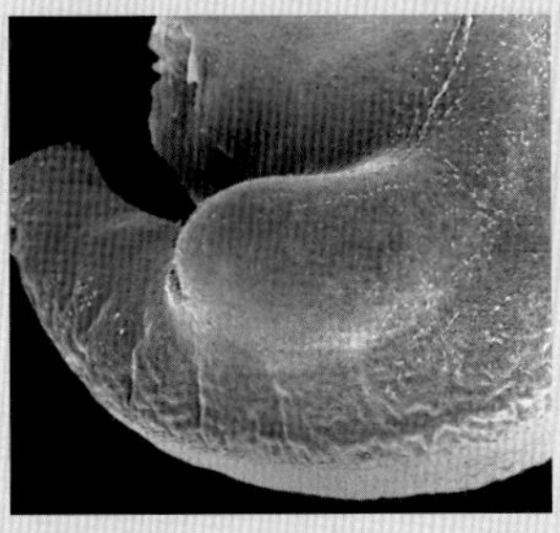

1.5.02

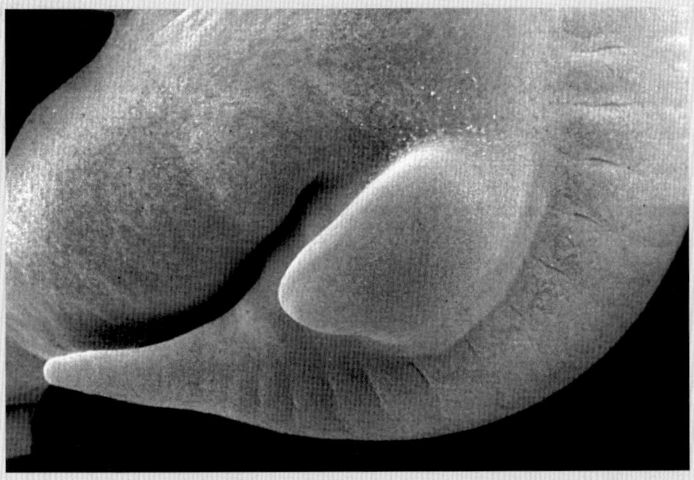

1.5.03 / 04

Embryos in the middle and at the end of week 5.

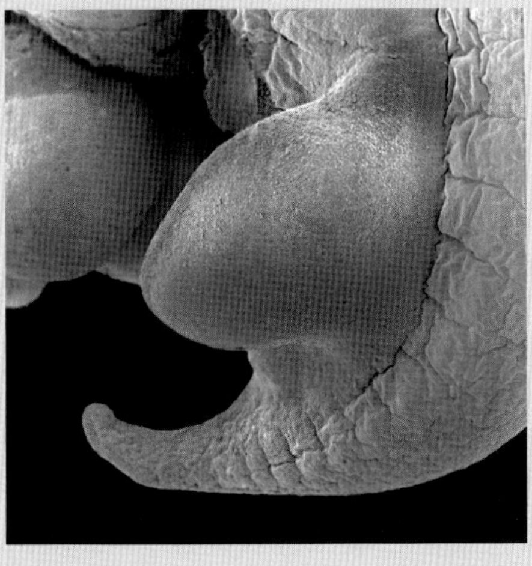

1.5.05

Embryo at the end of week 5.

Week 6

1.5.01–1.5.30
Developmental stages of the lower limb.

1.5.01–1.5.09
Lateral-left view of the left lower limb.

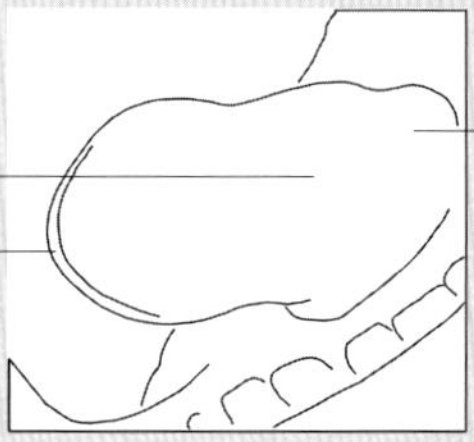

Week 7

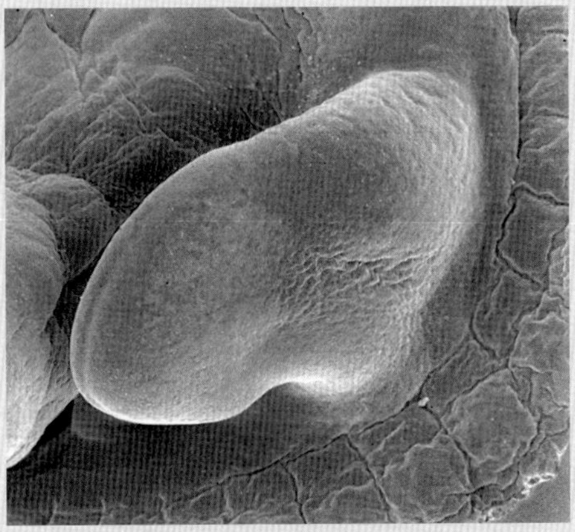

1.5.06 Embryos in the middle of week 6.

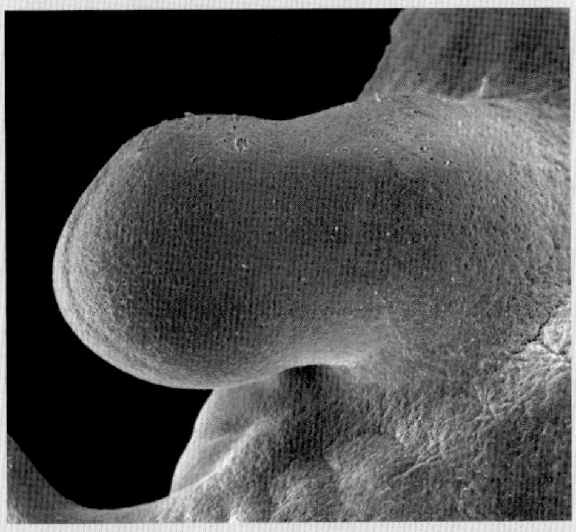

1.5.07

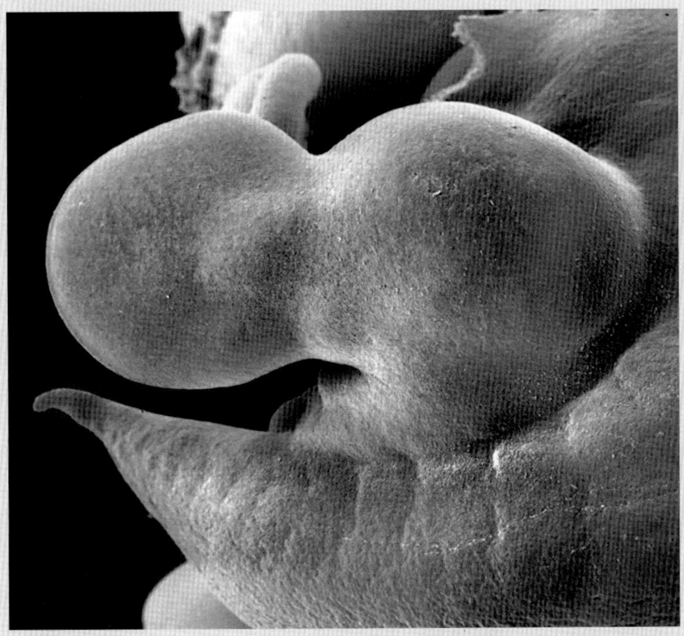

1.5.08 Same embryo as at week 7.

Week 7

1.
External Aspects

1.5
The Development of the Lower Limb

1.5.01–1.5.30
Developmental stages of the lower limb.

1.5.01–1.5.09
Lateral-left view of the left lower limb.

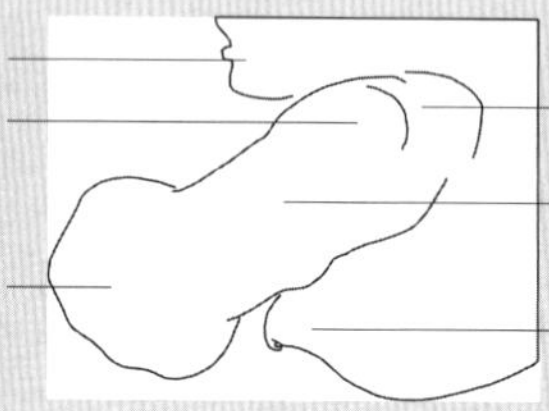

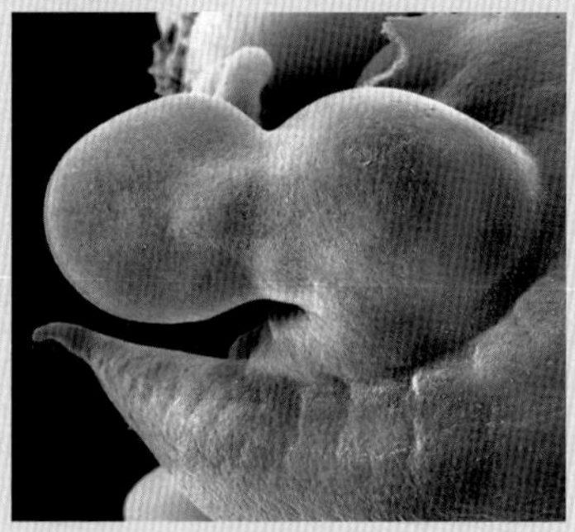

1.5.08a Same embryo as at week 5.

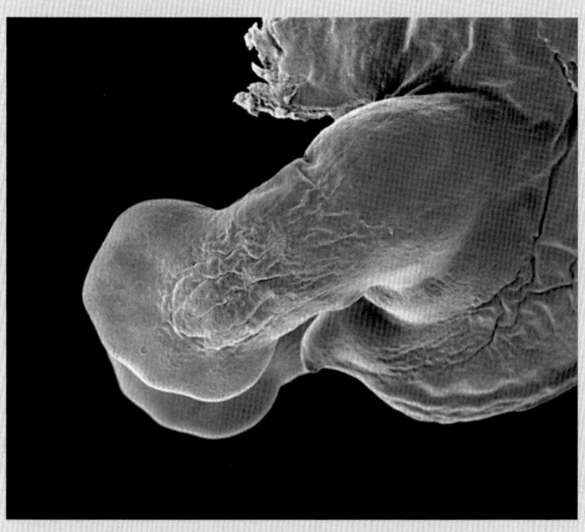

1.5.09 Embryo at the end of week 7.

15:1

Week 8

1.5.01–1.5.30
Developmental stages of the lower limb.

1.5.10–1.5.13
Plantar aspects of the left foot.

Week 9

Week 10

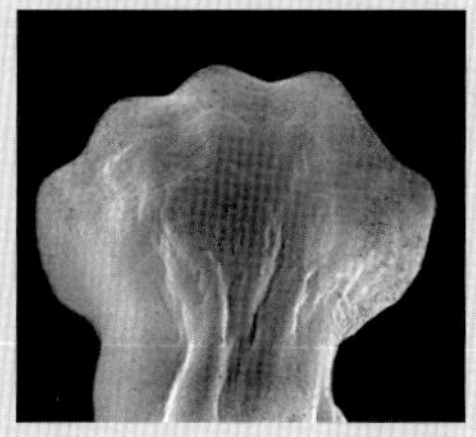

1.5.10 Embryo at week 8.

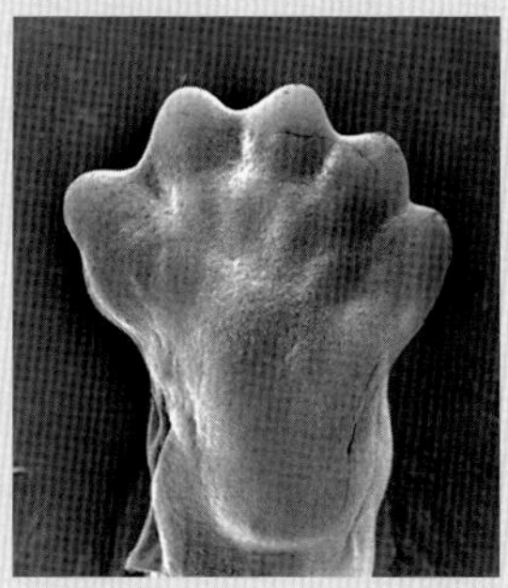

1.5.11 Embryo at the end of week 8.

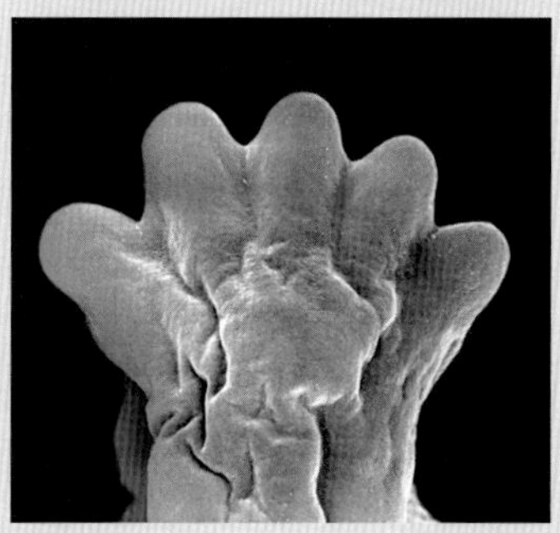

1.5.12 Embryo at week 9.

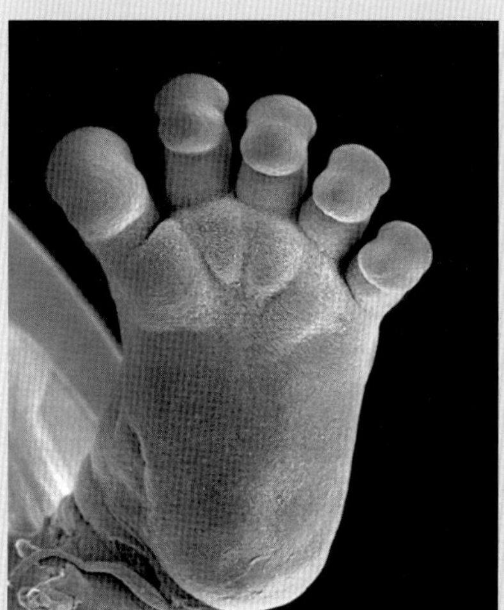

1.5.13 Embryo at week 10.

1.
External Aspects

1.5
The Development of the Lower Limb

1.5.01–1.5.30
Developmental stages of the lower limb.

1.5.14–1.5.22
Stages of the development of the lower limb: dorsal (1.5.14–1.5.19, 1.5.21), cranial (1.5.20), and caudal (1.5.22) views.

Week 5

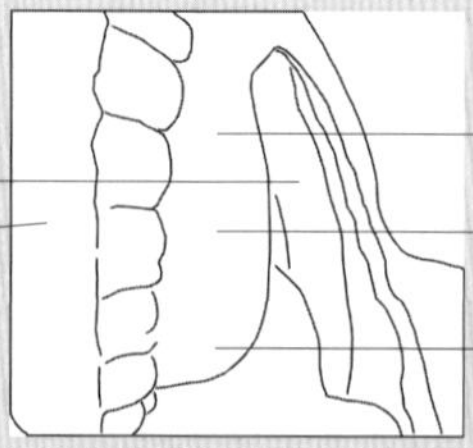

Week 5 / 6

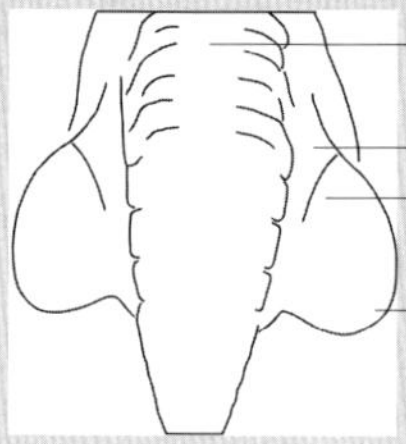

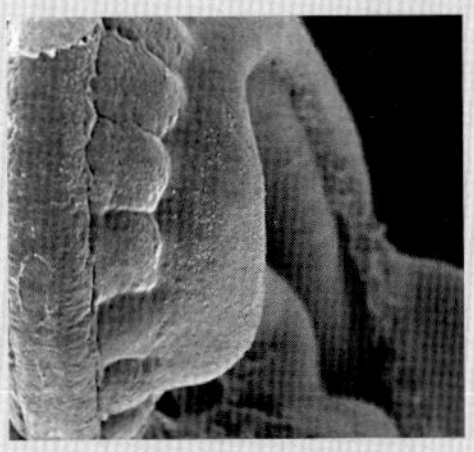

1.5.14 Embryo at the beginning of week 5.

1.5.15 Embryos at the end of week 5.

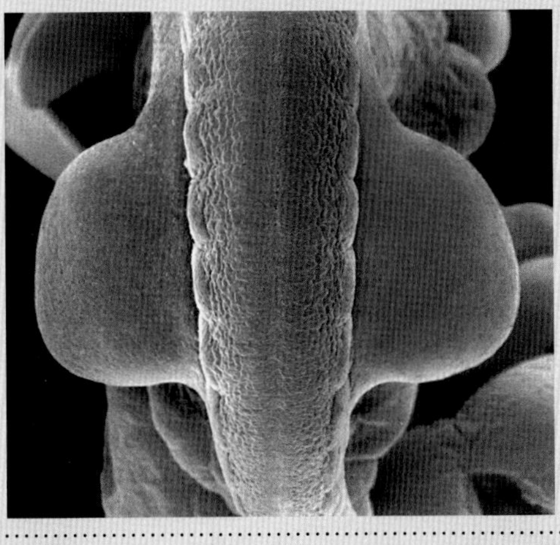

1.5.16

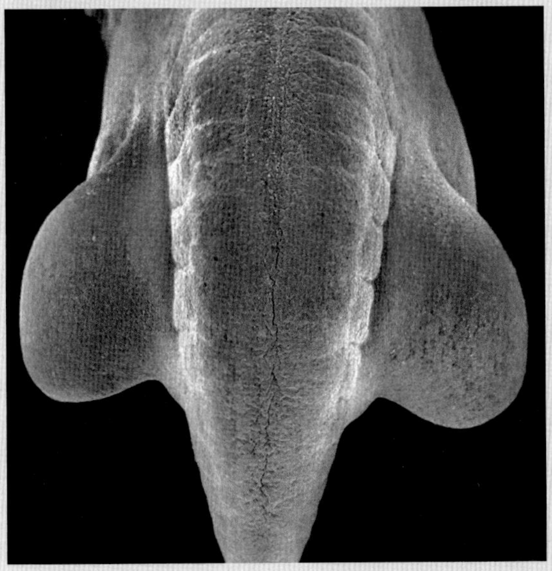

1.5.17 Same embryo as at the end of week 5/ early in week 6.

Week 5 / 6

1.
External Aspects

1.5
The Development of the Lower Limb

Week 6

1.5.01–1.5.30
Developmental stages of the lower limb.

1.5.14–1.5.22
Stages of the development of the lower limb: dorsal (1.5.14–1.5.19, 1.5.21), cranial (1.5.20), and caudal (1.5.22) views.

Week 7

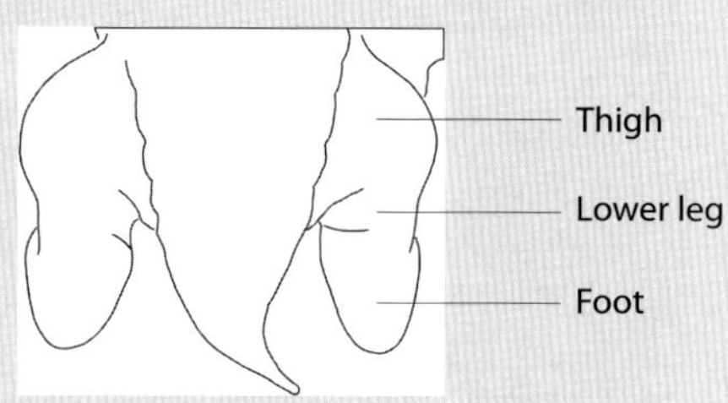

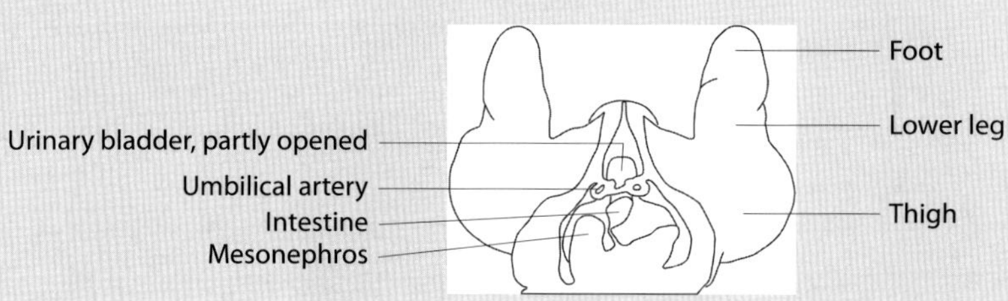

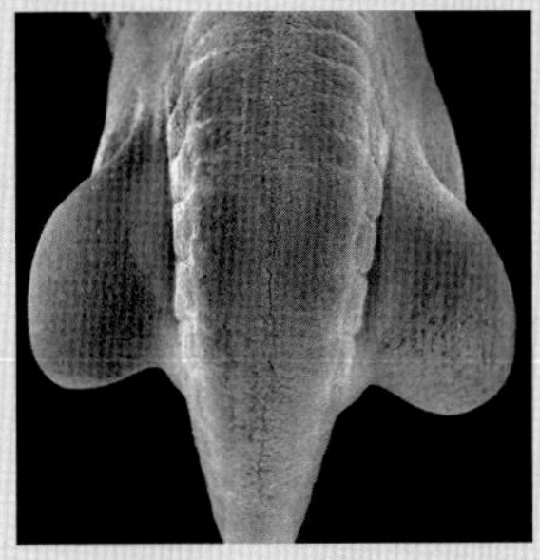

1.5.17a Same embryo as at the end of week 5/ early in week 6.

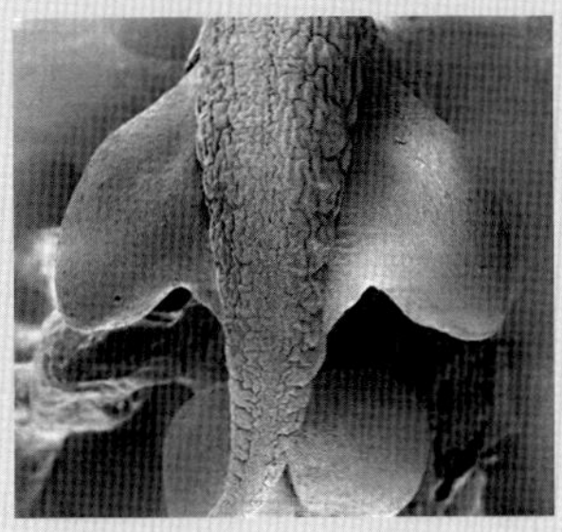

1.5.18 Embryo at week 6.

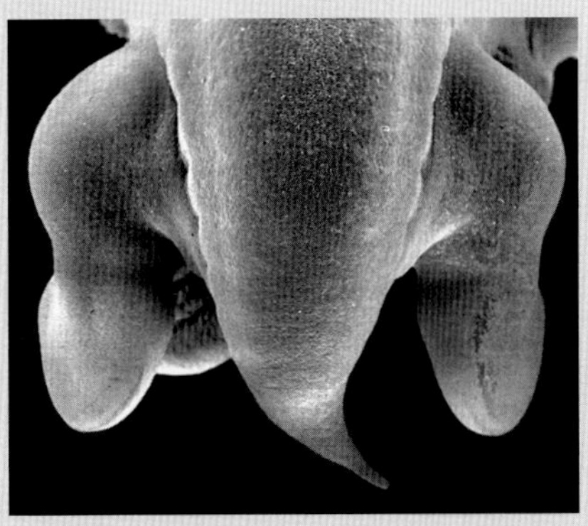

1.5.19 Embryo at week 7.

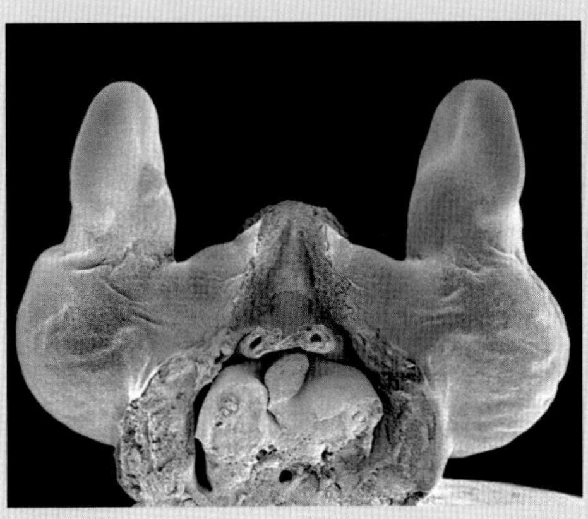

1.5.20 Embryo at week 7. Cranial aspect.

Week 7

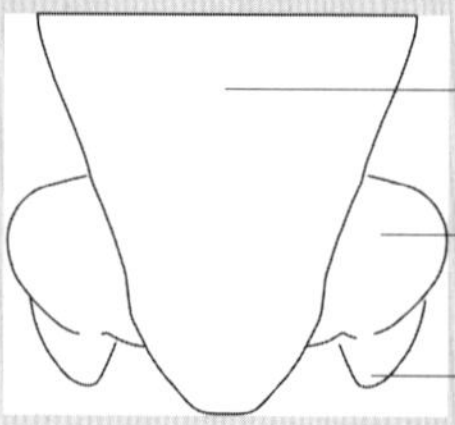

1.5.01–1.5.30
Developmental stages of the lower limb.

1.5.14–1.5.22
Stages of the development of the lower limb: dorsal (1.5.14–1.5.19, 1.5.21), cranial (1.5.20), and caudal (1.5.22) views.

Week 9

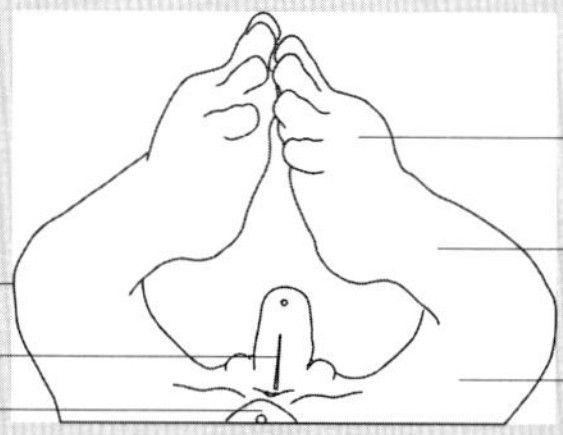

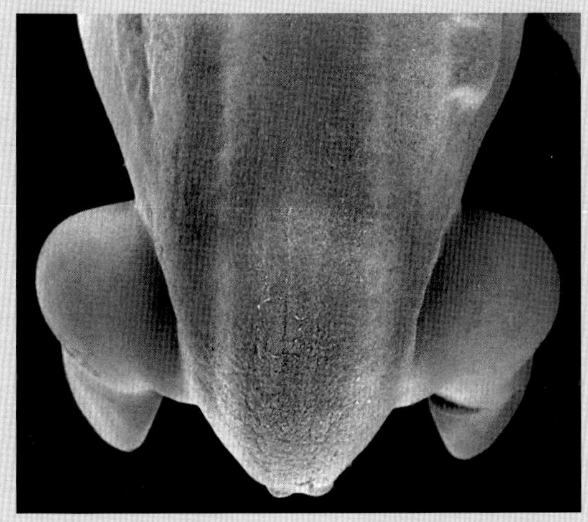

1.5.21 Embryo at week 7.
Dorsal aspect.

1.5.22 Embryo at week 9.
Caudal aspect.

Week 5 / 6

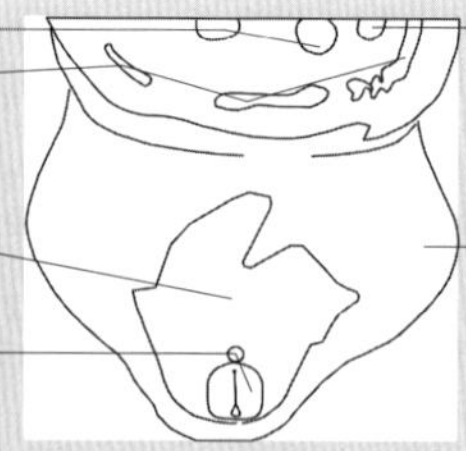

1.5.01–1.5.30

Developmental stages of the lower limb.

1.5.23–1.5.30

Stages of the development of the lower limb. Ventral view.

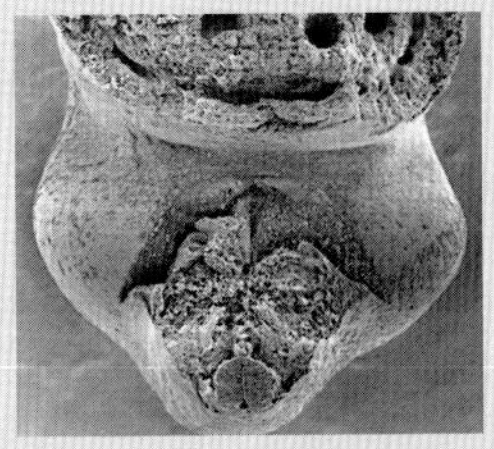

1.5.23

Embryos early, in the middle of, and late in weeks 5 and 6.

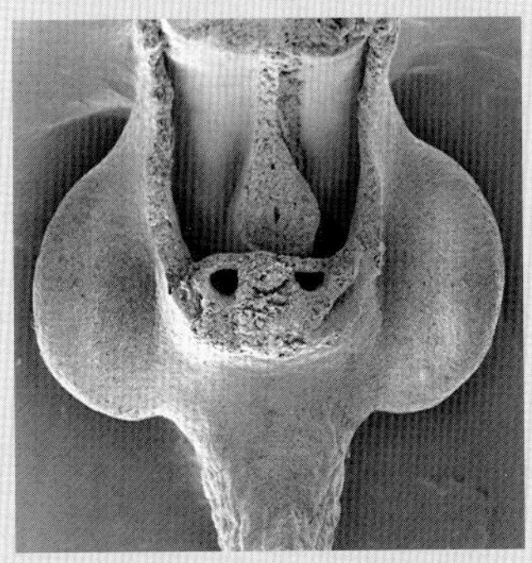

1.5.24

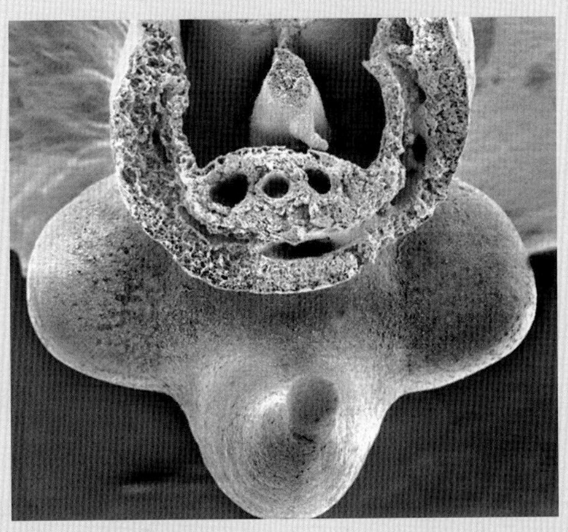

1.5.25

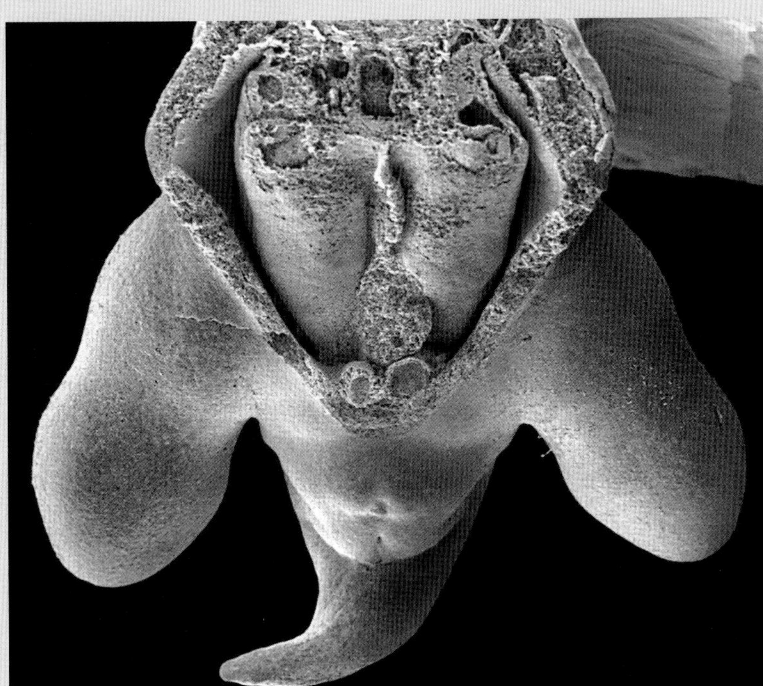

1.5.26

1.5.01–1.5.30
Developmental stages of the lower limb.

1.5.23–1.5.30
Stages of the development of the lower limb. Ventral view.

Week 6

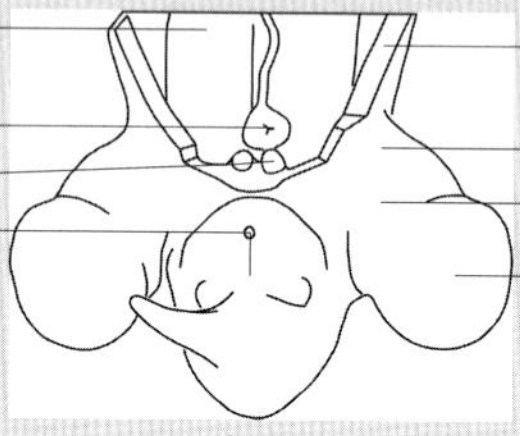

Week 7

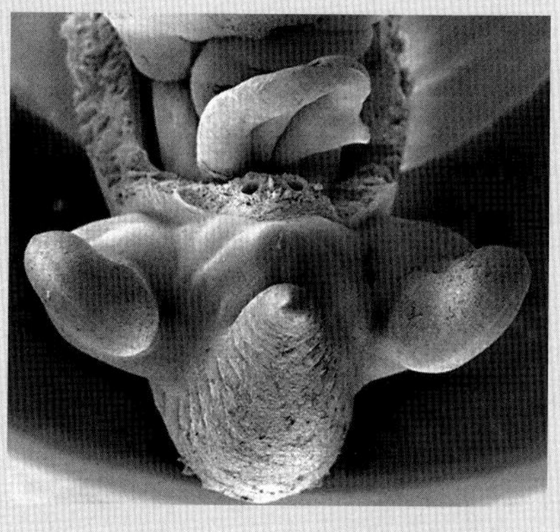

Week 8

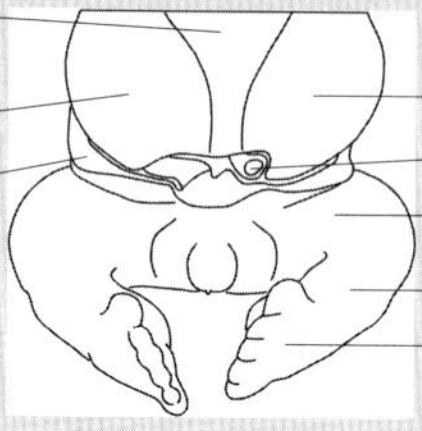

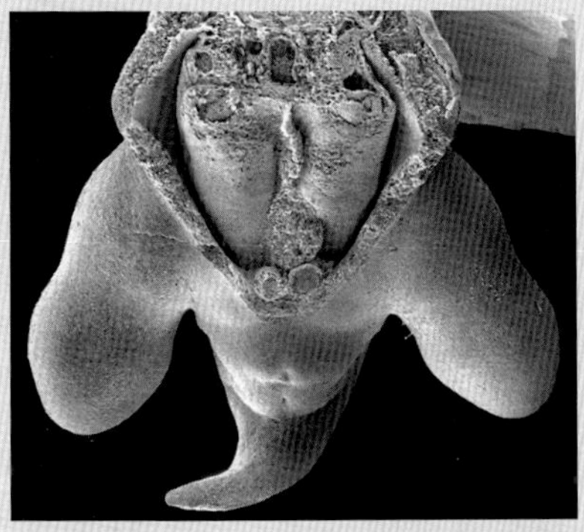

1.5.26a Embryos at week 6.

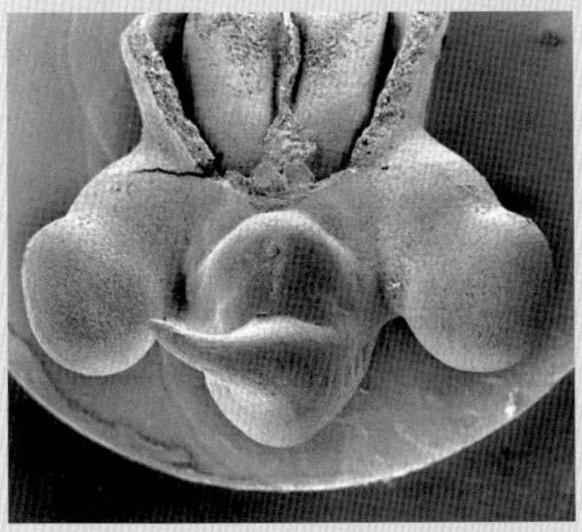

1.5.27

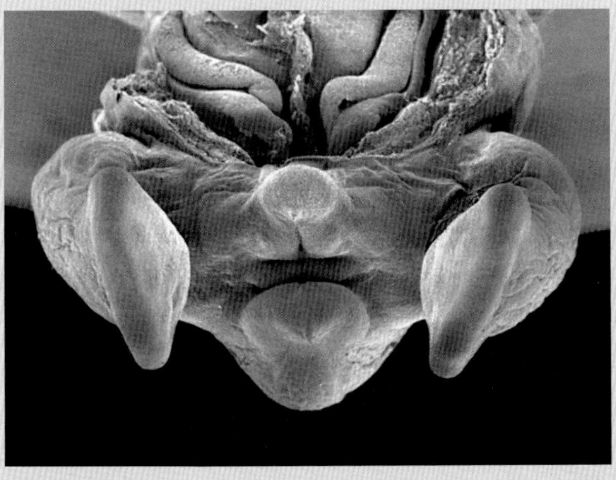

1.5.28 / 29 Embryos at week 7.

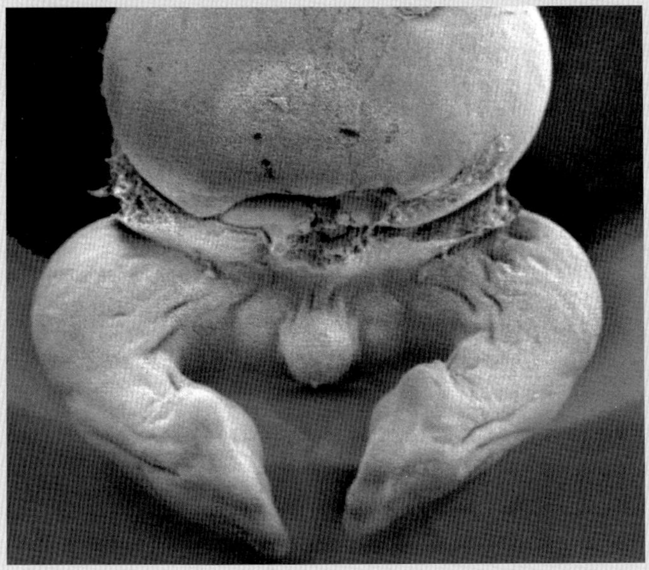

1.5.30 Embryo at week 8.

1.
External Aspects

1.6

The Development of the External Genitalia

In the ventral pelvic region a median furrow appears in week 5 where the endodermal outlet of the urinary and intestinal tracts, which are not yet separated, have contact to the ectoderm. This urethral groove is bordered, by paired elevations which later on will enclose the urethra (fig. 1.6.01–1.6.04). In week 6, the upper rims of the urethral folds form the swelling of the genital tubercle (fig. 1.6.03–1.6.05) which becomes elevated, growing in a caudal direction (fig. 1.6.06–1.6.08). When the endodermal outlets of the urinary and intestinal tracts are separated internally, the future anus is separated from the urethral groove by a bar. The opening of the anus laterally abuts the so-called anal swellings (fig. 1.6.22–1.6.26).

Up to week 7, the external genitalia are said to exhibit an 'indifferent stage' because the morphologically striking differences between male und female have either not yet been realized or are not yet known. As a first sign of sexual differentiation, in week 7 the urethral groove of female embryos remains short, whereas in male embryos the urethral groove extends up to the middle of the genital tubercle just beneath the epithelial tag of the glans (see fig. 1.6.15 and 1.6.16, 1.6.19 and 1.6.20).

In the following phase of development the so-called genital swellings, the anlage of the scrotum (or the greater labia), appear in week 8 (fig. 1.6.21). The formation of the definitive male and female characteristics does not begin before the 3rd month of gestation and is, therefore, beyond the scope of this atlas.

1.6.01–1.6.26
Developmental stages of the external genitalia.

1.6.01–1.6.08
Position of the developing external genitalia.

Week 5 / 6

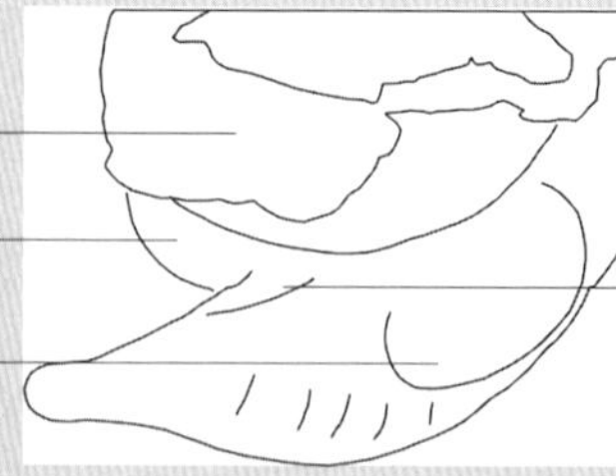

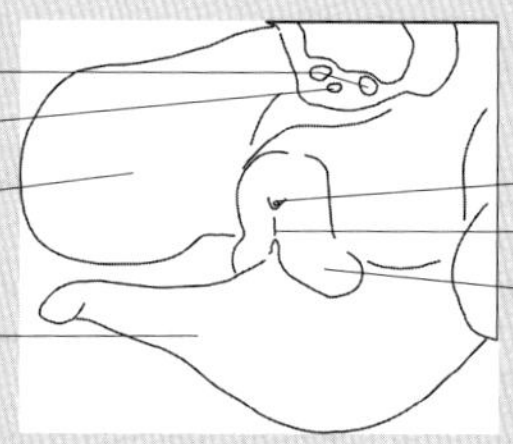

Week 6

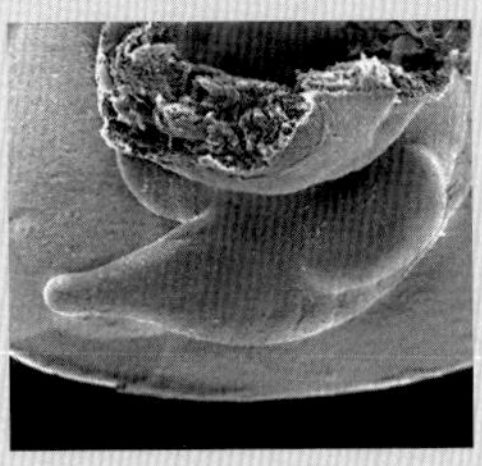

1.6.01

Embryos at week 5 and early in week 6. Ventral-left aspect.

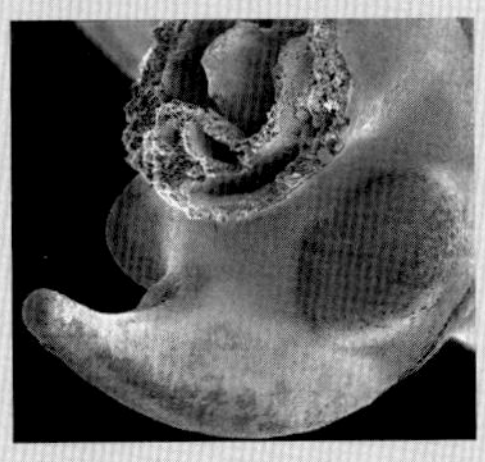

1.6.02

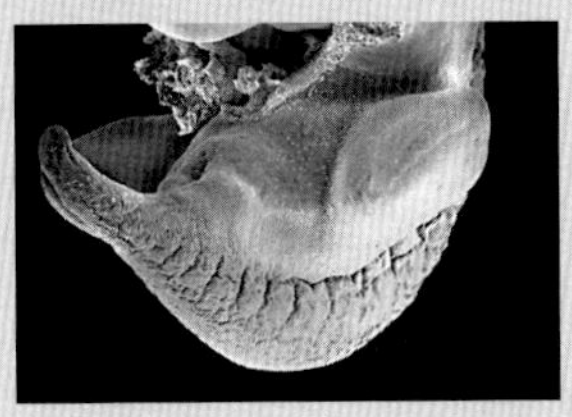

1.6.03

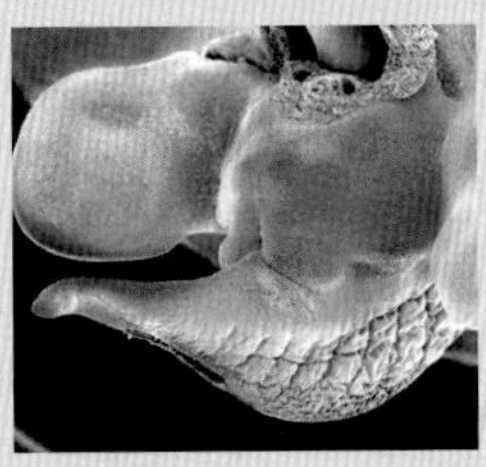

1.6.04

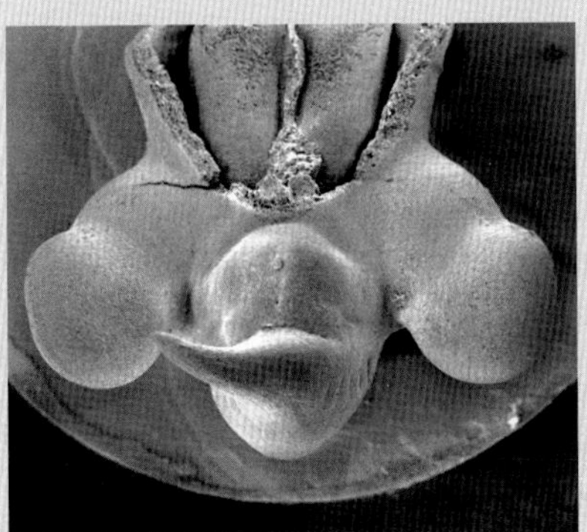

1.6.05

Embryo at week 6. Ventral aspect.

Week 6

Week 7

1.6.01–1.6.26
Developmental stages of the external genitalia.

1.6.01–1.6.08
Position of the developing external genitalia.

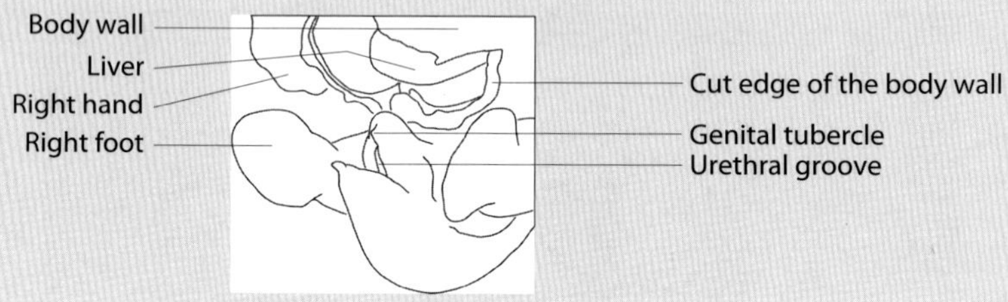

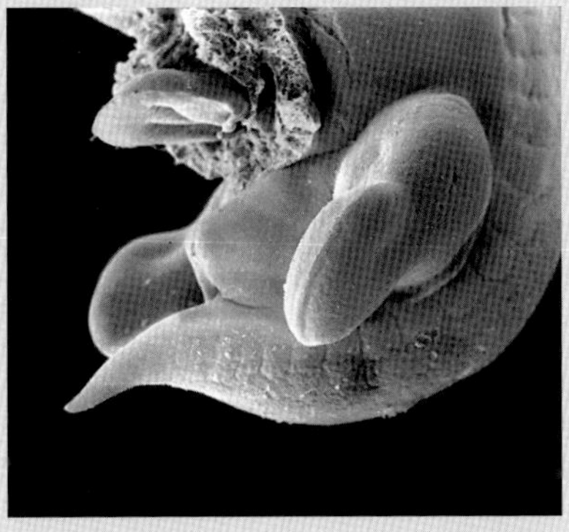

1.6.06 Embryo at the end of week 6. Ventral-left aspect.

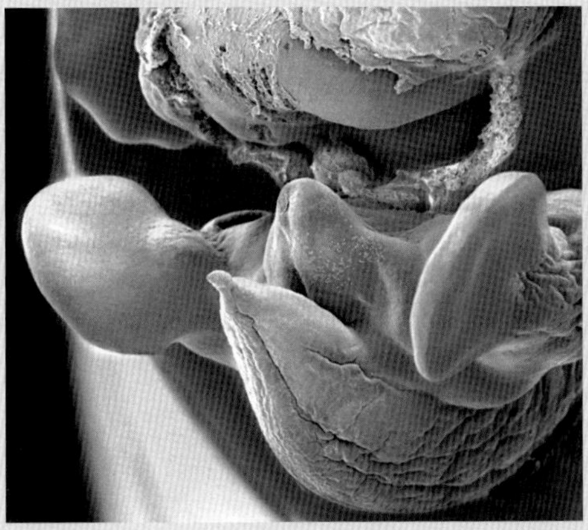

1.6.07 Embryo in the middle of week 7. Ventral-left aspect.

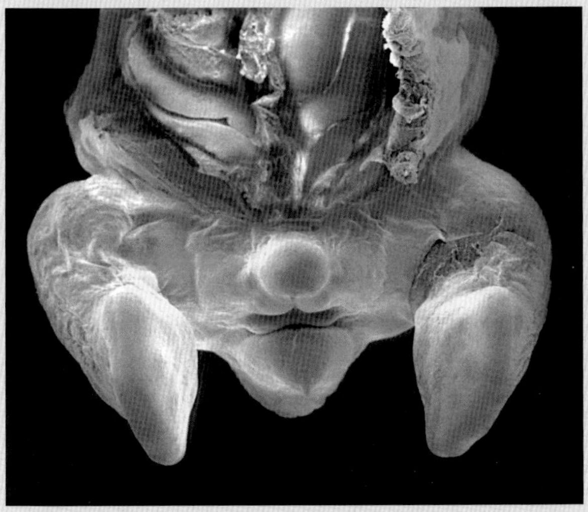

1.6.08 Embryo at the end of week 7. Ventral aspect.

1.6.01–1.6.26
Developmental stages of the external genitalia.

1.6.09–1.6.21
External genitalia.

Week 5

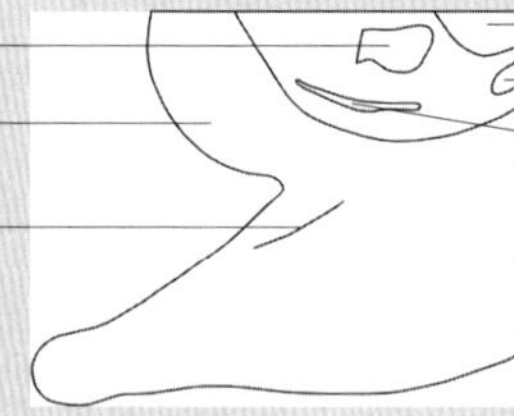

Week 6

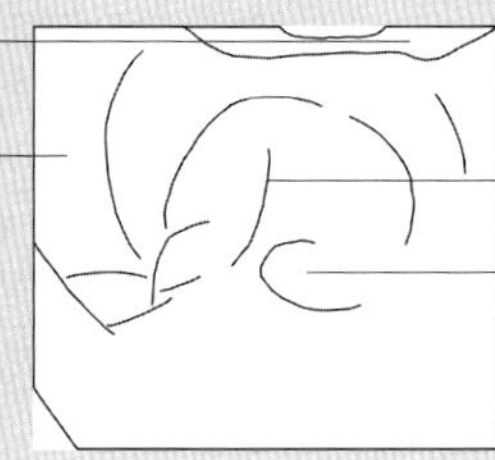

Week 6 / 7

1.6.09

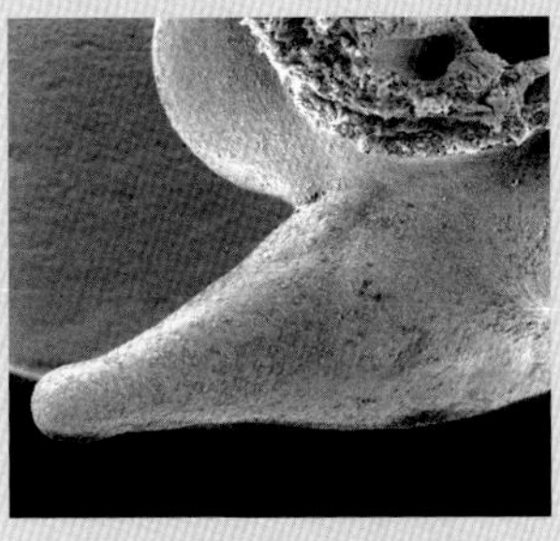

Embryos at week 5. Ventral-left view.

1.6.10

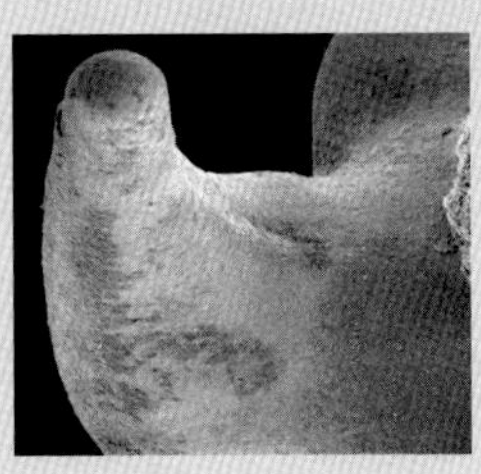

1.6.11

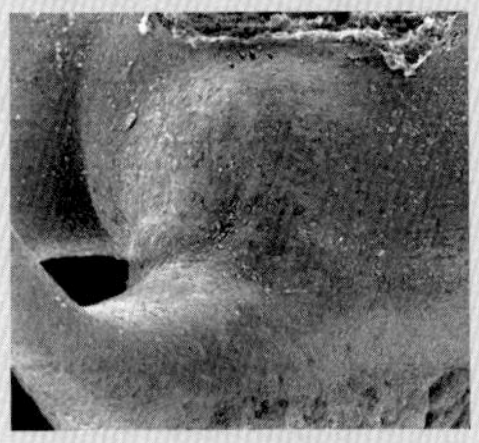

Embryos early in and in the middle of week 6.

1.6.12

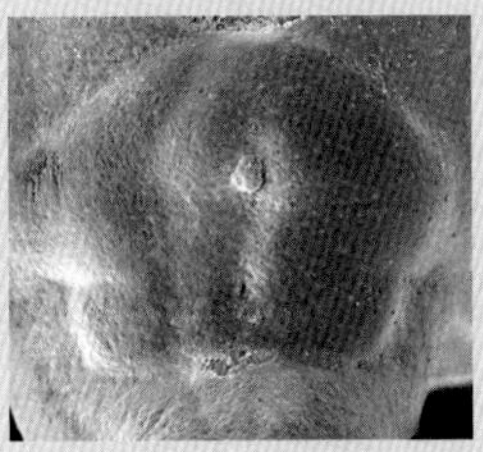

1.6.13

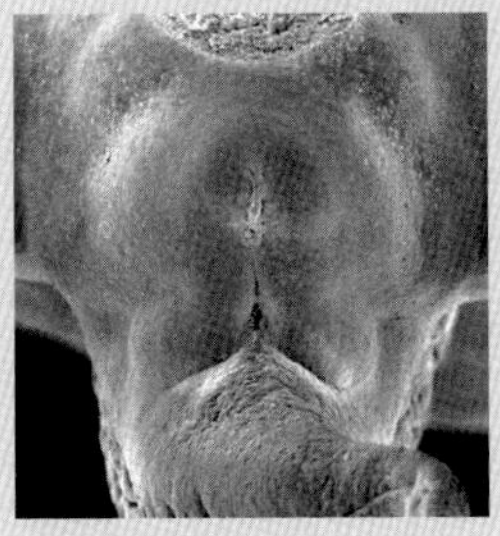

Embryos at the end of week 6 and early in week 7.

1.6.14

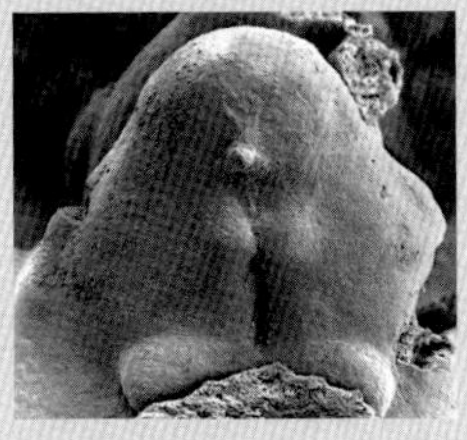

Week 7

1.6.01–1.6.26
Developmental stages of the external genitalia.

1.6.09–1.6.21
External genitalia.

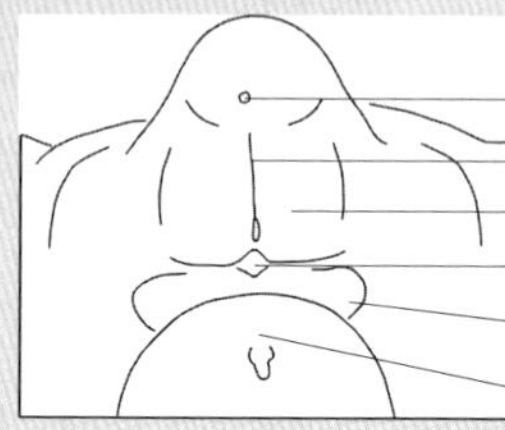

Week 8

1.6.15

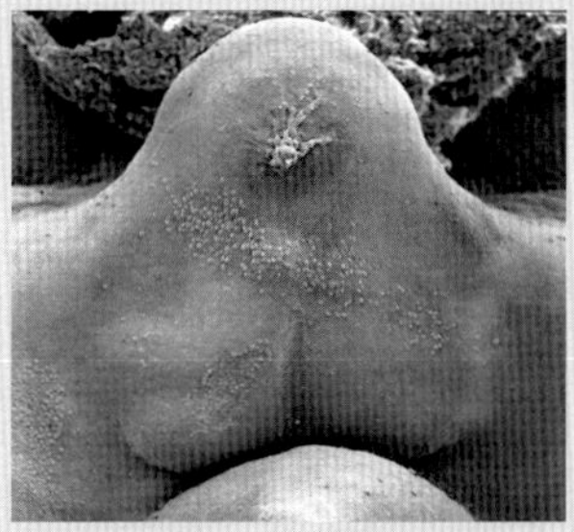

Embryos in the middle and at the end of week 7.

1.6.16

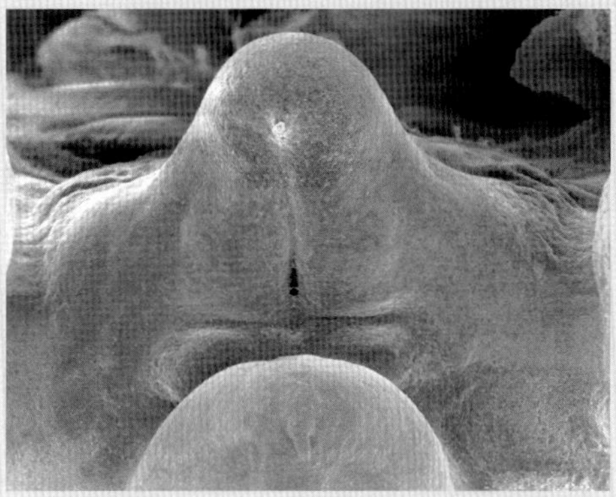

1.6.17

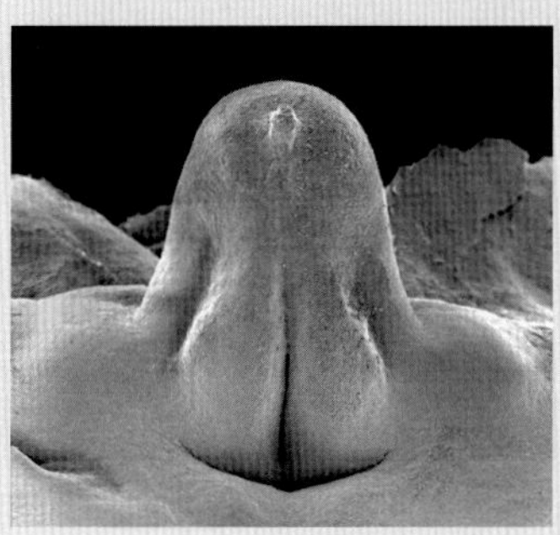

Embryos early in week 8.

1.6.18

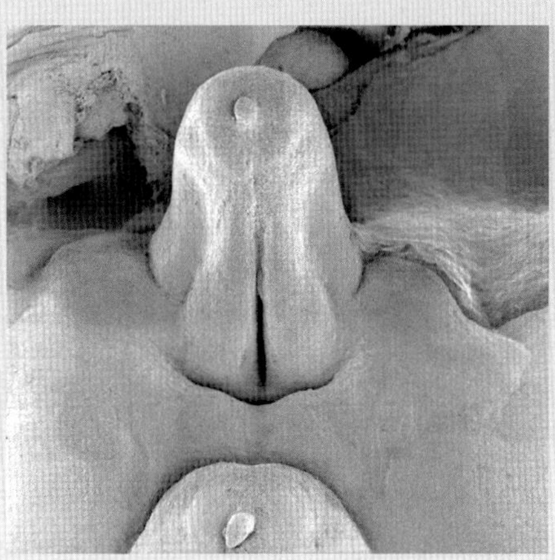

Week 8 / 9

1.6.01–1.6.26
Developmental stages of the external genitalia.

1.6.09–1.6.21
External genitalia.

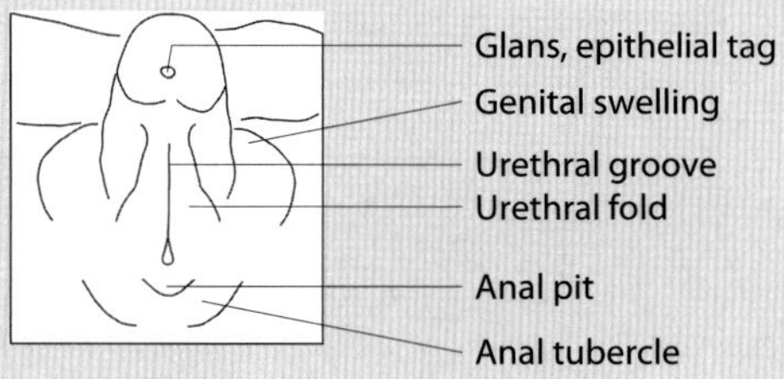

1.6.19

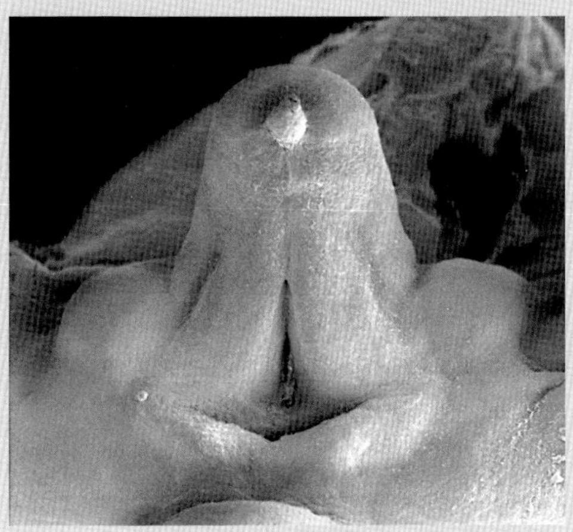

Embryos at the beginning and in the middle of week 8 and early in week 9.

1.6.20

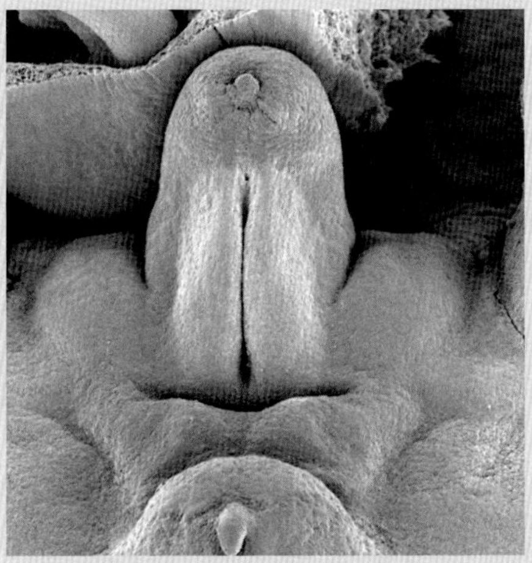

1.6.21

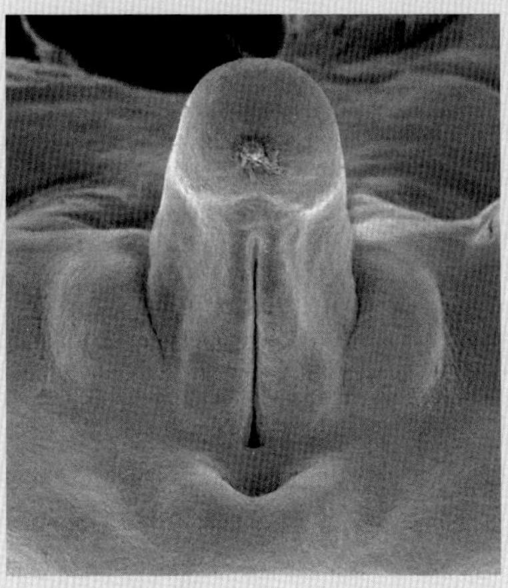

1.6.01–1.6.26
Developmental stages of the external genitalia.

Week 9

30:1

60:1

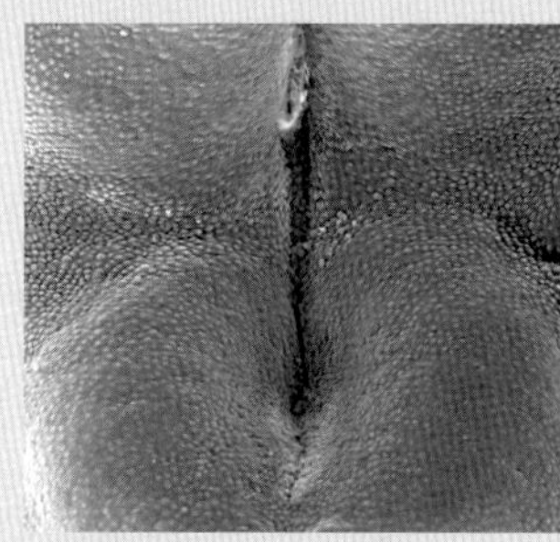

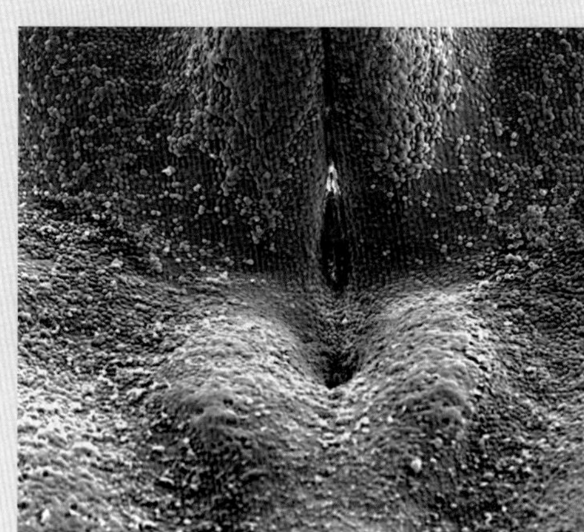

1.6.22

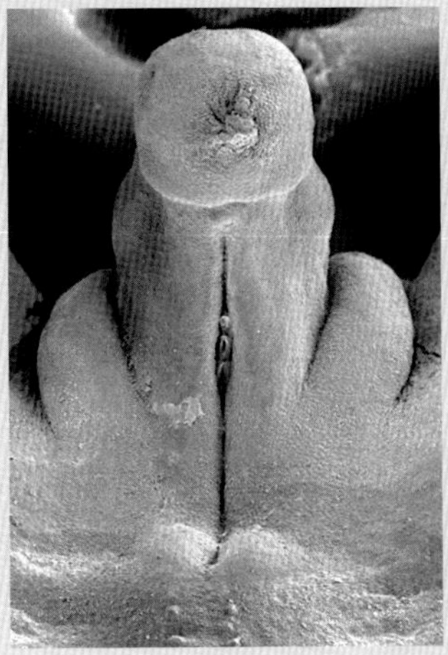

Stages of the development of the perineum. Embryos at week 9.

1.6.23 / 24

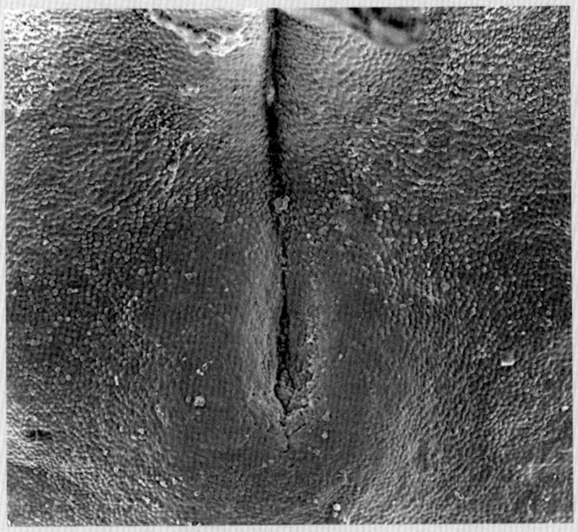

1.6.25 / 26

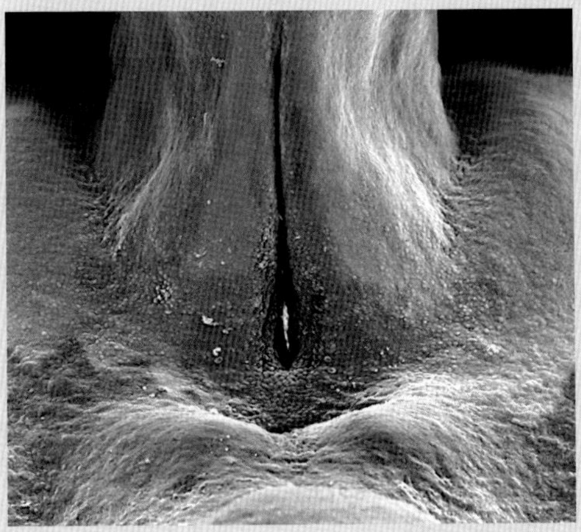

2

Endodermal Organs in the Head and Neck

2.1
The Development of the Walls of the Oral Cavity

The floor of the oral cavity is formed by the inner surface of the first two pharyngeal arches (fig. 2.1.01). The third, fourth, and sixth arches contribute to the pharynx (fig. 2.1.02, 2.1.03) and to the pharyngeal part of the tongue (fig. 2.1.24–2.1.26). The roof of the early oral cavity is formed by the epithelium covering the ventrally bent brain (fig. 2.3.01, 2.3.02). Following origination of the palatal processes and their fusion they form the definite roof of the oral cavity (see chapter 2.3).

The interior lateral circumferences of the right and the left first pharyngeal arches increase and form the lateral lingual swellings (fig. 2.1.02, 2.1.03). In the centre between the circumference of the right and the left first pharyngeal arches, a median swelling arises, the tuberculum impar (fig. 2.1.02–2.1.04). The furrows between these swellings disappear and consequently the anterior (oral) part of the tongue is formed (fig. 2.1.03–2.1.06, 2.1.11–2.1.18).

The posterior (pharyngeal) part of the tongue is formed by the second and the third pharyngeal arches (fig. 2.1.24–2.1.26). The median parts of these arches swell out and form the hypopharyngeal eminence (fig. 2.1.11, 2.1.12).

With the descent of the heart, the pharyngeal part of the tongue also descends, thus realizing the angularity between its oral and pharyngeal parts (fig. 2.1.23 and 2.2.22). By growing in a cranial direction, the tongue is folded away from the floor of the oral cavity (fig. 2.1.13–2.1.18), and by growing in a ventral direction, the tip of the tongue is freed from the floor of the oral cavity and becomes freely mobile (fig. 2.1.06–2.1.10, 2.1.14–2.1.18, 2.1.21–2.1.23).

The Origin of the Deciduous Teeth

In week 7, epithelial thickenings inside the upper and the lower lips grow into the underlying mesenchyme to form an external arch, the vestibular lamina and an internal arch, the dental lamina (fig. 2.1.27). The vestibular lamina will develop into the vestibular sulcus behind the lips, while the dental lamina gives rise to the teeth.

In each dental lamina ten buds are formed; the tooth germs successively arise in aboral directions (fig. 2.1.28, 2.1.29). Later on, the end of the tooth germ folds inward in an oral direction, thus forming the two-layered bell-shaped enamel organ (fig. 2.1.30). The development of the materials of the teeth, the dentine, the enamel and the periodontium and also the origin of the permanent teeth is during the fetal and the postnatal period and they are therefore beyond the scope of this atlas.

The Origin of the Thyroid Gland

The thyroid gland is formed by a depression in the midline of the valley between the first and the second pharyngeal arches, the foramen caecum (fig. 2.1.01). This bud develops into an endodermal pouch growing downward into the mesenchyme of the floor of the mouth (fig. 2.1.33, 2.1.34). During the descent of the heart, the thyroid gland descends to the neck and its connection to the tongue is reduced to the thyroglossal duct (fig. 2.1.35) which partially may remain as the pyramidal lobe. During its descent, the thyroid gland is crooked and forms two lobes connected by the isthmus.

2.1

Abbreviations

II–IV	pharyngeal arch II–IV
1	medial lingual swelling
2	lateral lingual swelling
3	tongue
4	epiglottis
5	laryngotracheal groove

Week 5

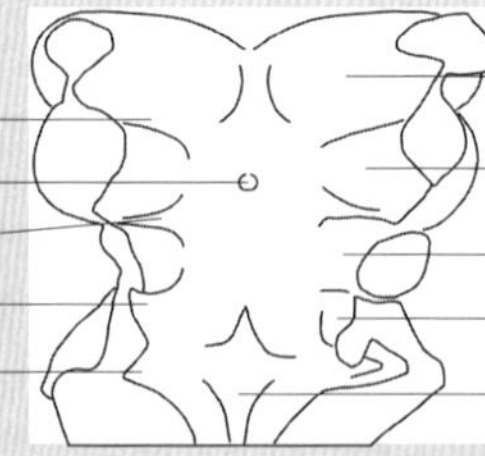

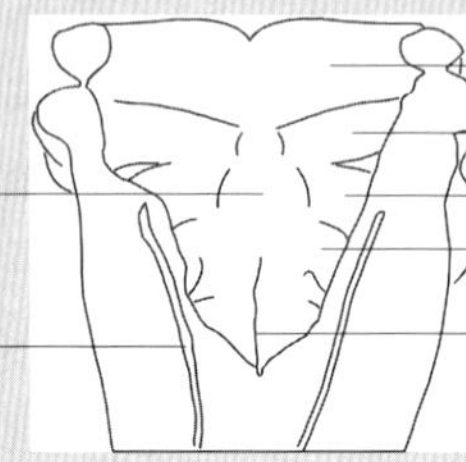

Week 6

2.1.01–2.1.37
Developmental stages of the walls of the oral cavity.

2.1.01–2.1.10
The floor of the mouth and the tongue. The brain and the upper jaw have been removed. Cranial view.

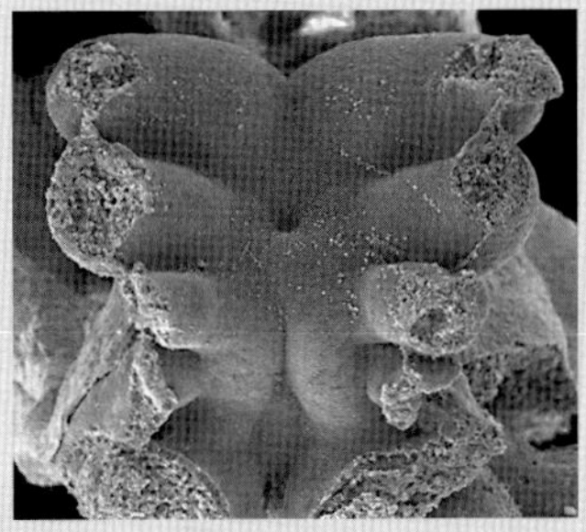

2.1.01 Embryos at week 5.

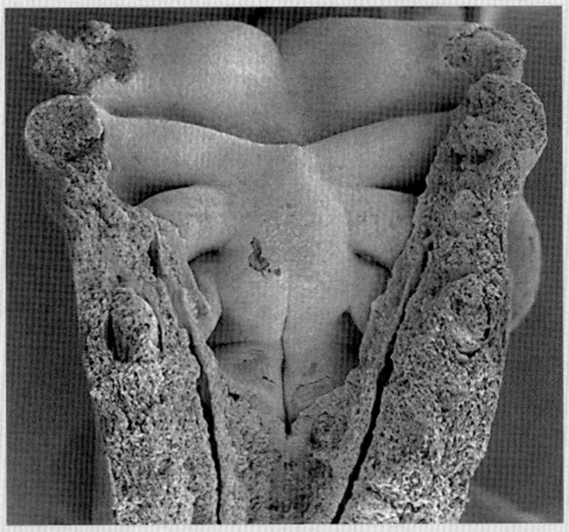

2.1.02

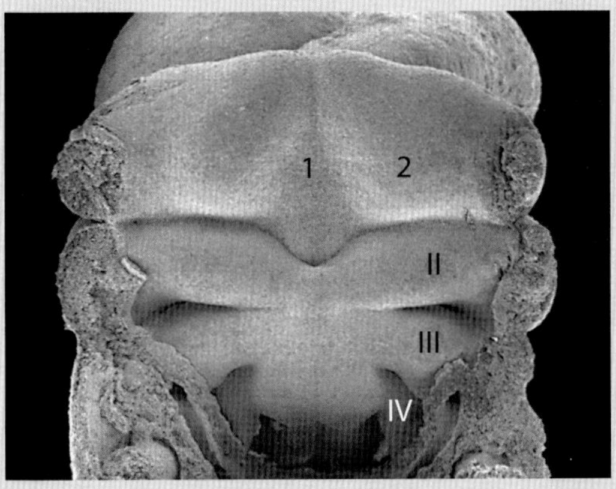

2.1.03 Embryos at week 6.

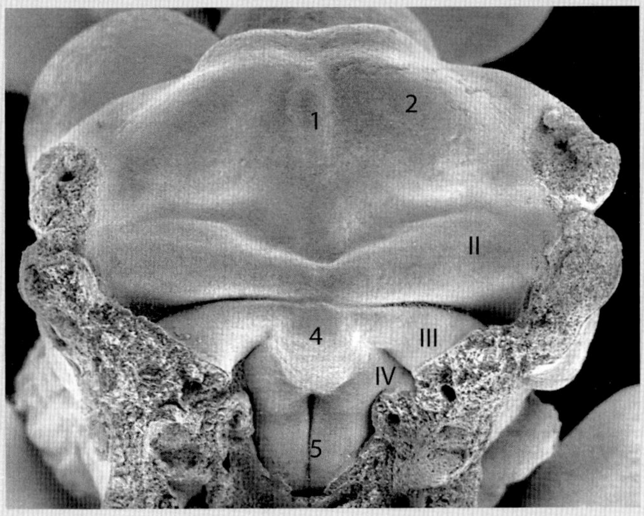

2.1.04

40:1

Week 6

2.
Endodermal Organs in the Head and Neck

2.1
The Development of the Walls of the Oral Cavity

2.1.01–2.1.37
Developmental stages of the walls of the oral cavity.

2.1.01–2.1.10
The floor of the mouth and the tongue. The brain and the upper jaw have been removed. Cranial view.

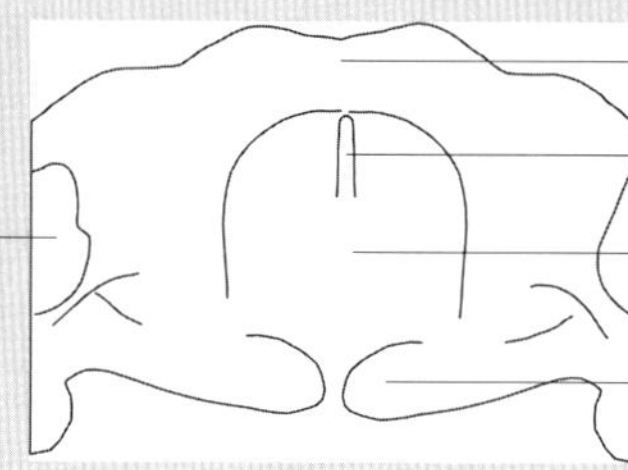

Week 7

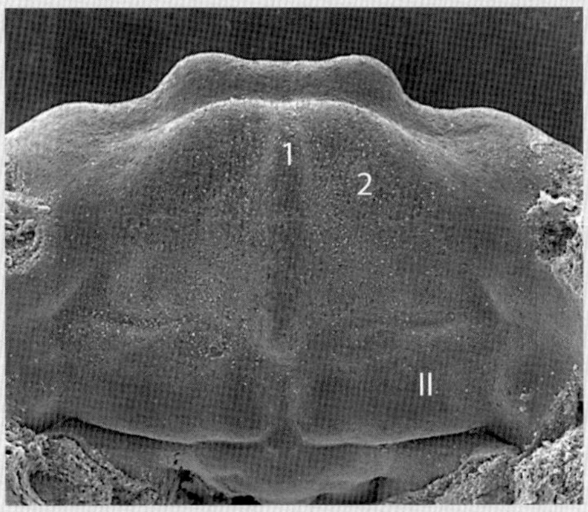

2.1.05 Embryos at the end of week 6.

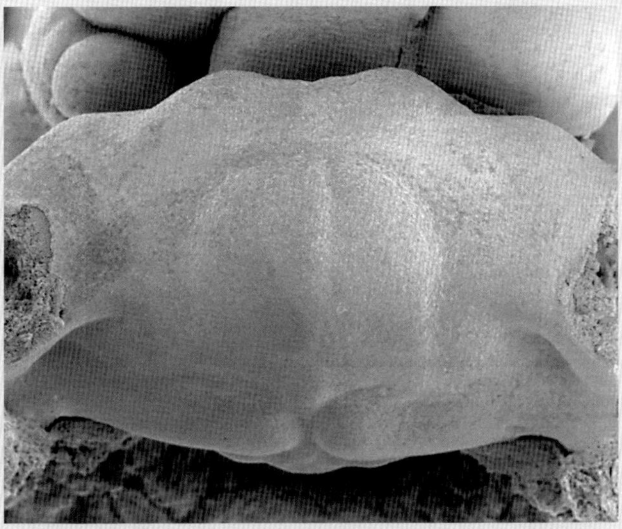

2.1.06

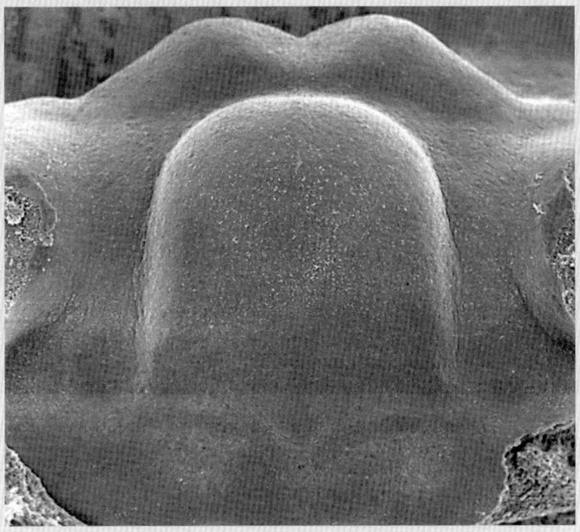

2.1.07 Embryo at week 7.

Week 8

2.1.01–2.1.37
Developmental stages of the walls of the oral cavity.

2.1.01–2.1.10
The floor of the mouth and the tongue. The brain and the upper jaw have been removed. Cranial view.

Week 9

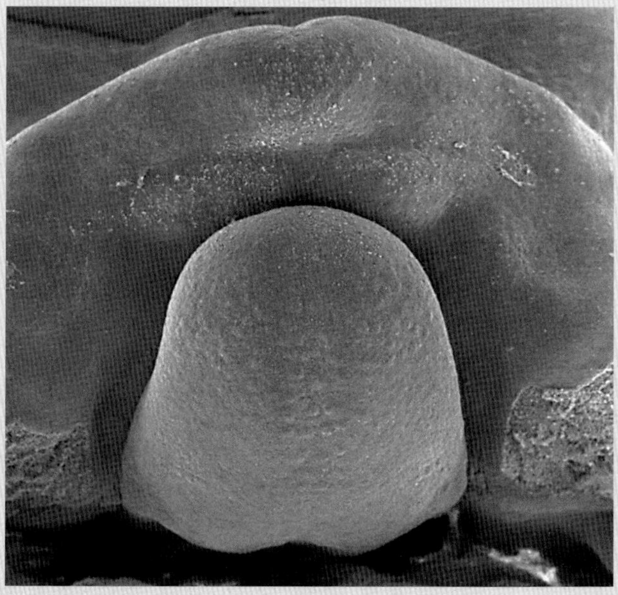

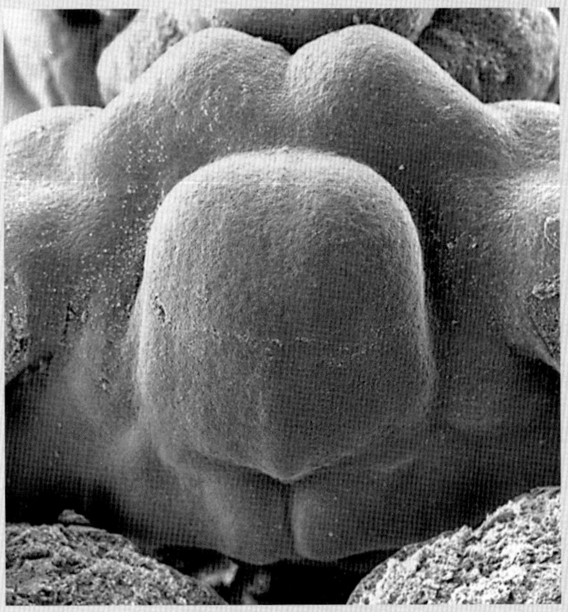

2.1.08 Embryo at week 8.

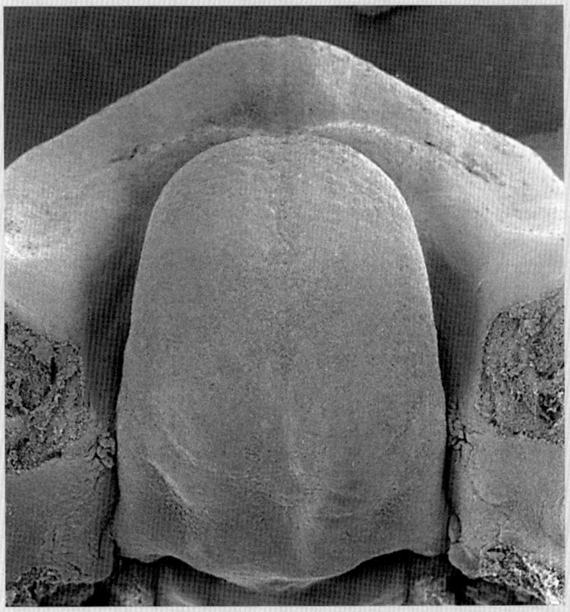

2.1.09 / 10 Embryos at week 9.

2.1.01–2.1.37

Developmental stages of the walls of the oral cavity.

2.1.11–2.1.18

The floor of the mouth and the tongue. The upper head and the upper jaw have been removed. Ventral and ventral-cranial view, respectively.

Week 5

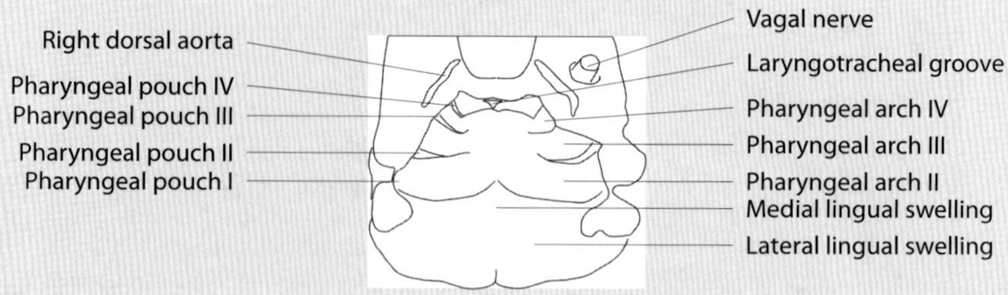

Week 6

Week 7

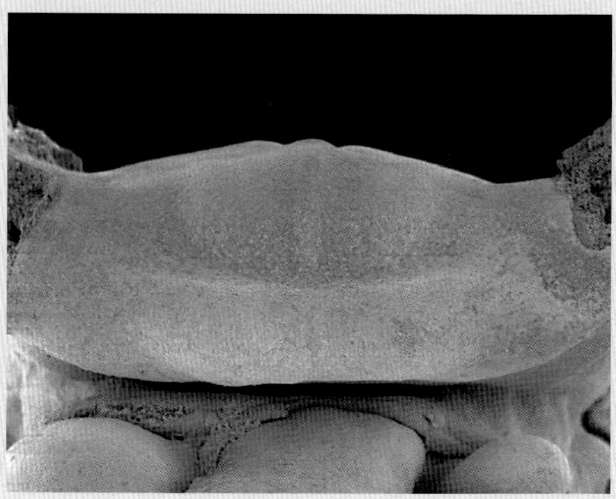

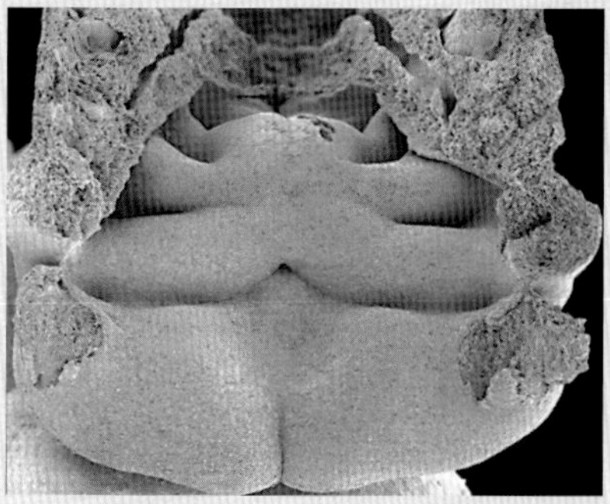

2.1.11 Embryo at week 5.

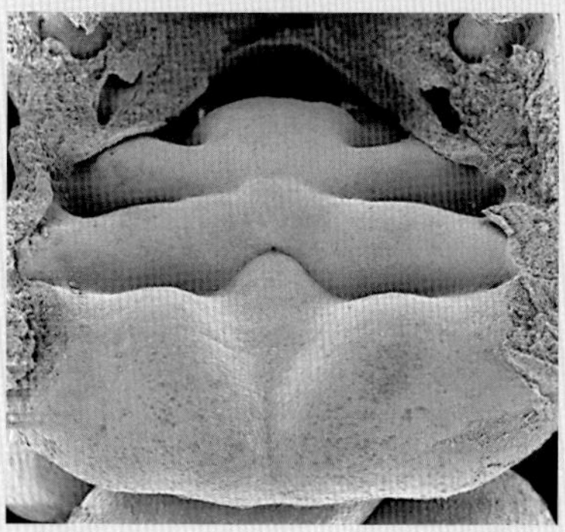

2.1.12 Embryos at week 6.

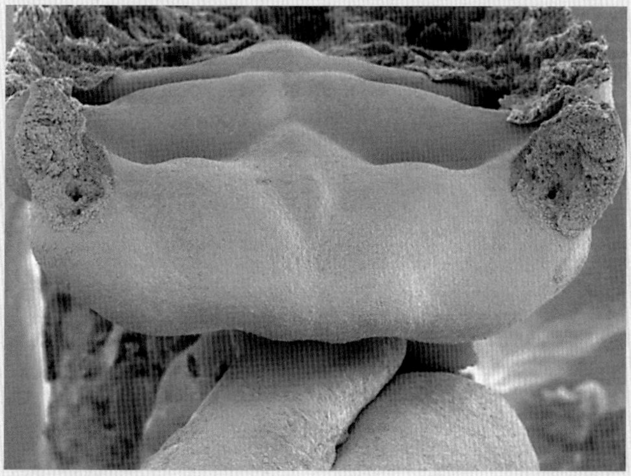

2.1.13

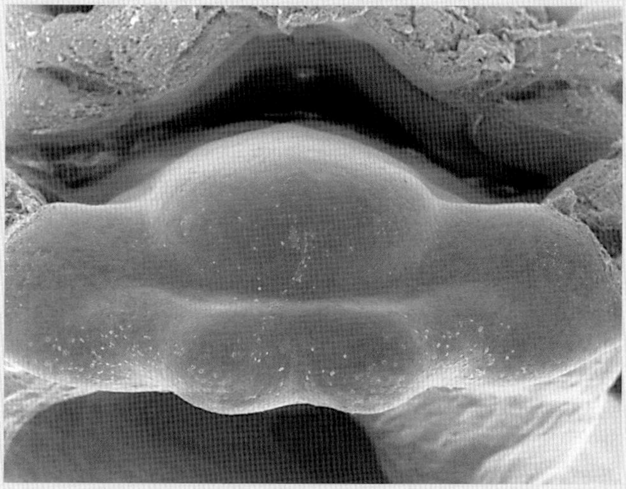

2.1.14 / 15 Embryos at week 7.

2.
Endodermal Organs in the Head and Neck

2.1
The Development of the Walls of the Oral Cavity

2.1.01–2.1.37
Developmental stages of the walls of the oral cavity.

2.1.11–2.1.18
The floor of the mouth and the tongue. The upper head and the upper jaw have been removed. Ventral and ventral-cranial view, respectively.

40:1

Week 8

Week 9

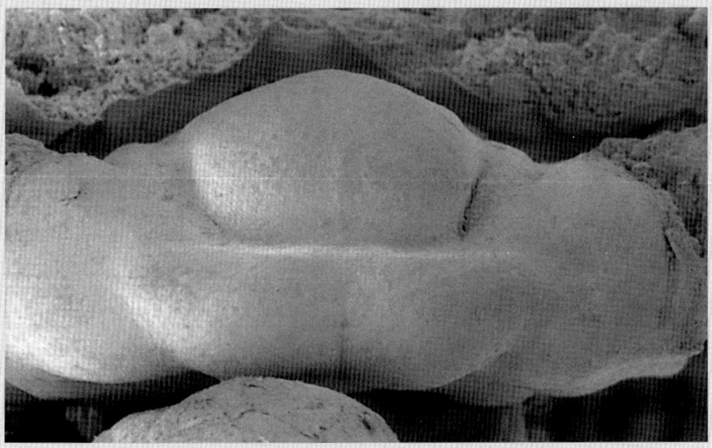

2.1.16 Embryos at week 8.

2.1.17

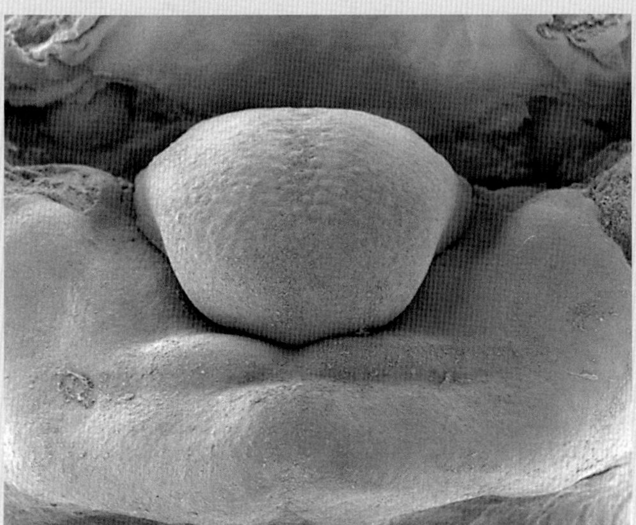

2.1.18 Embryo at week 9.

2.1.01–2.1.37

Developmental stages of the walls of the oral cavity.

2.1.19–2.1.23

The floor of the oral cavity and the tongue. The upper head and the upper jaw have been removed. Ventral-left view.

Week 6

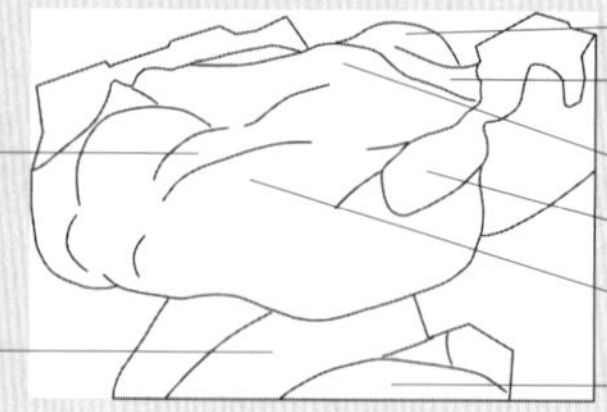

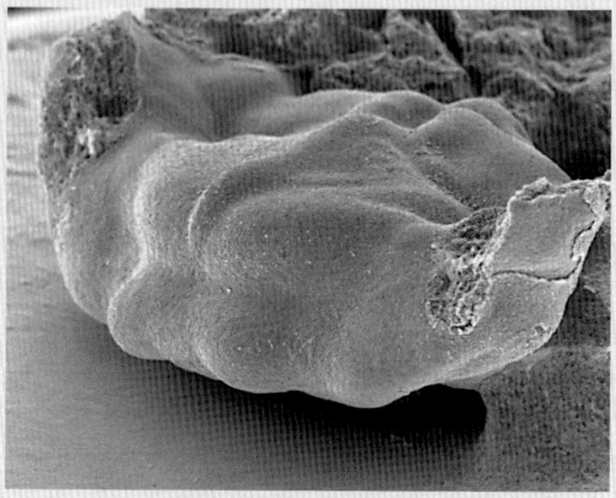

Week 7 / 8

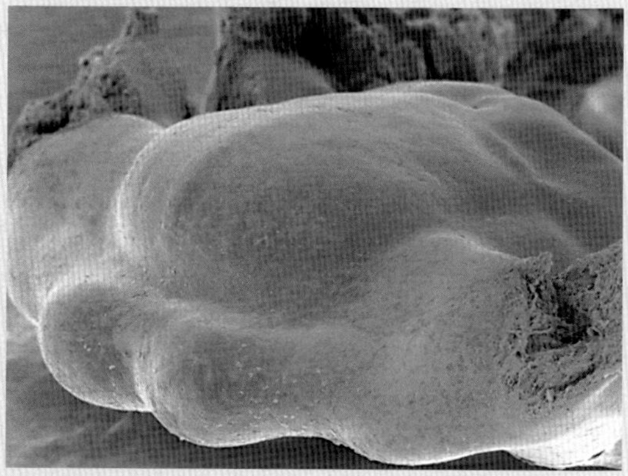

2.1.19

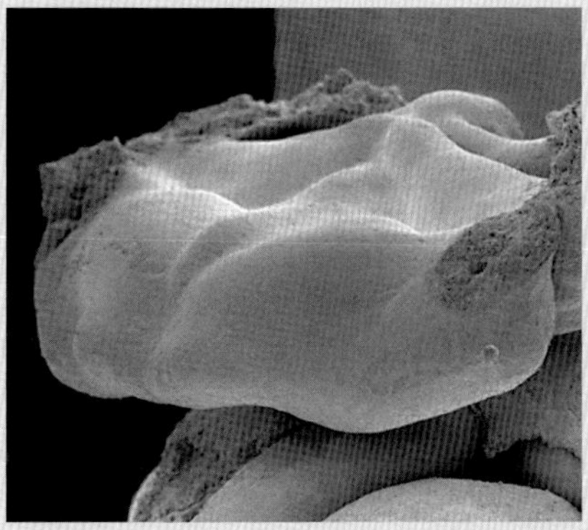

Embryos at week 6.

2.1.20 / 21

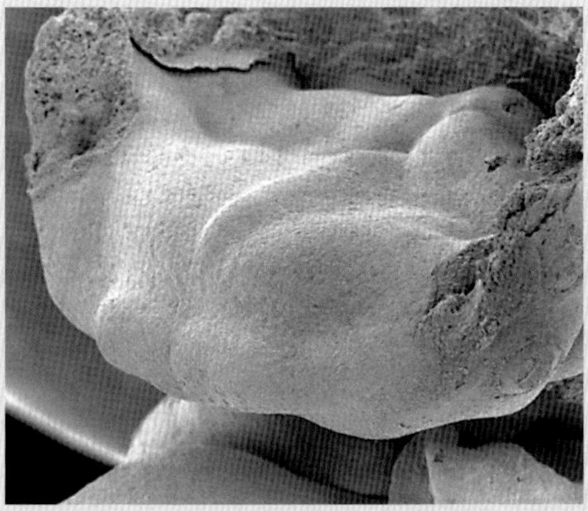

2.1.22 / 23

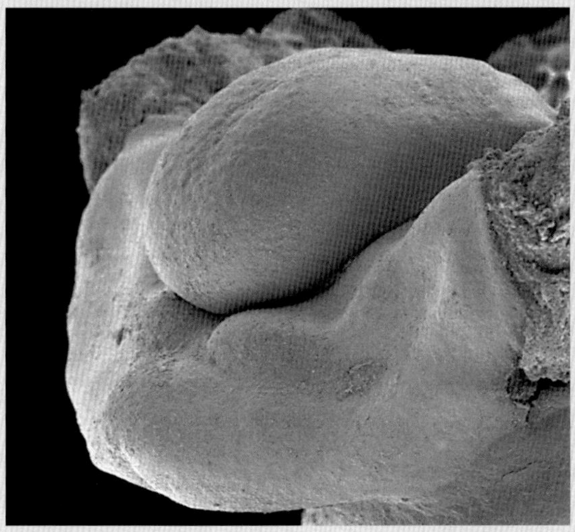

Embryos at weeks 7 and 8.

2.1.01–2.1.37
Developmental stages of the walls of the oral cavity.

2.1.24–2.1.26
The pharyngeal part of the tongue. Dorsal view.

Week 6

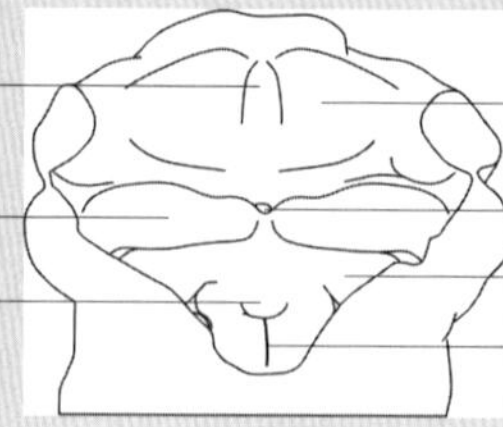

Week 7

Week 8

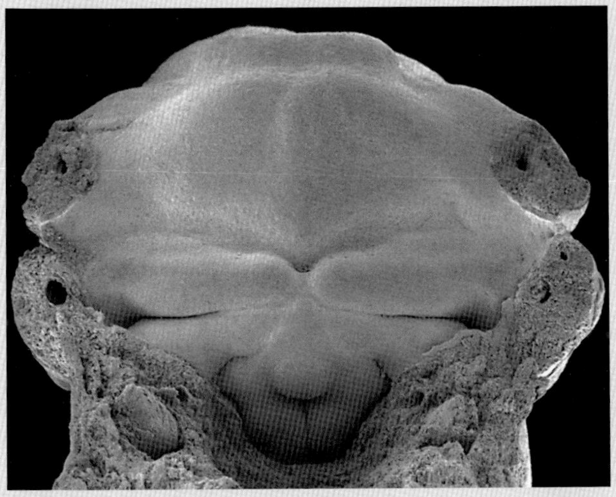

2.1.24 Embryo at week 6.

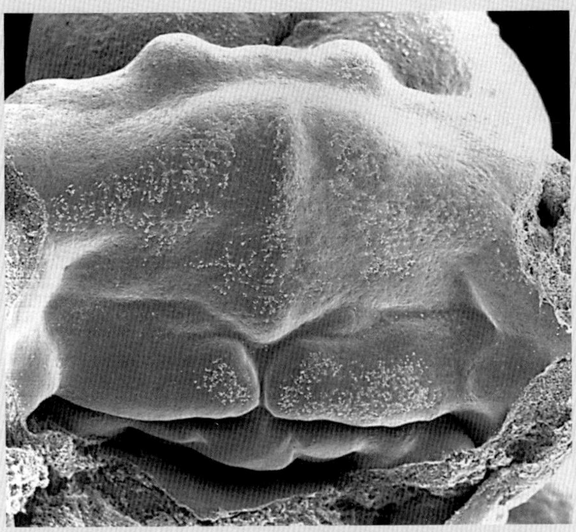

2.1.25 Embryo at week 7.

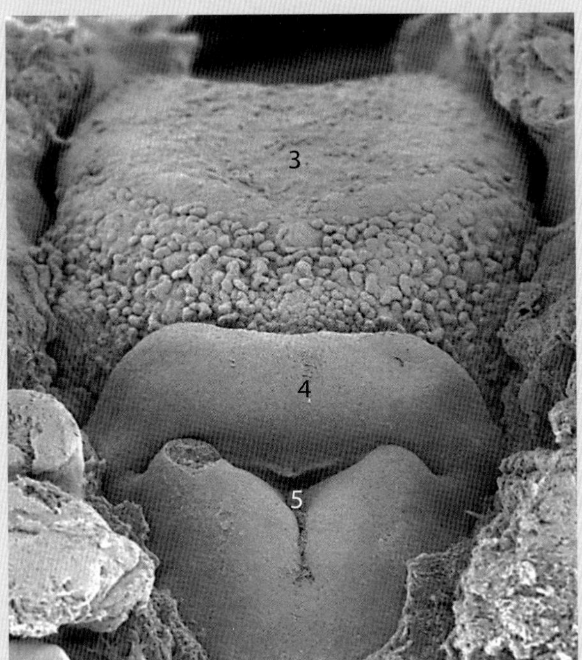

2.1.26 Embryo at week 8.

40:1

Week 8

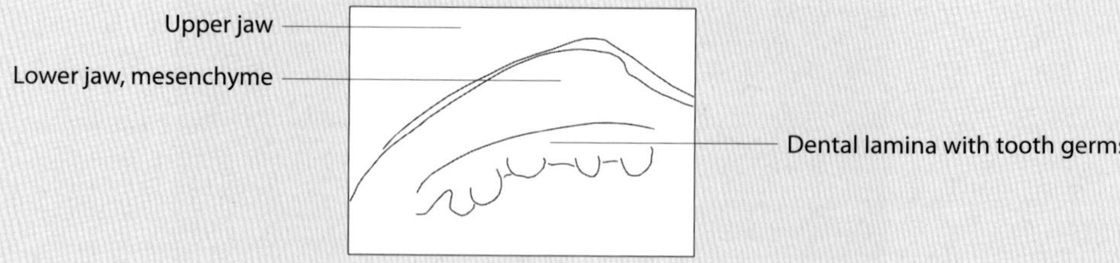

80:1

Week 9

2.1.01–2.1.37

Developmental stages of the walls of the oral cavity.

2.1.27–2.1.30

Early stages in the development of the teeth.

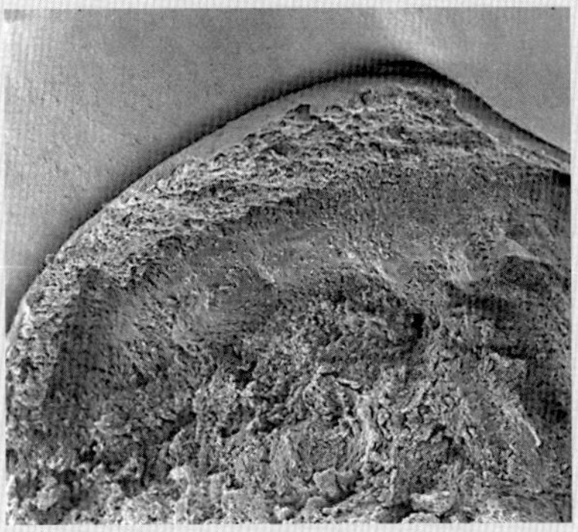

2.1.27 Right dental lamina with early tooth germs (incisor and premolar) of the lower jaw, seen from the ventral-caudal side. The vestibular lamina has been removed. Embryo at week 8.

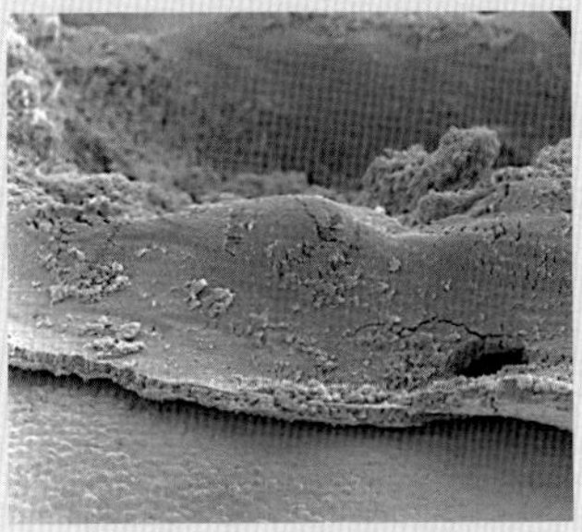

2.1.28 Tooth germ of a premolar in the upper jaw. Ventral view. Embryo at week 8.

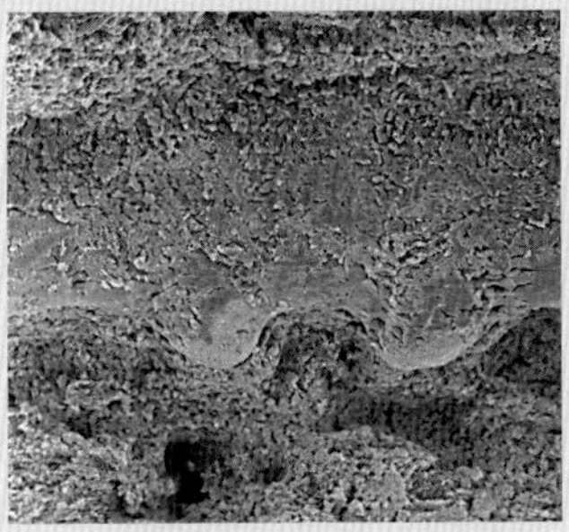

2.1.29 Incisor tooth germs of the upper jaw. Ventral view. Embryo at week 8.

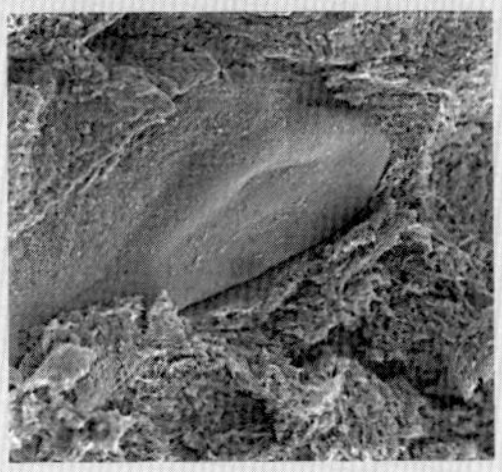

2.1.30 Early premolar bell-shaped enamel organ of the right upper jaw. Caudal-ventral view. Embryo at week 9.

2.
Endodermal Organs in the Head and Neck

2.1
The Development of the Walls of the Oral Cavity

2.1.01–2.1.37
Developmental stages of the walls of the oral cavity.

2.1.31–2.1.35
Early stages in the development of the thyroid gland.

40:1

Week 5 / 7

80:1

Week 5

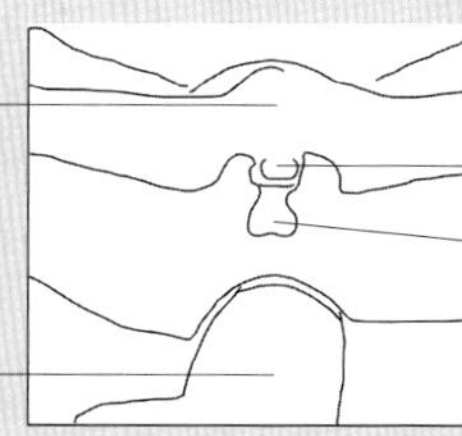

160:1

80:1

Week 7

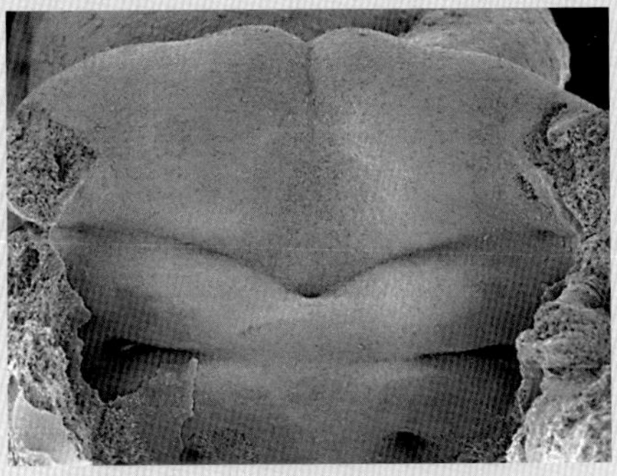

2.1.31 Position of the foramen caecum. Cranial view. Embryos at weeks 5 and 7.

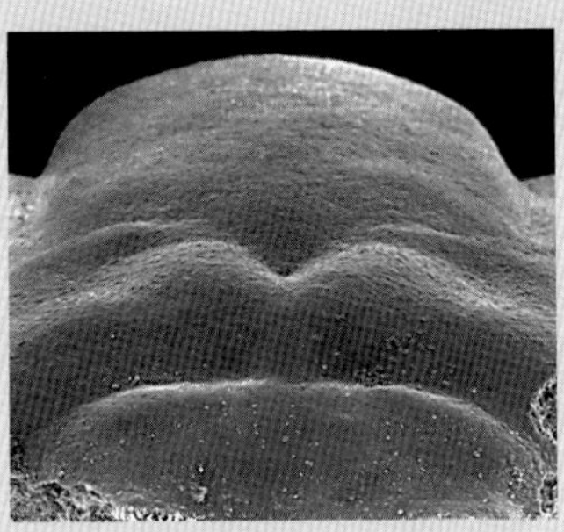

2.1.32

2.1.33 Early bud of the thyroid gland exposed ventrally. Ventral view.

2.1.34

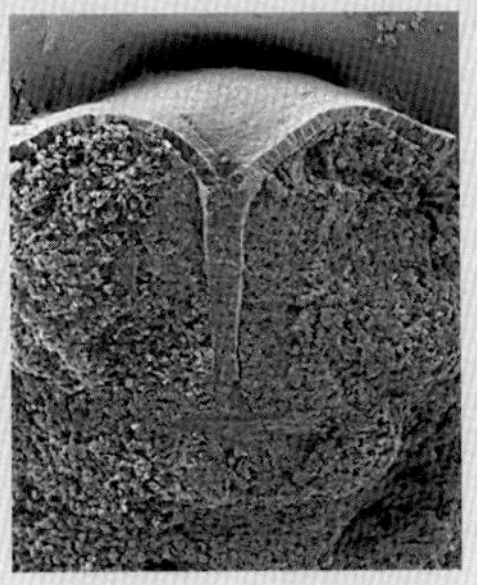

2.1.35 Foramen caecum and thyreoglossic duct exposed ventrally. Ventral view.

2.1.01–2.1.37

Developmental stages of the walls of the oral cavity.

2.1.36 / 37

Salivary glands.

Week 8

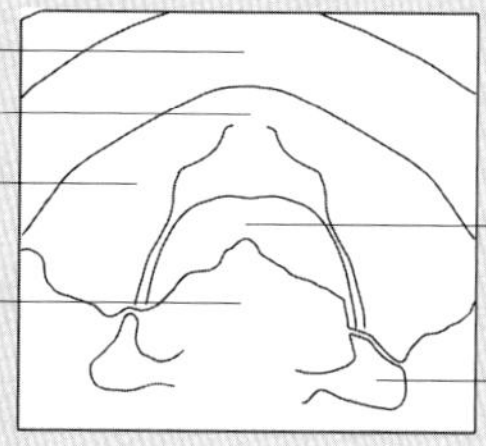

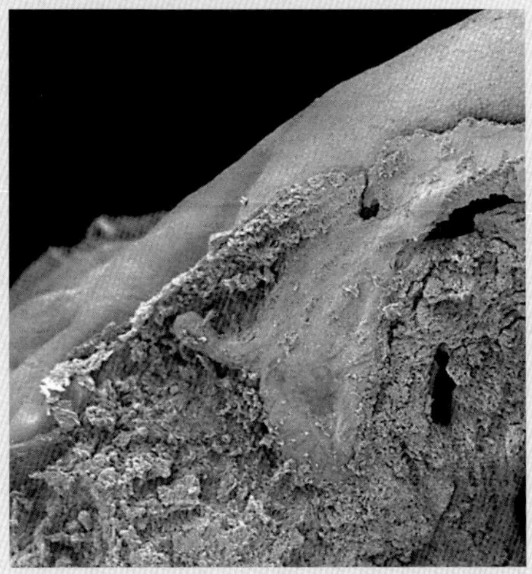

2.1.36 Bud of the right parotic gland. The connective tissue underlying the floor of the mouth and the gland has been removed. Caudal view. Embryo at week 8.

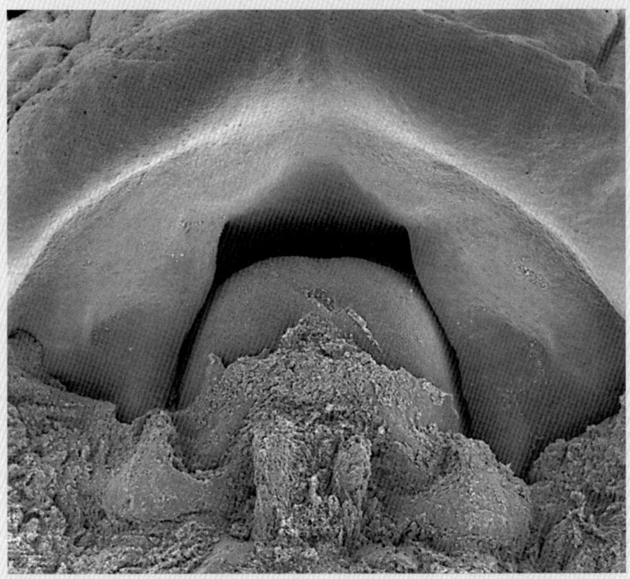

2.1.37 Bud of the sublingual and submandibular glands. Caudal view.

2.2 The Development of the Nose

The formation of ectodermal thickenings (fig. 2.2.01) within the right and left confines of the frontal and the lateral head is the beginning of the development of the external nose and also the nasal cavities. The centre of a nasal placode deepens and forms a nasal pit (fig. 2.2.02, 2.2.03), whereas the margins of the pit are folded up and form the medial and the lateral nasal prominences (fig. 2.2.04–2.2.08 and 1.2.08–1.2.12).

The nasal pit deepens and grows inward and forms a nasal groove or sac (fig. 2.2.10–2.2.15) which in its most dorsal portion caudally comes into contact with the roof of the mouth (fig. 2.2.16). The adjacent epithelial layers of the nasal sac and of the roof of the oral cavity rupture, thus producing a communication between the oral cavity and the nasal sac, the primary choana (fig. 2.2.17–2.2.19). The nasal sac, now termed nasal duct, elongates together with the primary choana in a dorsal direction (fig. 2.2.19–2.2.21, 2.2.24–2.2.26) until the primary choana extends up to the nasopharynx (fig. 2.2.26).

The right and the left nasal ducts are the future nasal cavities. The median tissue delimited by the two nasal ducts laterally is the nasal septum in which its cartilaginous lamina is formed (fig. 2.2.28). The separation of the nasal cavities from the oral cavity is realized by the elevation of the palatal processes (see chapter 2.3 for details). When the palatal processes reach their horizontal position they fuse with each other and also with the free border of the nasal septum (fig. 2.2.29–2.2.31) and form the floor of the nasal and also the roof of the oral cavity.

2.2

Abbreviations

1	palatal process
2	primary palate
3	upper jaw
4	nasal septum
5	auditory tube
6	hard palate
7	pharynx

Week 5

2.2.01–2.2.31
Developmental stages of the nose.

2.2.01–2.2.09
Development of the external aspects of the nose.

Week 6

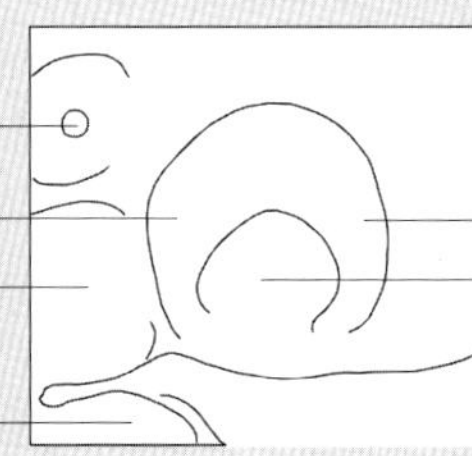

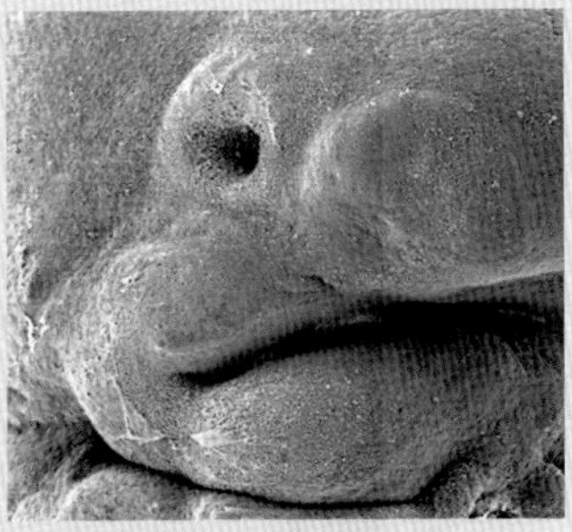

2.2.01

Right nasal placode. Lateral view. Embryos at week 5.

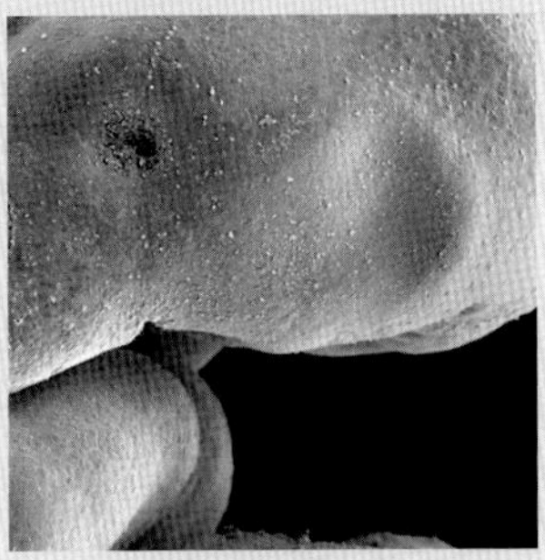

2.2.02

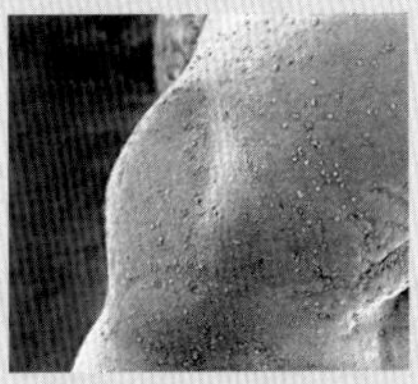

2.2.03

Right nasal placode. Ventral view. Embryo at week 5.

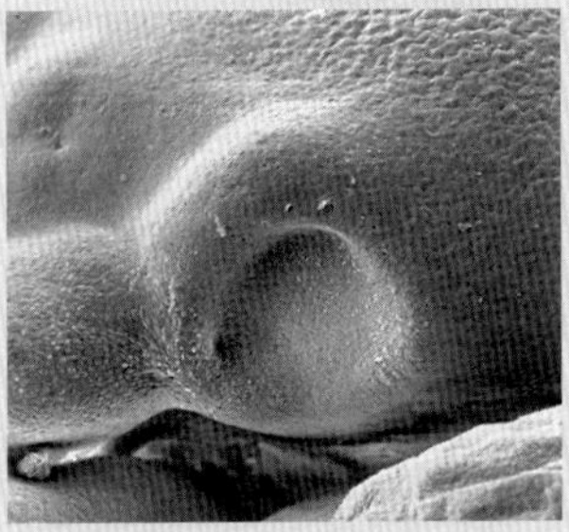

2.2.04

The right nasal pit deepens into the nasal groove. Embryos at week 6.

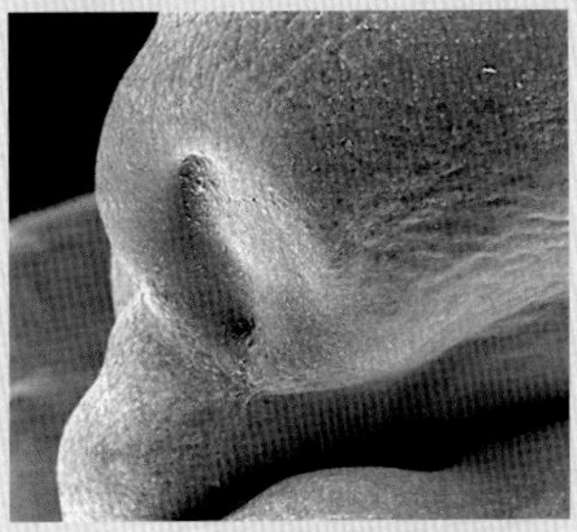

2.2.05

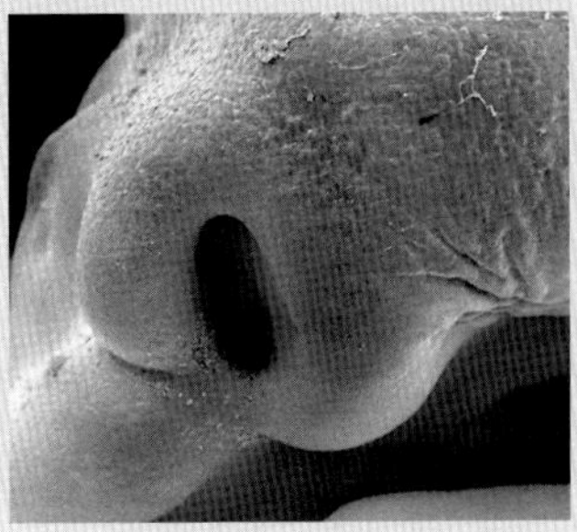

2.2.06

Week 7

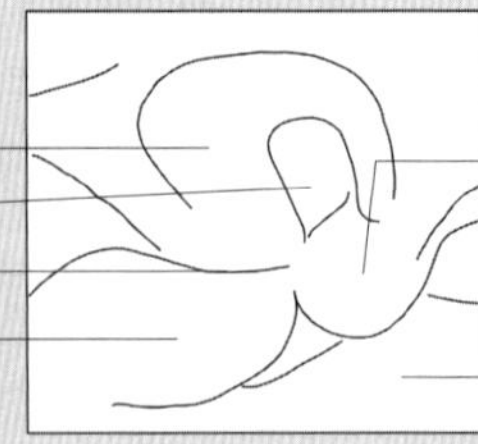

2.2.01–2.2.31
Developmental stages of the nose.

2.2.01–2.2.09
Development of the external aspects of the nose.

Week 8

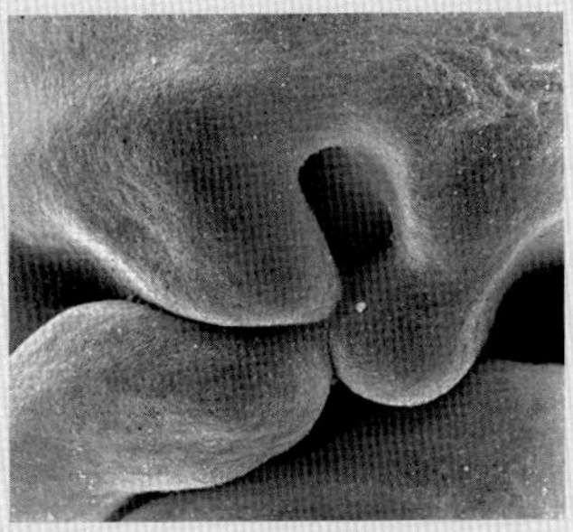

2.2.07

The nasal swellings arise around the nasal groove. Embryos at week 7.

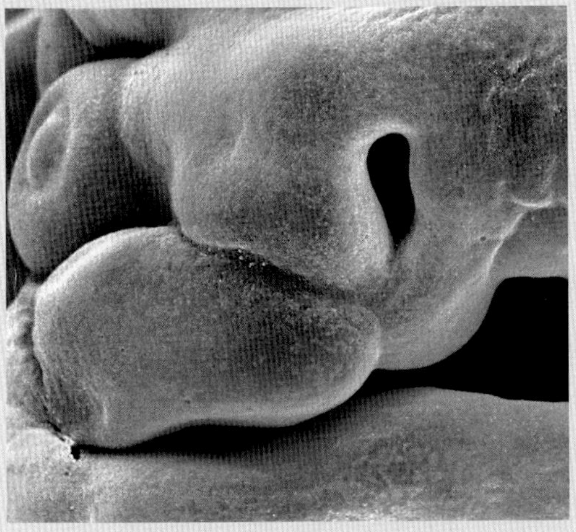

2.2.08

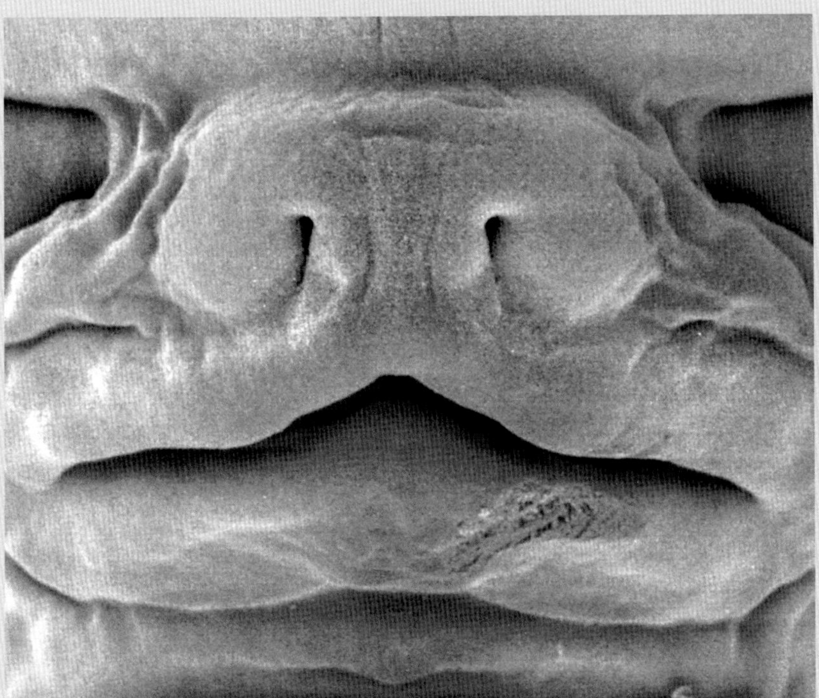

2.2.09

The apparently unpaired external nose has become prominent. Embryo at week 8.

2.2.01–2.2.31
Developmental stages of the nose.

2.2.10–2.2.23
Development of the internal aspects of the right nasal cavity. The nasal grooves/nasal ducts have been exposed and partially fenestrated to demonstrate the lumen.

Week 5

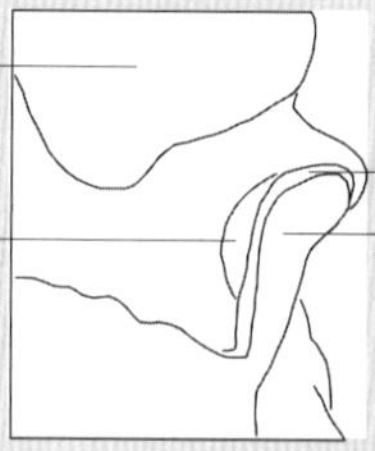

Week 6

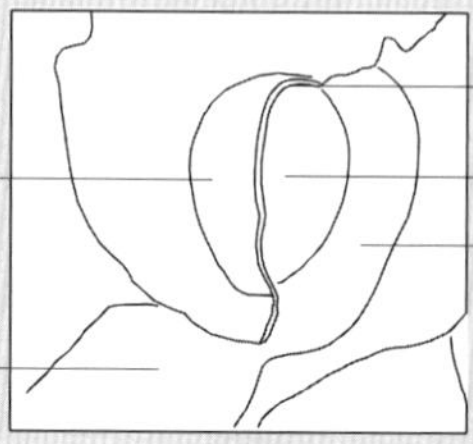

Week 7

2.2.10

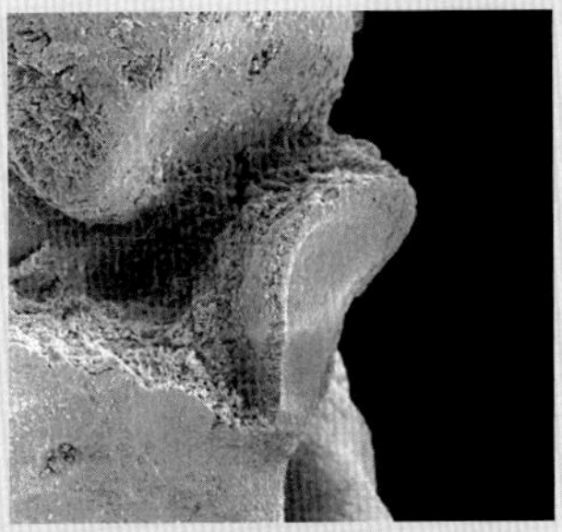

Right nasal placode. Lateral view. Embryo at week 5.

2.2.11

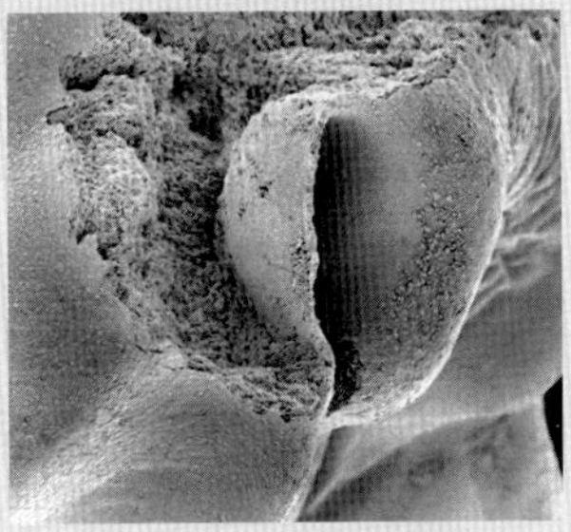

The nasal groove deepens into the nasal duct. Embryos at week 6.

2.2.11 / 12
Lateral views.

2.2.13
Ventral view.

2.2.12

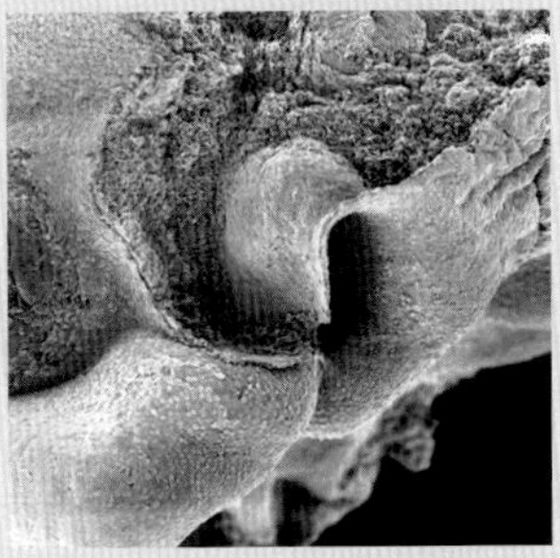

2.2.13

2.2.14

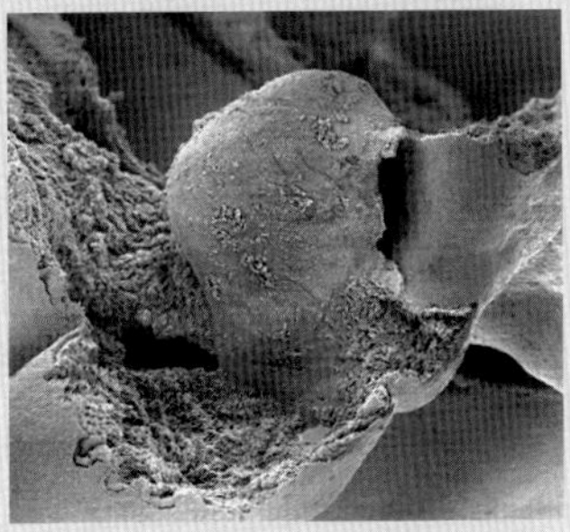

The epithelial lateral wall of the right nasal duct has been exposed. Lateral view. Embryos at week 7.

2.2.15

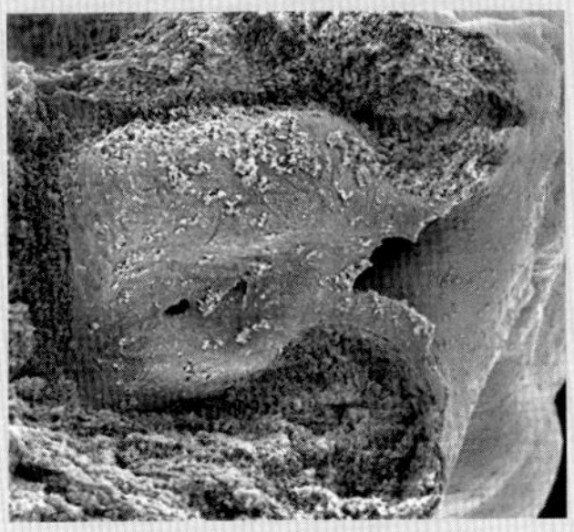

Week 7

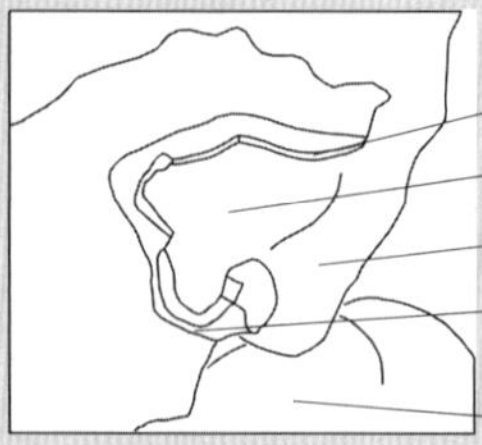

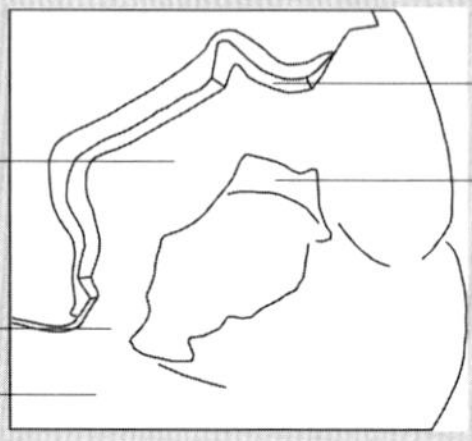

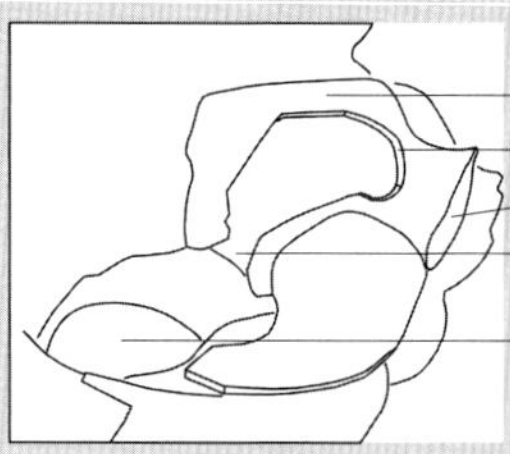

2.2.01–2.2.31
Developmental stages of the nose.

2.2.10–2.2.23
Development of the internal aspects of the right nasal cavity. The nasal grooves/nasal ducts have been exposed and partially fenestrated to demonstrate the lumen.

2.2.16–2.2.19
The lateral epithelial wall of the right nasal duct has been removed. Embryos at week 7.

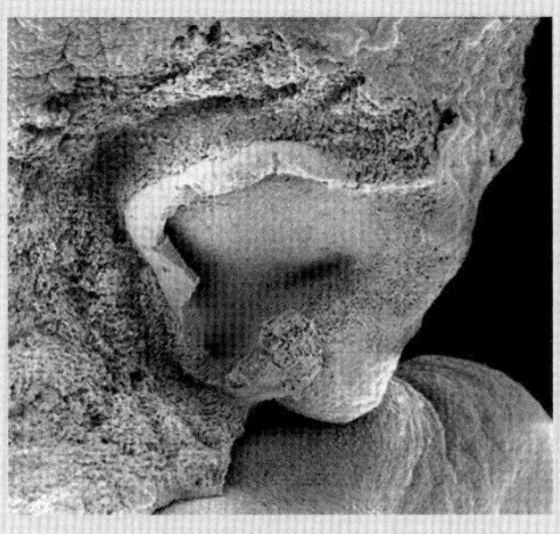

2.2.16

The nasal groove still does not show any communication with the oral cavity.

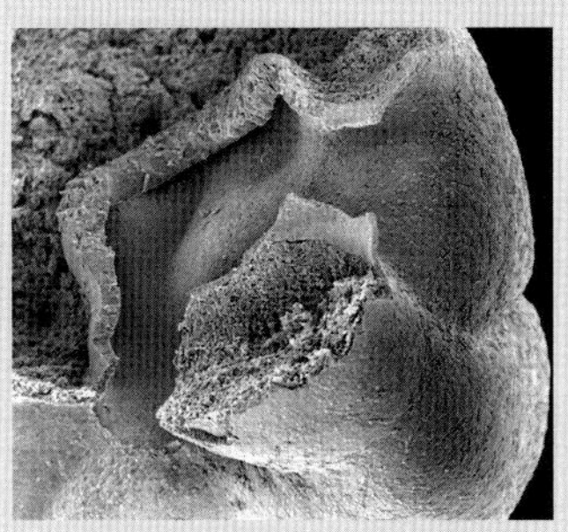

2.2.17

The epithelial contact between nasal and oral epithelia is torn, thus forming the primary choana.

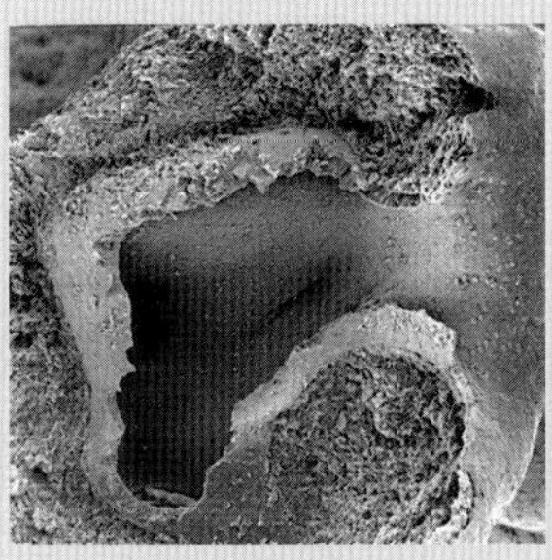

2.2.18

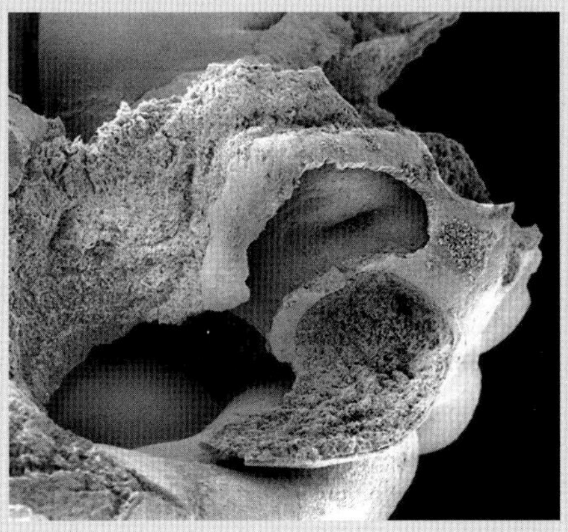

2.2.19

Week 8

2.2.01–2.2.31
Developmental stages of the nose.

2.2.10–2.2.23
Development of the internal aspects of the right nasal cavity. The nasal grooves/nasal ducts have been exposed and partially fenestrated to demonstrate the lumen.

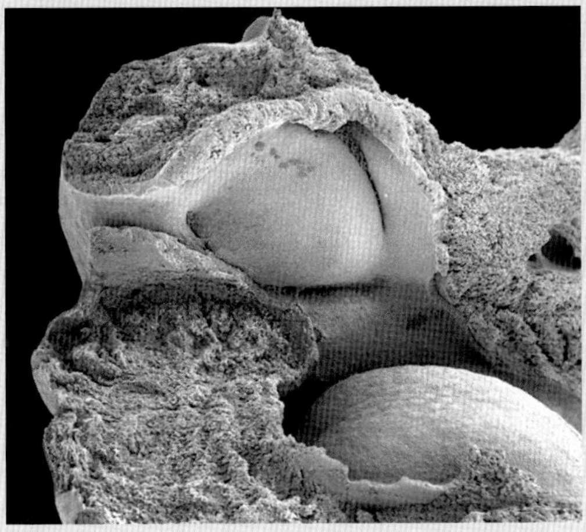

2.2.20 Upon removal of the left nasal cavity and the medial wall of the right nasal cavity, the lateral wall, exhibiting the first conchal bud, is shown. Left view. Embryo at week 8.

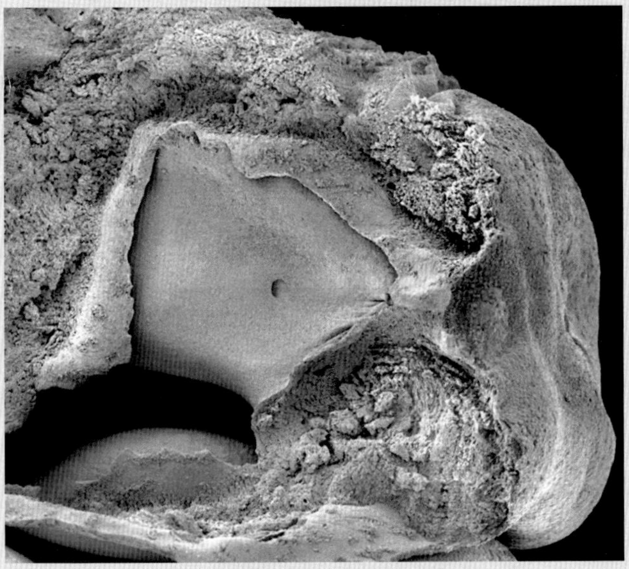

2.2.21 Upon removal of the lateral wall of the right nasal cavity, the medial wall can be seen, exhibiting the nasal septum and the organ of Jacobson. Right view. Embryo late in week 8.

Week 8

2.2.01–2.2.31
Developmental stages of the nose.

2.2.10–2.2.23
Development of the internal aspects of the right nasal cavity. The nasal grooves/nasal ducts have been exposed and partially fenestrated to demonstrate the lumen.

2.2.22 / 23
Developmental stages of the conchae. Embryos at week 8.

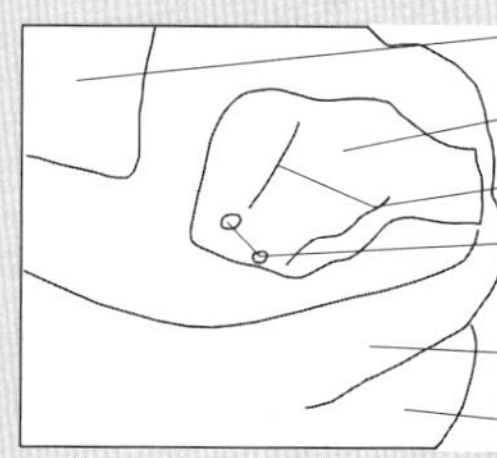

2.2.22 Upon removal of the left nasal cavity and the medial wall of the right nasal cavity, the lateral wall with the conchae can be seen. Left view. More advanced stage than the embryo in figure 2.2.20.

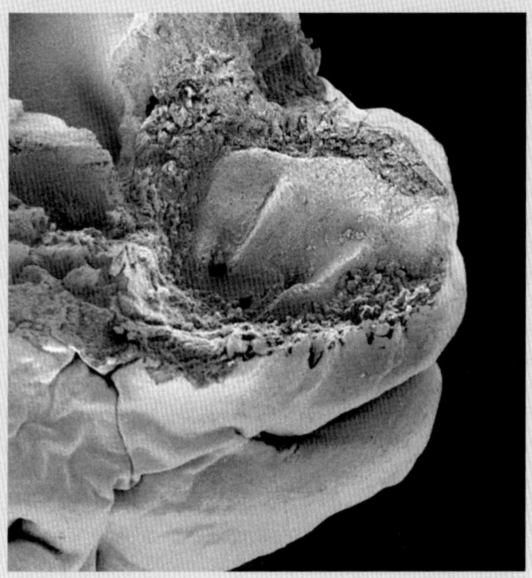

2.2.23 The epithelial lateral wall of the right nasal cavity shows the foldings of the developing conchae. Right view.

2.2.01–2.2.31
Developmental stages of the nose.

2.2.24–2.2.26
Developmental stages of the elongation of the primary choana. The lower jaw has been removed. Caudal view.

Week 7

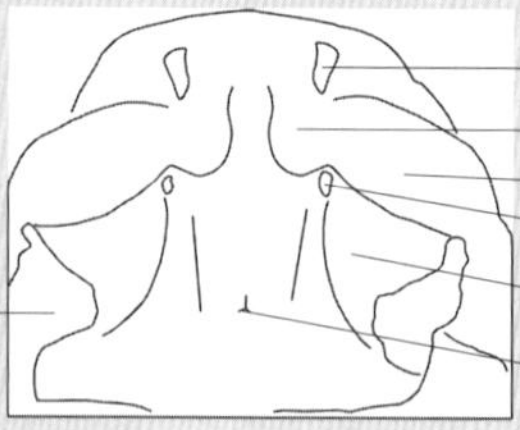

Week 8

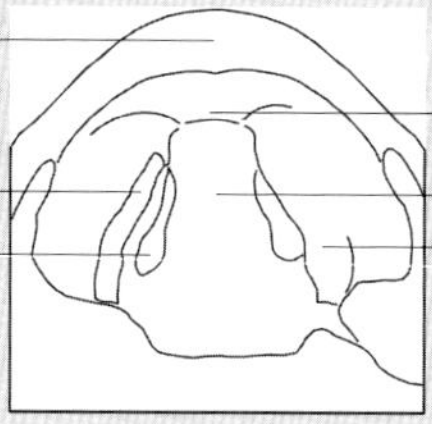

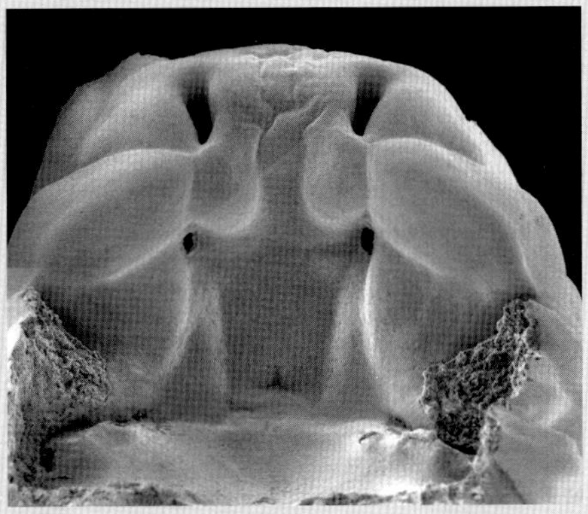

2.2.24 Embryo at week 7.

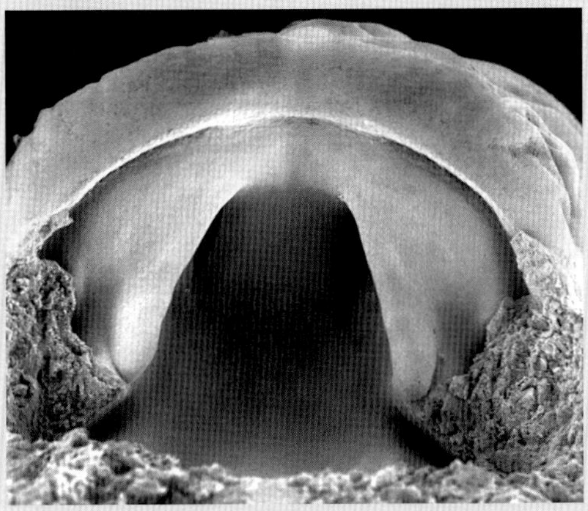

2.2.25 Embryo at week 8.

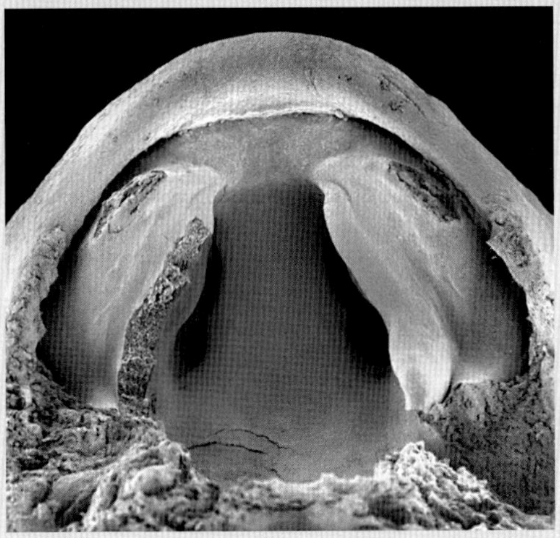

2.2.26 The right palatine process has been partly resected to show the elongation of the primary choana. Embryo late in week 8.

Week 7

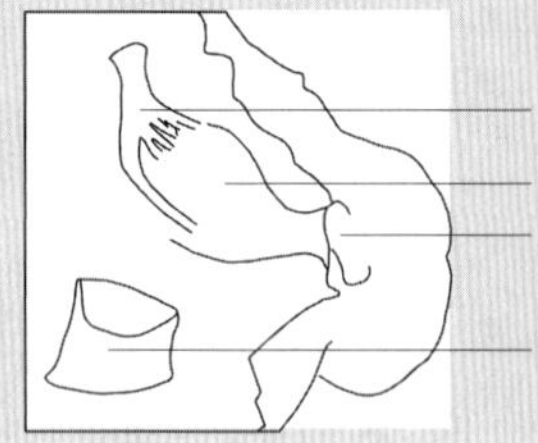

Week 8

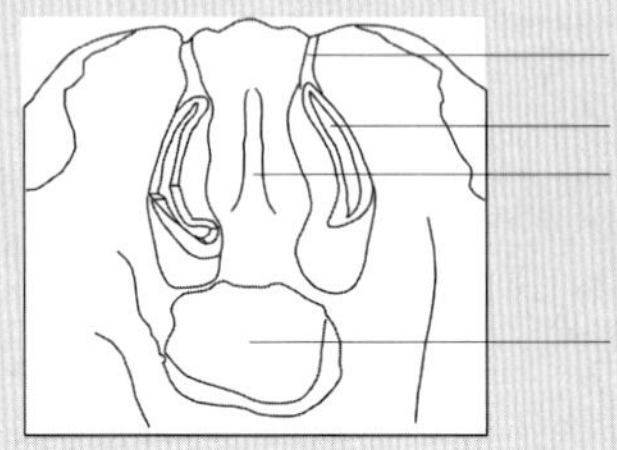

2.2.01–2.2.31
Developmental stages of the nose.

2.2.27–2.2.31
Developmental stages of the nasal septum.

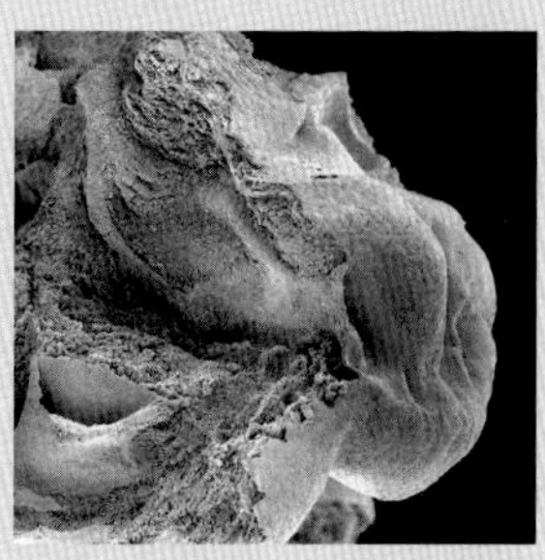

2.2.27 The right nasal cavity has been removed to show the epithelial medial wall of the left nasal cavity with the medial nasal nerves. Right view. Embryo at week 7.

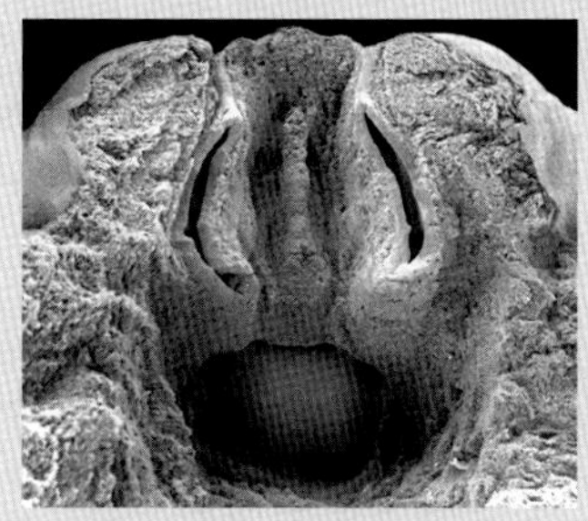

2.2.28 Both nasal cavities have been opened cranially. The connective tissue between the nasal cavities has been removed except the cartilage of the nasal septum. Cranial-dorsal view. Embryo at week 8.

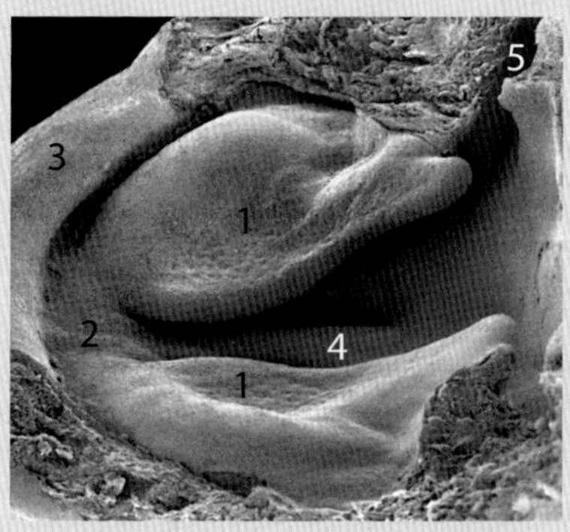

2.2.29 Caudal and lateral-right view of the palatine processes and the lower free border of the nasal septum. Elevation of the palatine process has started ventrally, thus establishing the contact between it and the nasal septum. Embryo at week 8.

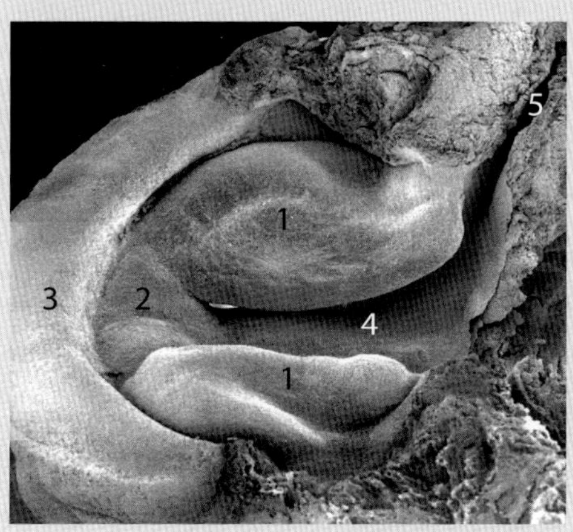

2.2.30 Caudal and lateral-right view of the palatine processes and the lower free border of the nasal septum. Elevation of the palatine process is advanced in the dorsal direction, thus completing the contact between it and the nasal septum. Embryo at week 8.

Week 10

2.2.01–2.2.31
Developmental stages of the nose.

2.2.27–2.2.31
Developmental stages of the nasal septum.

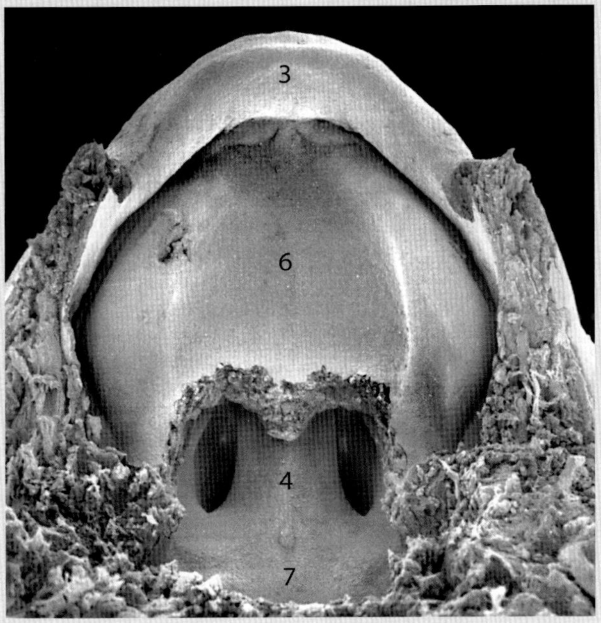

2.2.31 Dorsal and caudal view of the palate and the nasal septum. The fusion of the palatine processes with each other and the free border of the nasal septum has been completed. The soft palate has been removed. Embryo at week 10.

2.3 The Development of the Palate

The roof of the early oral cavity is formed by the epithelium covering the ventrally bent brain (fig. 2.3.01, 2.3.02). The final form of the roof develops from the primary palate and the palatal processes. The primary palate corresponds to the region between the early still roundish primary choanae and comprises the inside portions of the medial nasal prominences (fig. 2.3.03, 2.3.05). The palatal processes arise from the lateral wall of the oral cavity and inside the maxillary eminences (fig. 2.3.03). They grow out in a vertical direction and form processes which abut the tongue laterally (fig. 2.3.04–2.3.07).

When the head takes on an oblong form, the tongue descends and the palatal processes reach a horizontal position (fig. 2.3.09). They grow in a medial direction and their free borders meet in the midline and fuse, thus forming the definite palate (secondary palate) above the tongue (fig. 2.3.12, 2.3.13). The fusion of the palatal processes starts ventrally and continues dorsally (fig. 2.3.12, 2.3.14).

The hard palate is formed by the anterior part, the soft palate by the most posterior portions of the fused palatal processes (fig. 2.3.14). The junction of the primary palate with the palatal processes, the secondary palate, is not realized by fusion but by merging. Figures 2.3.10 and 2.3.11 show that the processes of palatal growth and the descent of the tongue are not necessarily right-left symmetrical.

2.3

Abbreviations

1	palatal process, still vertical
2	palatal process, nearly horizontal
3	palatal processes, fusing
4	hard palate
5	soft palate
6	primary palate
7	nasal septum
8	tongue
9	upper jaw

2.3.01–2.3.14
Developmental stages of the palate.

Week 4

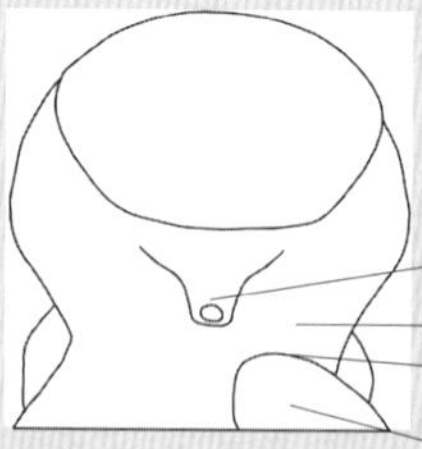

Week 5 / 7

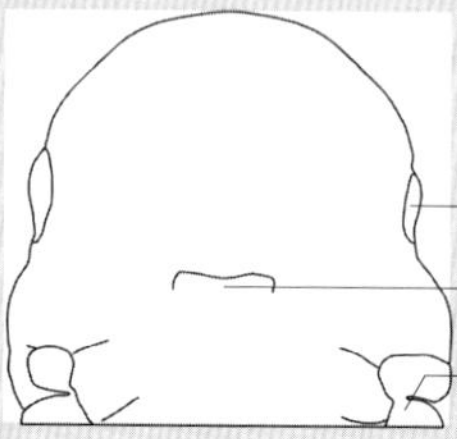

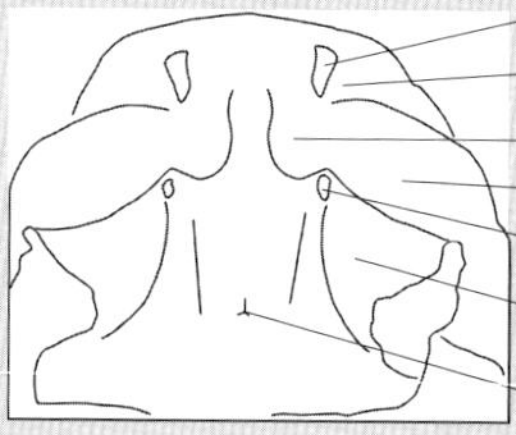

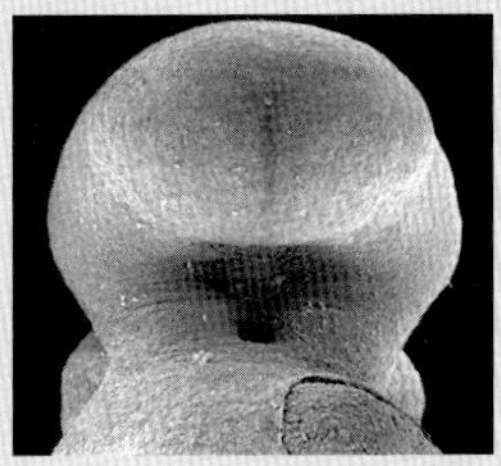

2.3.01 Oral cavity. Ventral view. Embryo at week 4.

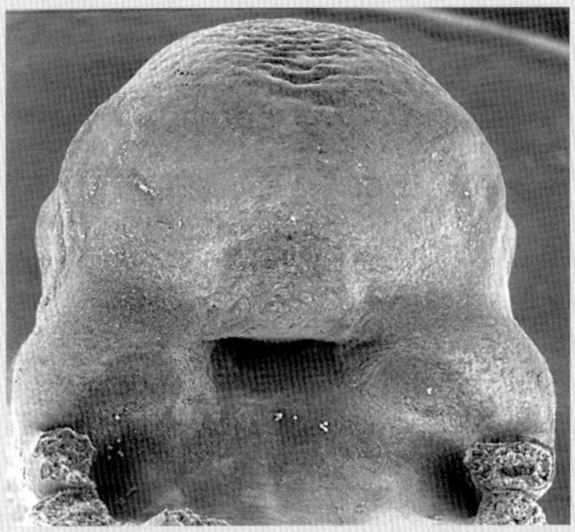

2.3.02 Oral cavity. Caudal view. The lower jaw has been removed. Embryos at weeks 5 and 7.

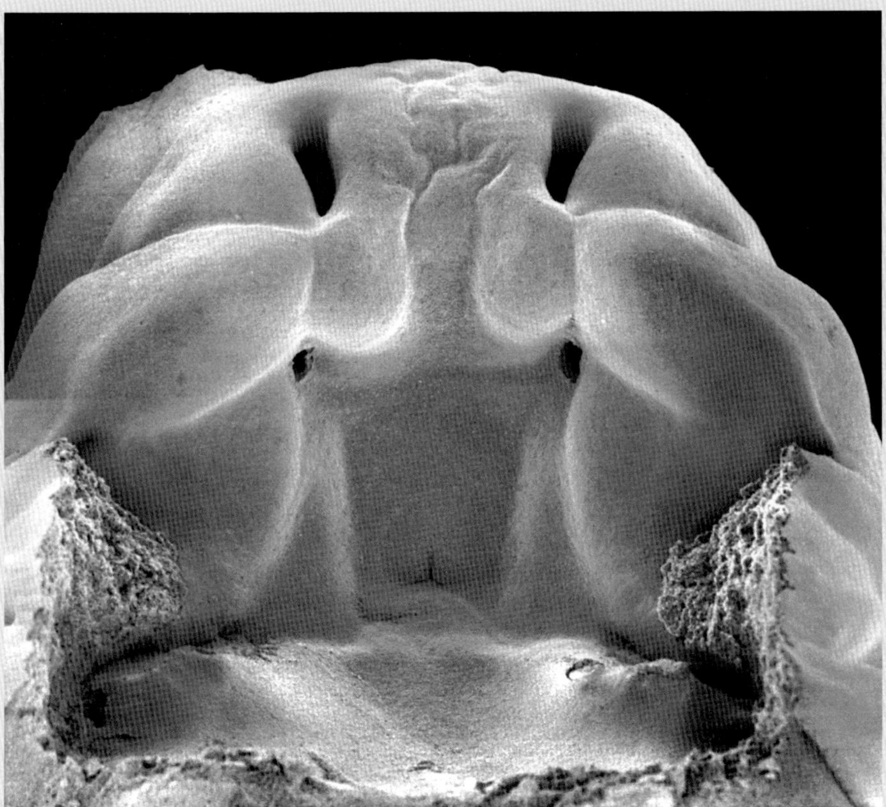

2.3.03

Week 7

2.
Endodermal Organs in the Head and Neck

2.3
The Development of the Palate

2.3.01–2.3.14
Developmental stages of the palate.

Week 8

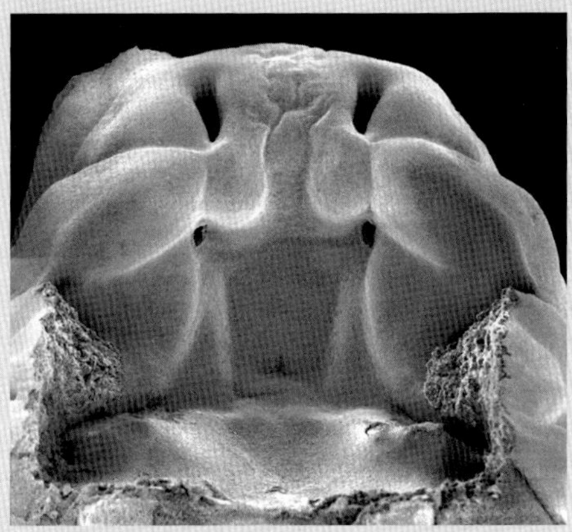

2.3.03a Same embryo as in figure 2.3.03.

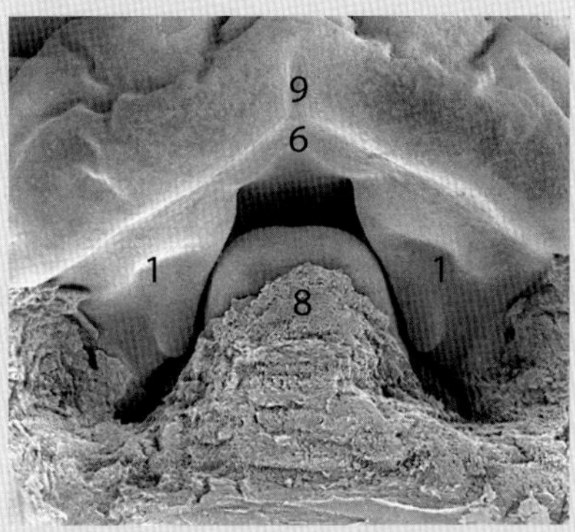

2.3.04 The lower jaw and the floor of the oral cavity except the tongue have been removed. Caudal view. Embryo late in week 7.

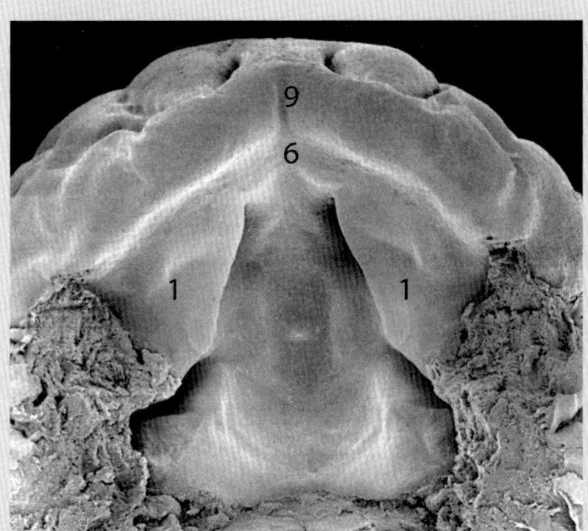

2.3.05 Same embryo as in figure 2.3.04. After removal of the tongue, the future nasal cavity is visible. Caudal view.

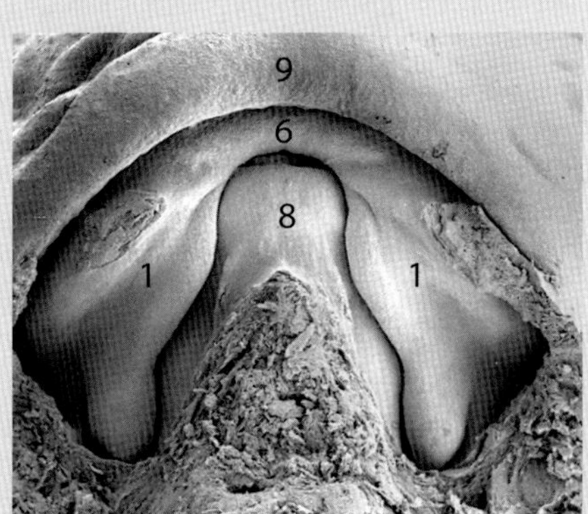

2.3.06 The lower jaw and the floor of the oral cavity except the tongue have been removed. Caudal view. Embryo at week 8.

Week 8

2.3.01–2.3.14
Developmental stages of the palate.

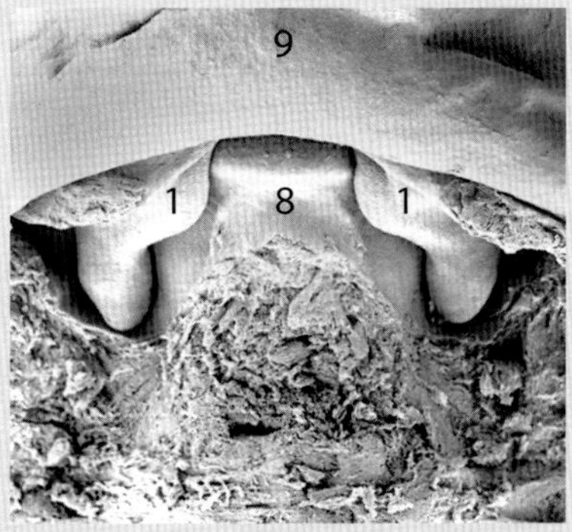

2.3.07 Same embryo and the same state of dissection as in figure 2.3.06. The ventral view shows the position of the tongue and the palatal processes. Ventral view.

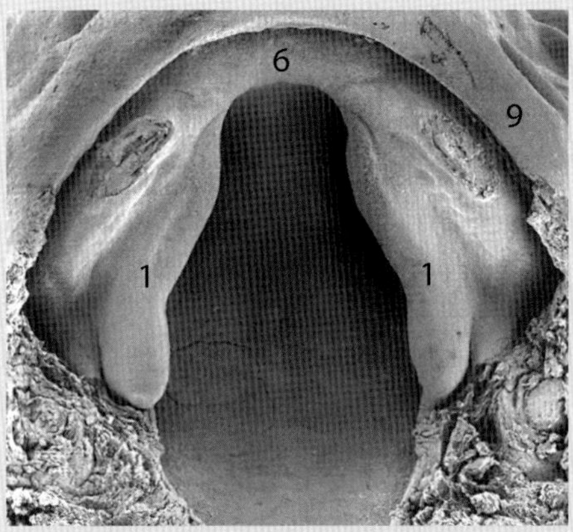

2.3.08 Same embryo as in figure 2.3.07. After removal of the tongue, the palatal processes and the future nasal cavity are visible. Caudal view.

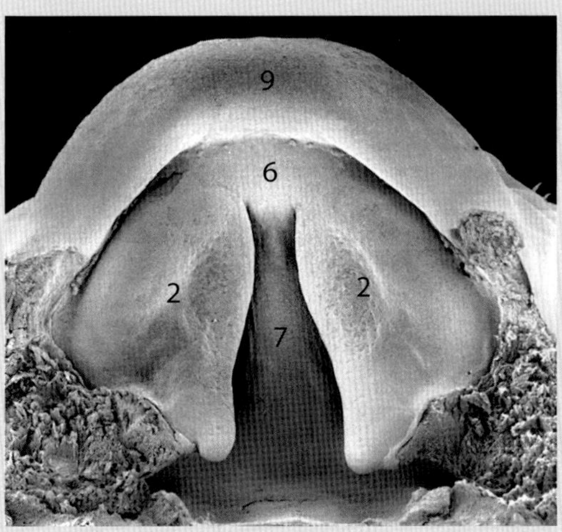

2.3.09 The lower jaw and the tongue have been removed. More advanced elevation of the palatal processes. Embryo late in week 8. Caudal view.

Week 8

2.3.01–2.3.14
Developmental stages of the palate.

Week 9

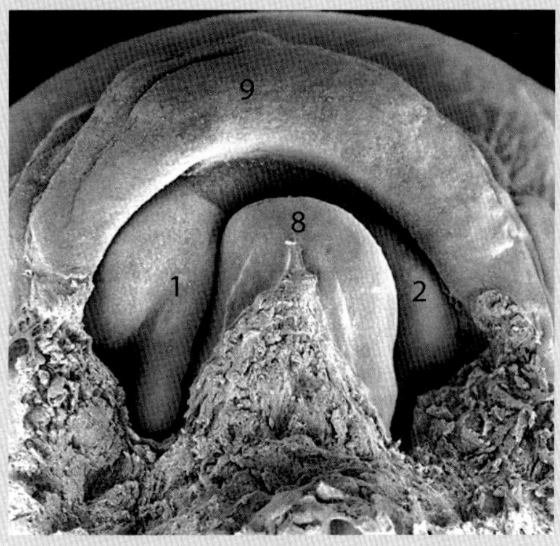

2.3.10 In this embryo the elevation of the palatal processes is not right-left symmetrical. Embryo late in week 8 (see fig. 2.4.04). Due to the more advanced elevation of the left palatal process, the tongue is presented in an oblique position. Caudal view.

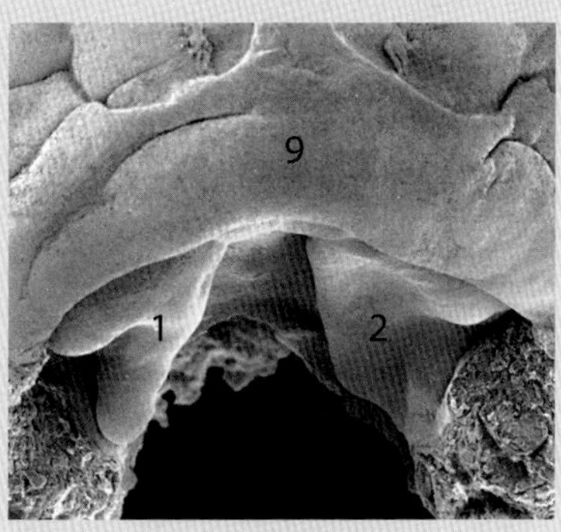

2.3.11 Same embryo as in figure 2.3.10. The tongue has been removed to show the asymmetrical positions of the palatal processes. Ventral view.

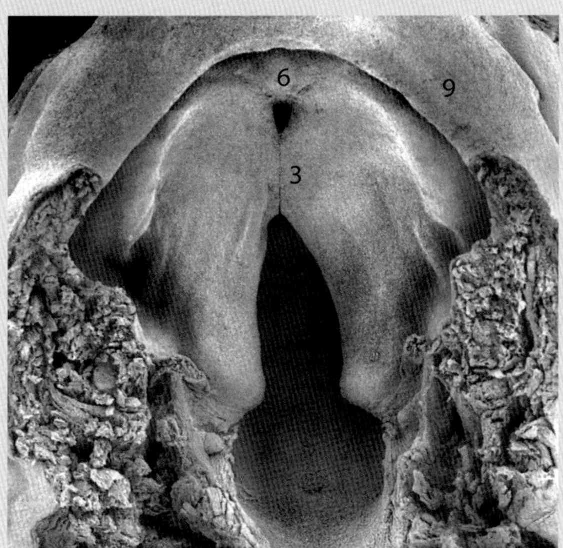

2.3.12 The lower jaw and the tongue have been removed. The elevation of the palatal processes is advanced and their fusion has begun ventrally. Embryo at week 9. Caudal view.

20:1

Week 10

2.3.01–2.3.14
Developmental stages of the palate.

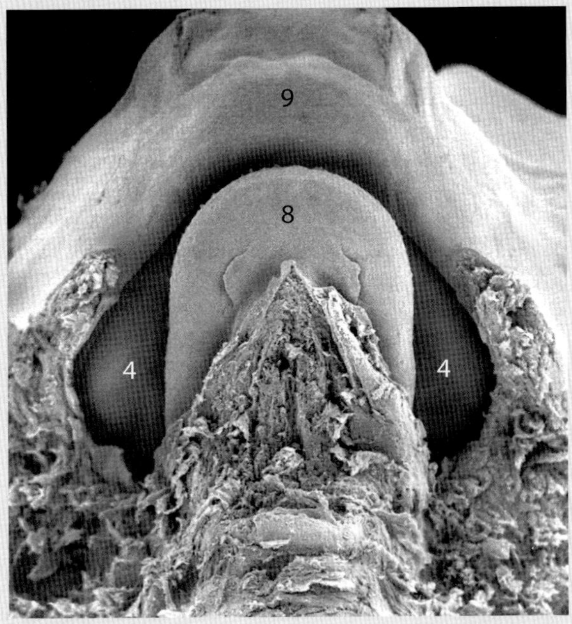

2.3.13 Embryo at week 10. Caudal view. The lower jaw and the floor of the oral cavity, except the tongue, have been removed. The palate is partly hidden behind the tongue.

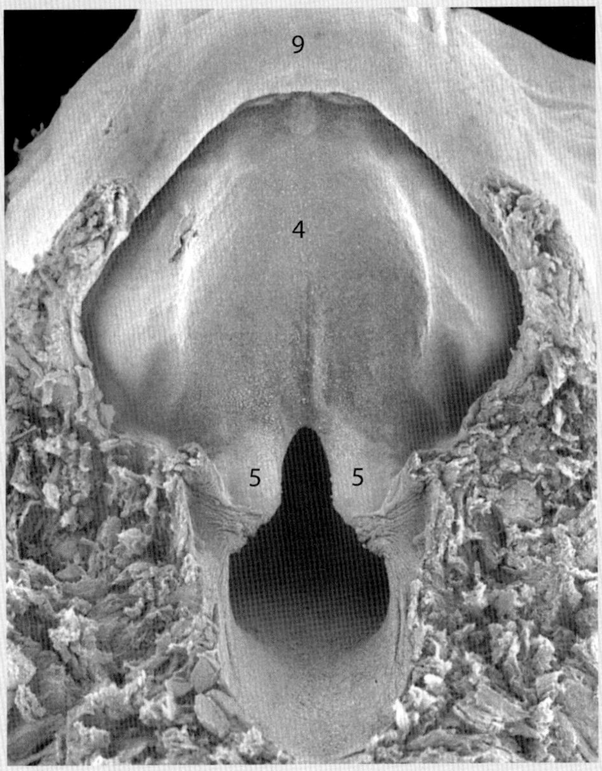

2.3.14 Embryo at week 10. Caudal view. After removal of the tongue, the hard and the soft palate are visible.

2.4 The Development of the Pharynx

The Auditory Tube

The auditory tube joins the upper portion of the pharynx (nasopharynx) to the middle ear. It develops from the first pharyngeal pouch (fig. 2.4.03, 2.4.04, 2.4.07, 2.2.29, 2.2.30, 6.3.24). Its orifice is situated immediately behind the palatal processes and is, therefore, hidden from a ventral view into the oral cavity behind the soft palate (fig. 2.4.05).

The Hypopharynx

The relief of the hypopharynx is formed by the internal surfaces of the pharyngeal arches III and IV (fig. 2.4.07, 2.4.08). The median parts of these arches swell out and form the hypopharyngeal eminence (fig. 2.4.09, 2.4.10). The caudal border of the hypopharynx is formed by the sixth arches (fig. 2.4.07–2.4.09) which eventually take part in the formation of the arytenoid swellings of the larynx (fig. 2.4.09, 2.4.10).

The position of the pharyngeal arch arteries in relation to the pharyngeal arches is shown in figures 2.4.12–2.4.14.

2.4

Abbreviations

II	pharyngeal arch II
III	pharyngeal arch III
IV	pharyngeal arch IV
1	medial lingual swelling
2	lateral lingual swelling
3	tongue
4	epiglottis
5	laryngotracheal groove
6	palatal process
7	auditory tube
8	pharynx

20:1

Week 5

Week 7

Week 8

2.
Endodermal Organs in the Head and Neck

2.4
The Development of the Pharynx

2.4.01–2.4.14
Developmental stages of the pharynx.

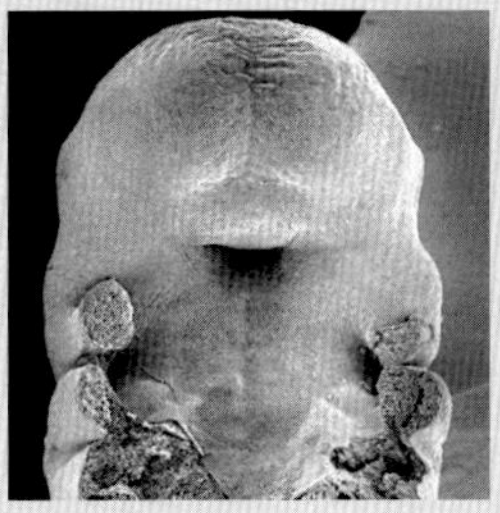

2.4.01

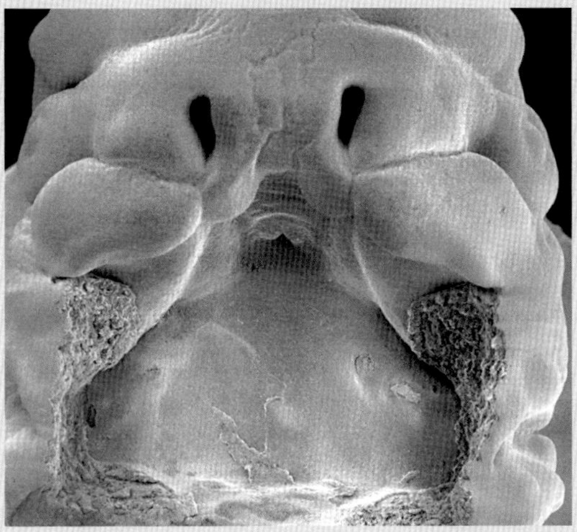

2.4.02

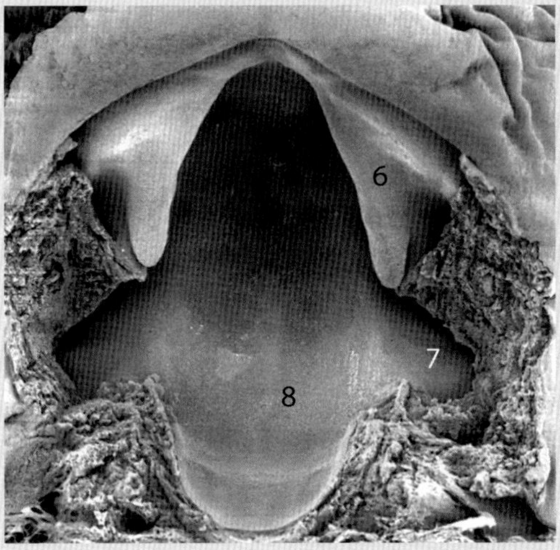

2.4.03

Oral cavity and pharynx. The lower jaw and the floor of the oral cavity have been removed. Embryos at weeks 5, 7 and 8. Ventral views.

2.4.01–2.4.14
Developmental stages of the pharynx.

20:1

Week 8

Week 10

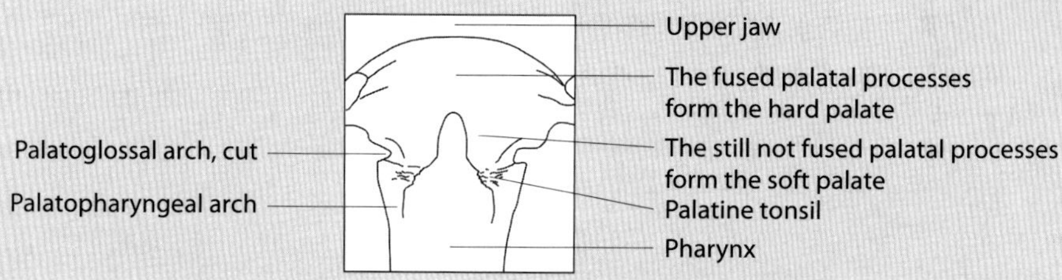

80:1

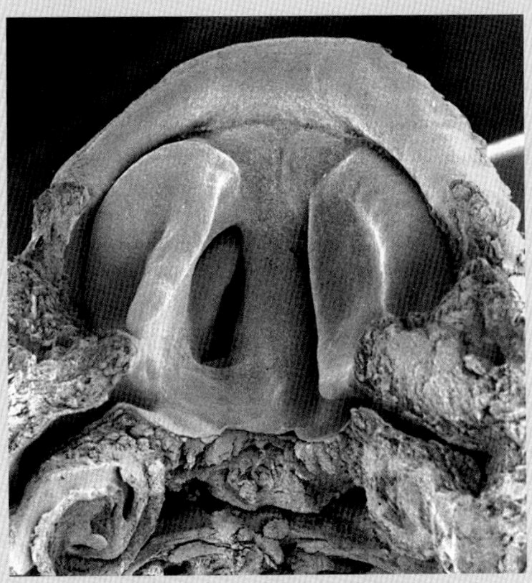

2.4.04 The position of the orifice of the auditory tube. Same embryo as in figures 2.3.10 and 2.3.11. The lower jaw and the floor of the oral cavity have been removed. Caudal view.

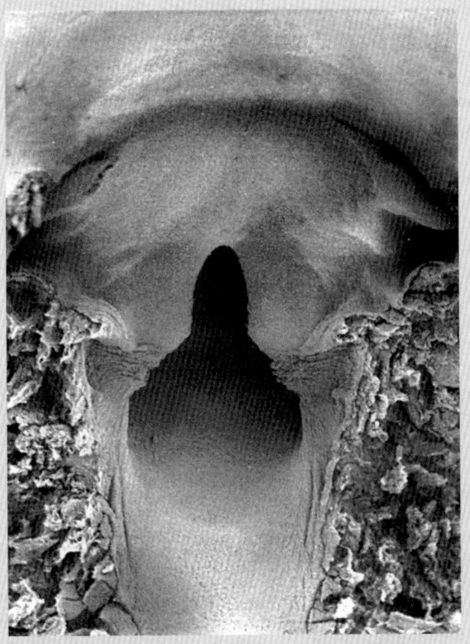

2.4.05 Ventral view of the hard and the soft palate and the palatine tonsil, the palatopharyngeal and the palatoglossal arches. The lower jaw and the floor of the oral cavity have been removed. Embryo at week 10.

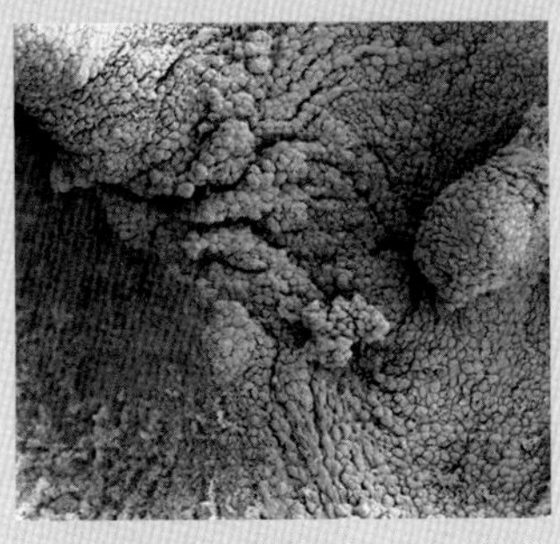

2.4.06 Anlage of the palatine tonsil. Embryo at week 10. Ventral view.

2.4.01–2.4.14
Developmental stages of the pharynx.

Week 5

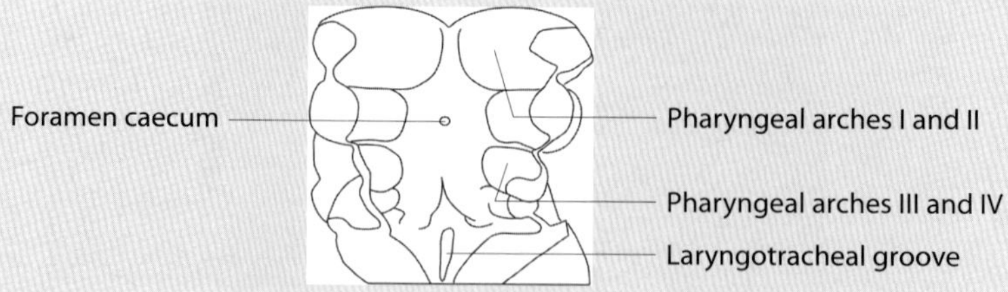

Week 5 / 6

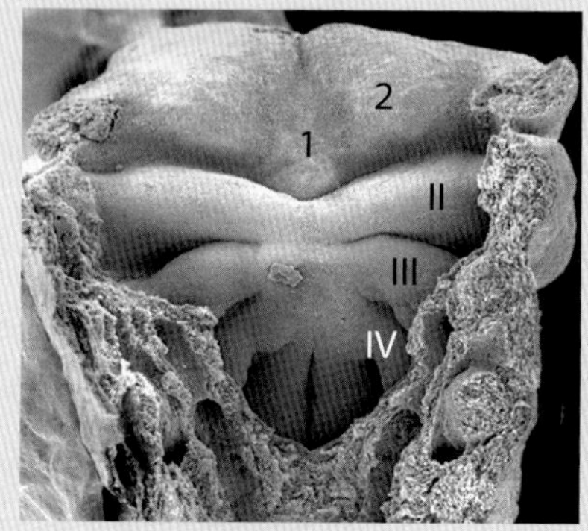

Week 6 / 7

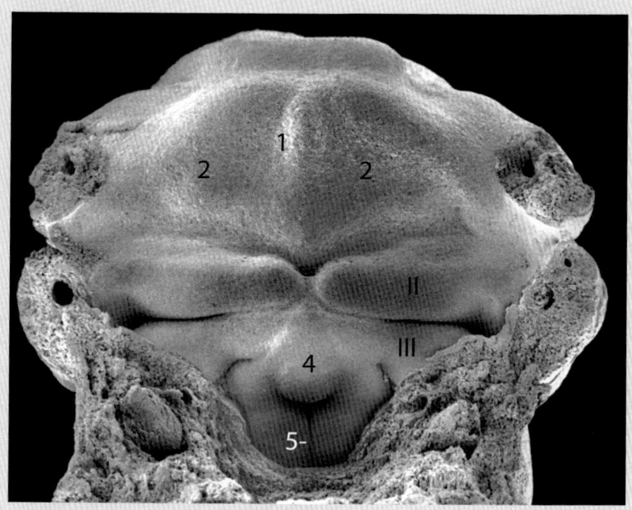

2.4.07

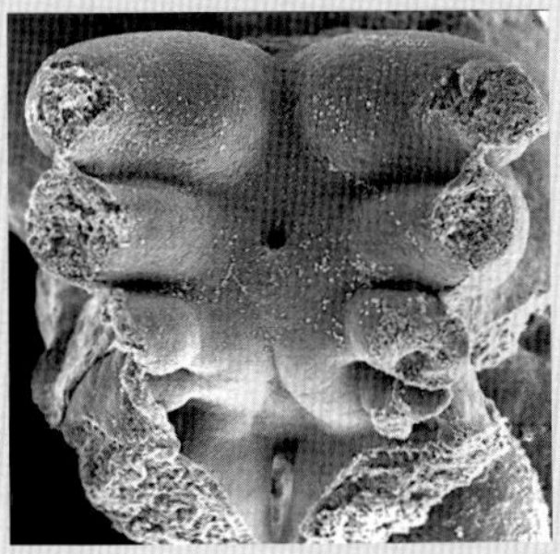

Cranial-dorsal view of the pharyngeal region of the tongue and the pharynx. Embryos at weeks 5 (2.4.07, 2.4.08), 6 (2.4.09, 2.4.10) and 7 (2.4.11).

2.4.08 / 09

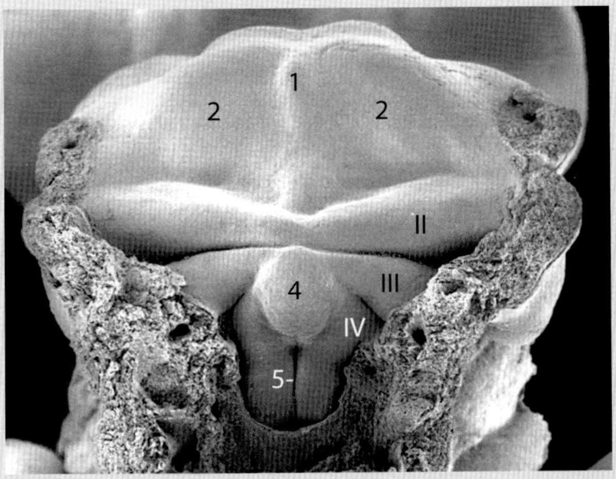

2.4.10 / 11

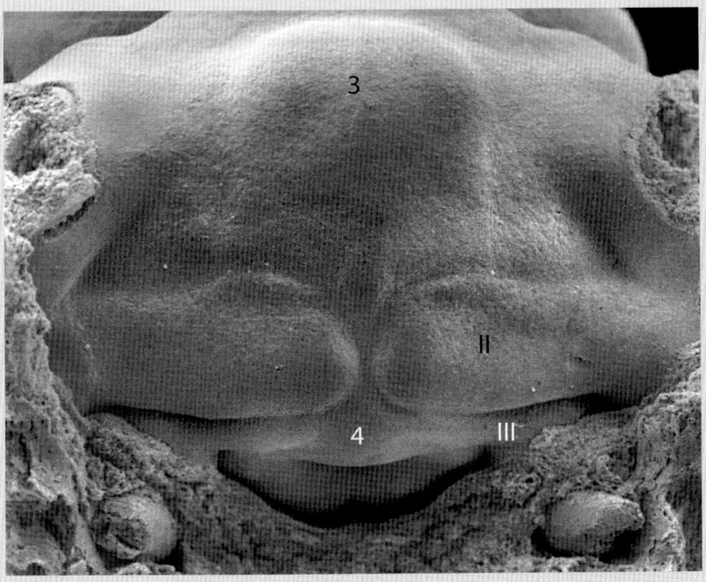

2.4.01–2.4.14
Developmental stages of the pharynx.

Week 5

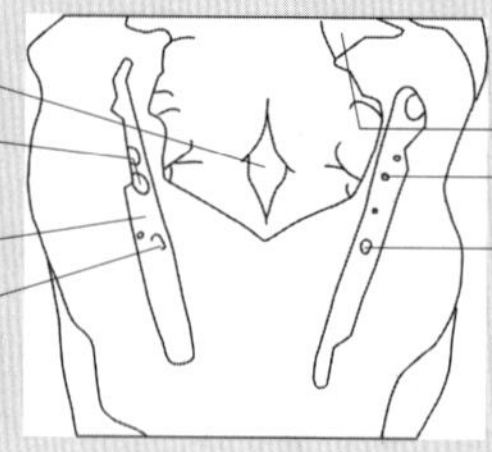

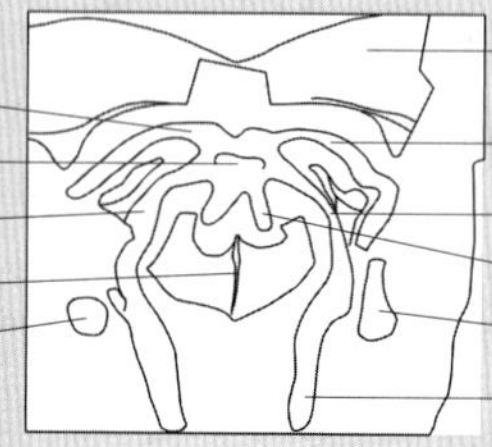

Week 6

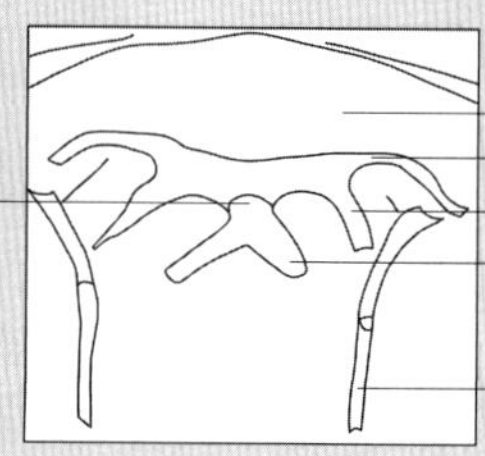

2.4.12

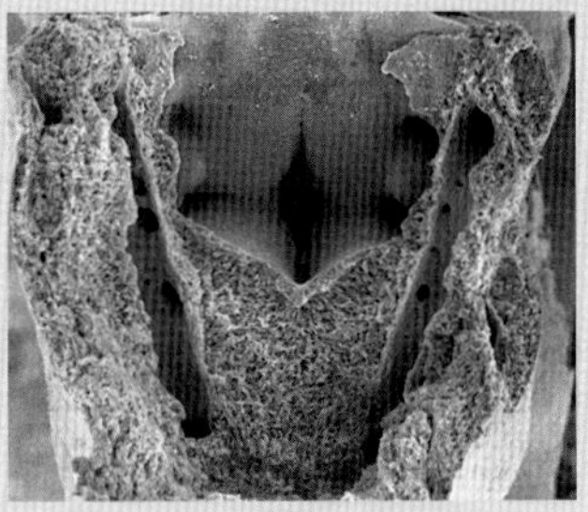

Positional relationships of the pharyngeal arch arteries. Dorsal view. Embryos at weeks 5 (2.4.12, 2.4.13) and 6 (2.4.14).

2.4.13

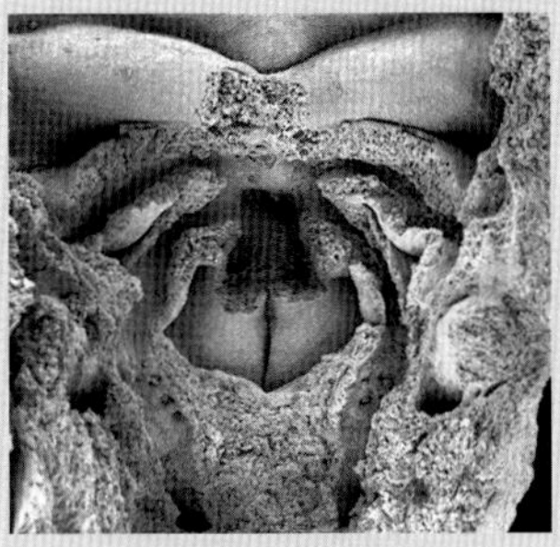

2.4.14

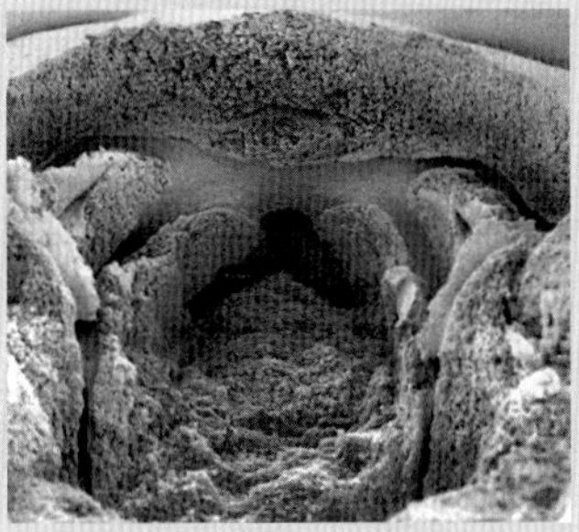

2.5

The Development of the Larynx

The larynx develops in the region where the lung bud branches as a diverticulum from the foregut (fig. 2.5.01–2.5.04). The orifice of the early trachea is laterally bordered by the 6th pharyngeal arches (fig. 2.5.05, 2.5.06) which eventually take part in forming the arytenoid swellings (fig. 2.5.07, 2.5.08). Thus the glottis is formed as a narrow rift in the sagittal direction.

The posterior region of the hypopharyngeal swelling grows out to form the epiglottis (fig. 2.6.09–2.6.11 and 2.1.04). The opposite epithelia of the glottis become apposed and form an epithelial lamina (fig. 2.5.12). Only in the 3rd month is the epithelial lamina separated and the lumen of the glottis is re-established.

2.5

Abbreviations

I	pharyngeal arch I
II	pharyngeal arch II
III	pharyngeal arch III
IV	pharyngeal arch IV
1	medial lingual swelling
2	lateral lingual swelling
3	foramen caecum
4	epiglottis
5	laryngotracheal groove
6	arytenoid swelling
8	pharynx
9	oesophagus
10	trachea

2.5.01–2.5.13
Developmental stages of the larynx.

Week 4

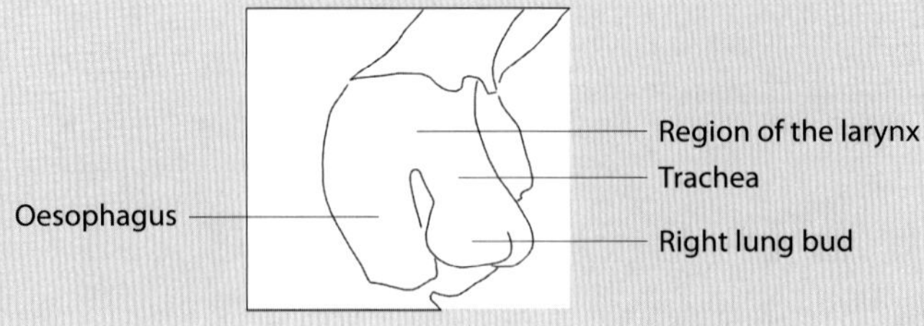

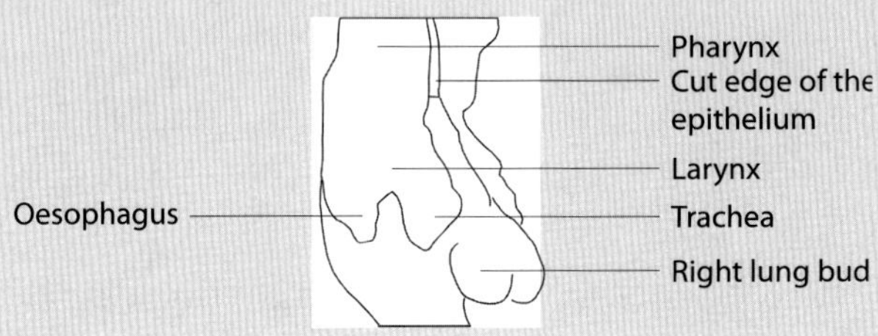

Week 5

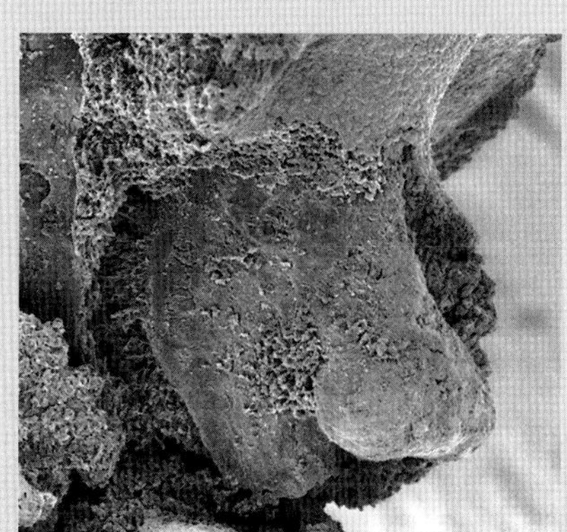

2.5.01

External and internal aspects of the early laryngeal region and the lung buds. Embryos at weeks 4 (2.5.01, 2.5.02) and 5 (2.5.03, 2.5.04). Ventral-left view.

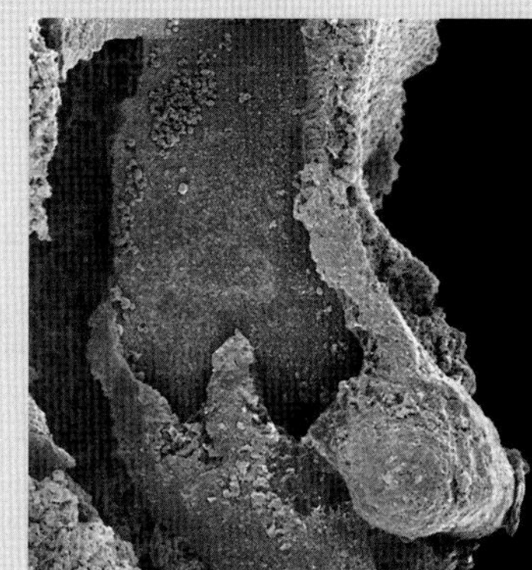

2.5.02

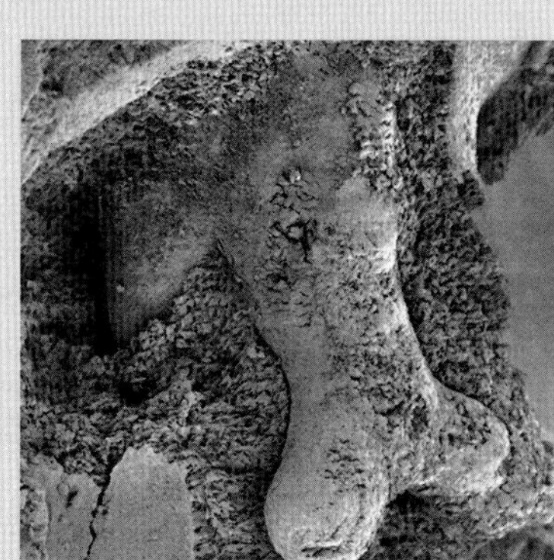

2.5.03

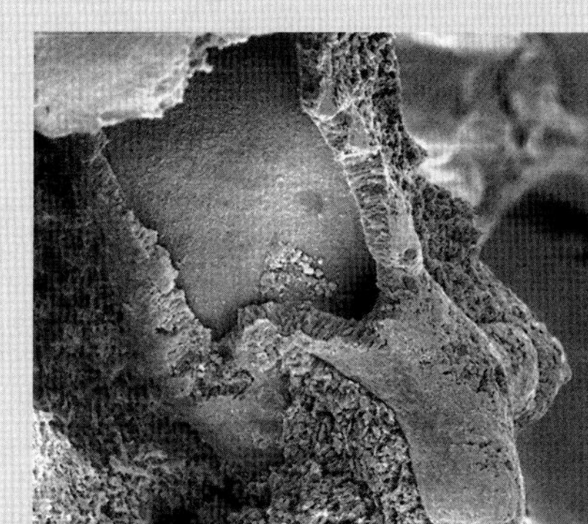

2.5.04

Week 5

2.5.01–2.5.13
Developmental stages of the larynx.

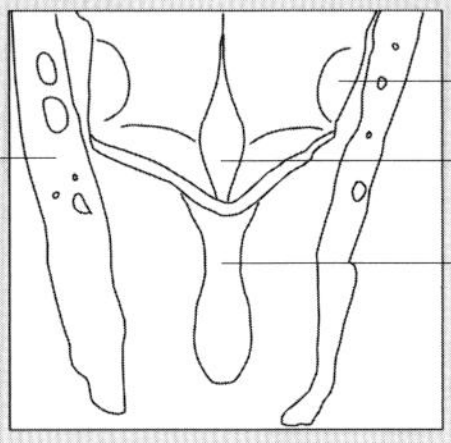

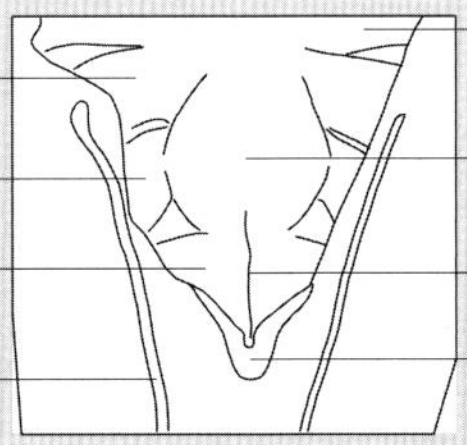

Week 6

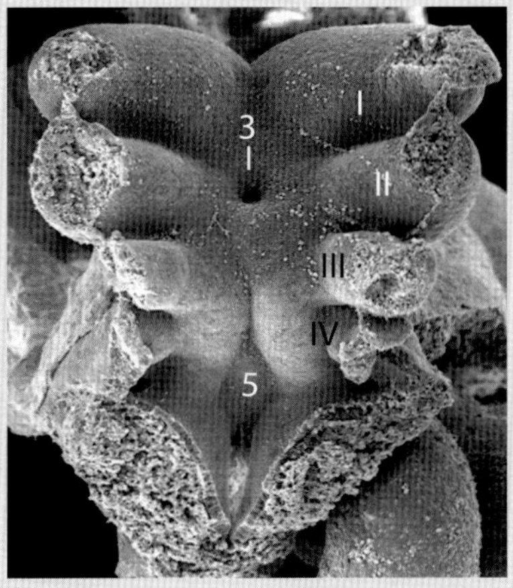

2.5.05

Dorsal view of the hypopharynx and the branching off of the larynx. Embryos at weeks 5 (2.5.05–2.5.07) and 6 (2.5.08).

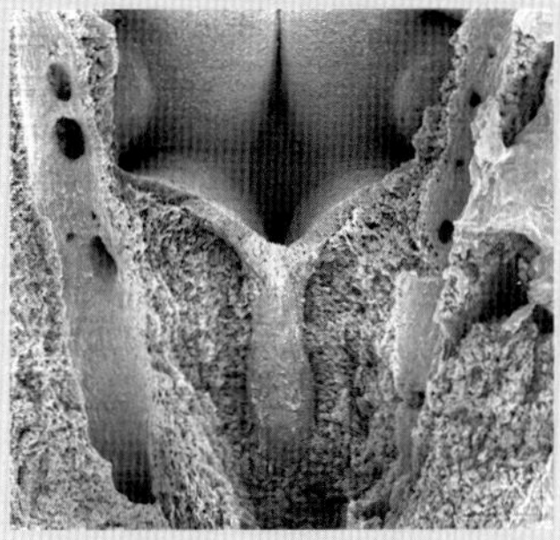

2.5.06

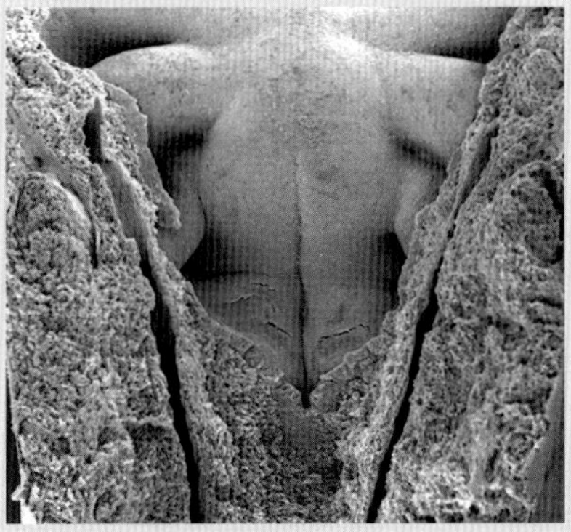

2.5.07

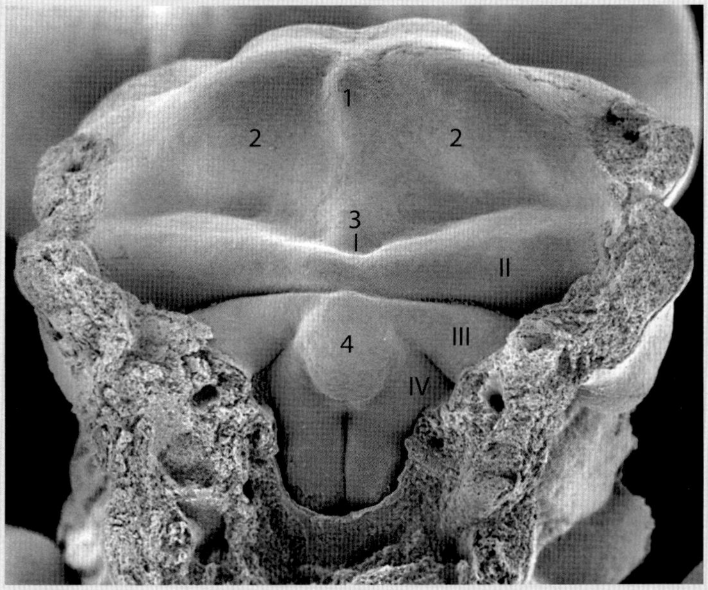

2.5.08

Week 6

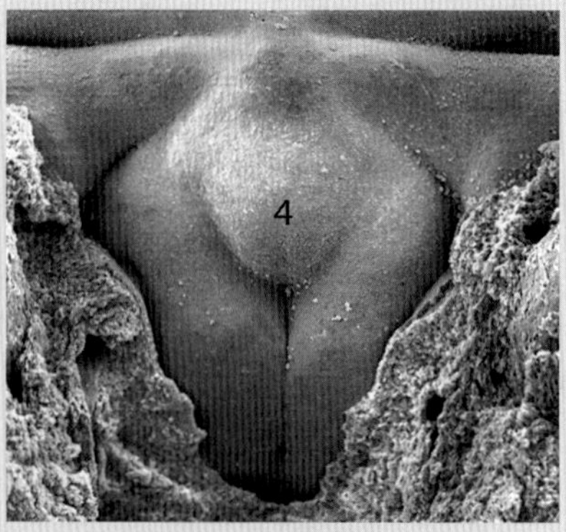

2.5.01–2.5.13
Developmental stages of the larynx.

Week 7

Week 6

Week 8

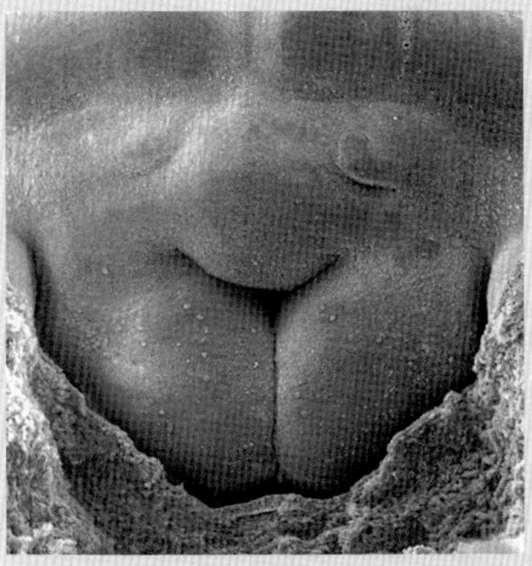

2.5.09 / 10

Cranial view of the laryngotracheal groove. Embryos late in week 6. Individual variations in embryos at the same developmental stage.

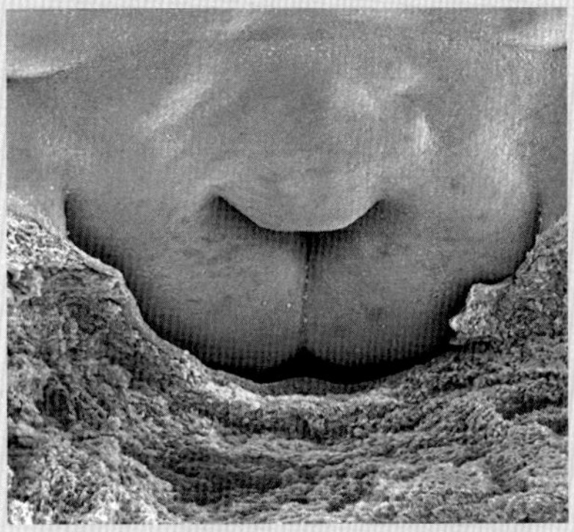

2.5.11

Cranial view of the laryngotracheal groove. Embryo at week 7.

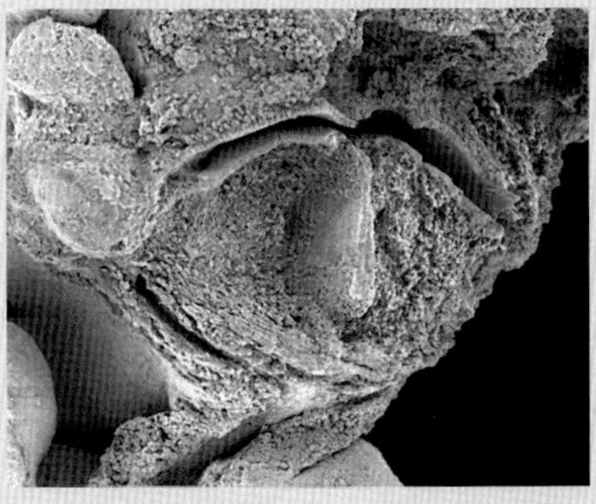

2.5.12

Cranial view of the larynx and the pharynx. Embryo at week 6.

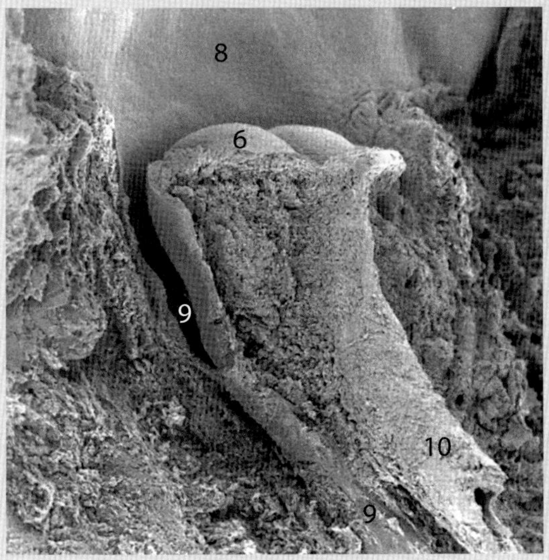

2.5.13

Ventral-right view of the pharynx, the larynx and the oesophagus. Embryo at week 8.

2.6
The Position of the Thymus and the Parathyroid Glands

The thymus develops from the endoderm of the ventral outpocketings of the third pharyngeal pouch and from the ectoderm of the third pharyngeal groove. Beginning at about week 7, the thymus descends and eventually reaches its definite position in the superior mediastinum.

The parathyroid glands develop from the endoderm of the dorsal outpocketings of the third and the fourth pharyngeal pouches.

Week 5 / 6

2.6.01–2.6.06
Position of the thymus.

2.6.01 / 02
Lateral left view of the exposed epithelial buds of the anlage of the thymus and the superior parathyroid. Same embryo at the end of week 5/ beginning of week 6.

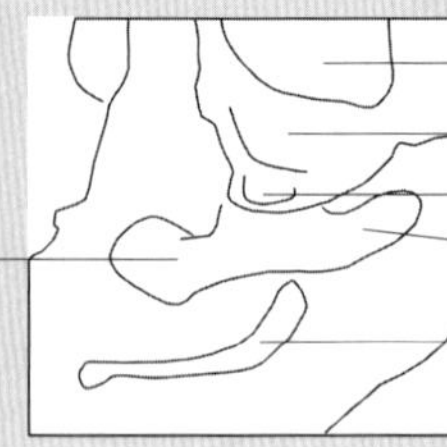

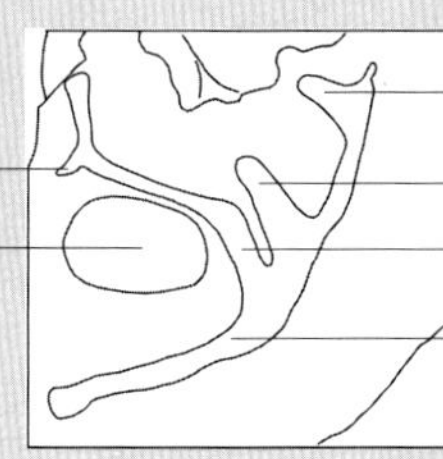

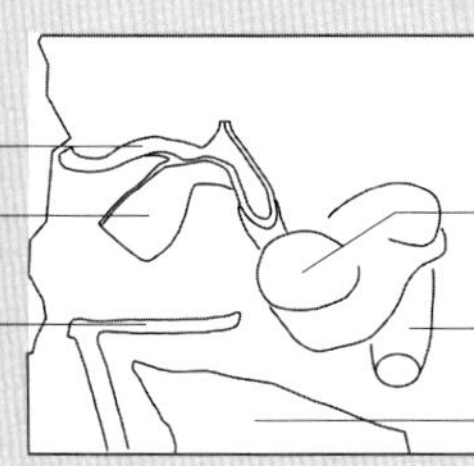

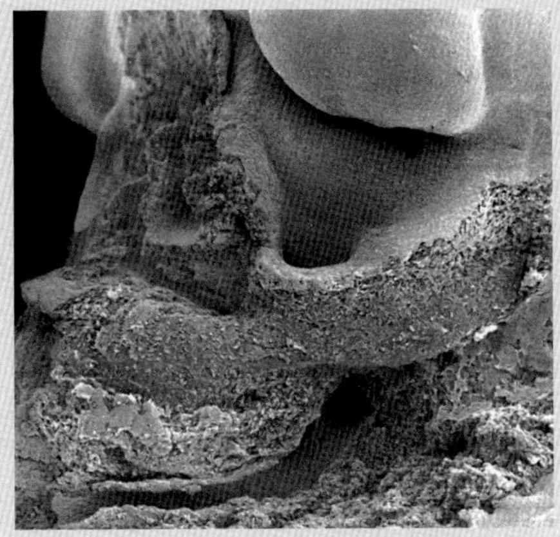

2.6.01 Positional relationships to the third pharygeal arch.

2.6.02 The connection of the thymus to the pharyngeal pouch is cut. The superior parathyroid has been removed.

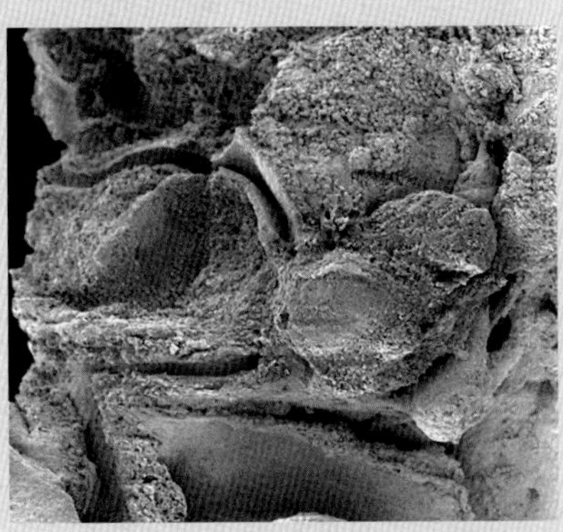

2.6.03 Ventral-left view of the position of the thymus and the parathyroid. The most ventral portion of the thymus has been removed. Embryo at the end of week 6.

2.6.01–2.6.06
Position of the thymus.

100:1

Week 6

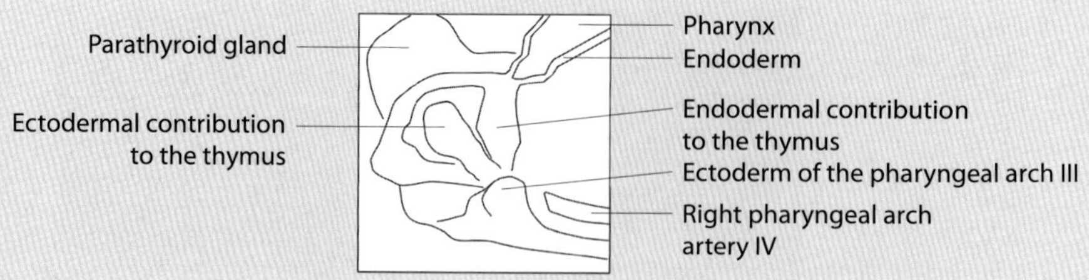

200:1

50:1

Week 8

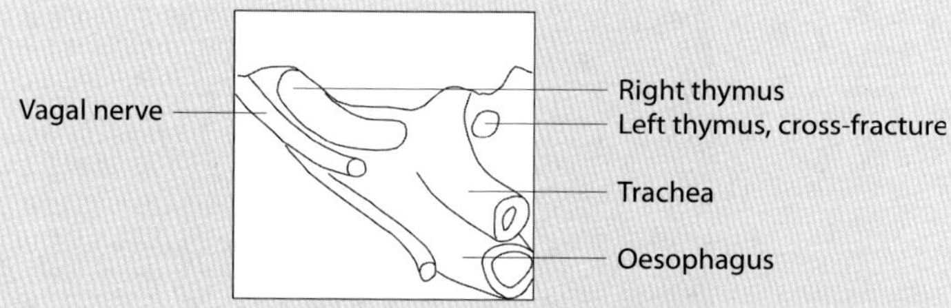

2.6.04 Ventral-right view of the right thymus. The superficial surrounding tissue has been removed to show the ectodermal part. Embryo at the end of week 6.

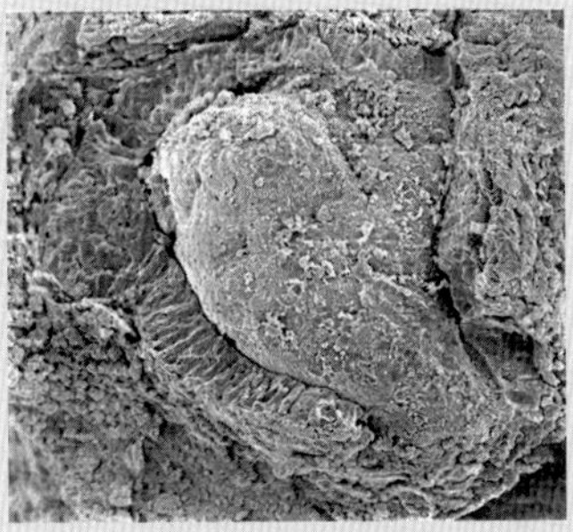

2.6.05 Same region at a higher magnification to show the structure of the thymus anlage.

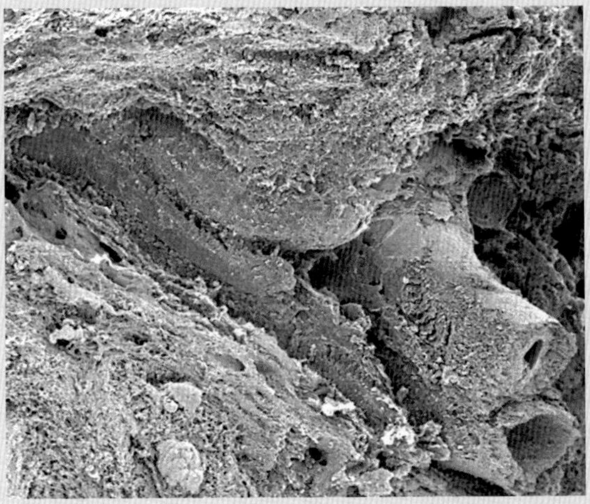

2.6.06 Ventral-right view of the thymus. Embryo at week 8.

3

Organs in the Thorax

3.1 The Development of the Heart

External Form

By the end of week 3 of gestation, the heart arises as a straight tube at the level of the neck. Immediately after its origin the contractile motions of the heart begin. Soon after its appearance, the straight tube begins to loop (fig. 3.1.01–3.1.05), and due to this looping the cardiac regions of inflow tract, ventricular region and outflow tract are established. Beginning in week 4, the heart descends in a caudal direction until it reaches its final position at the thoracic level at about weeks 6 and 7.

In week 4, the ventricular region expands and the first indentation of the interventricular sulcus becomes visible (fig. 3.1.09, 3.1.11, 3.1.13). During week 5 the interventricular sulcus deepens, the initially small right ventricle enlarges and the right and left atrium expand and become visible in an anterior view (fig. 3.1.19, 3.1.20).

Due to the kinking of the outflow tract, its left and right portions are deformed differentially during week 8, the left portion conducting the blood from the right ventricle (pulmonary flow path) and the right portion conducting the blood from the left ventricle (aortic or systemic flow path) (fig. 3.1.19–3.1.24). The auricles of the atria encompass the outflow tract (fig. 3.1.22, 3.1.23).

Beginning in weeks 8/9, the major coronary vessels are established (fig. 3.1.24, 3.1.29) and the coronary circulation arises.

Interior of the Heart: The Cardiac Septation

In week 4, the young cardiac loop resembles a tube which is curved in an intricate way. It pumps the blood from the inflow to the outflow tract in an undivided flow path. In the heart of the adult, however, there are two separate flow paths, the pulmonary and the systemic (aortic), which are completely separated under normal conditions. Additionally, each of these flow paths is organized into two chambers, an inflow chamber, the atrium, and an outflow chamber, the ventricle.

The transformation of the undivided flow path into two definite paths is realized by a system of septa which arise at the different levels of the loop. These septa join up to build a partition wall which divides the undivided bloodstream into the pulmonary and the systemic stream.

There are two kinds of septa, primarily paired mesenchymal and unpaired muscular septa.

The process of septation occurs in the following manner: the paired septa arise as superior and inferior atrioventricular septa (cushions) between the inflow tract (atria) and ventricles (fig. 3.1.30–3.1.32); and also between the ventricular region and the outflow tract as right and left conus septum (fig. 3.1.33, 3.1.34); thirdly, they arise as superior and inferior truncus septa on the boundary between upstream and downstream regions of the outflow tract (fig. 3.1.35).

The opposing septa in each region grow into the lumen almost at right angles to the bloodstream, their free edges come into contact and eventually they fuse to form a uniform unpaired septum (e.g. fig. 3.1.36, 3.1.37).

Meanwhile, the primarily unpaired muscular septa are formed. Starting from the external cardiac wall they fold up into the lumen of the inflow tract as interatrial septum (fig. 3.1.45–3.1.47) and into the lumen of the ventricular region as the interventricular septum (fig. 3.1.36, 3.1.37). The more complicated septation in the atria will be described later.

In a second phase of septation the septa of the neighbouring levels connect to each other in a direction parallel to the bloodstream: the atrial septum connects with the atrioventricular septa (cushions) (fig. 3.1.50, 3.1.51). The inferior atrioventricular cushion connects to the posterior margin of the interventricular septum (fig. 3.1.38) and the anterior margin of the interventricuar septum connects to the left portion of the conus septum (fig. 3.1.37, 3.1.38).

The right conus septum connects via the superior and the inferior atrioventricular cushions to the posterior margin of the interventricular septum (fig. 3.1.36, 3.1.39, 3.9.40). Before the definite fusion of these septa in the ventricular region is realized, a foramen between the right and the left ventricular flow paths remains open (fig. 3.1.39–3.9.41). This interventricular foramen is closed by the middle or the end of the 3rd month.

In the outflow tract, the conus septa connect to the truncus septa in the following way: the left conus septum connects to the inferior truncus septum (fig. 3.1.37) and the right conus septum connects to the superior truncus septum (fig. 3.1.36, 3.1.38).

Septation is completed by the aorticopulmonary septum which arises in the aortic sac between the systemic and pulmonary pharyngeal arch arteries (fig. 3.1.42, 3.1.43). It grows in an upstream direction and connects with its limbs to the superior and the inferior truncus septa (fig. 3.1.44).

Atrial septation is a little more intricate because there is a second muscular septum (septum secundum) independent of the already mentioned septum (primum). As noted above, the septum primum arises first by ingrowth from dorsal into the atria (fig. 3.1.45–3.1.47). At the base of the heart, its limbs connect to the atrioventricular cushions (fig. 3.1.50, 3.1.51).

During the ingrowth of the septum primum, below its free rim a communication between the atria remains open: the foramen primum (fig. 3.1.47). Simultaneously the septum primum is perforated in its upper parts, thus establishing a second communication between the atria, the foramen secundum (fig. 3.1.48–3.1.50, 3.1.52).

The foramen primum becomes reduced in size until it is eventually closed during the 3rd month. The foramen secundum increases in size until it reaches about half of the area of the septum primum.

No earlier than in the 3rd month does the second interatrial septum arise, the septum secundum. It arises to the right of the septum primum and grows from ventral to dorsal (fig. 3.1.53). The septum secundum remains incomplete and almost covers the free rim of the septum primum.

In this way the final embryonic and fetal communication between the atria, the foramen ovale is established. It is bordered by the superior rim of the septum primum and the inferior rim of the septum secundum. The foramen ovale is closed shortly after birth by the increasing pressure in the left atrium.

There are extreme individual variations in the moment of realization of the second phase of septation, in which the septa of the neighbouring regions connect to each other. Additionally, the origin of the septa in the first phase of septation is not realized in the order described. It was not the goal of the description to give a time schedule of the septation, but to give an outline of the septal topography.

Missing or incomplete or dystopic development of the septa and an abnormal connection of septa may well be the formal cause of a great variety of cardiac malformations.

3.1 Abbreviations

la	left atrium
lv	left ventricle
ra	right atrium
rv	right ventricle
1	sinus venosus
2	ventricular loop
3	interventricular sulcus
4	outflow tract, conus
5	outflow tract, truncus
6	pulmonary trunk
7	ascending aorta
8	aortic arch
9	atrioventricular sulcus
10	anterior descending branch of left anterior coronary artery
11	anterior leaflet of the tricuspid valve
12	septal leaflet of the tricuspid valve
13	anterior papillary muscle
14	posterior papillary muscle
15	supraventricular crest
16	pulmonary semilunar valve
17	common cardinal vein
18	hepatic vein
19	superior cardinal vein
20	right sinus valve
21	left sinus valve

3.1.01–3.1.29
Development of the external form of the heart. Right, ventral, and left views.

Week 4

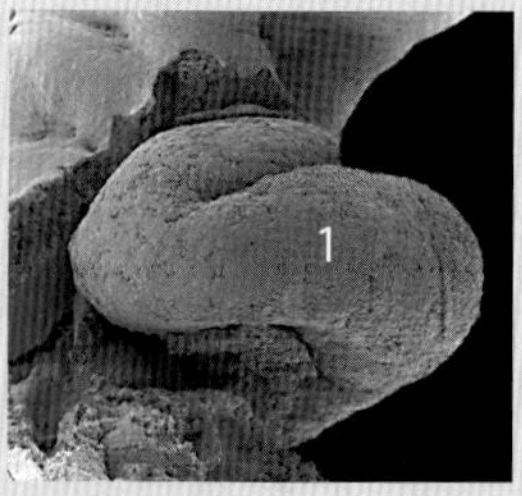

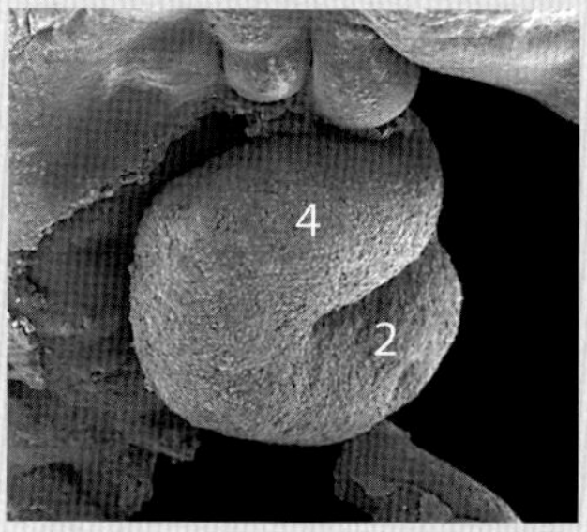

Week 5

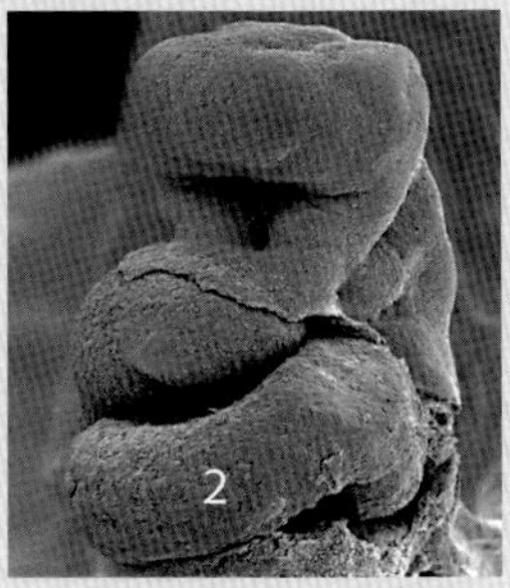

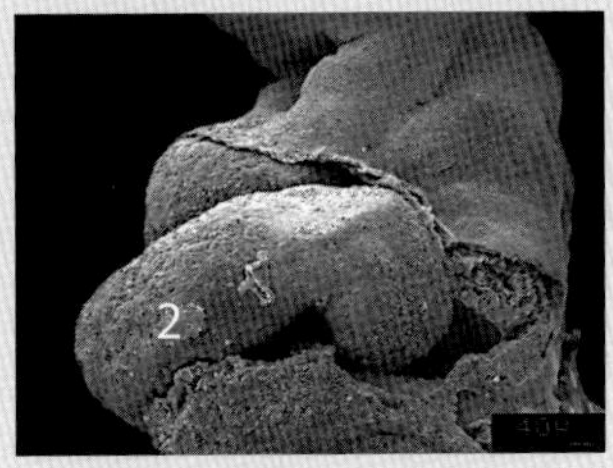

3.1.01 / 02 / 03

Same embryo
at week 4.
Right, ventral, and left
views.

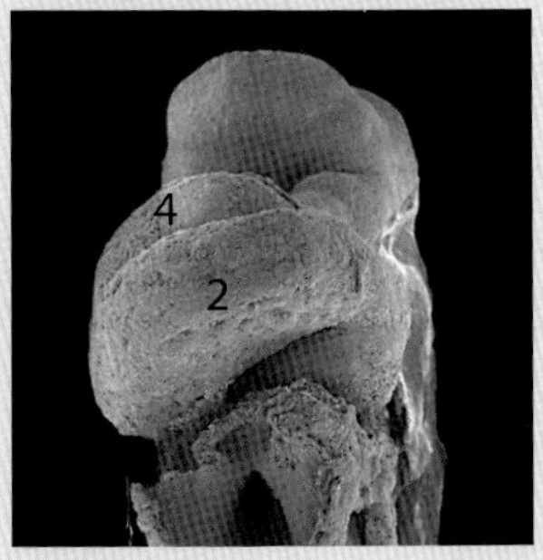

3.1.04

Embryos at week 4.
Ventral view.

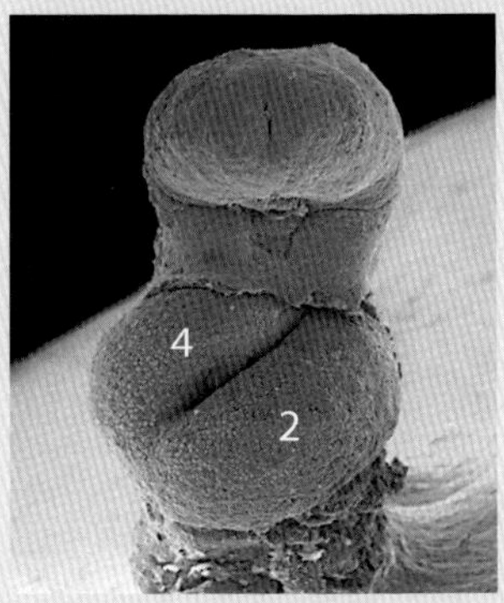

3.1.05

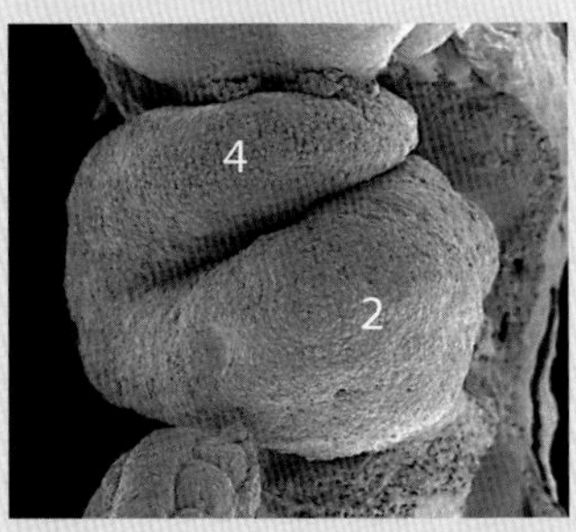

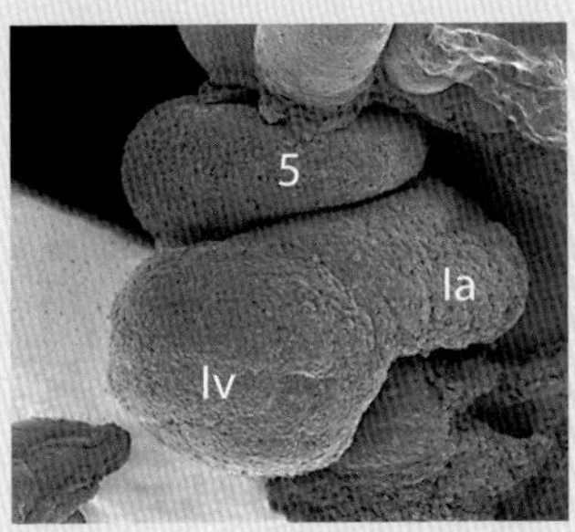

3.1.06 / 07 / 08

Same embryo at the end
of week 4.
Right, ventral, and left
views.

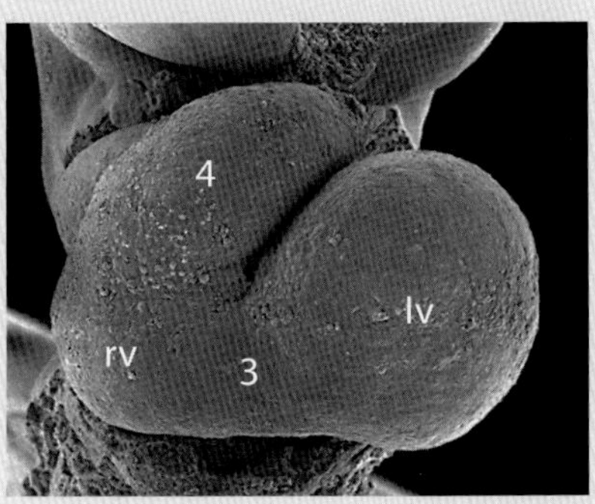

3.1.09

Embryo at week 5.
Ventral view.

40:1

Week 5

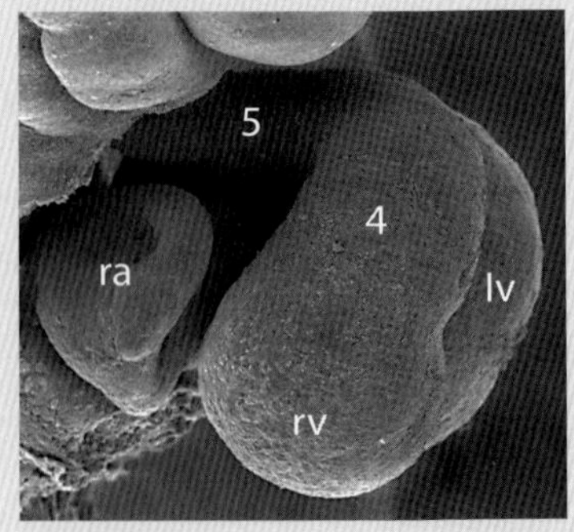

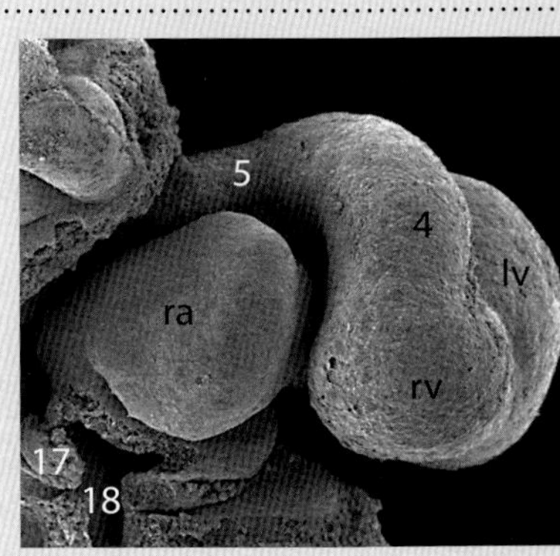

Week 6

3.1.01–3.1.29

Development of the external form of the heart. Right, ventral, and left views.

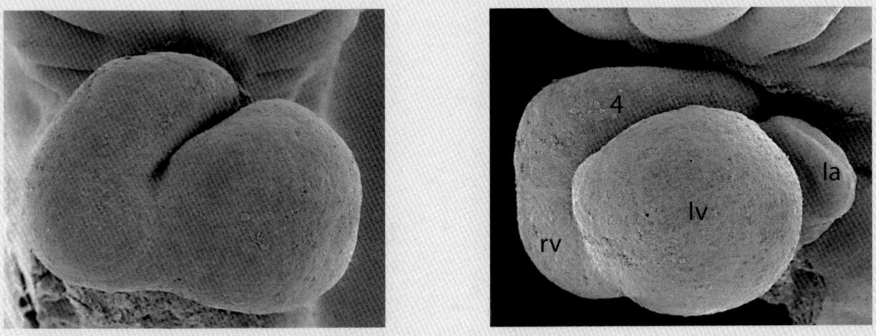

3.1.10 / 11 / 12

Same embryo at week 5.
Right, ventral, and left views.

3.1.13

Embryo at week 5.
Ventral view.

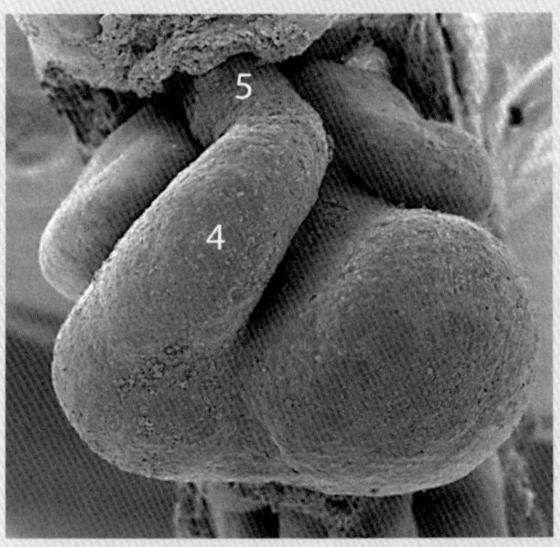

3.1.14

Embryo at the end of week 5.
Ventral-cranial view.

3.1.15 / 16 / 17

Same embryo at the end of week 5.
Right, ventral, and left views.

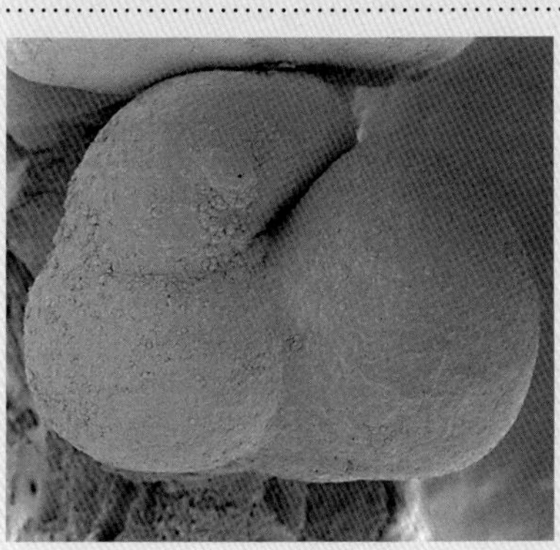

3.1.18

Embryo at week 6.
Ventral view.

Week 6

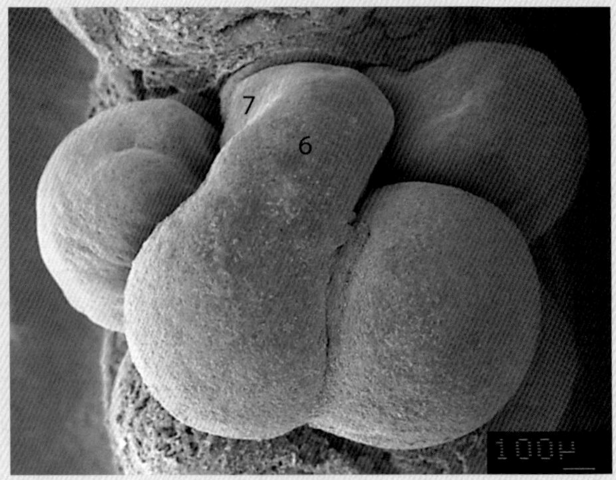

3.1.01–3.1.29

Development of the external form of the heart. Right, ventral, and left views.

Week 7

3.1.19 / 20

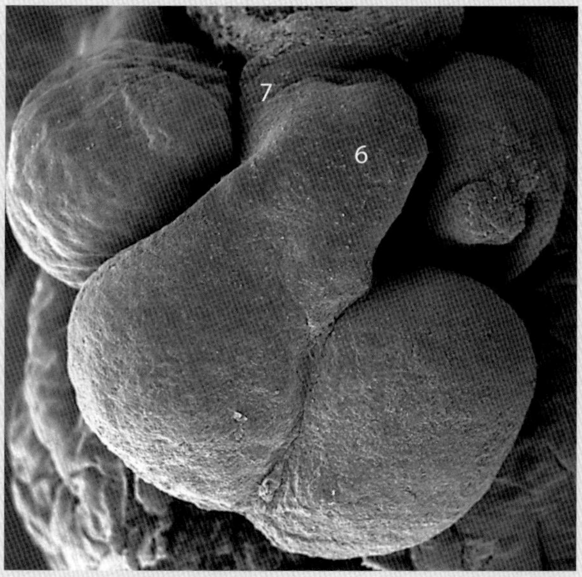

3.1.19
Embryo early in week 6.

3.1.20
Embryo at the end of week 6.
Ventral view.

3.1.21

Embryo at week 7.
Ventral view.

Week 7 / 8 / 9

3.1.01–3.1.29

Development of the external form of the heart. Right, ventral, and left views.

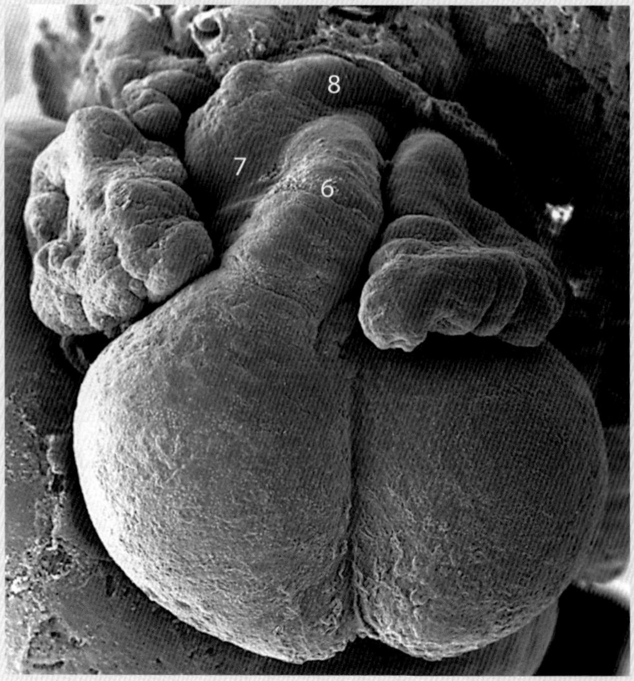

3.1.22 / 23

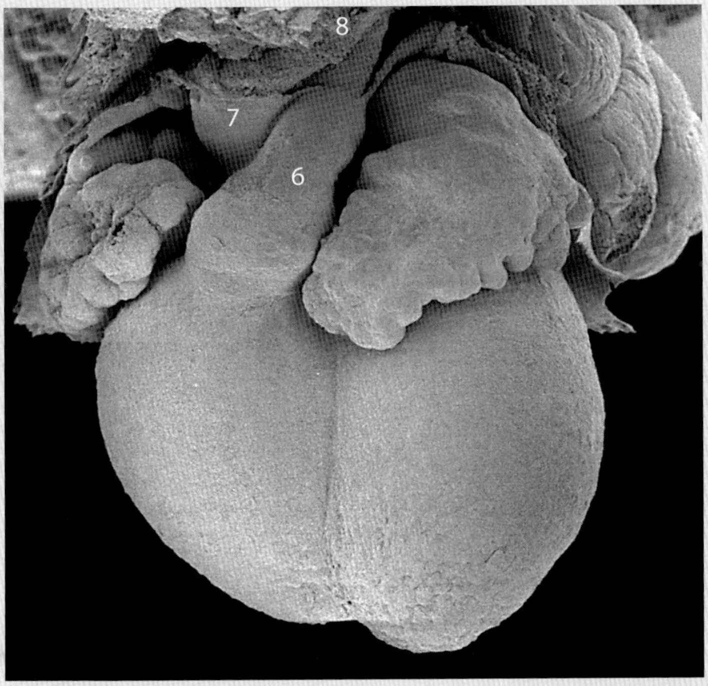

Embryos at weeks 7, 8 and 9.
Ventral views.

3.1.24

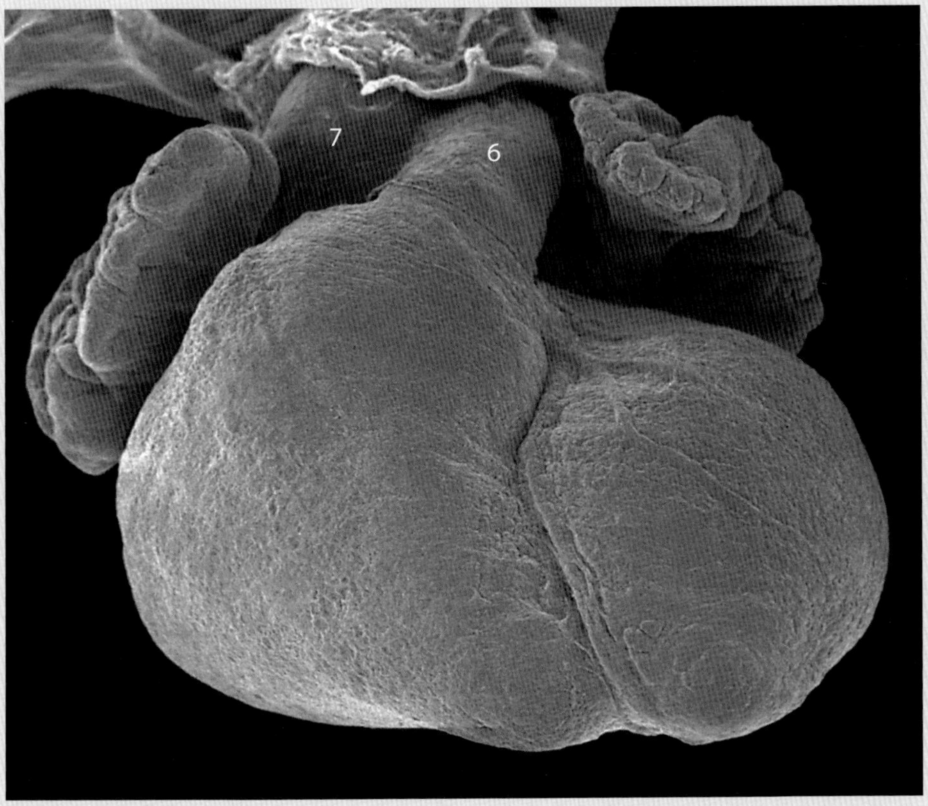

Weeks 5 / 6 / 8

3.1.01–3.1.29

Development of the external form of the heart. Right, ventral, and left views.

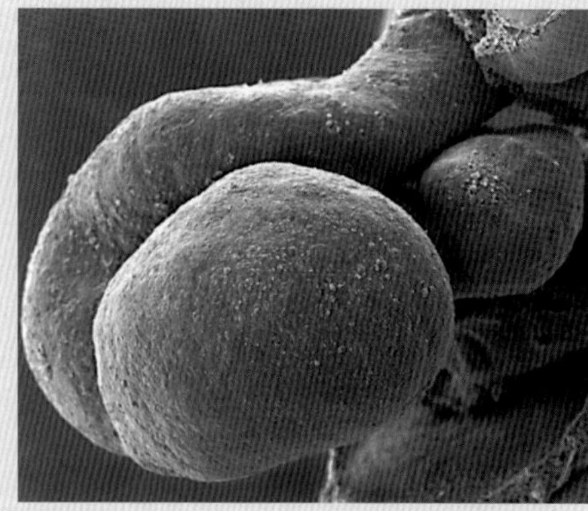

3.1.25 / 26

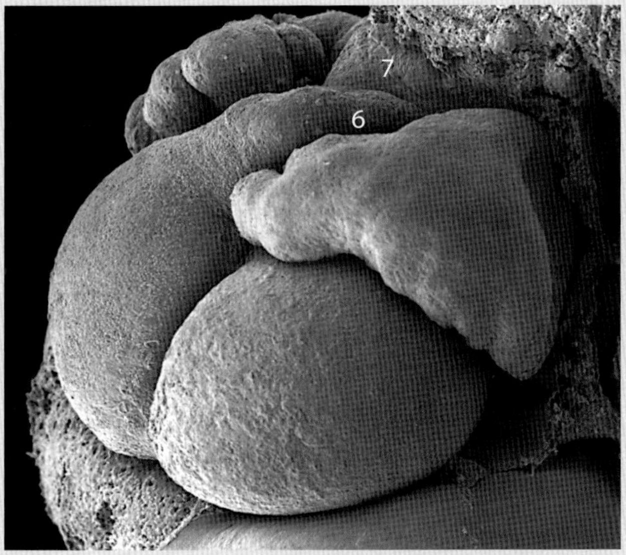

Embryos at weeks 5, 6 and 8.
Left views.

3.1.27

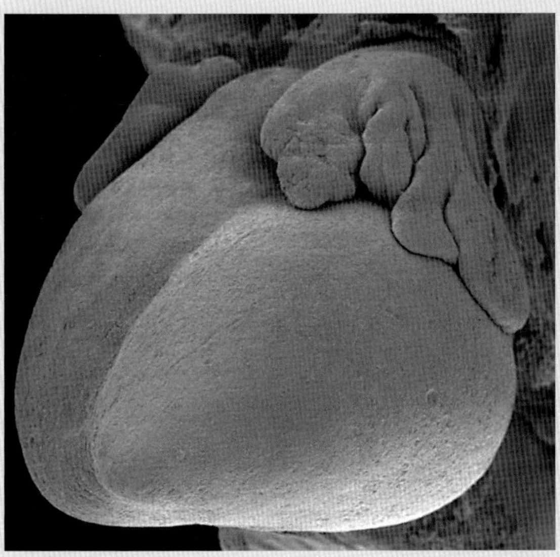

3.1.01–3.1.29

Development of the external form of the heart. Right, ventral, and left views.

3.1.28 / 29

Embryos at weeks 6 and 9.
Right views.

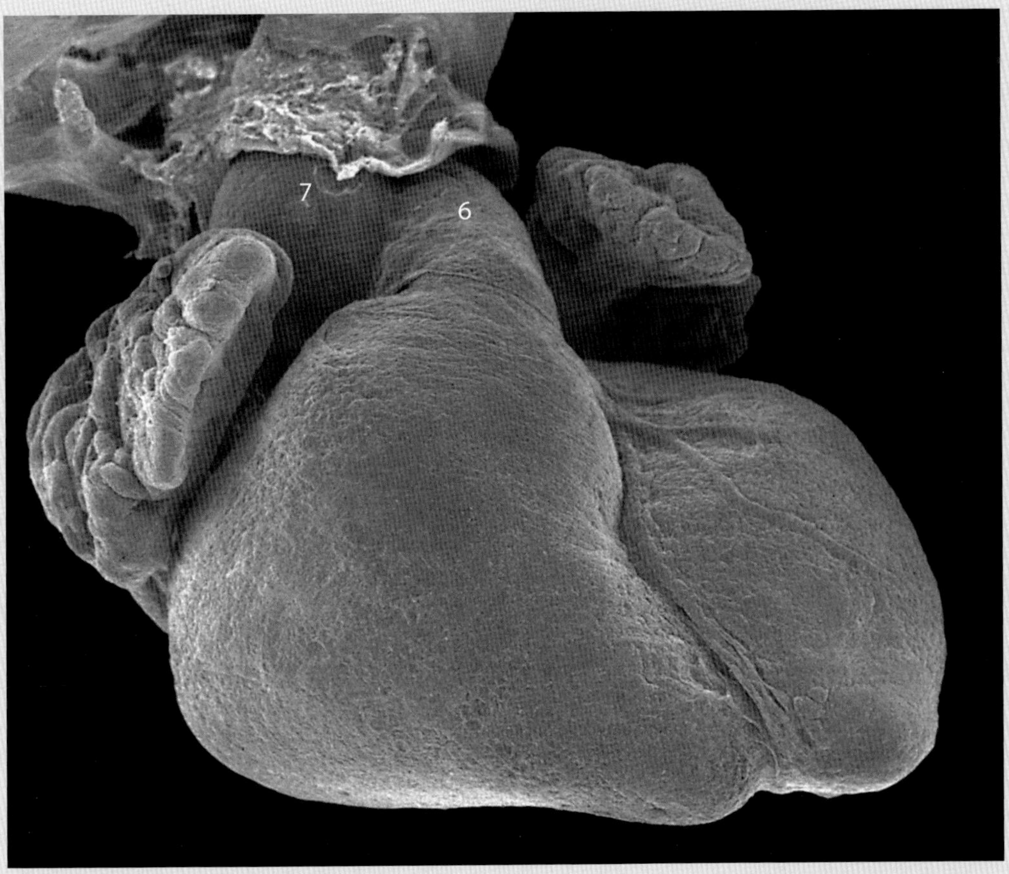

3.1.30–3.1.41
Developmental stages of the cardiac ventricular septation.

Week 6

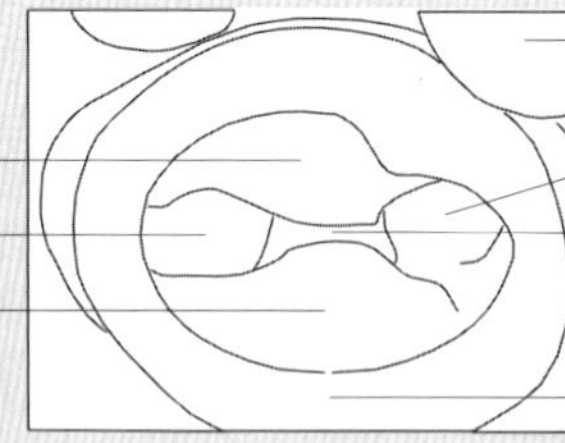

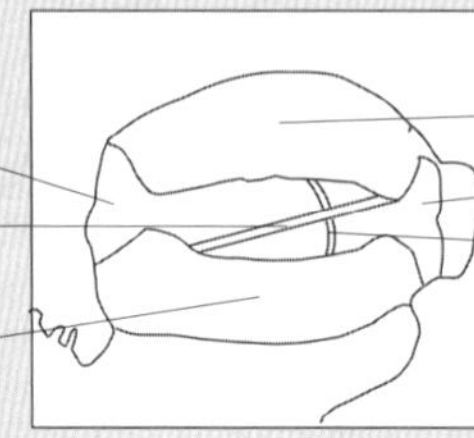

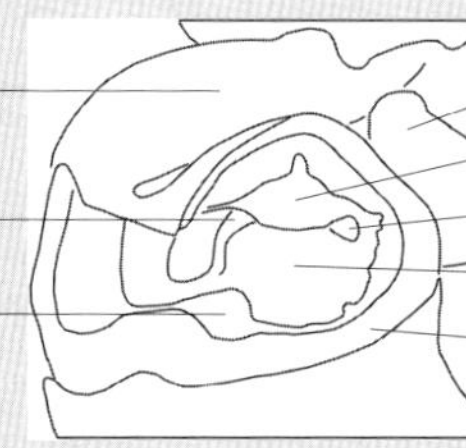

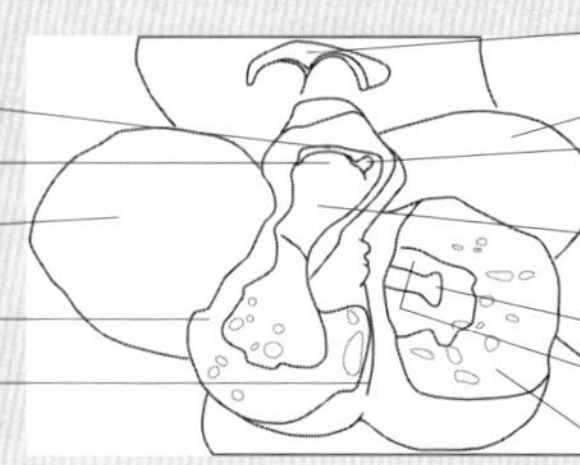

3.1.30

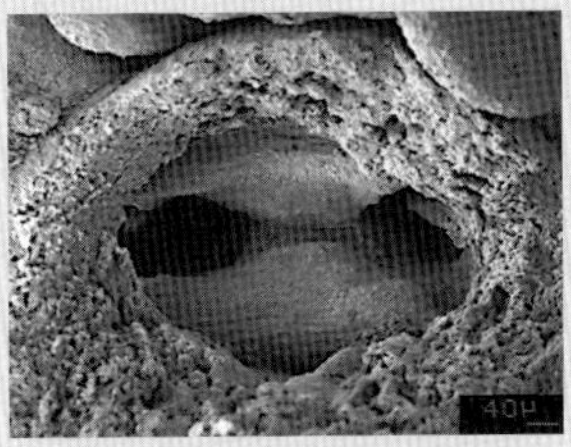

Caudal view of different stages of the atrioventricular cushions.
Embryos at week 6.

3.1.31

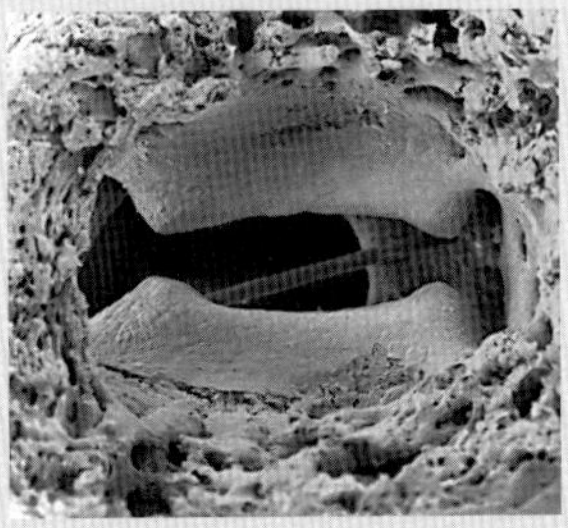

3.1.32

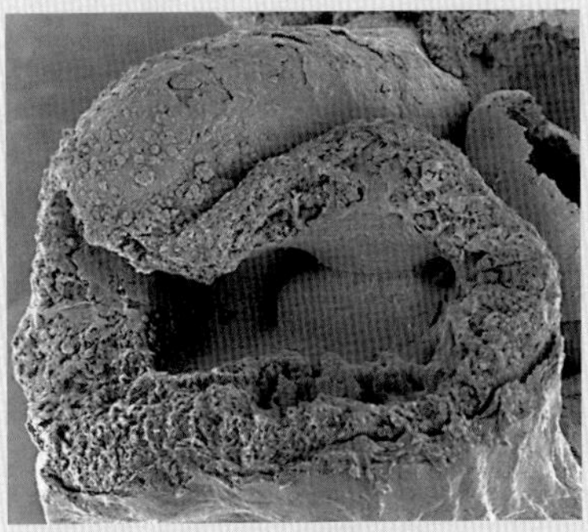

3.1.33 / 34

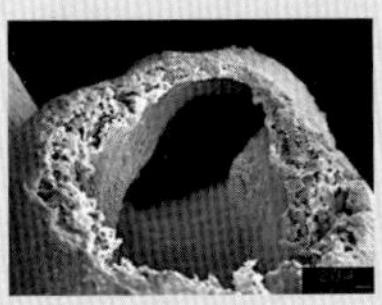

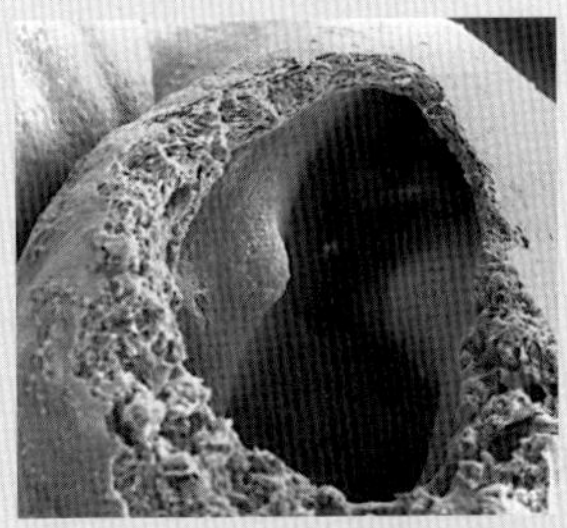

Position of the conus (3.1.33, 3.1.34) and the truncus (3.1.35) septa.
View in a downstream direction.
Embryos at week 6.

3.1.35

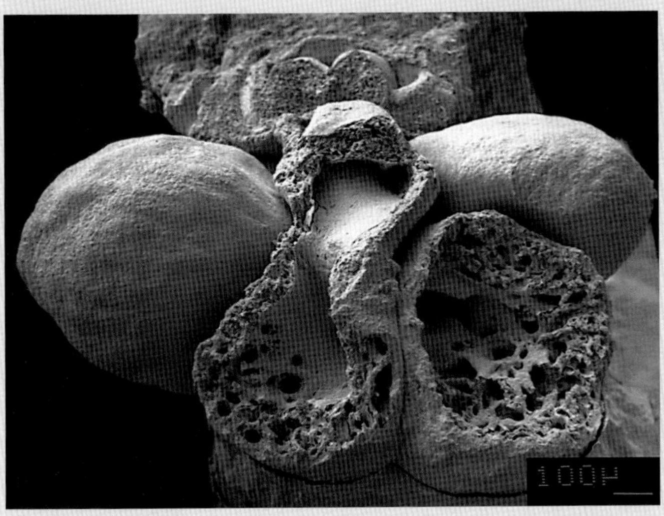

Week 7

3.1.30–3.1.41
Developmental stages of the cardiac ventricular septation.

3.1.36–3.1.40
Stages of the connection of the conus septa with the ventricular septum and the truncus septa. Different devopmental stages of the closure of the interventricular foramen. Ventral-right view of the right ventricle. Embryos at week 7.

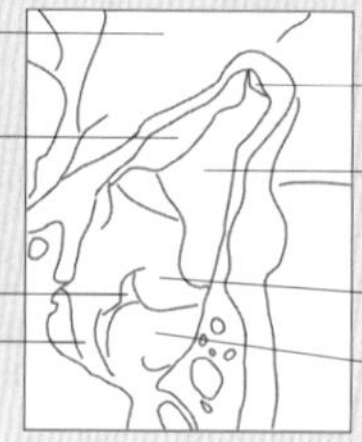

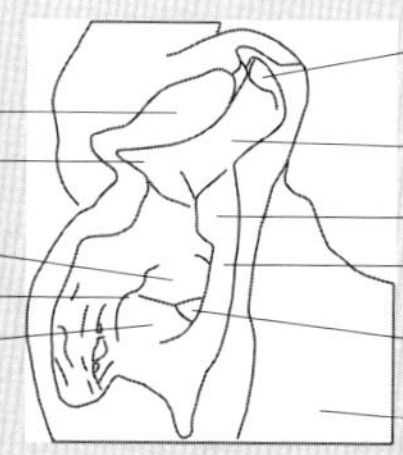

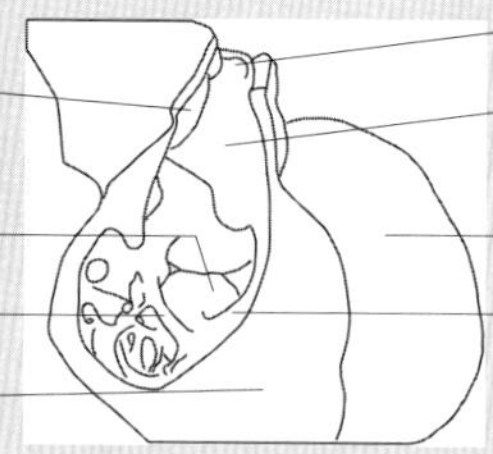

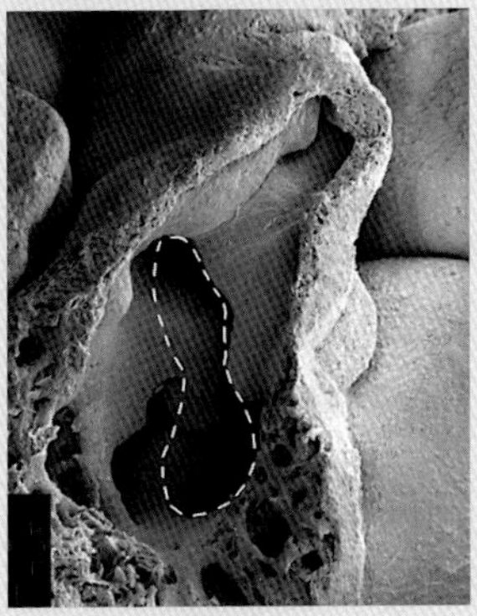

3.1.36 The broken line indicates the boundary of the interventricular foramen still wide open. Whereas the left portion of the interventricular septum has already been formed, the septal protrusions forming the right portion have not yet advanced to the corresponding portion of the broken line.
(Compare with the more advanced stage in fig. 3.1.39.)

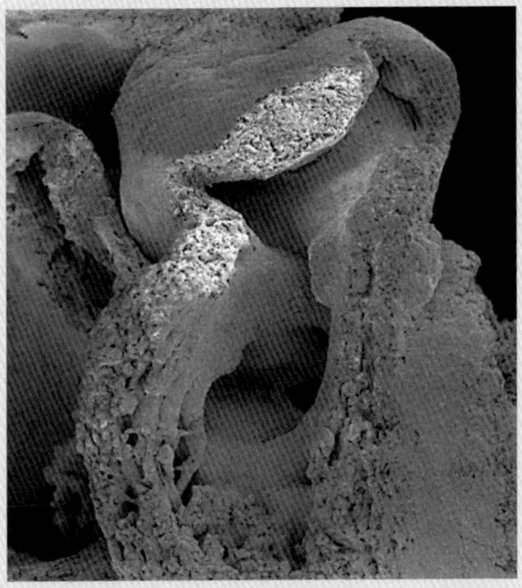

3.1.37 Connection of the interventricular septum via the left conus septum to the inferior truncus septum. The corresponding connection of the right conus septum to the superior truncus septum has been removed.

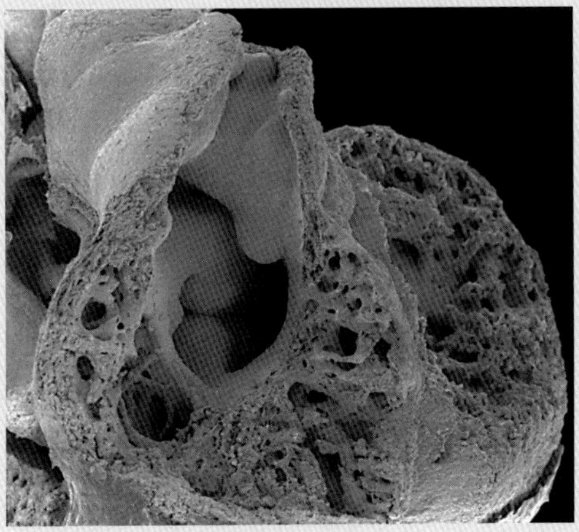

3.1.38 Different stages of the development of the interventricular foramen.

3.1.30–3.1.41

Developmental stages of the cardiac ventricular septation.

3.1.36–3.1.40

Stages of the connection of the conus septa with the ventricular septum and the truncus septa. Different devopmental stages of the closure of the interventricular foramen. Ventral-right view of the right ventricle. Embryos at week 7.

Week 7

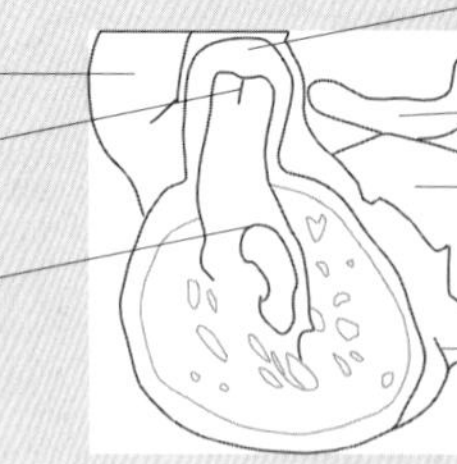

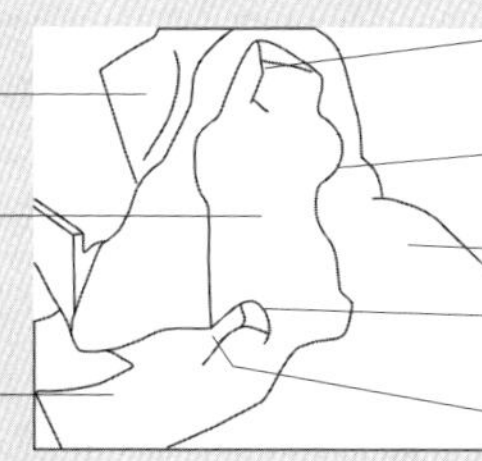

Week 8

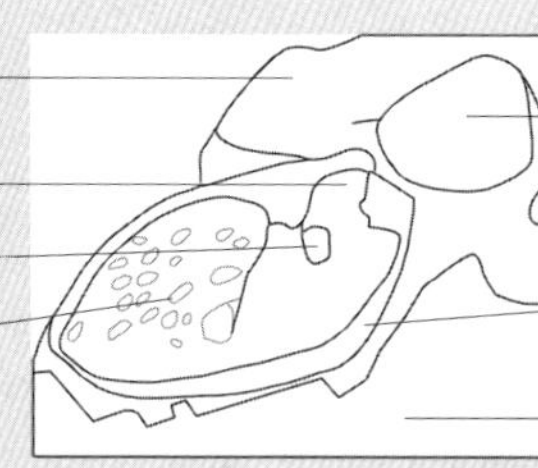

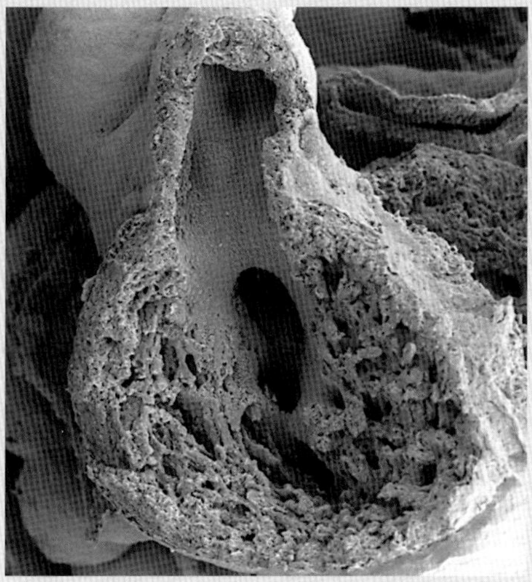

3.1.39 Different stages of the development of the interventricular foramen.

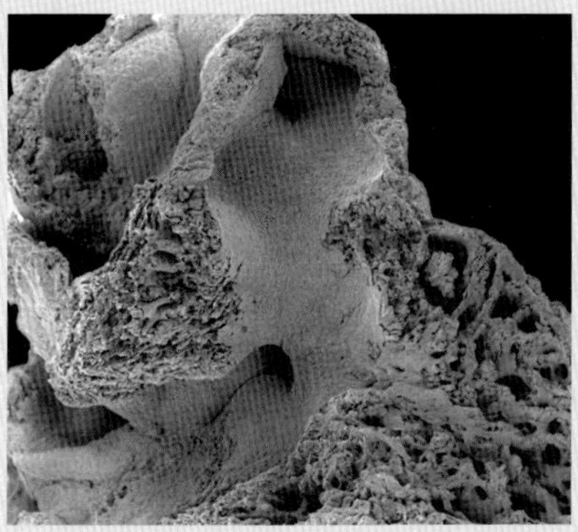

3.1.40

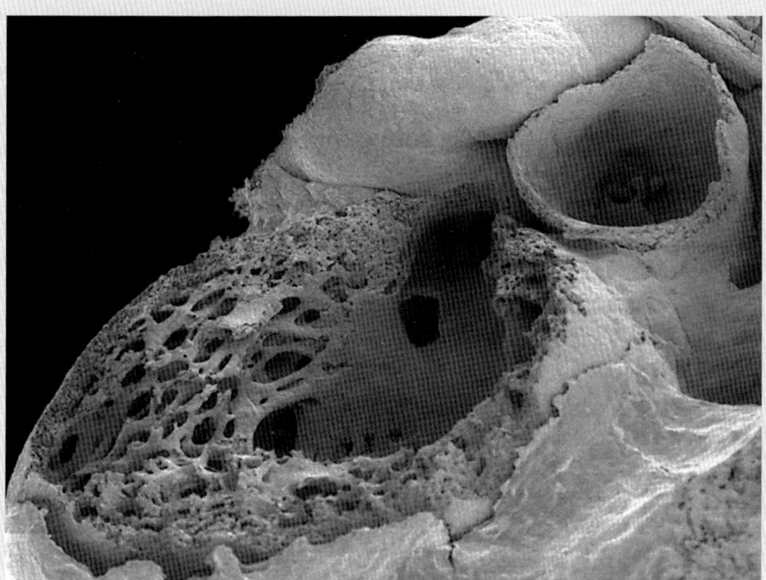

3.1.41 Left view into the left ventricle shows the interventricular foramen and the outflow tract of the left ventricle (aortic conus). Embryo at week 8.

Weeks 5 / 6 / 7

3.1.42–3.1.44
Septation of the truncus (distal outflow tract) and the aortic sac. Embryos at weeks 5, 6 and 7.

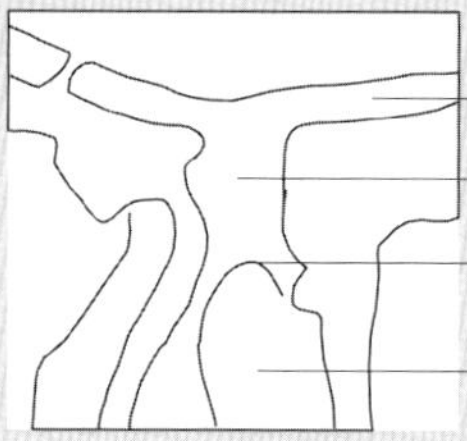

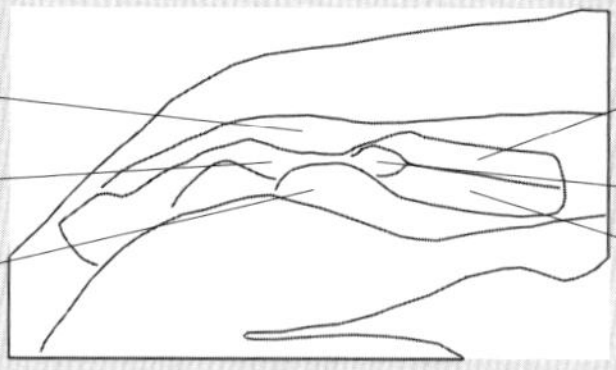

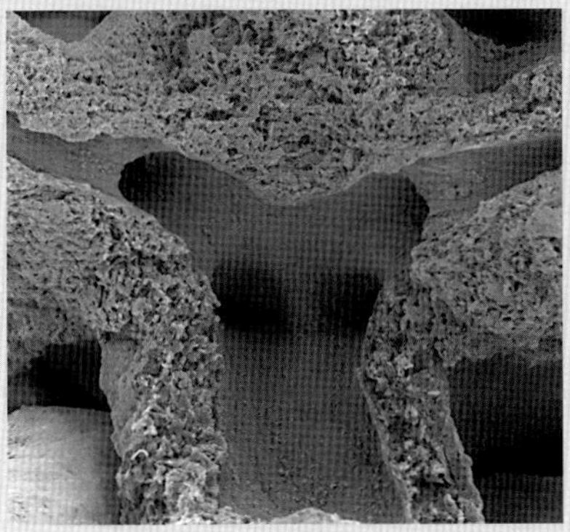

3.1.42 View in a downstream direction into the aortic sac and the origin of the third, fourth and sixth aortic arch arteries.

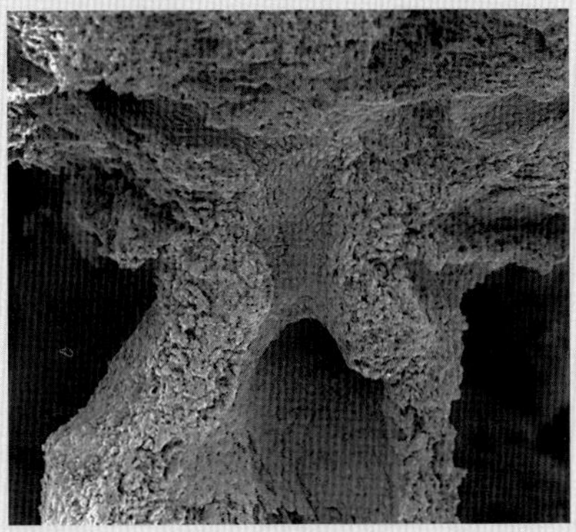

3.1.43 Same view as in figure 3.1.42. The more protruded aorticopulmonary septum now separates the pulmonary from the systemic flow path, which gives rise to the third and fourth aortic arch arteries.

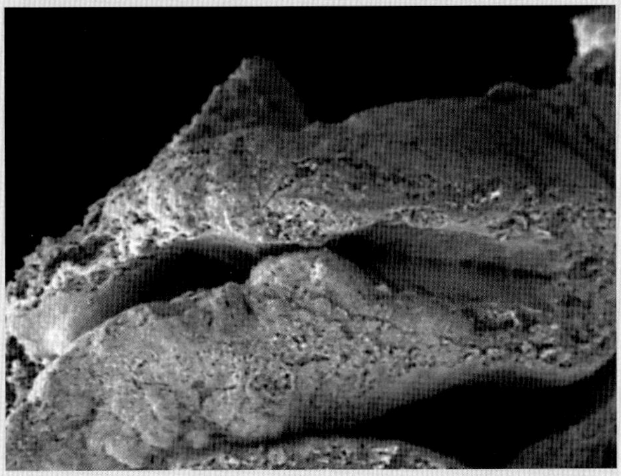

3.1.44 Lateral left view of the distal truncus. The connection of the aortico-pulmonary septum to the truncus septa is nearly complete, and separates the pulmonary from the systemic flow paths.

Week 5

3.1.45–3.1.56
Septation of the atria.

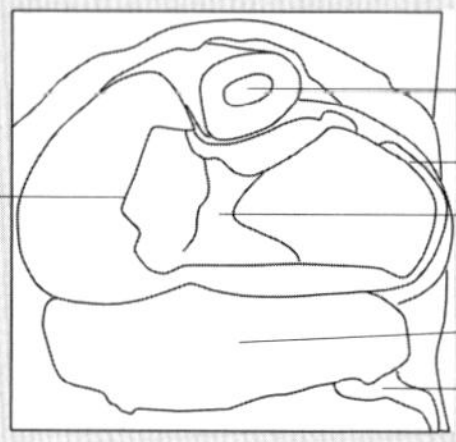

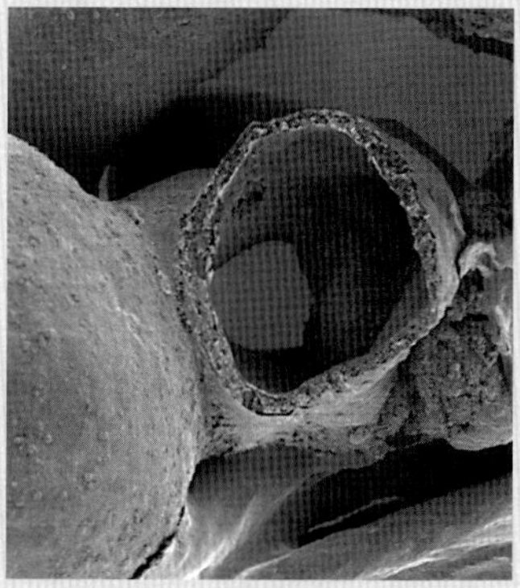

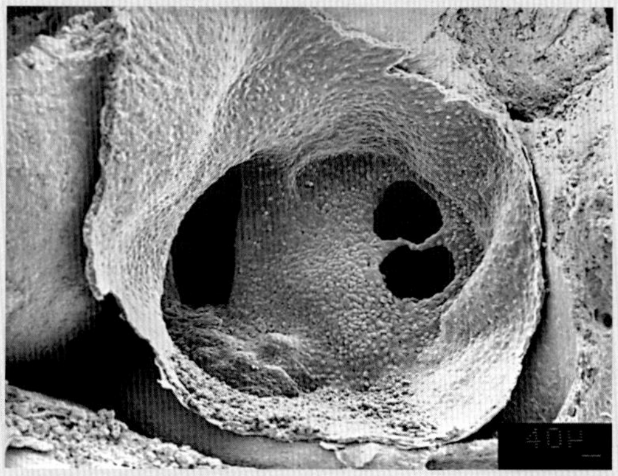

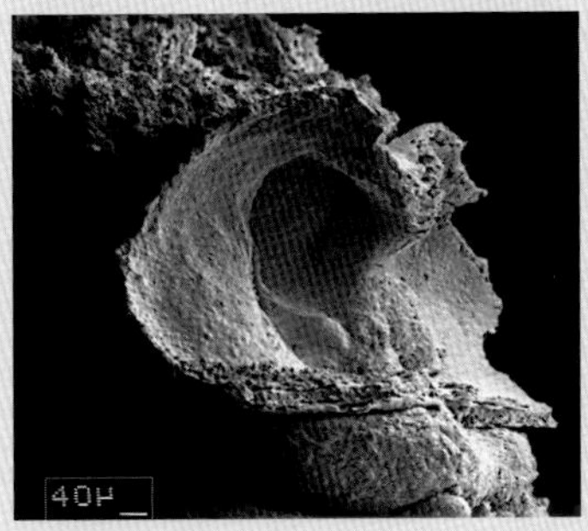

3.1.45

Origin of the primary interatrial septum. Ventral-right and ventral view. Embryos early in week 5.

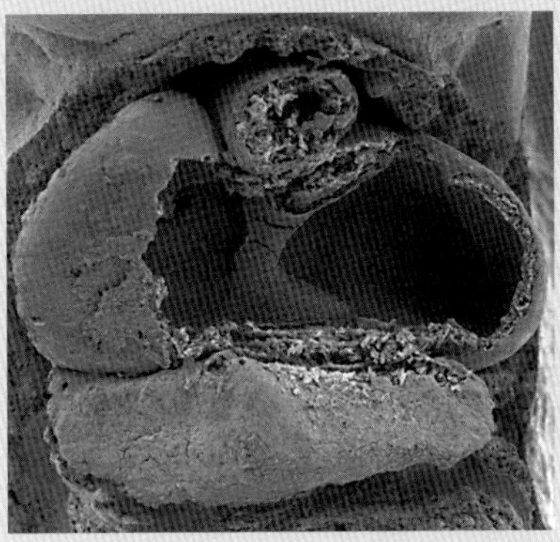

3.1.46

3.1.47 / 48

Lateral (left) view of the opened left atrium.

3.1.47
The wide open interatrial foramen primum.

3.1.48
The first indication of the interatrial foramen secundum. The first foramen is hidden beneath the atrial wall.

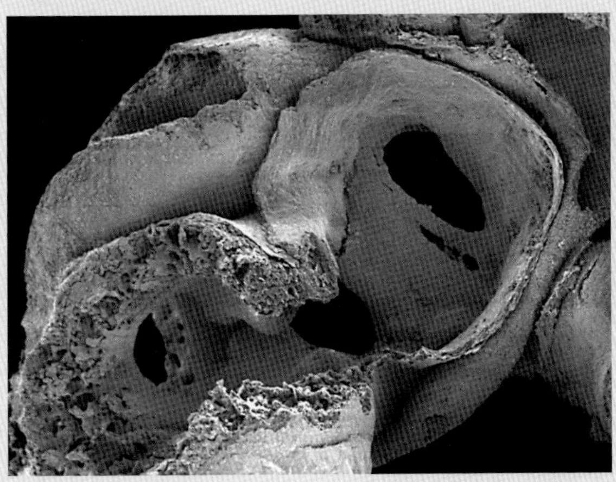

3.1.49 / 50

Lateral (left) view of the opened left atrium.

The interatrial foramen secundum has become enlarged.

3.1.45 3.1.56
Septation of the atria.

Week 7

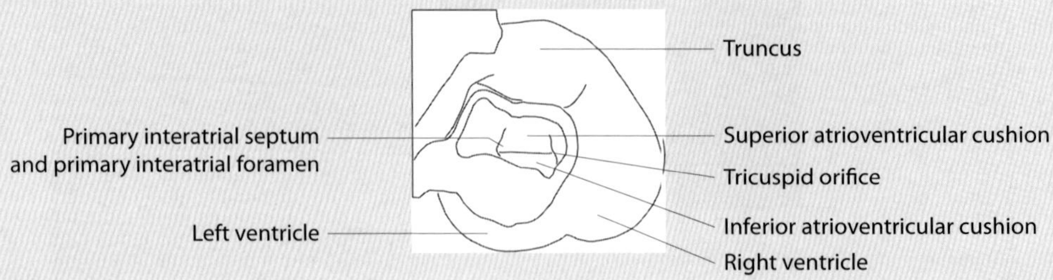

Week 8

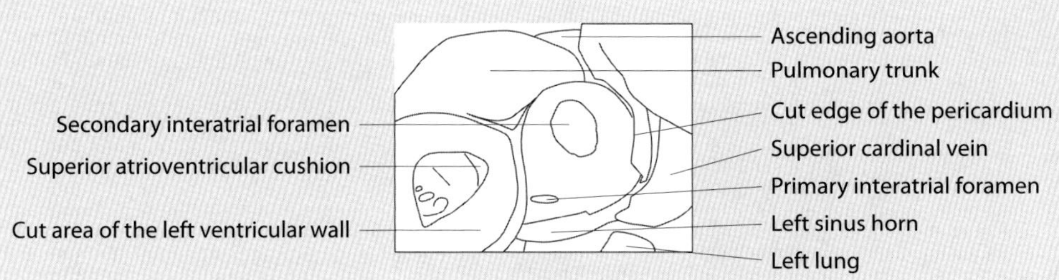

Week 10

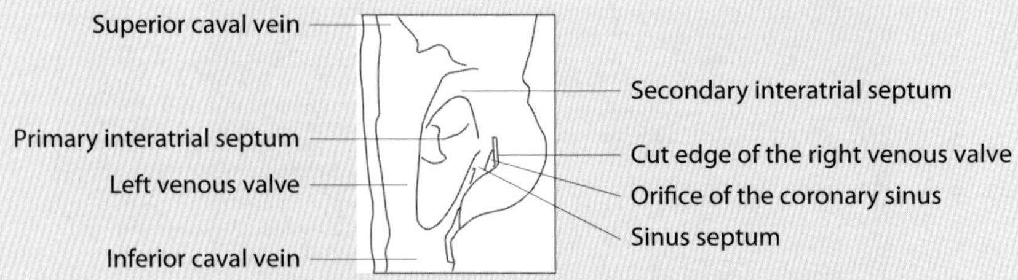

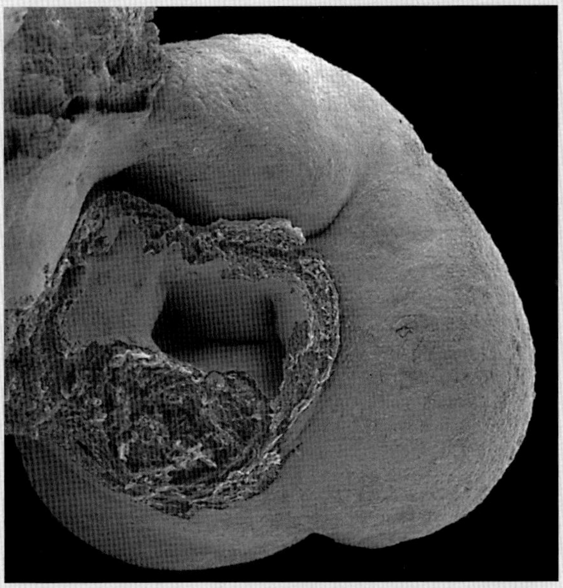

3.1.51 Closure of the primary interatrial foramen.

Cranial-right view of the atrioventricular cushions and the primary interatrial foramen. The walls of the right atrium have been removed.
Embryo at week 7.

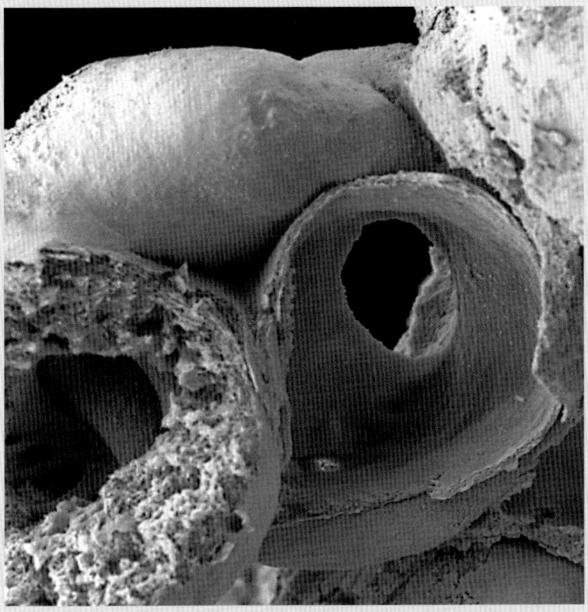

3.1.52 Closure of the primary interatrial foramen.

Left view of the left atrium. The secondary foramen has become enlarged (see fig. 3.1–50), the primary foramen is now only a very small opening.
Embryo at week 8.

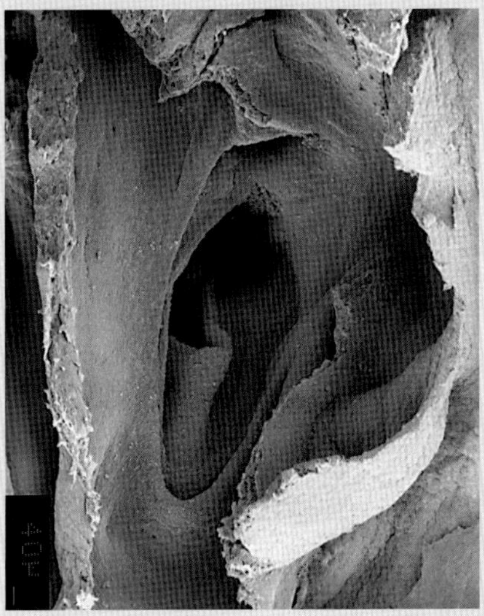

3.1.53 Origin of the secondary interatrial septum and the foramen ovale. Right view of the right atrium.
Embryo at week 10.

Week 6

3.1.45–3.1.56
Septation of the atria.

3.1.54–3.1.56
Orifice of the common, the superior and the inferior cardinal veins and the hepatic vein into the right atrium, and the sinus valves. Consecutive dissections of an embryo at week 6.

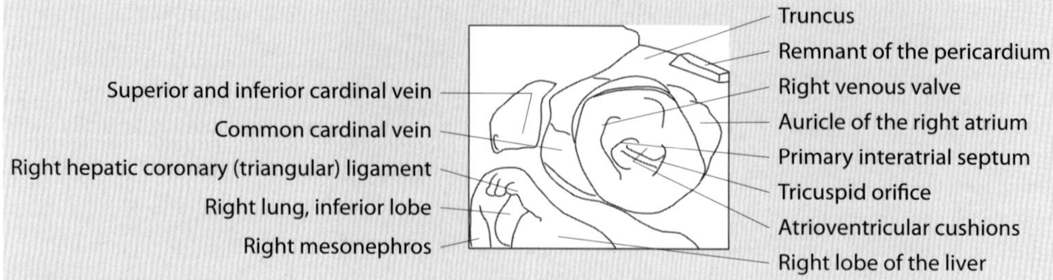

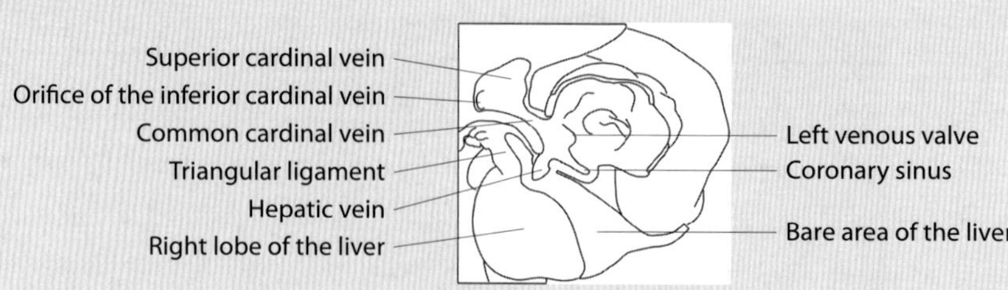

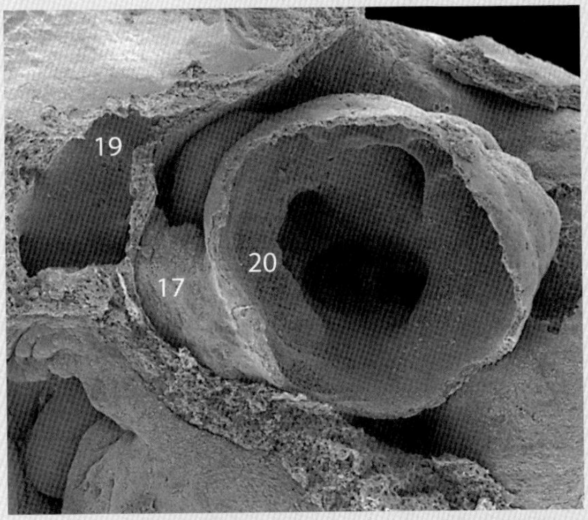

3.1.54 The right atrium has been opened to show the right venous valve.

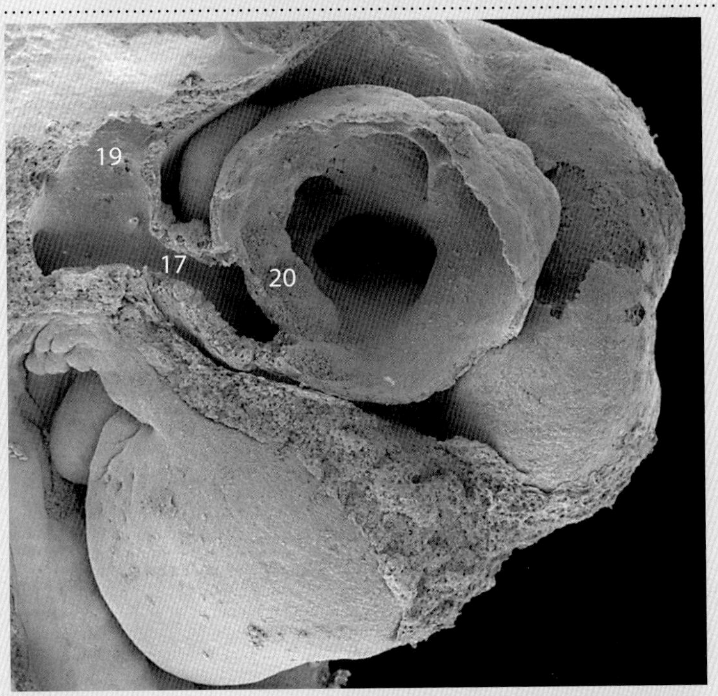

3.1.55 The superior and common cardinal veins have been opened.

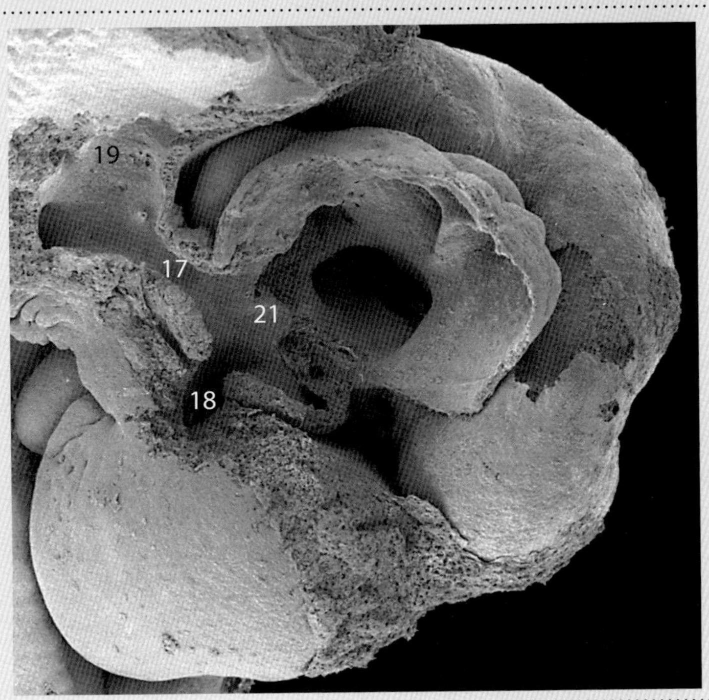

3.1.56 The right sinus valve and its origin from the atrial wall are removed to show the left sinus valve and the opening of the hepatic (future inferior caval vein) into the right atrium.

3.1.57–3.1.68
Developmental stages of the cardiac valves.

3.1.57–3.1.60
Development of the mitral valve.

Week 5

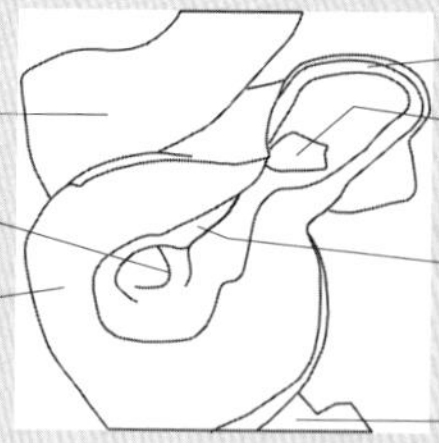

Week 7

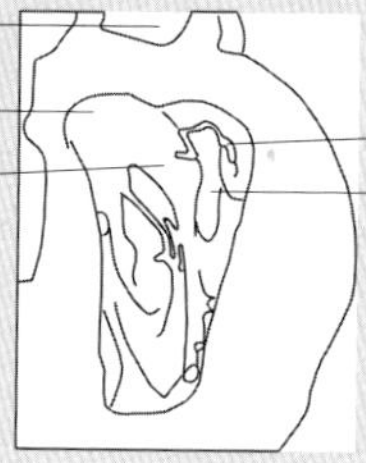

Week 9

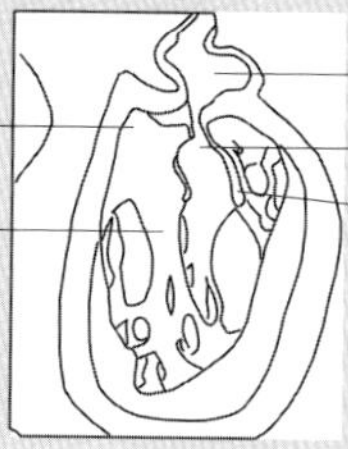

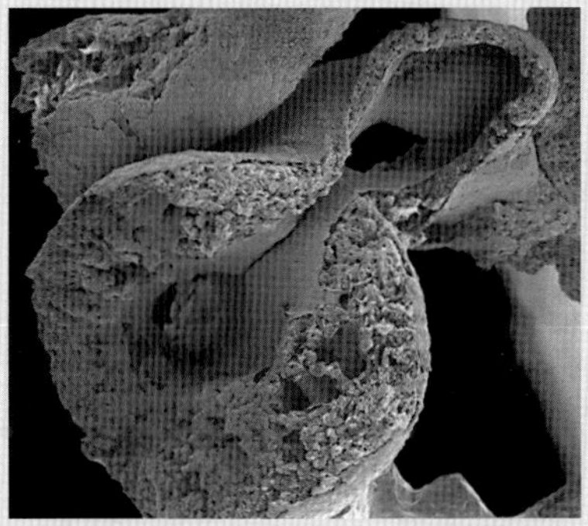

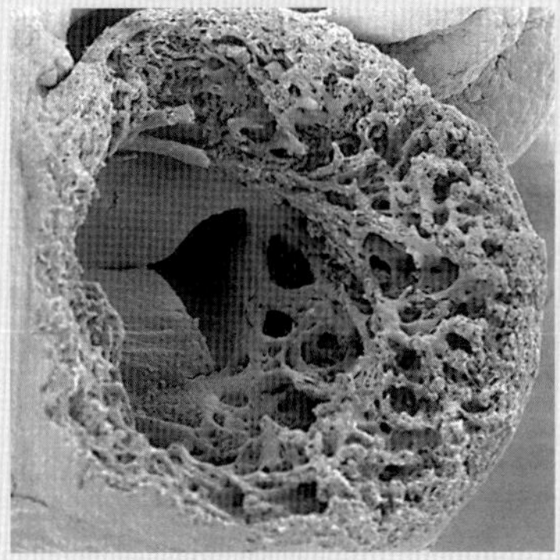

3.1.57 / 58

3.1.57
Lateral (left) view of the heart.
Embryo at week 5.

3.1.58
Caudal view of the opened left ventricle.
Embryo at week 5.

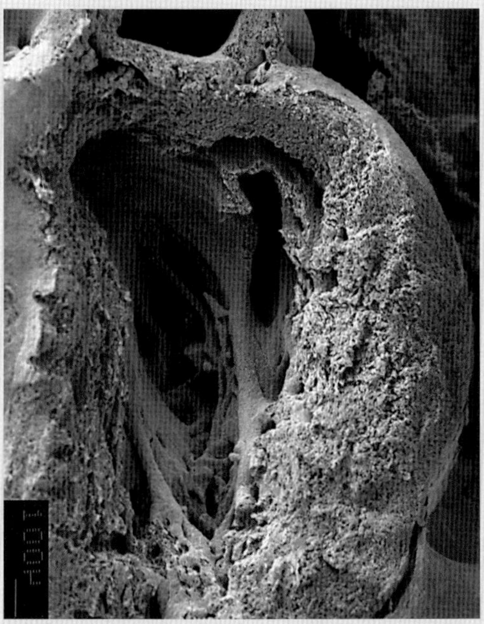

3.1.59

Ventral and caudal view of the opened left ventricle.
Embryo at week 7.

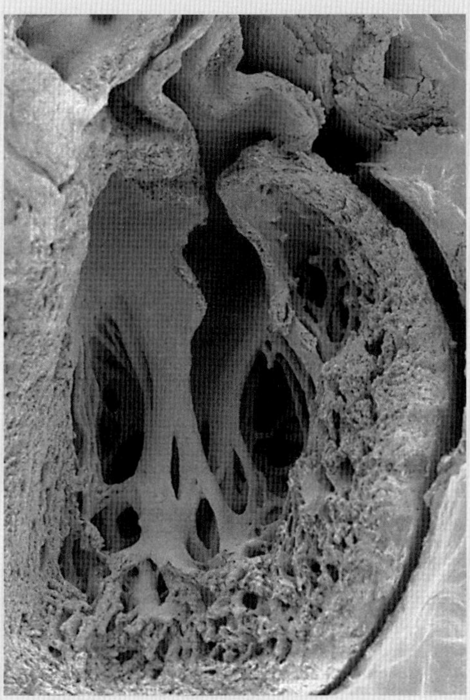

3.1.60

Ventral and caudal view of the opened left ventricle.
The leaflets of the mitral valve have been partly removed to show the connection to the left atrium.
Embryo at week 9.

3.1.57–3.1.68
Developmental stages of the cardiac valves.

3.1.61–3.1.64
Development of the tricuspid valve.

Week 6

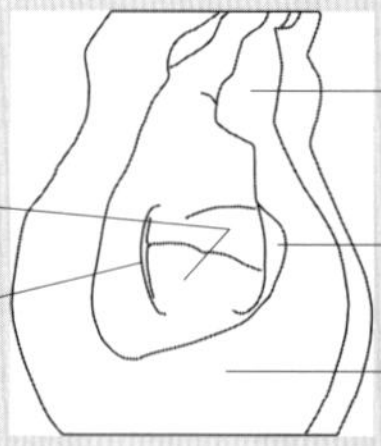

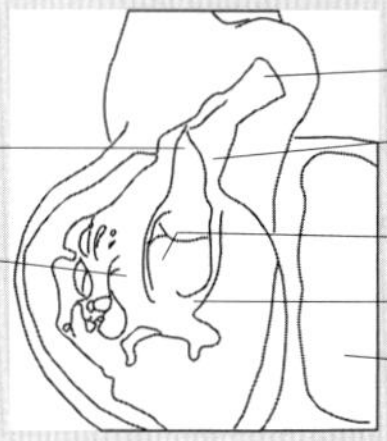

Week 9

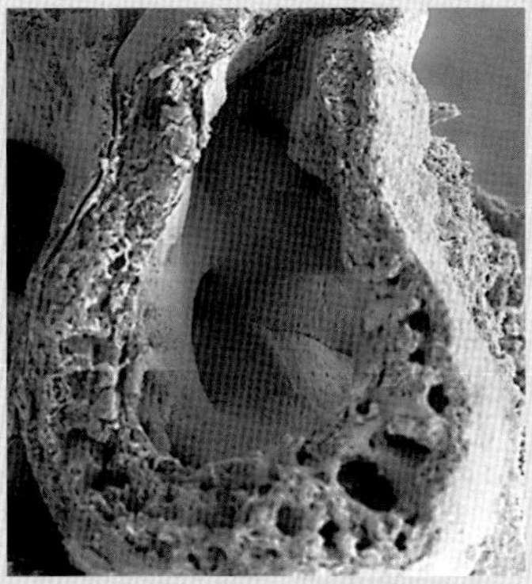

3.1.61 Ventral-right view of the right ventricle. Embryo at week 6.

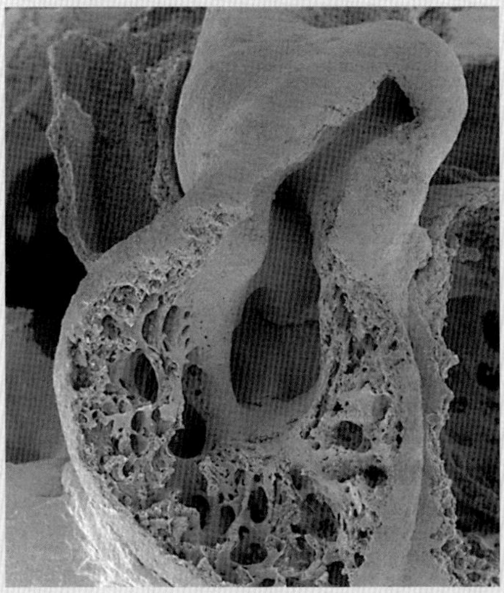

3.1.62 Ventral-right view of the right ventricle. Embryo at the end of week 6.

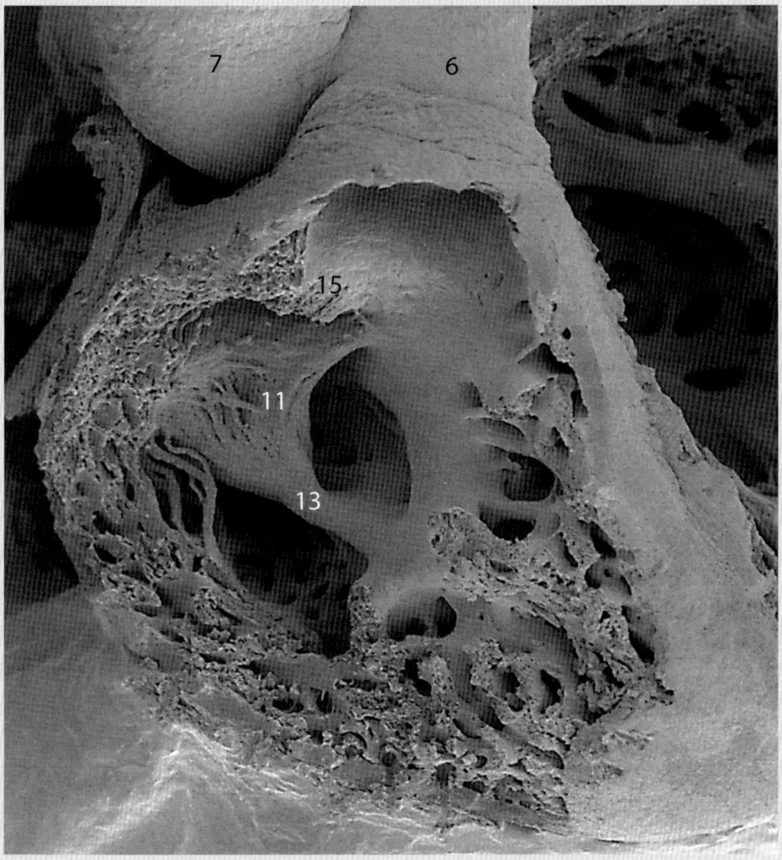

3.1.63 Ventral-right view of the right ventricle. Embryo at the end of week 9.

50:1

Week 9

3.1.57–3.1.68
Developmental stages of the cardiac valves.

3.1.65–3.1.68
Development of the pulmonary valve.

100:1

Weeks 7 / 8

Week 9

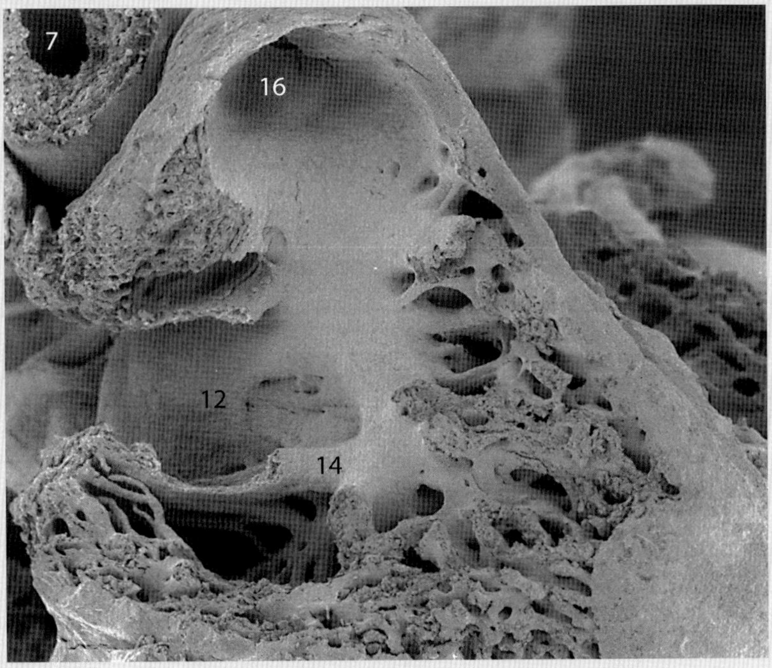

3.1.64

Same embryo as in figure 3.1.63. Ventral-right view of the right ventricle. The mural leaflet of the tricuspid valve has been removed to show the septal leaflet.

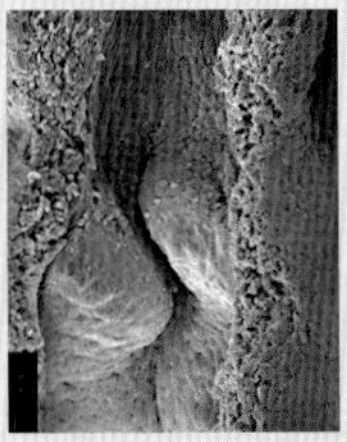

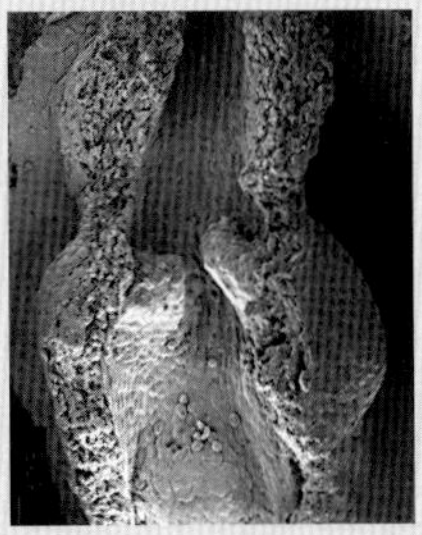

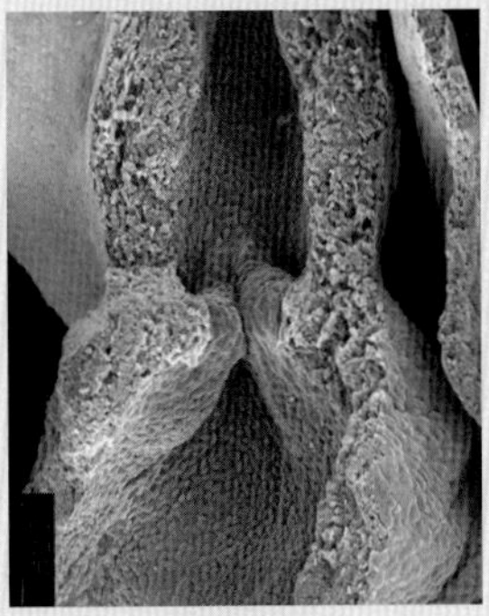

3.1.65 / 66 / 67

Ventral view of the opened and vertically adjusted pulmonary trunk. Due to the opening of the trunk, the anterior leaflet has been removed. Embryo at week 7, late in week 7 and at week 8.

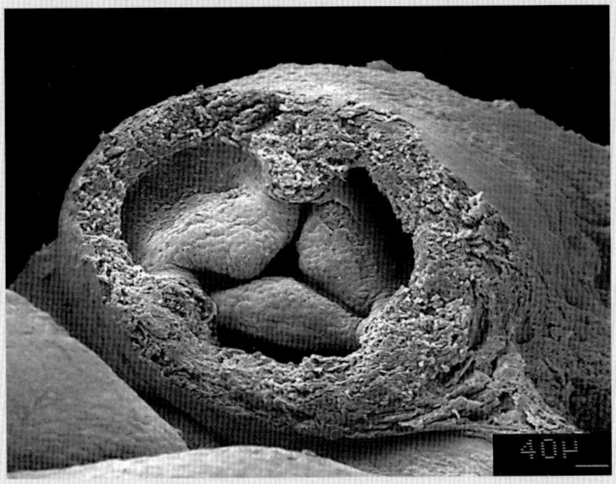

3.1.68

Cranial view of the pulmonary valve. Embryo at week 9.

3.1.69–3.1.91
Developmental aspects of the aortic arch complex.

Week 5

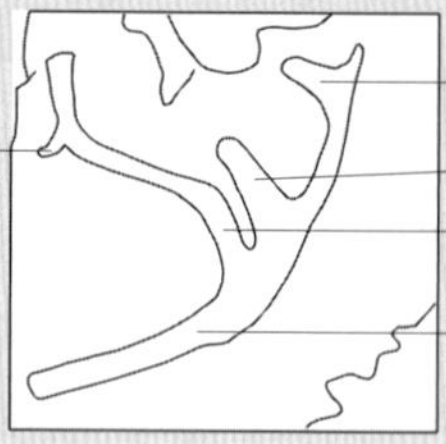

Week 6

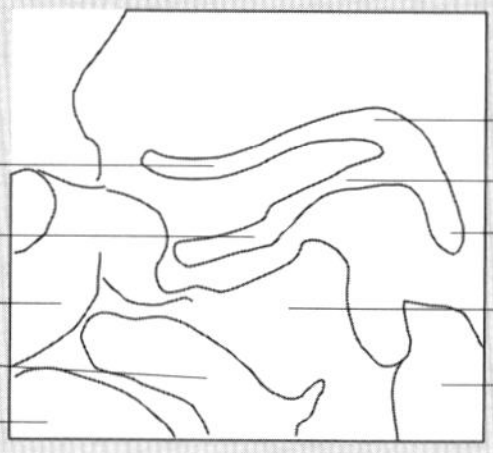

3.1.69

Ventral-left view of the opened left pharyngeal arch arteries and the aorta. Embryo at week 5.

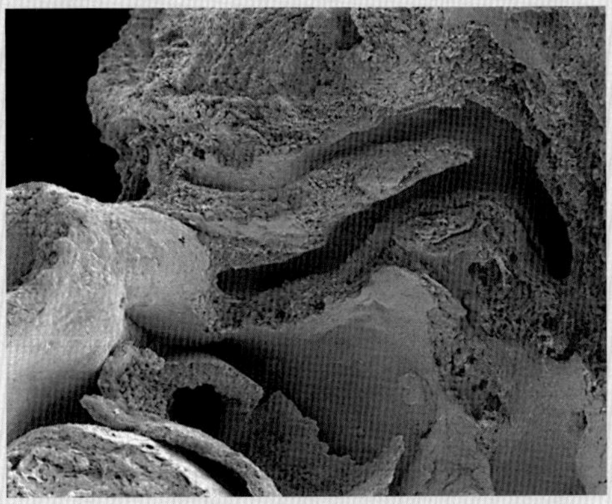

3.1.70

Ventral-left view of the left fourth aortic arch artery (ascending aorta and aortic arch), the left sixth aortic arch artery (pulmonary trunk and arterial duct), and the descending aorta. Embryo at week 6.

3.1.69–3.1.91

Developmental aspects of the aortic arch complex.

Week 6

Week 9

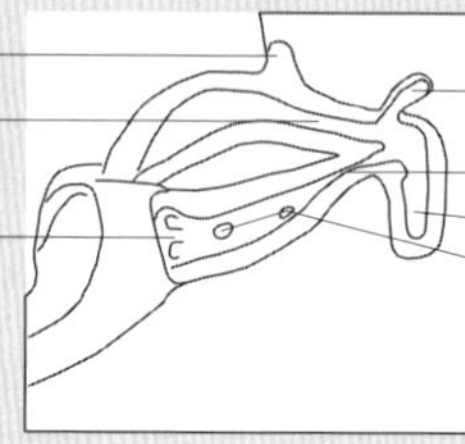

Week 10

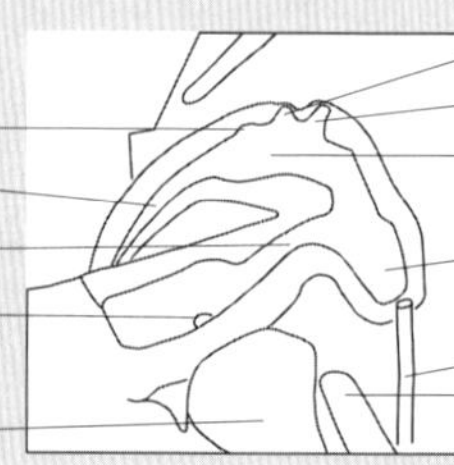

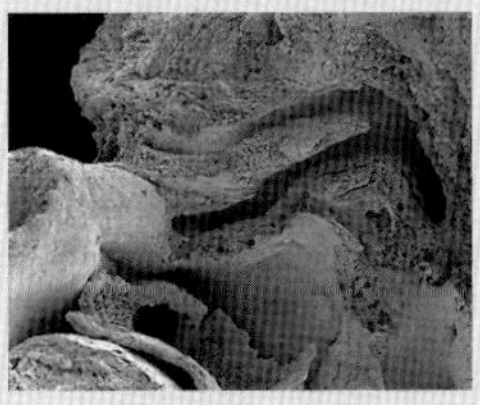

3.1.70a Same embryo as in figure 3.1.70.

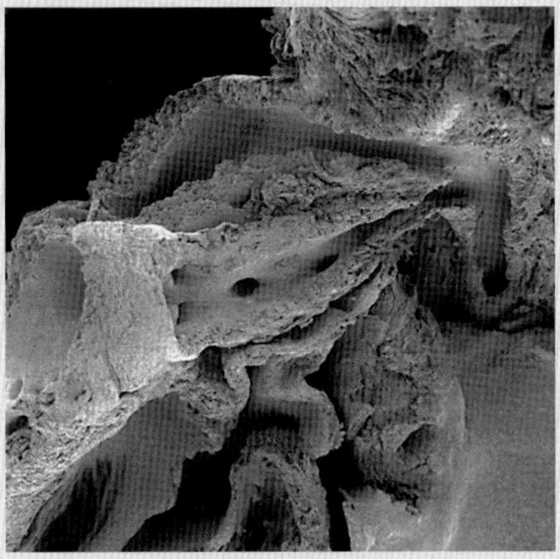

3.1.71 Ventral-left view of the left fourth aortic arch artery (ascending aorta and aortic arch), the left sixth aortic arch artery (pulmonary trunk and arterial duct) and the descending aorta. Embryo at week 9.

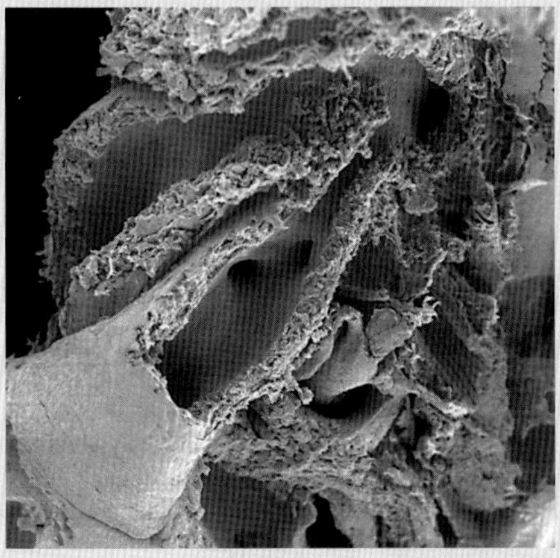

3.1.72 Ventral-left and caudal view of the pulmonary trunk with the orifices of the left and the right pulmonary arteries, and the arterial duct (ductus arteriosus), and the ascending aorta and the aortic arch (regarding the position of the orifices of the pulmonary arteries: see fig. 3.1.71). Embryo at week 10.

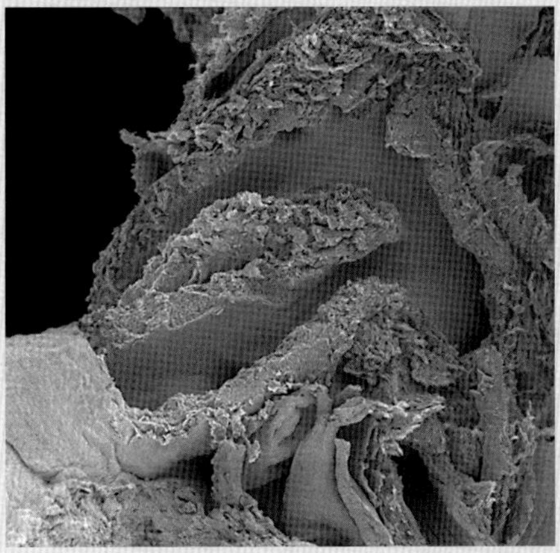

3.1.73 Same embryo as in figure 3.1.72. Ventral-left view.

3.1.69–3.1.91
Developmental aspects of the aortic arch complex.

Week 4

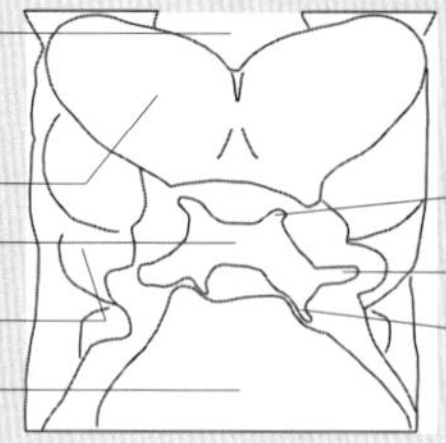

Week 5

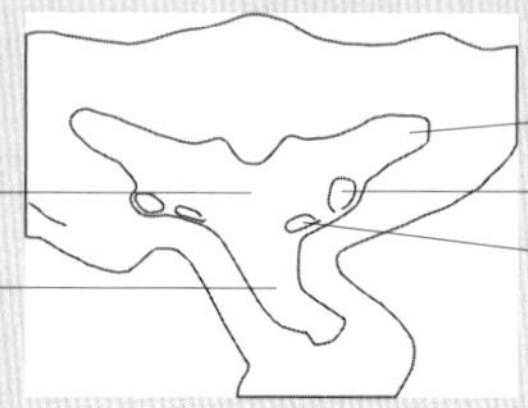

3.1.74–3.1.78
Septation of the aortic sac.
View in a downstream direction of the aortic sac and the exits of the pharyngeal arch arteries.

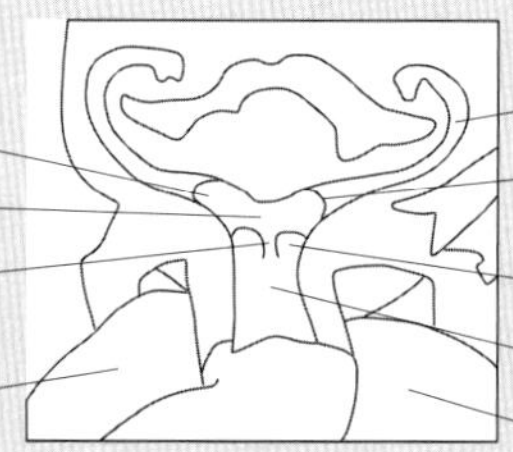

Week 6

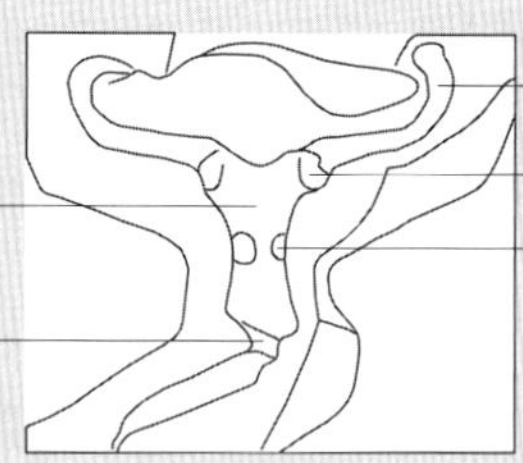

Week 5

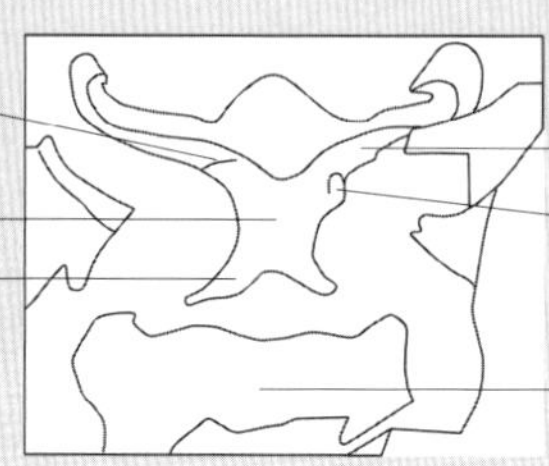

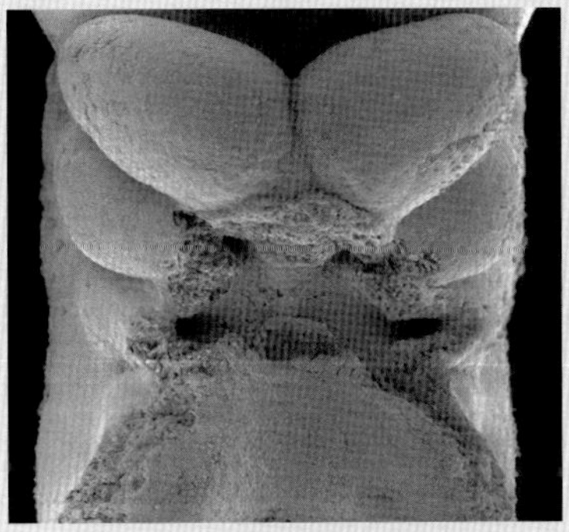

3.1.74 The exits of the sixth aortic arch arteries are not yet apparent.
Embryo at week 4.

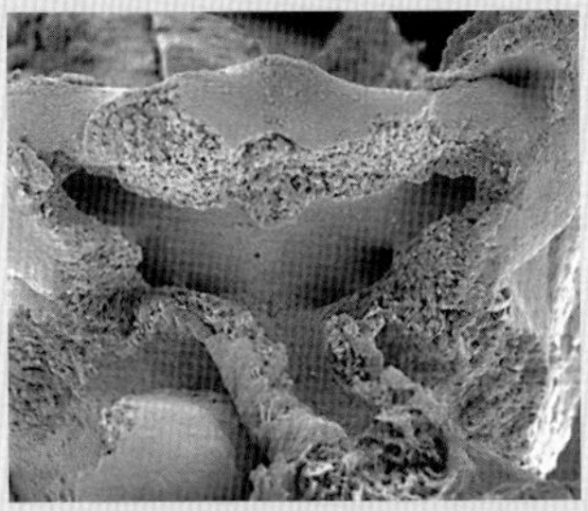

3.1.75 The exits of the sixth aortic arch arteries are situated laterally.
Embryo at week 5.

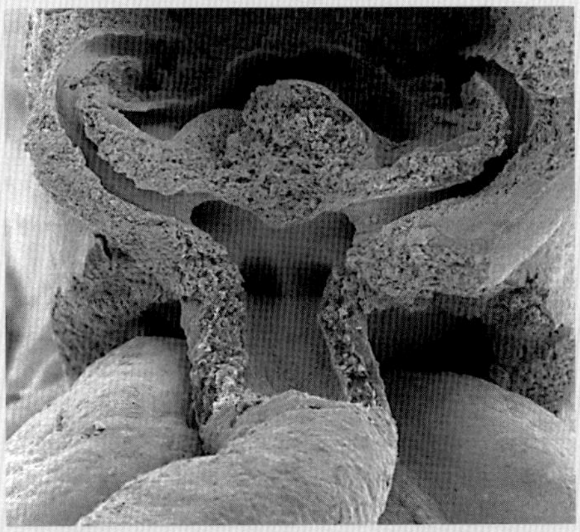

3.1.76 The exits of the sixth aortic arch arteries have approached the midline.
Embryo at week 5.

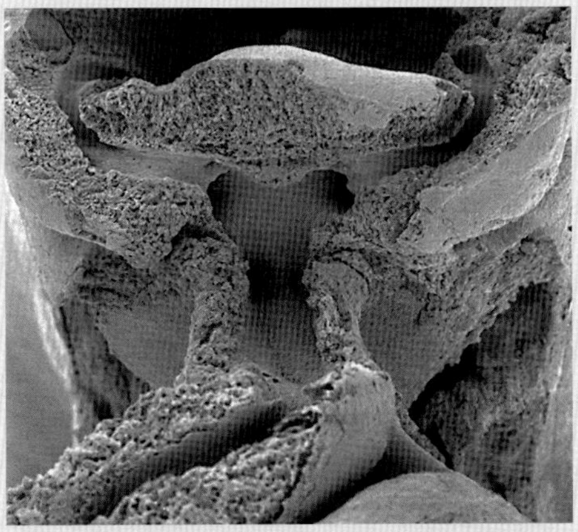

3.1.77 The septum separating the third and fourth aortic arch arteries (carotico-aortic septum) has approached the midline.
Embryo at week 6.

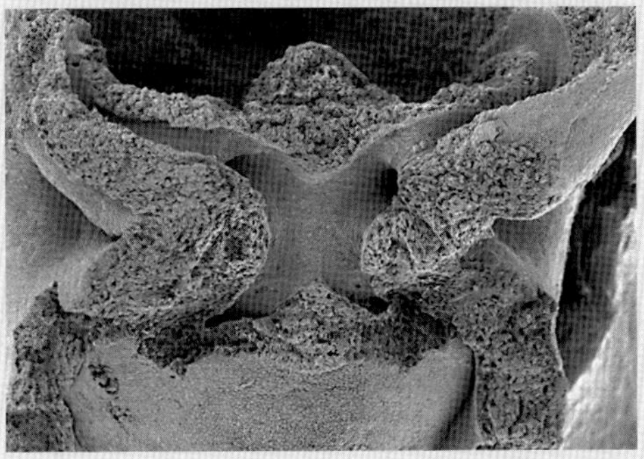

3.1.78 The distal outflow tract of the heart has been removed to show the proximal opened sixth aortic arch arteries.
Embryo at week 5.

3.1.69–3.1.91
Developmental aspects of the aortic arch complex.

3.1.79–3.1.84
Development of the aorticopulmonary septum.

Week 6

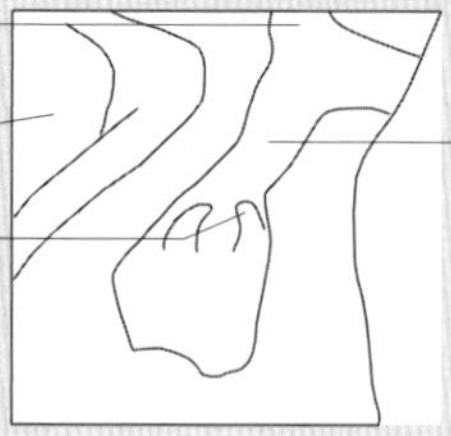

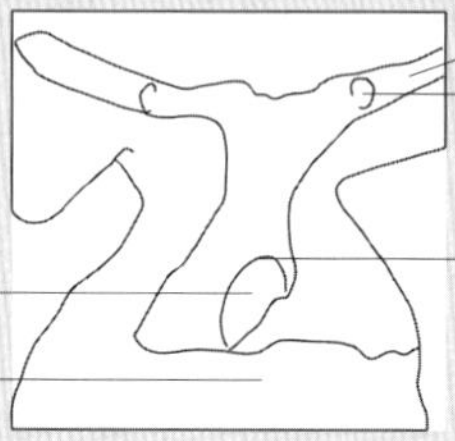

Week 5

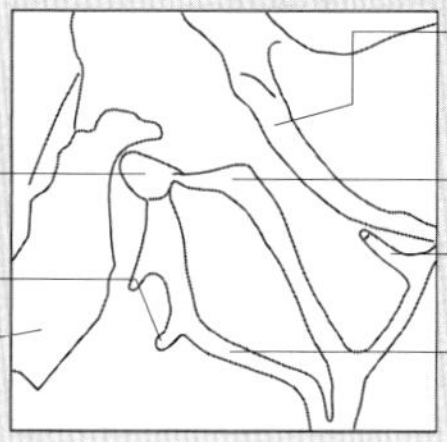

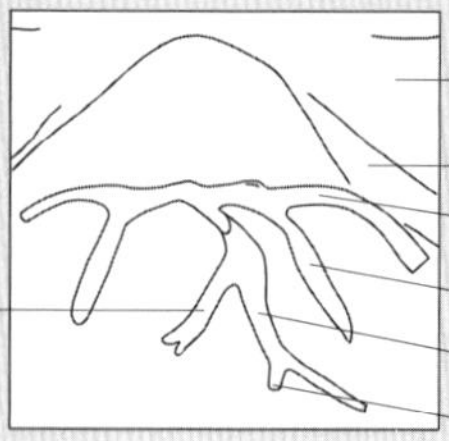

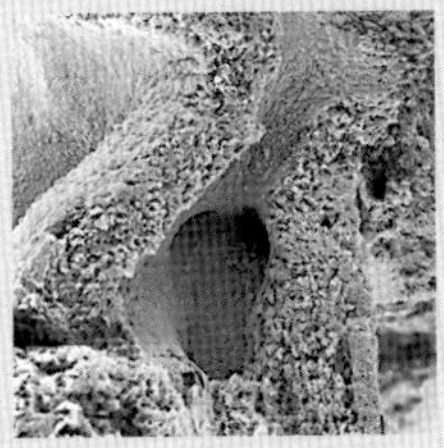

3.1.79 Less and more advanced stages of early development of the aorticopulmonary septum.
Embryos at week 6.

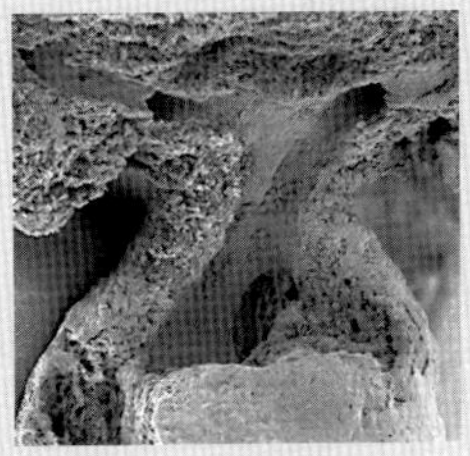

3.1.80

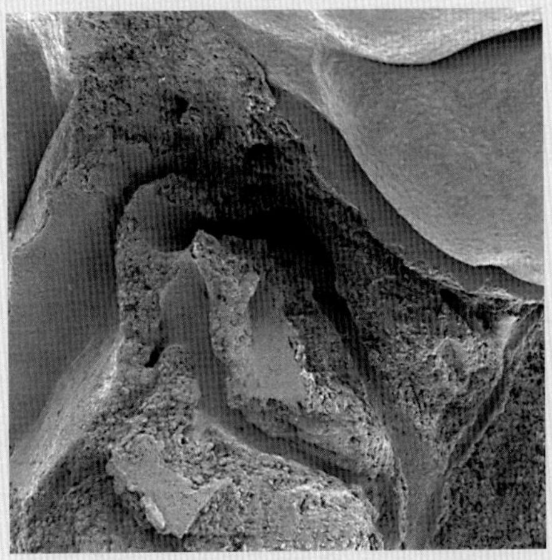

3.1.81 Ventral-left view of the left fourth and sixth aortic arch arteries.
Embryo late in week 5.

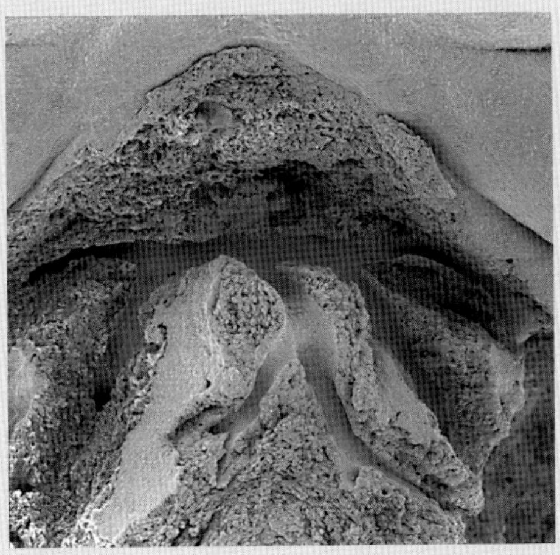

3.1.82 Ventral view of the sixth aortic arch arteries. The distal outflow tract of the heart has been removed.
Embryo at week 5.

3.1.69–3.1.91
Developmental aspects of the aortic arch complex.

3.1.79–3.1.84
Development of the aorticopulmonary septum.

Week 6

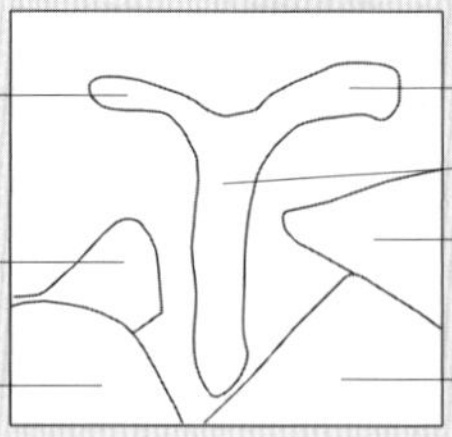

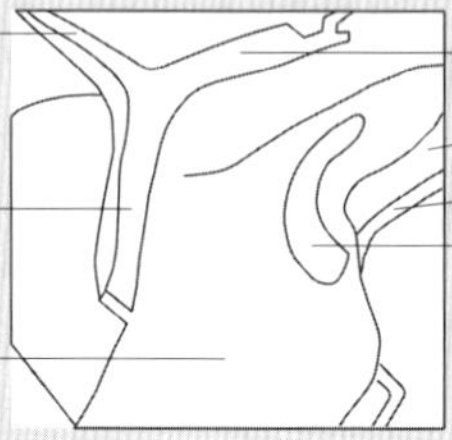

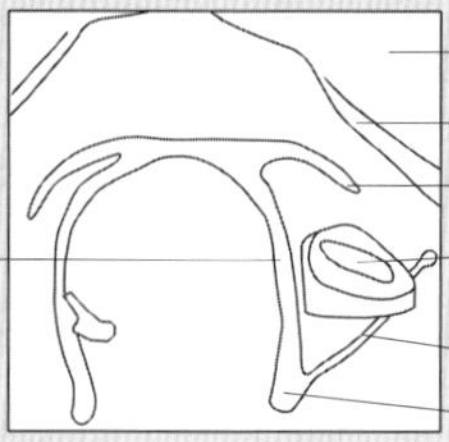

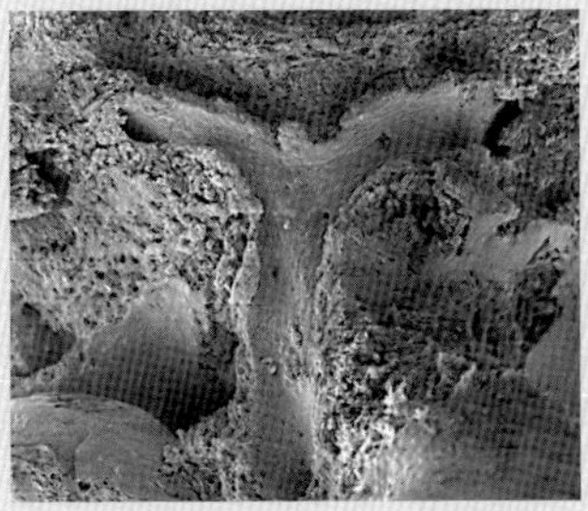

3.1.83 Cranial and ventral view of the aortic outflow tract (ascending aorta). The aorticopulmonary septum separates the systemic and the pulmonary flow paths completely.

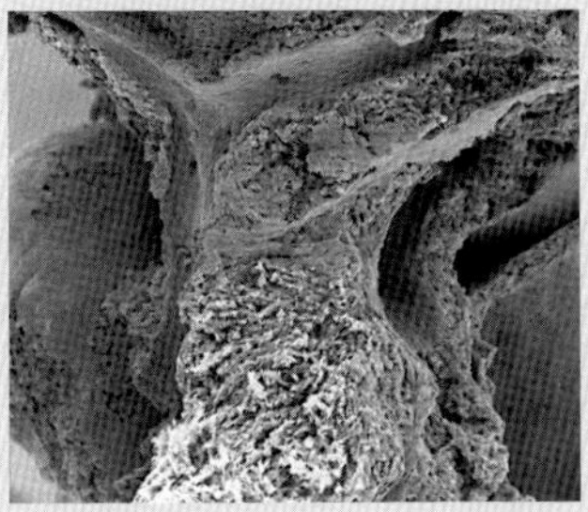

3.1.84 Same embryo as in figure 3.1.83. Cranial and ventral-left view shows the pulmonary trunk beneath the ascending aorta.

3.1.85 Ventral view of the positions of the third and fourth aortic arch arteries. The distal outflow tract of the heart has been removed.
Embryo at week 6.

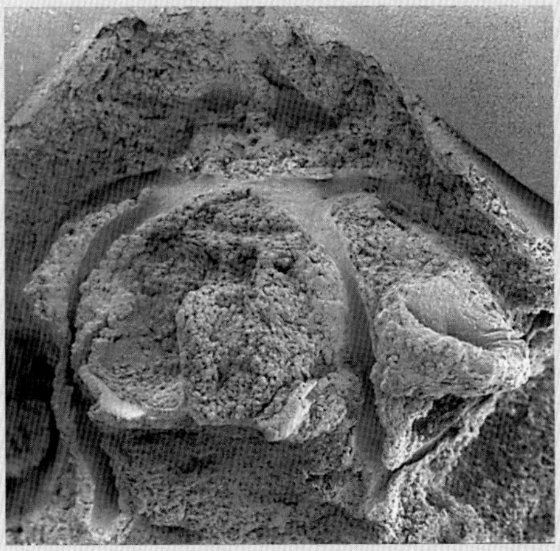

3.1.86 Ventral and left view of the positions of the third and fourth aortic arch arteries. The distal outflow tract of the heart has been removed.
Embryo at week 6.

3.1.69–3.1.91
Developmental aspects of the aortic arch complex.

3.1.87–3.1.91
The aortic arch complex in dorsal view.

60:1

Week 5

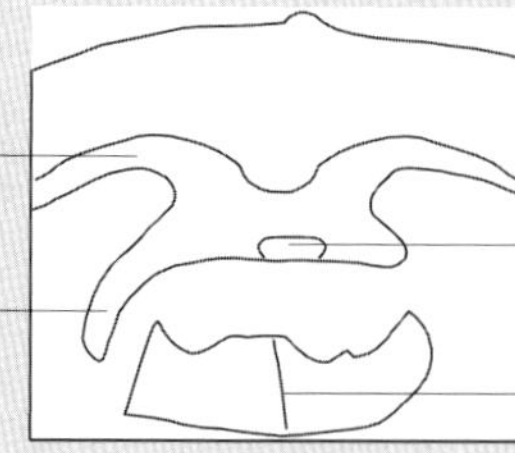

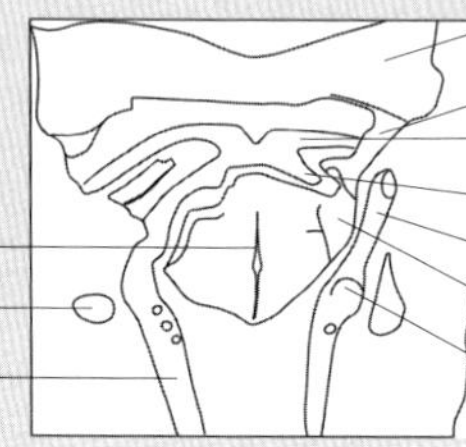

120:1

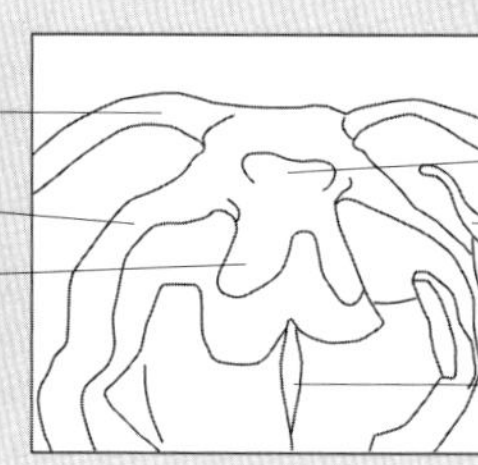

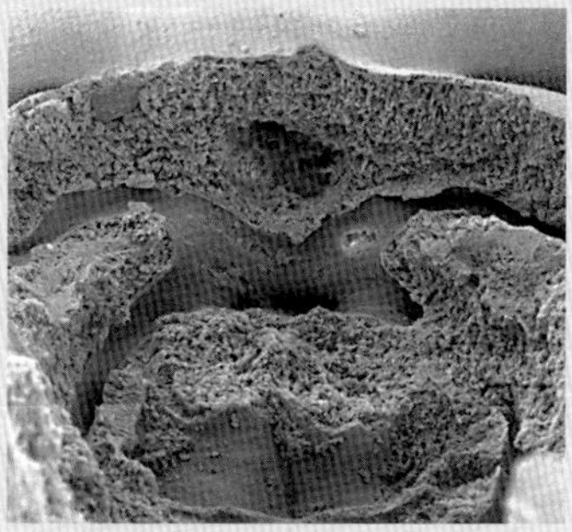

3.1.87 The aortic sac and the third and fourth aortic arch arteries.
Embryo at week 5.

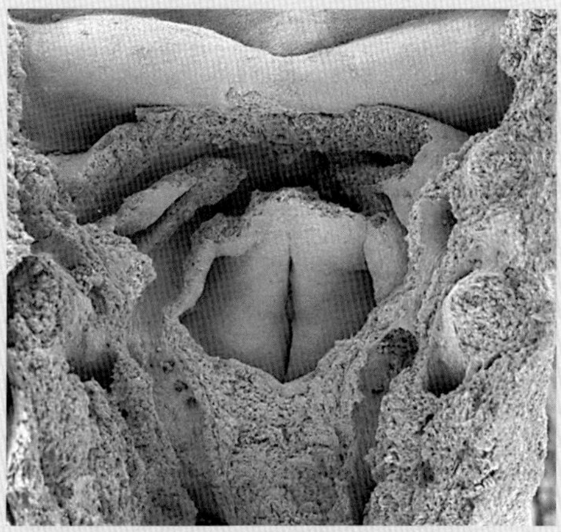

3.1.88 The third and fourth aortic arch arteries.
Embryo at week 5.

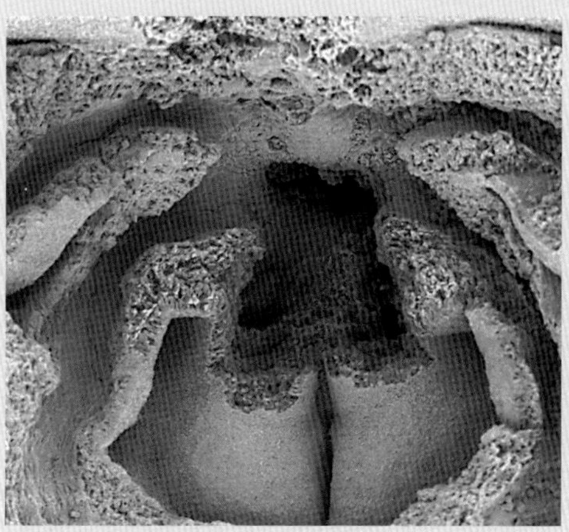

3.1.89 Same embryo as in figure 3.1.88. Parts of the sixth arches have been removed to show the exits of the sixth aortic arch arteries.

3.1.69–3.1.91
Developmental aspects of the aortic arch complex.

3.1.87–3.1.91
The aortic arch complex in dorsal view.

Week 6

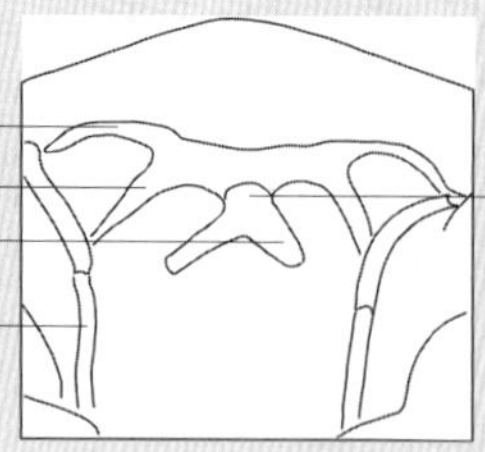

Week 5

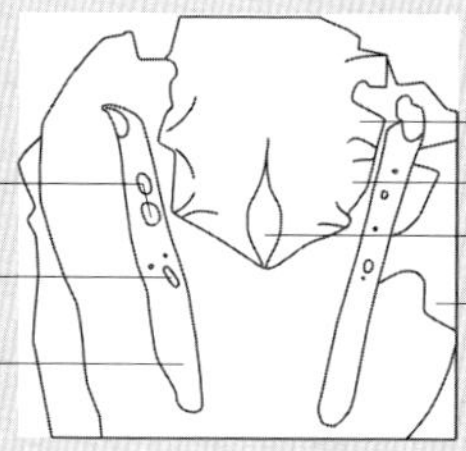

3.1.90 Aortic sac and third, fourth and sixth aortic arch arteries.
Embryo at week 6.

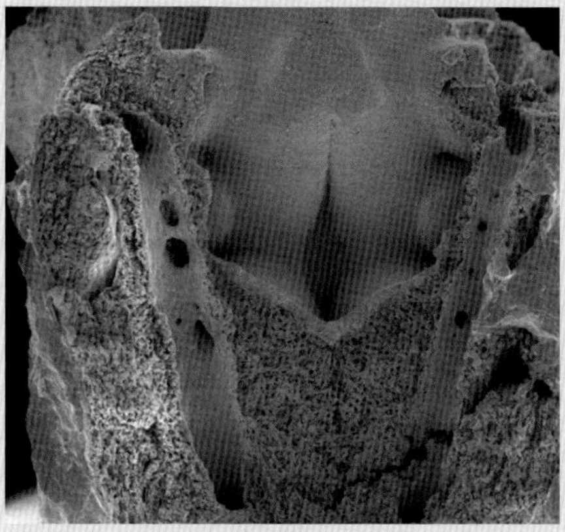

3.1.91 The paired dorsal aortae with the orifices of the pharyngeal arch arteries.
Embryo at week 5.

3.2 The Development of the Lungs

The lungs arise as an unpaired endothelial evagination from the ventral wall of the foregut, cranial from the anterior intestinal port (fig. 3.2.01–3.2.03). The evagination elongates in a caudal direction and forms the trachea. Its caudal end divides into the right and the left lung buds (fig. 3.2.04–3.2.06). The lung buds grow in caudal and lateral directions and form two bulges at the level of the pericardial-peritoneal canal (fig. 3.2.07, 3.2.14, 3.2.55–3.2.59).

Very early differences between the right and the left lung buds become apparent in their angle of ramification from the trachea and in their length (fig. 3.2.05, 3.2.06, 3.2.08, 3.2.09).

The elongated lung buds develop lobar buds, three in the right and two in the left lung (fig. 3.2.10–3.2.12). These lobar buds do not develop in the same plane but each grows in its characteristic direction (fig. 3.2.10, 3.2.11). Beginning at about week 6, the lobar buds ramify and form the segmental buds, which correspond to the adult segmental bronchi (fig. 3.2.49–3.2.54) whose ramifications produce an increasing number of secondary and tertiary buds (fig. 3.2.48, 3.2.54). The development of the functional alveoli does not begin before week 24 and ends postnatally.

The positional development of the lung is determined by its origin almost in the midline, dorsal to the heart, and its early position in the pericardial-peritoneal canal (fig. 3.2.58, 3.2.59). Starting from this position the lungs are transposed in lateral and caudal directions by the growth of the pericardioperitoneal canals which form the pleural cavities (fig. 3.2.13–3.2.33). The separation of the pleural cavities from the pericardial cavity occurs in week 6 (fig. 3.2.18, 3.2.29), the separation from the peritoneal cavity is realized by the pleuroperitoneal folds (fig. 3.2.61, 3.2.62). A small pleuroperitoneal communication may be preserved up to the 3rd month (fig. 3.2.63).

The differences in the development of the external form of the right and the left lung are shown in figures 3.2.33–3.2.48.

3.2

Abbreviations

d	diaphragm
il	intestinal loop
la	left atrium
lv	left ventricle
mn	mesonephros
ra	right atrium
rv	right ventricle
s	stomach
1	lung bud
2	superior lobe of the lung
3	middle lobe of the lung
4	inferior lobe of the lung
5	common cardinal vein
6	superior cardinal vein
7	inferior cardinal vein
8	superior caval vein
9	left sinus horn
10	triangular ligament
11	pleuroperitoneal canal, coelomic duct
12	conus
13	truncus
14	ductus venosus
15	duodenum
16	liver
17	hepatic vein

Week 4

3.2.01–3.2.12
Early developmental stages of the lung. The heart has been removed. Ventral view.

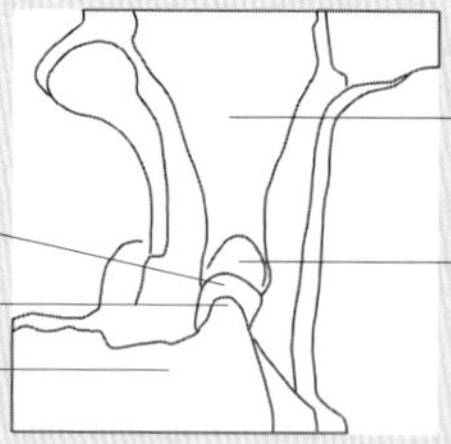

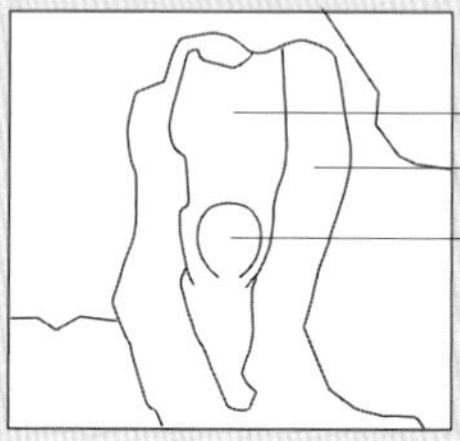

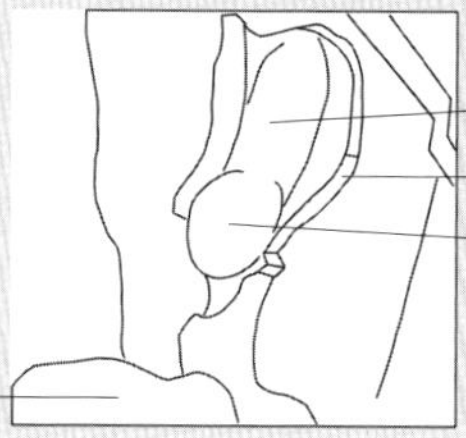

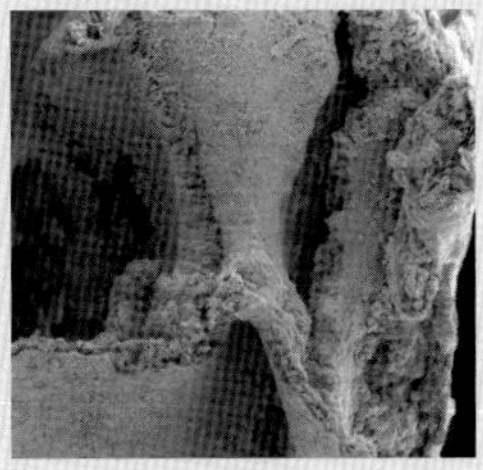

3.2.01

The lung bud is only a tiny epithelial outpocketing. Embryo at week 4.

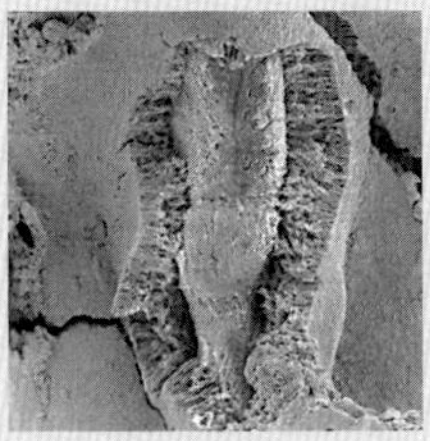

3.2.02

Ventral (fig. 3.2.02) and ventral-left (fig. 3.2.03) view of the lung bud. Same embryo at week 4, more advanced stage.

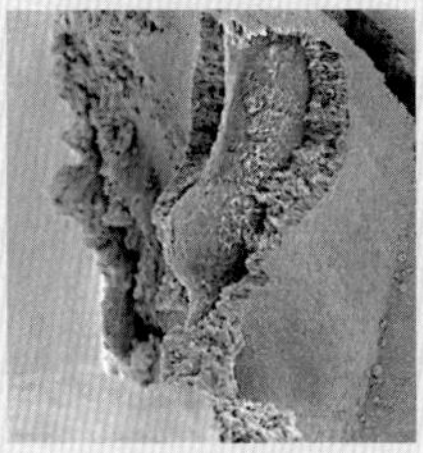

3.2.03

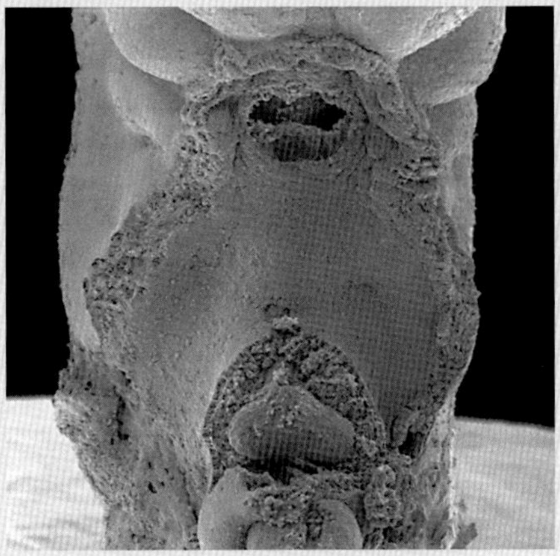

3.2.04

The heart has been removed. Dorsal wall of the pericardial cavity and first dichotomic ramifications of the bud into the anlage of the right and the left lung.
The overlying mesenchyme has been removed. Embryo at week 4.

Week 5

3.2.01–3.2.12
Early developmental stages of the lung. The heart has been removed. Ventral view.

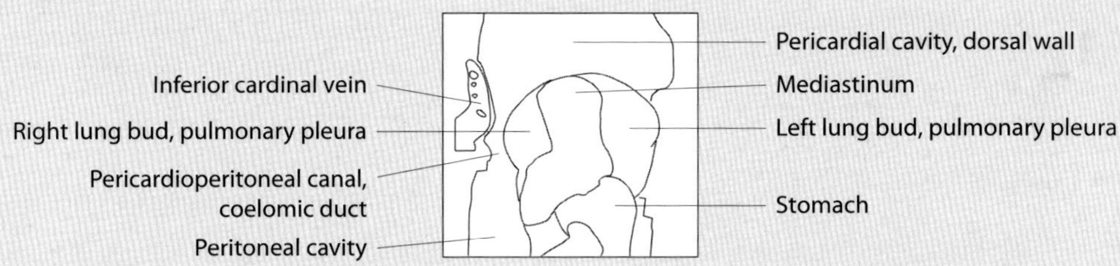

Week 6

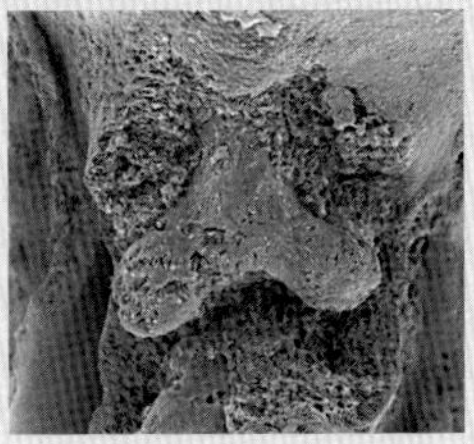

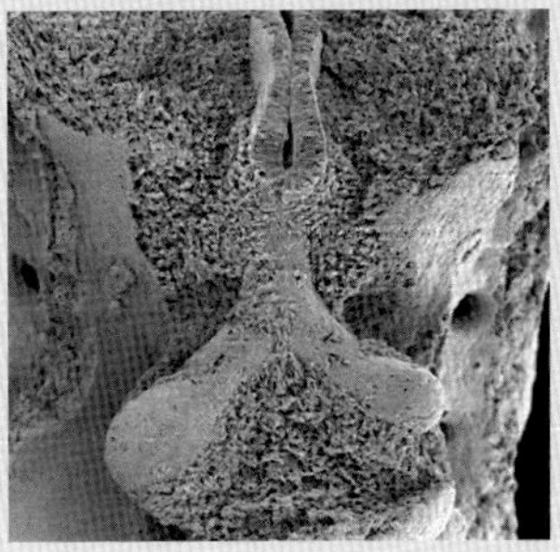

3.2.05 / 06

3.2.05
Right and left lung bud. The overlying mesenchyme and the epithelium of the serosa have been removed.
Embryo at week 5.

3.2.06
Right and left lung buds. The overlying mesenchyme and the epithelium of the serosa have been removed.
More advanced embryo at week 5.

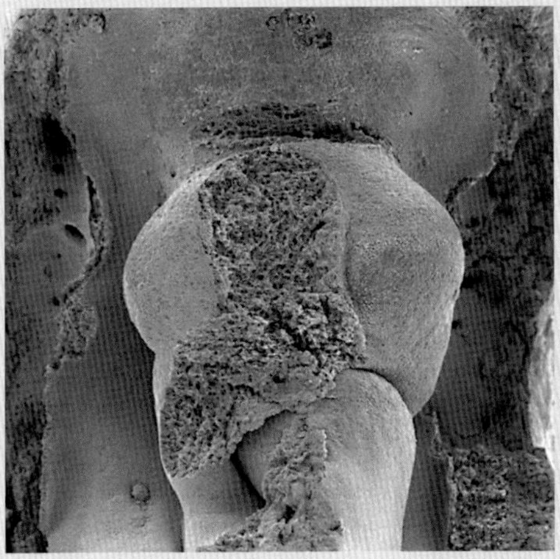

3.2.07

External aspect of the serosa epithelium that covers the lungs.
The developmental stage corresponds to the embryo of figure 3.2.06.

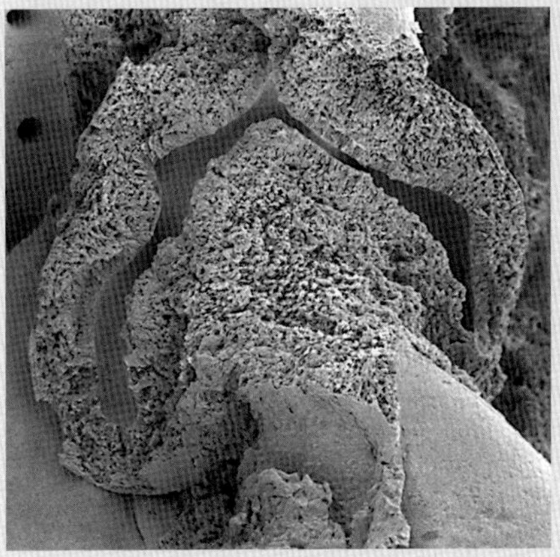

3.2.08 / 09

Fig. 3.2.08
The serosa and the mesenchyme overlying the endoderm have been removed. The lung anlagen have been opened.
Embryo at week 6.

Fig. 3.2.09
The lungs have been opened. The first indications of the developing lobes.
Embryo at week 6.

Week 6

3.2.01–3.2.12
Early developmental stages of the lung. The heart has been removed. Ventral view.

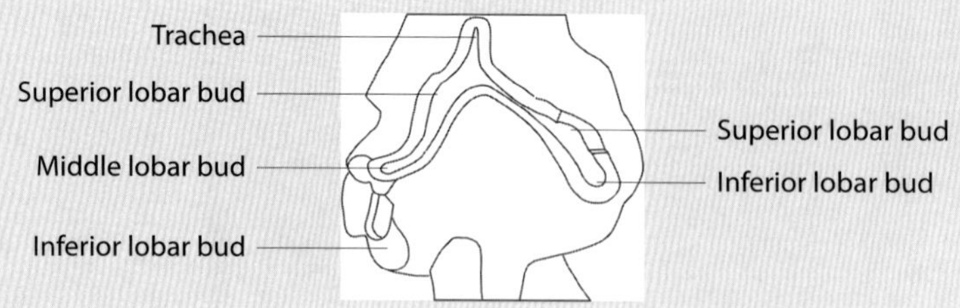

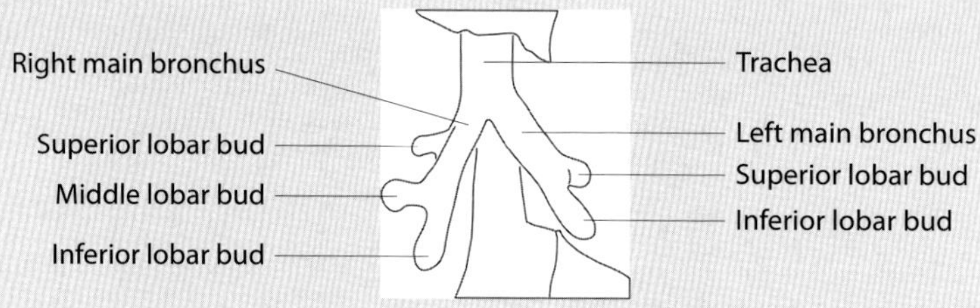

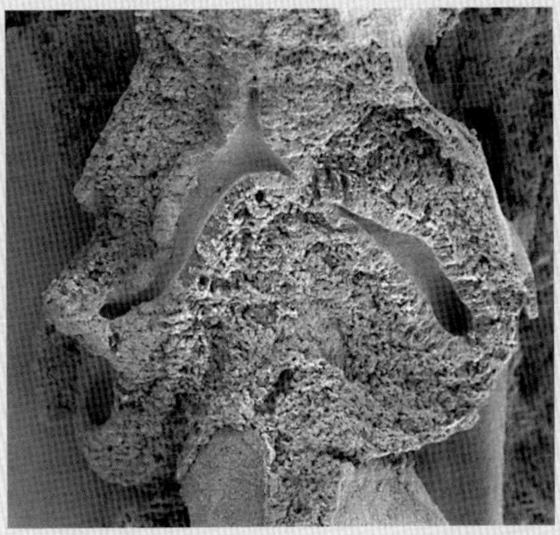

3.2.10 The lungs have been opened. The middle lobe of the right lung is more advanced.
Embryo at week 6.

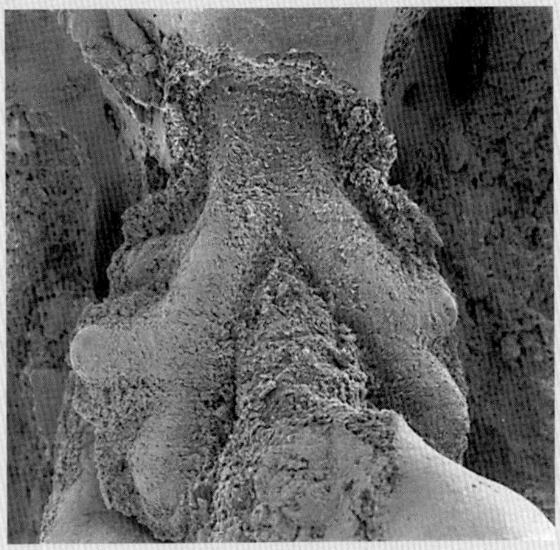

3.2.11 The external aspect of the endoderm of the lobes.
Embryo at week 6.

3.2.12 Advanced stage of lobe development.
Embryo at the end of week 6.

Week 5

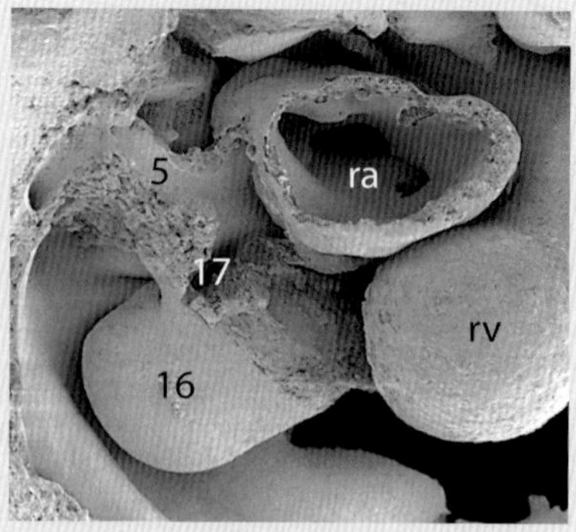

3.2.13–3.2.21

Positional development of the right lung. The adjacent pictures show different dissection steps of the same embryo. Pictures on the left show aspect of the opened peritoneal cavity. In the pictures on the right the structures overlying the lung have been removed to reveal the entire lung. Lateral (right) view.

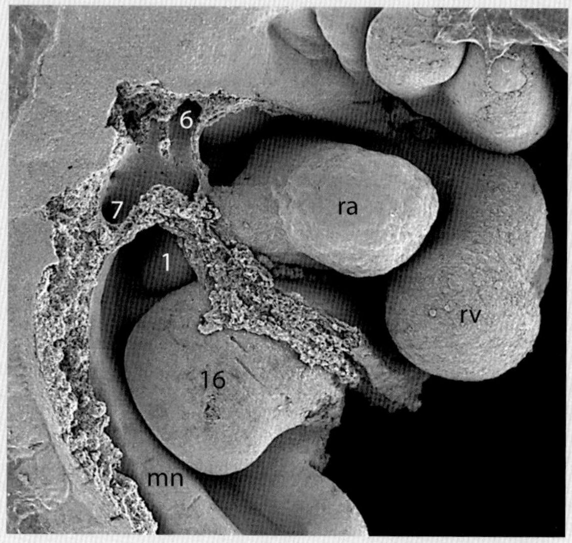

Week 6

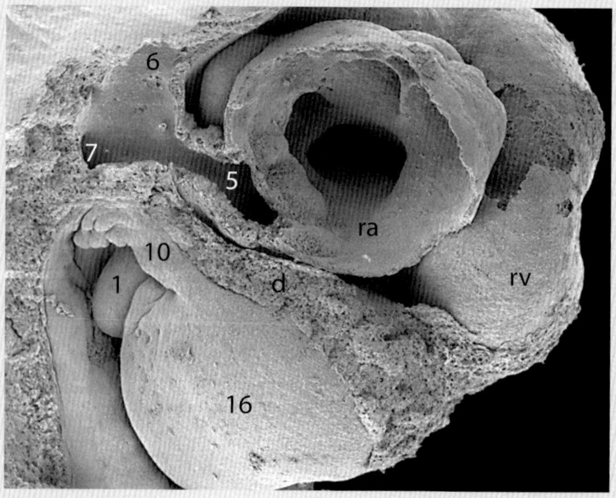

3.2.13 / 14

Embryos at week 5.

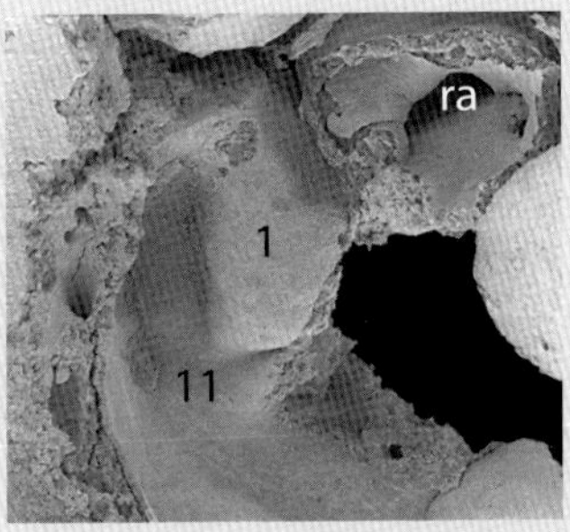

3.2.15 / 16

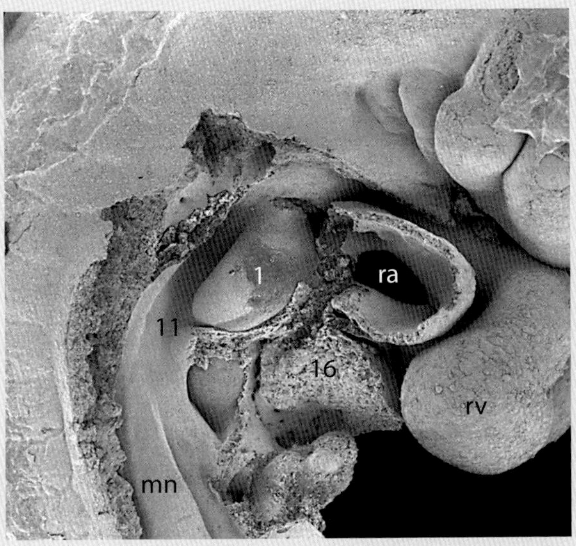

3.2.17 / 18

Embryos early and late in week 6.

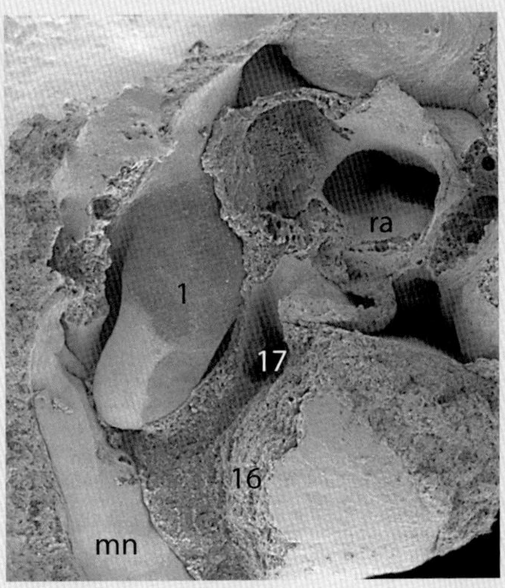

Week 6

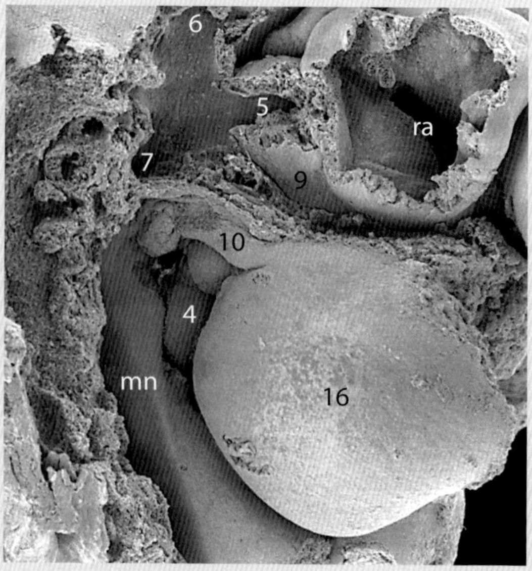

3.2.13–3.2.21
Positional development of the right lung. The adjacent pictures show different dissection steps of the same embryo. Pictures on the left show aspect of the opened peritoneal cavity. In the pictures on the right the structures overlying the lung have been removed to reveal the entire lung. Lateral (right) view.

Week 7

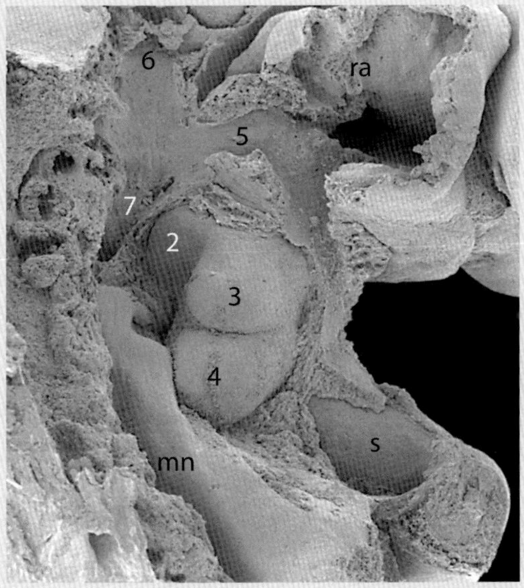

3.2.19 / 20 Embryos early and late in week 6.

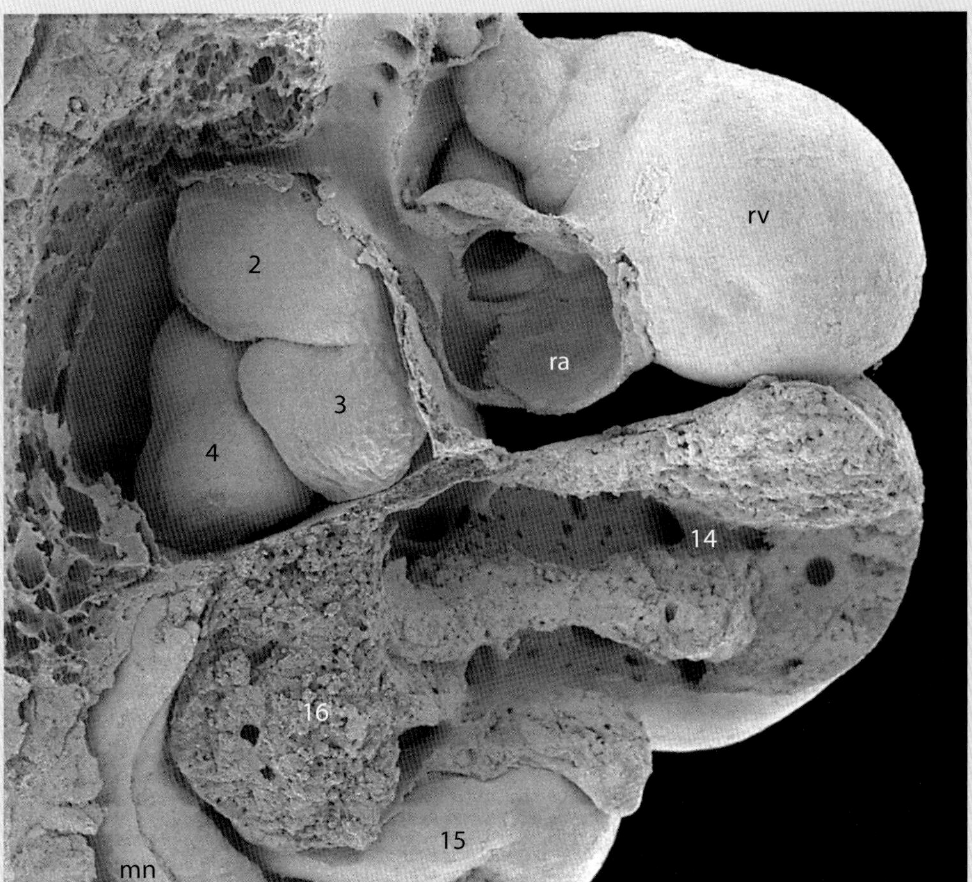

3.2.21 The lung has almost reached its adult position in the region between heart and liver. Only the costal pleura has been removed. The separation of the pleural and the peritoneal cavities is almost complete (see fig. 3.2.63).

Week 5

3.2.22–3.2.32
Positional development of the left lung. The adjacent pictures show different dissection steps of the same embryo. Pictures on the left show aspect of the opened peritoneal cavity. In the pictures on the right the structures overlying the lung have been removed to reveal the entire lung. Lateral (left) view.

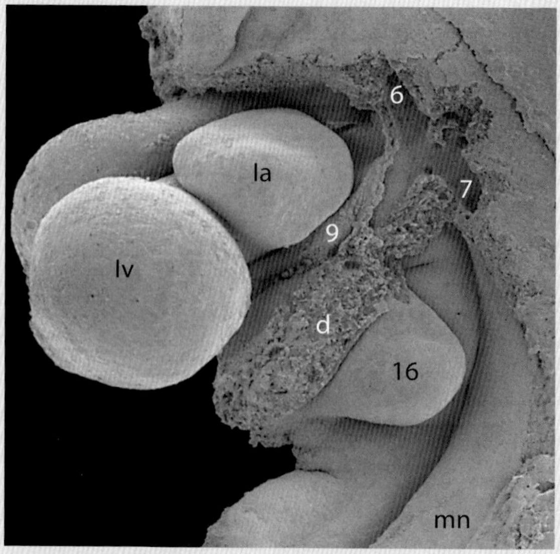

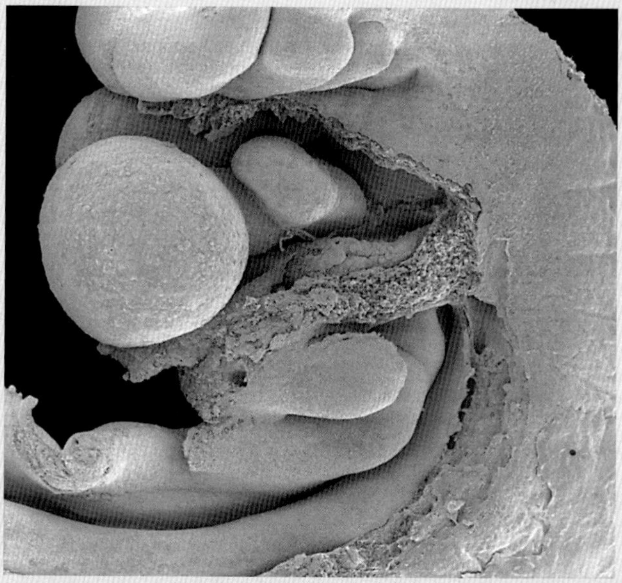

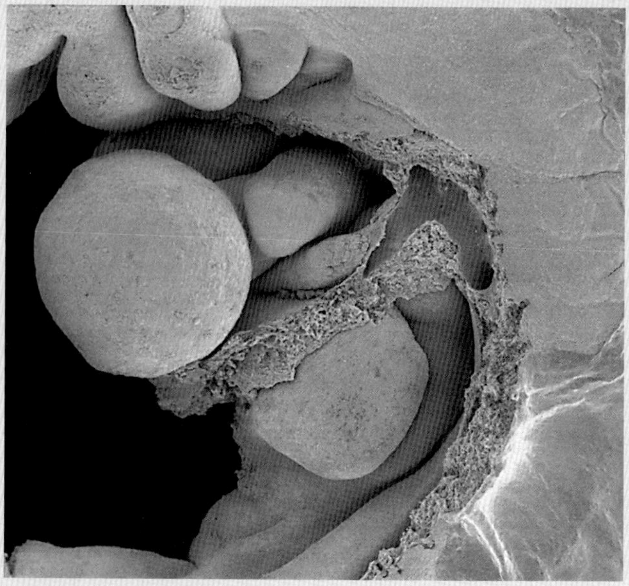

3.2.22 / 23

Embryos at week 5.

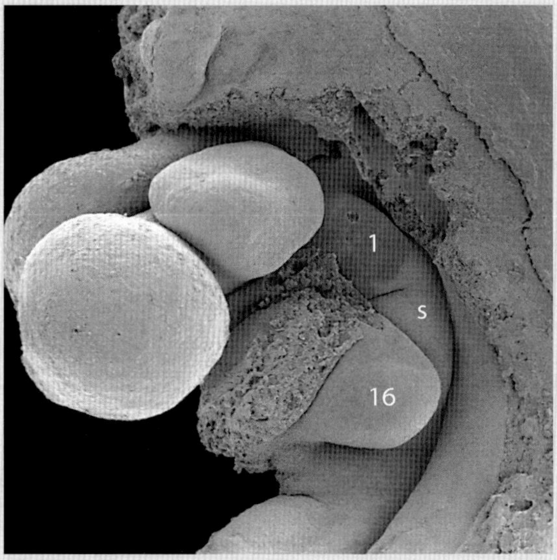

3.2.24 / 25

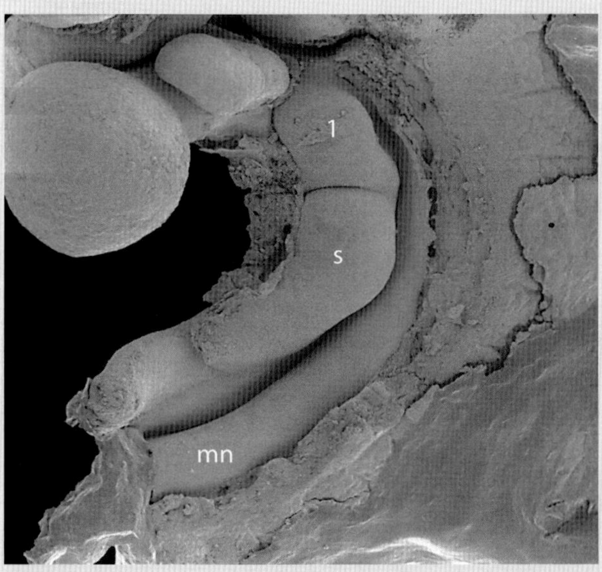

3.2.26 / 27

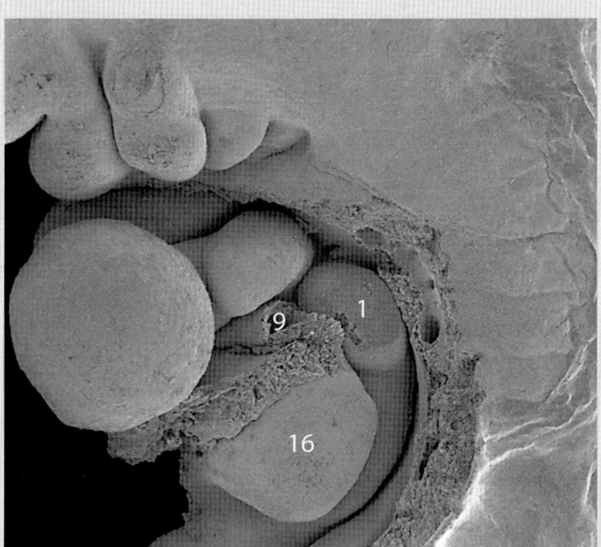

3.2.22–3.2.32

Positional development of the left lung. The adjacent pictures show different dissection steps of the same embryo. Pictures on the left show aspect of the opened peritoneal cavity. In the pictures on the right the structures overlying the lung have been removed to reveal the entire lung. Lateral (left) view.

Week 6

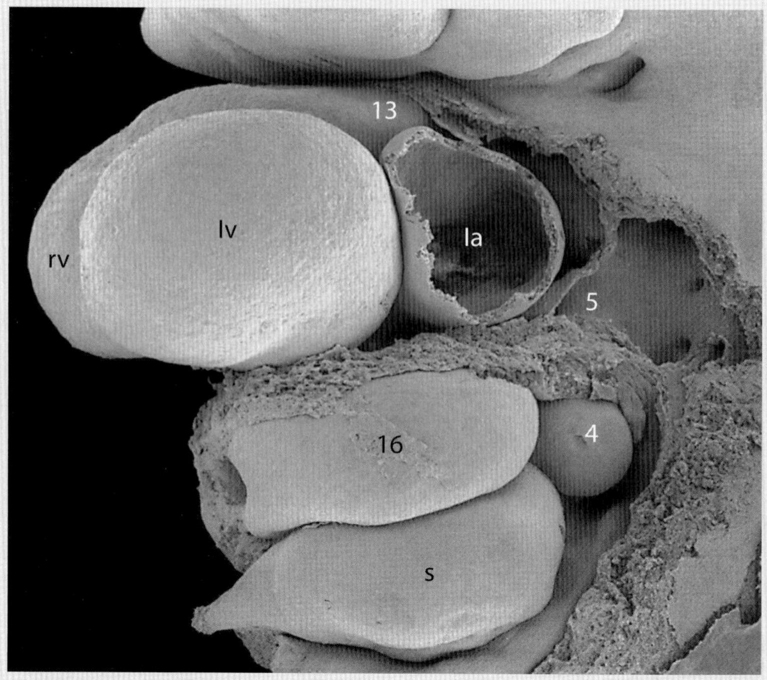

Week 7

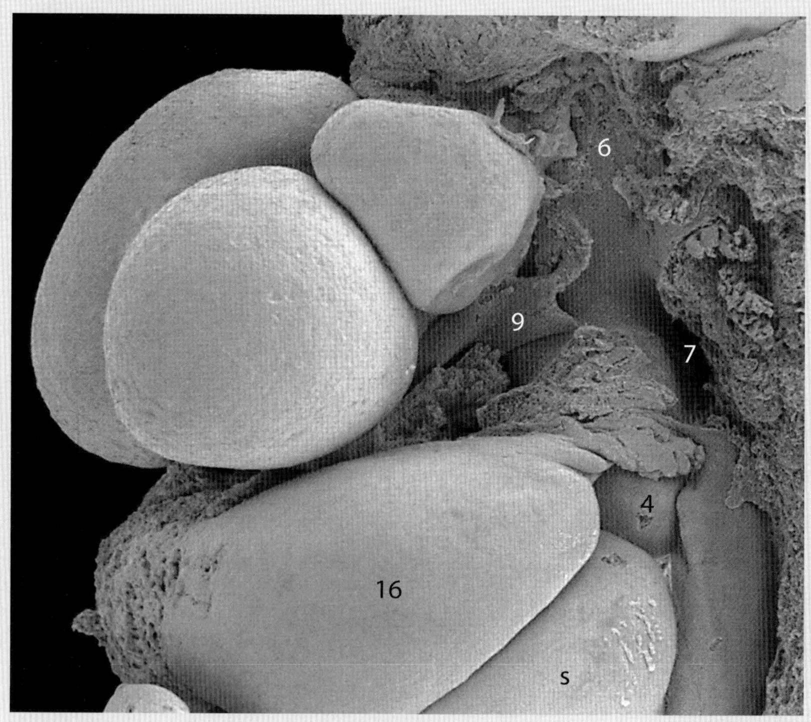

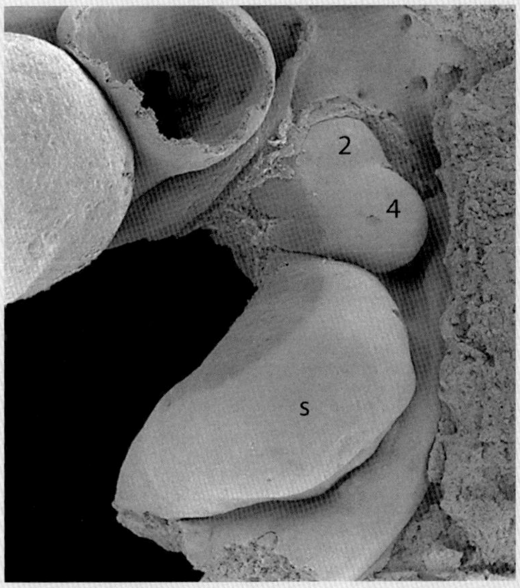

3.2.28 / 29

Embryo at week 6.

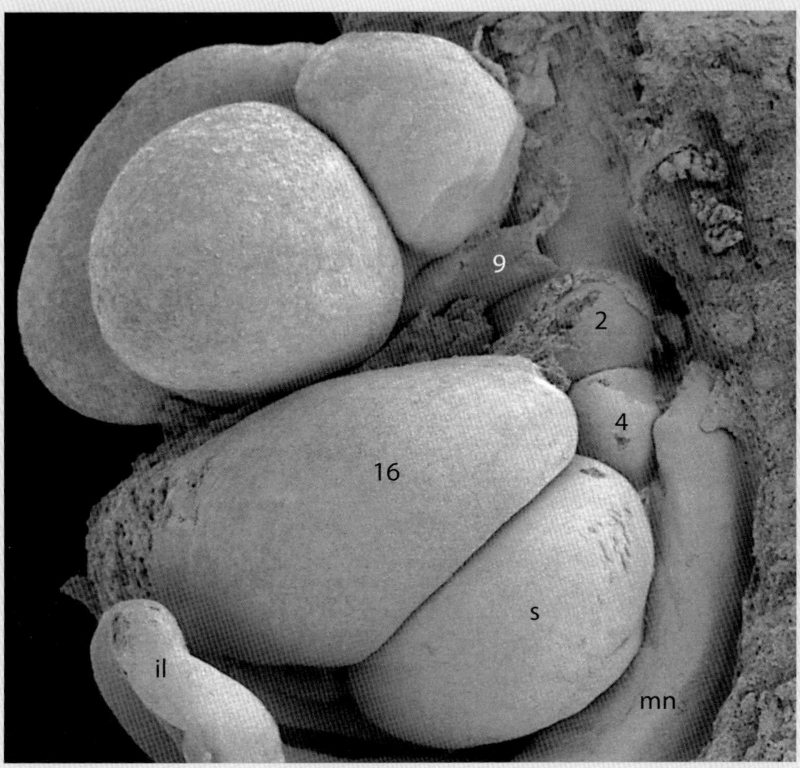

3.2.30 / 31

Embryo at week 7.

3.2.22–3.2.32

Positional development of the left lung. The adjacent pictures show different dissection steps of the same embryo. Pictures on the left show aspect of the opened peritoneal cavity. In the pictures on the right the structures overlying the lung have been removed to reveal the entire lung. Lateral (left) view.

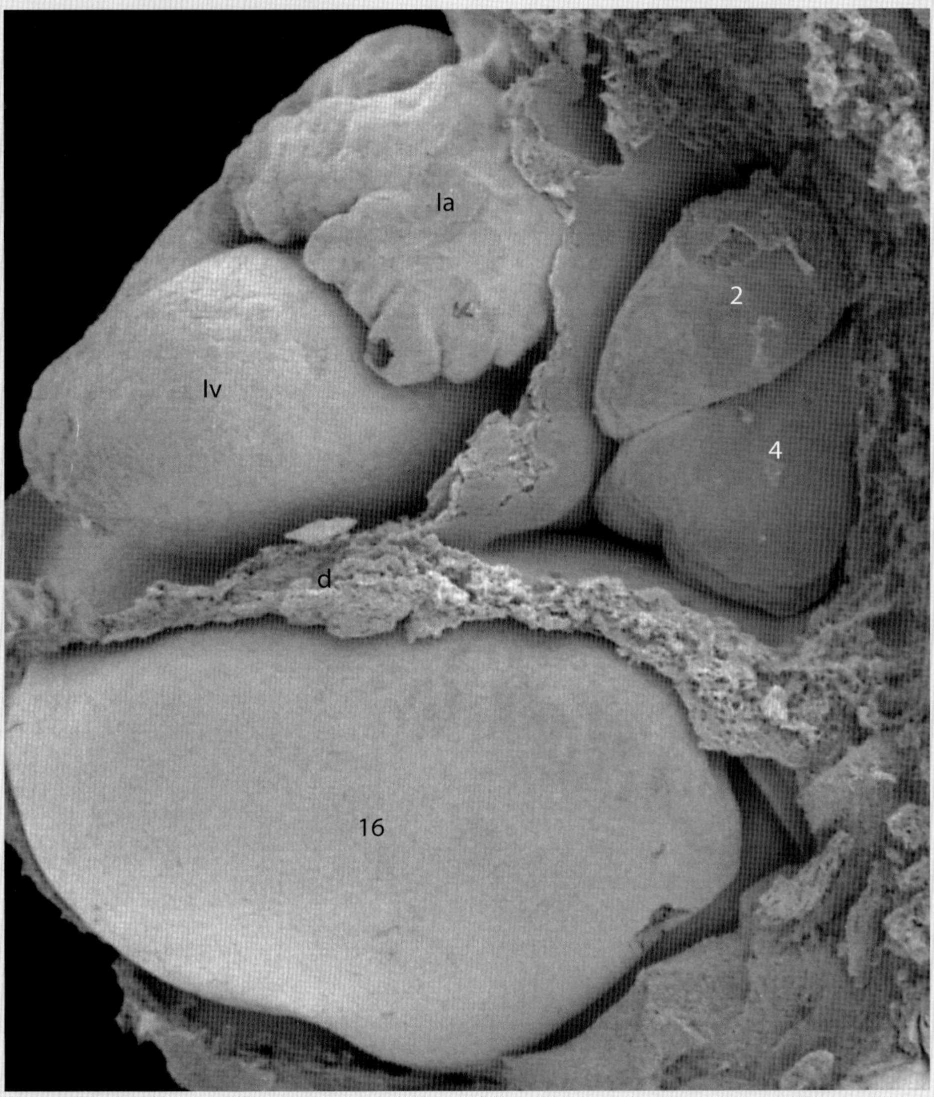

3.2.32 The lung has almost reached its adult position in the region between heart and liver. Only the costal pleura has been removed. The separation of the pleural and the peritoneal cavities is almost complete (see fig. 3.2.63).

3.2.33–3.2.48
Development of the external form of the lungs.
The pictures on the right show the right lung in the lateral (left) view. The pictures on the right show the left lung in the lateral (left) view.

Week 5

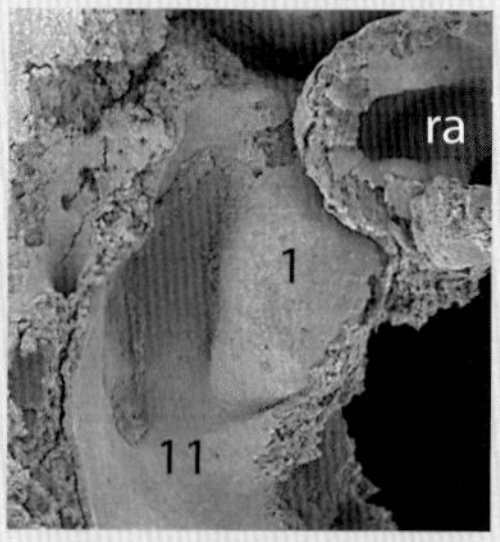

Week 6

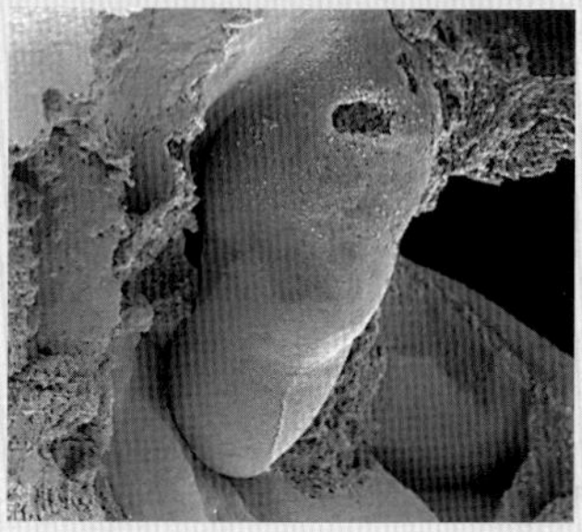

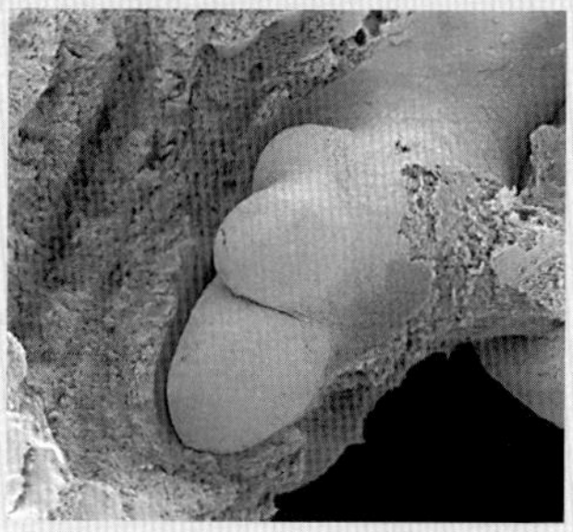

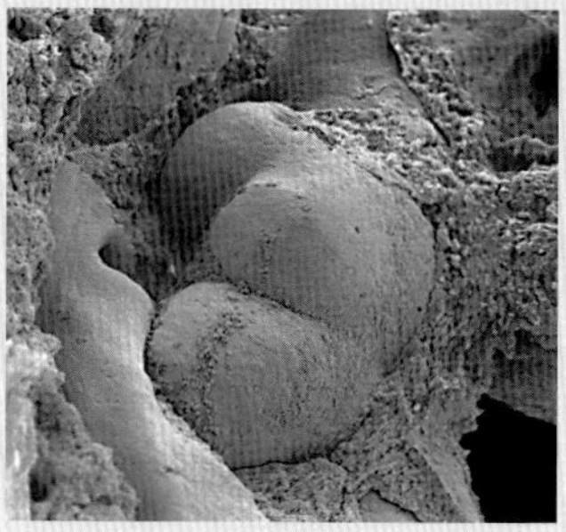

Week 7

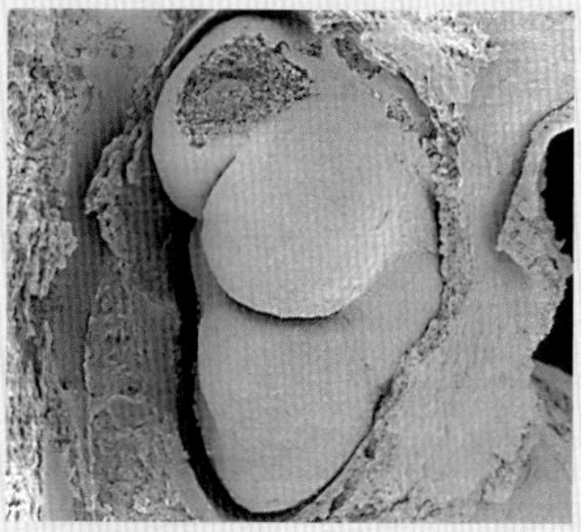

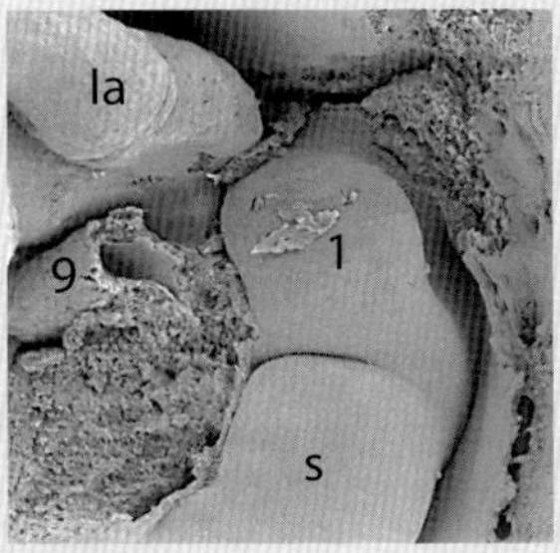

3.2.33 / 34

Embryo at week 5.

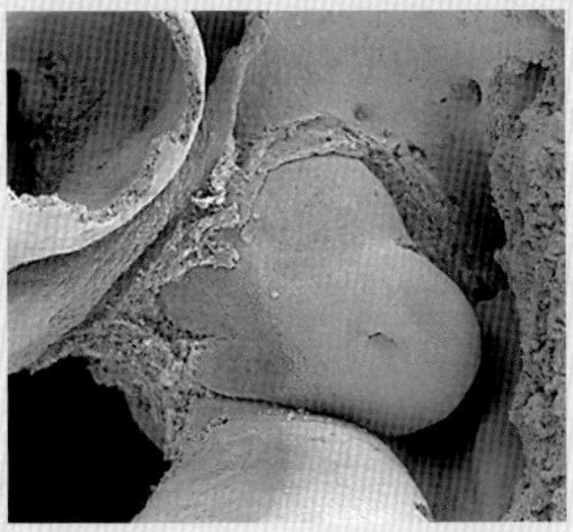

3.2.35 / 36

Embryo at week 6.

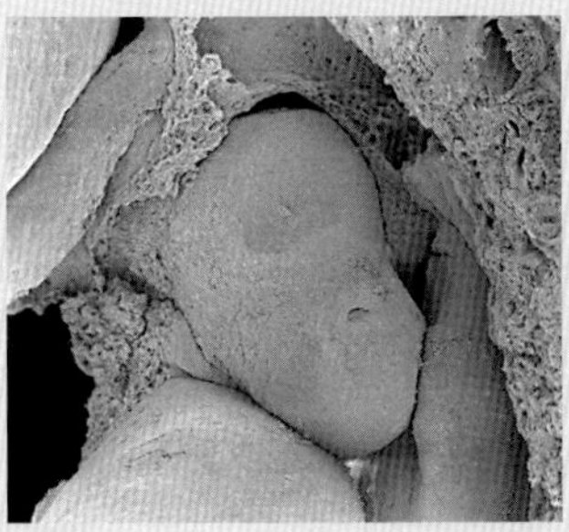

3.2.37 / 38

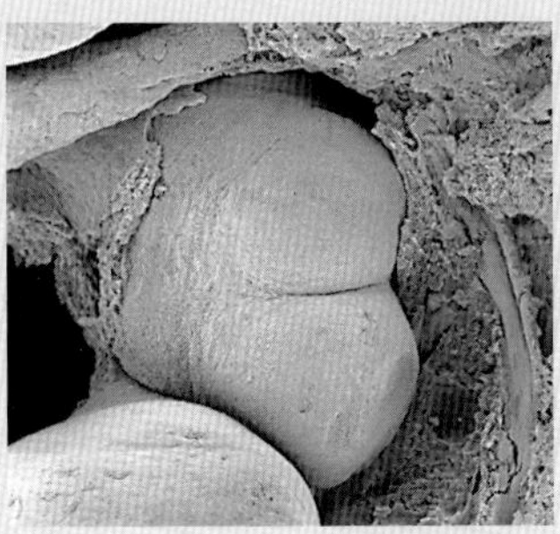

3.2.39 / 40

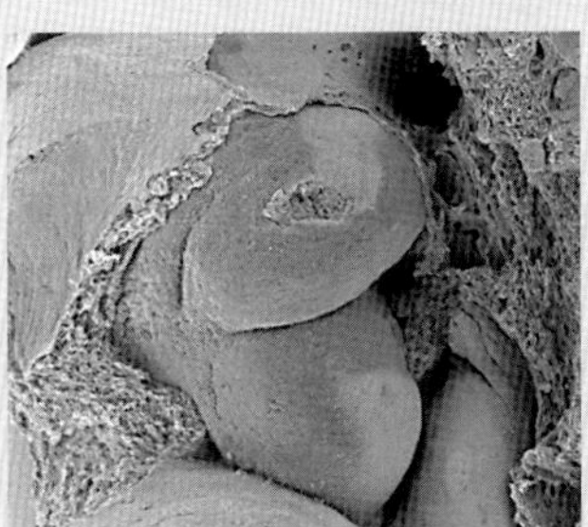

3.2.41 / 42

Embryo at week 7.

3.2.33–3.2.48

Development of the external form of the lungs.
The pictures on the left show the right lung in the lateral (right) view.
The pictures on the right show the left lung in the lateral (left) view.

Week 7

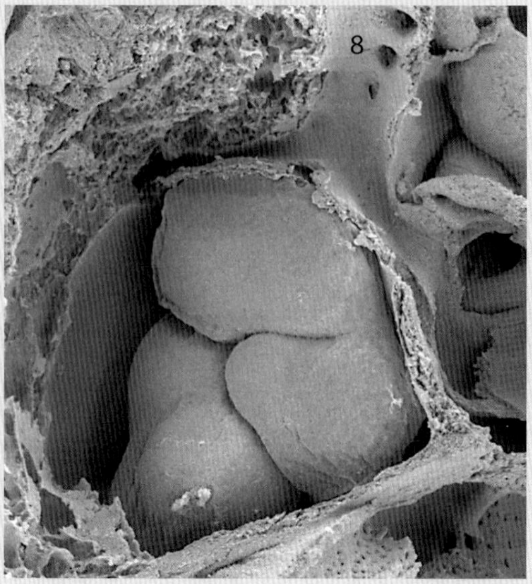

Week 8

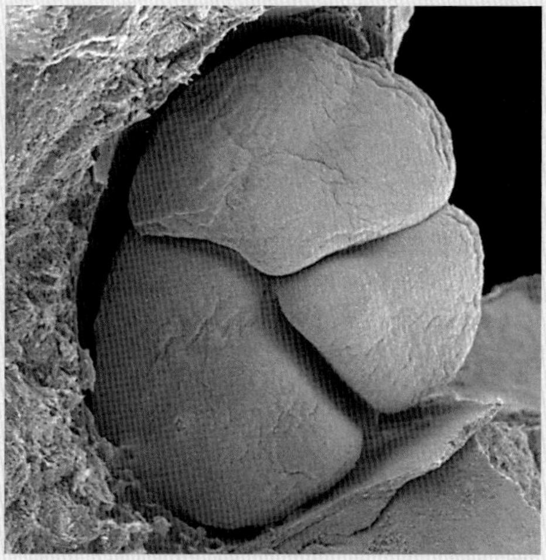

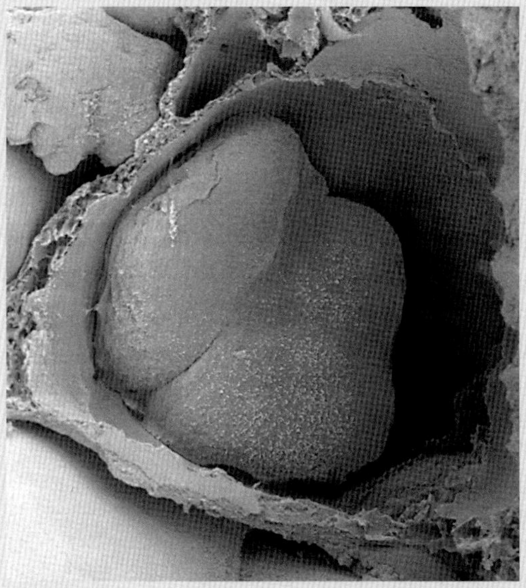

3.2.43 / 44 Embryo at week 7.

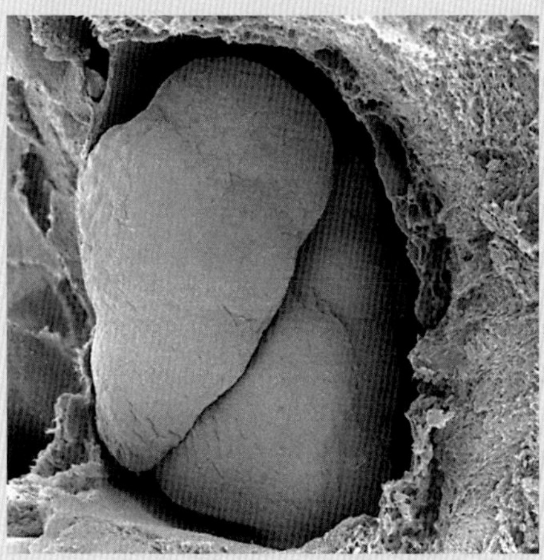

3.2.45 / 46 Embryo at week 8.

Week 9 / 10

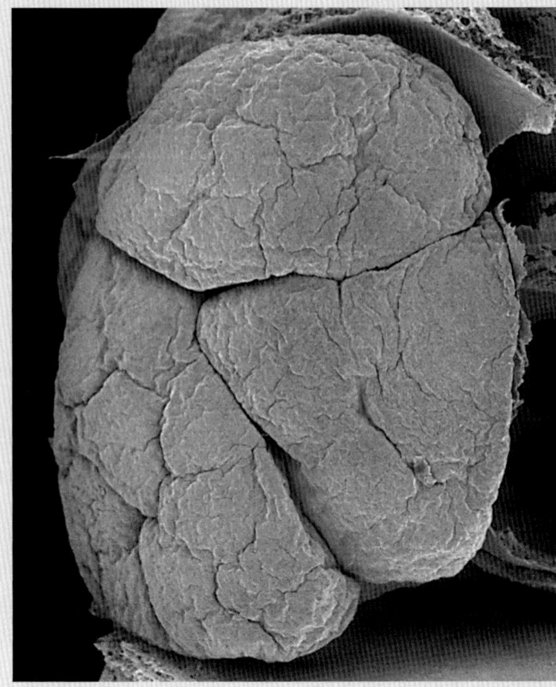

3.2.33–3.2.48

Development of the external form of the lungs.
The pictures on the left show the right lung in the lateral (right) view.
The pictures on the right show the left lung in the lateral (left) view.

3.2.47 / 48

3.2.47
Right lung. Lateral (right) view.
Embryo at week 9.

3.2.48
Left lung. Lateral (left) view.
Embryo late in week 9 or 10.

Week 7

Week 8

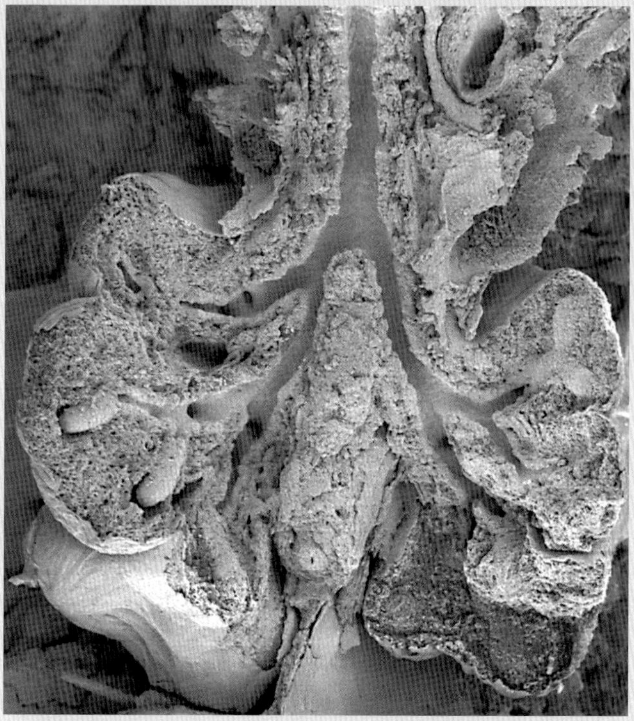

3.2.49–3.2.54
Early stages of alveolar development.

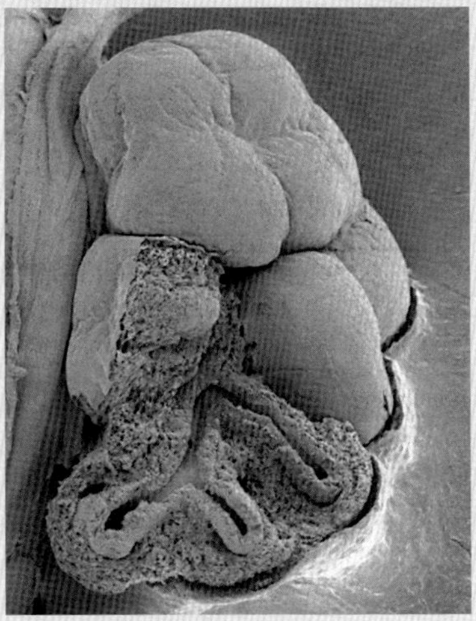

3.2.49 / 50

3.2.49
Ventral view of the first lobular alveoles.
Embryo at week 7.

3.2.50
Dorsal view of the right lung.
Embryo at week 7.

3.2.51 / 52 / 53

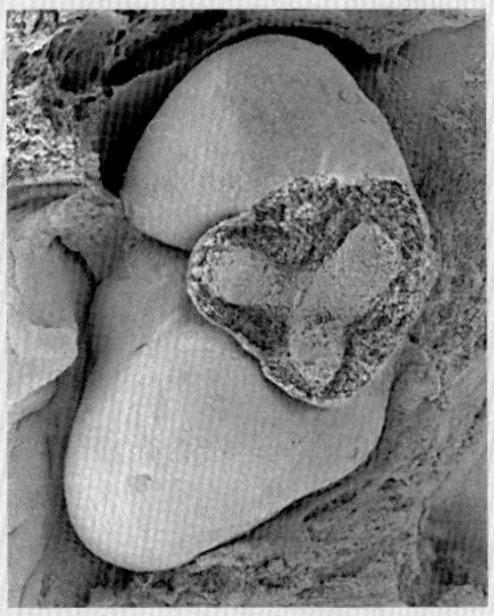

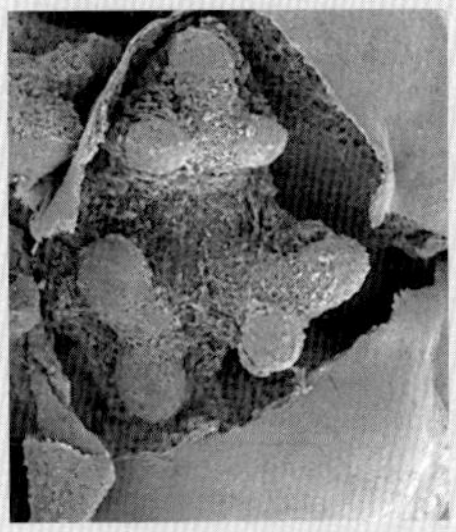

3.2.51
Ventral view of the lungs.
Embryo at week 8.

3.2.52
Lateral (right) view of the middle lobe of the right lung.
Embryo late in week 7.

3.2.53
Lateral (right) view of the middle lobe of the right lung.
Embryo at week 8.

3.2.54

Lateral (right) view of the superior and the middle lobe of the right lung.
Embryo at week 8.

30:1

3.2.55–3.2.63
Pleural cavity.

Week 4

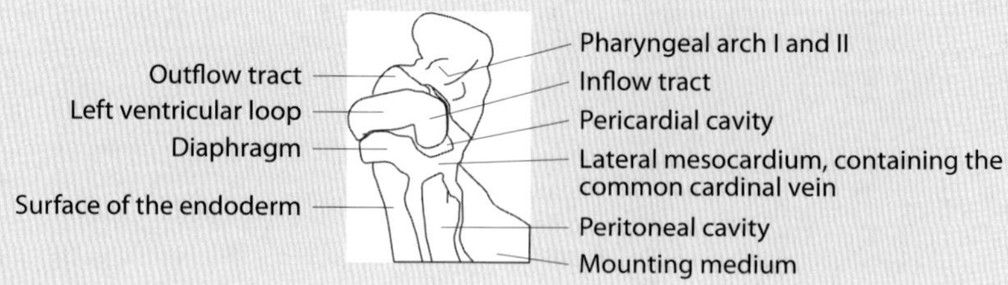

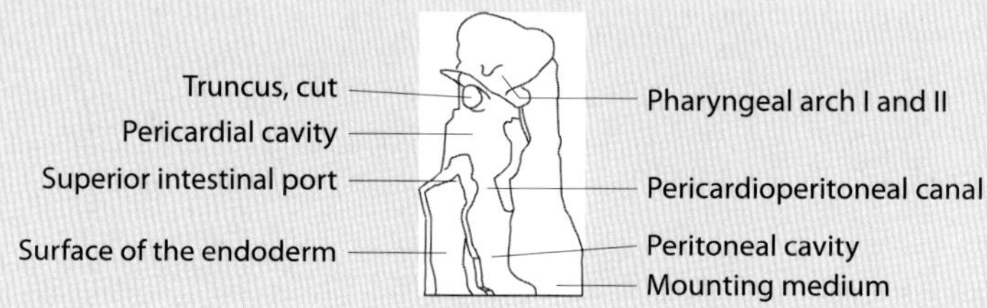

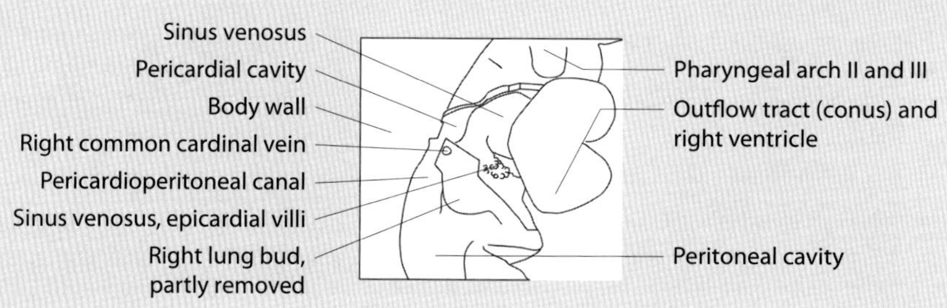

Week 5

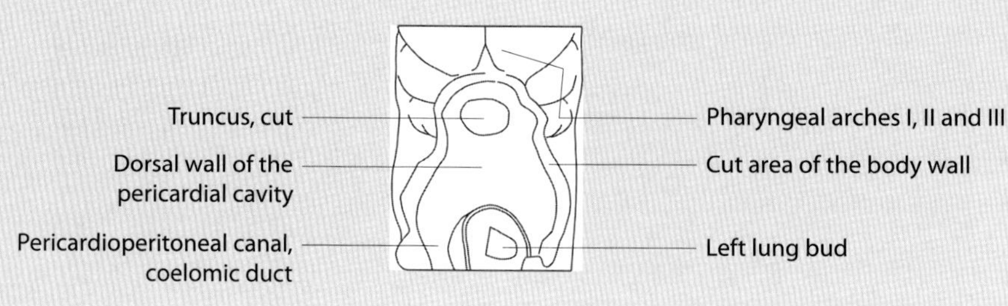

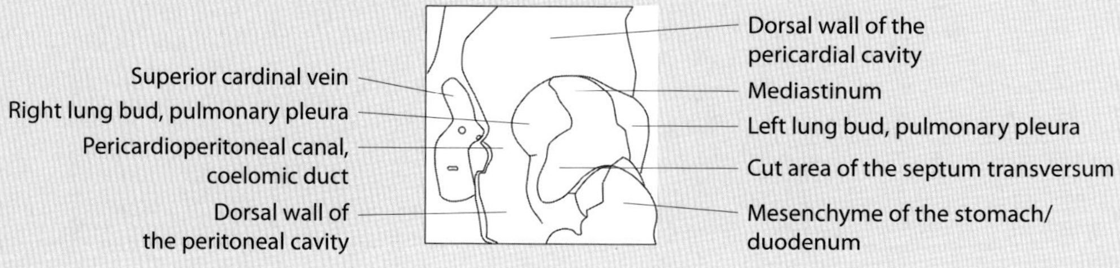

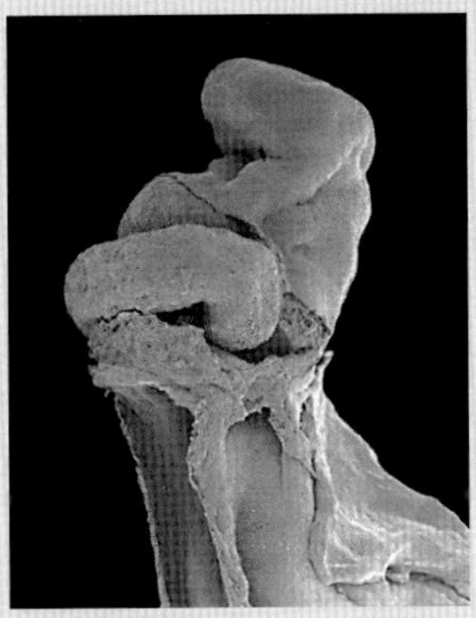

3.2.55 Lateral (left) view. The pericardial and coelomic cavities have been opened.
Embryo at week 4.

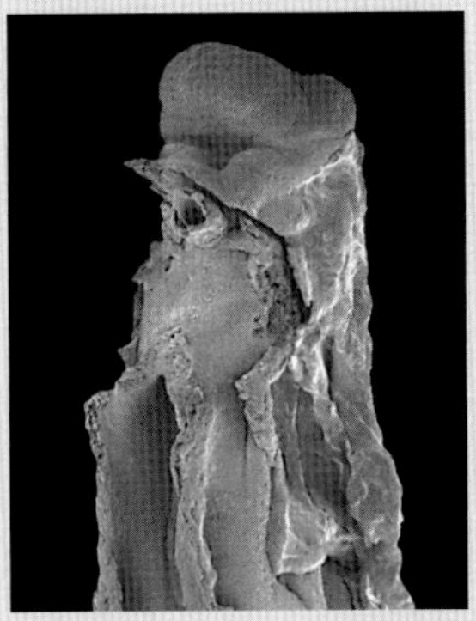

3.2.56 Same embryo as in figure 3.2.55. Ventral-left view. The heart has been removed. After the removal of the left common cardinal vein, the connection between the pericardial and the peritoneal cavities is visible: the left coelomic duct (pericardioperitoneal duct).

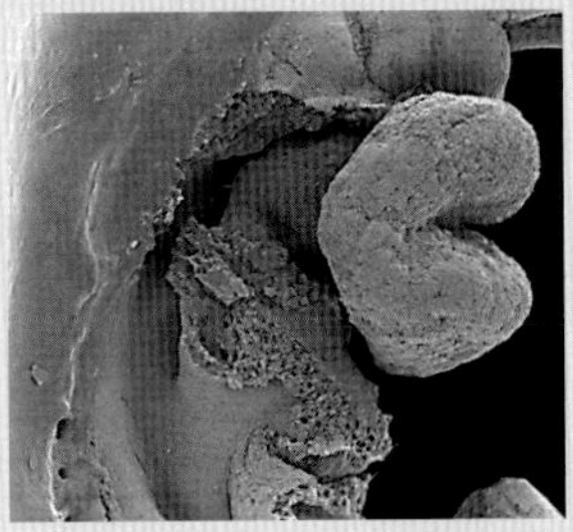

3.2.57 Lateral (right) view. The right common cardinal vein has been removed and the right coelomic duct in its positional relationship to the lung is shown.
Embryo at week 4.

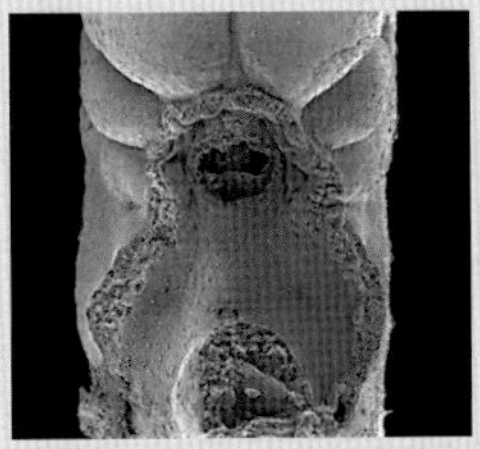

3.2.58 Ventral-right view. The heart and the common cardinal vein have been removed to show the positional relationship of the right lung to the coelomic duct.
Embryo at week 5.

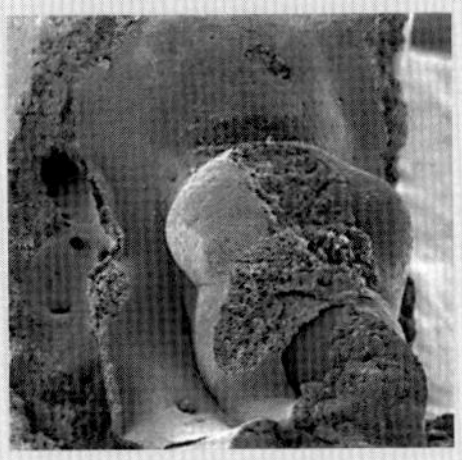

3.2.59

3.2.55–3.2.63
Pleural cavity.

Week 5

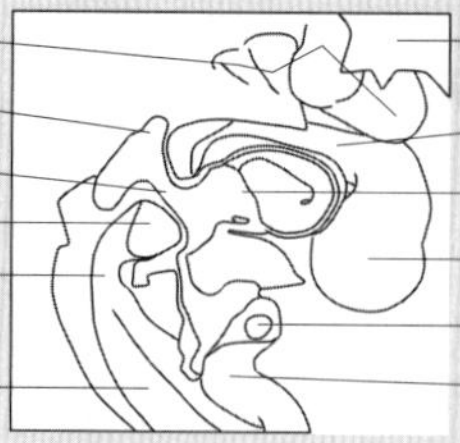

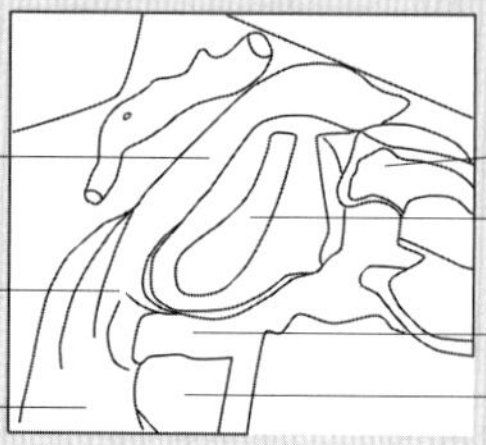

Week 6

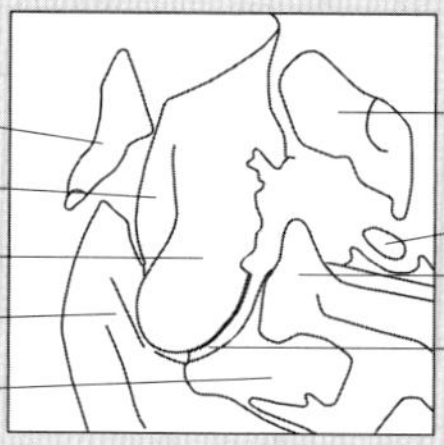

Week 8

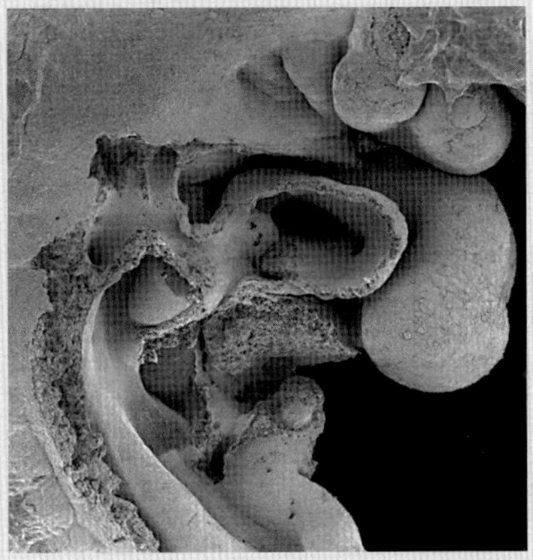

3.2.60 Lateral (right) view.
The part of the right lung that is not covered by the common cardinal vein.
Embryo at week 5.

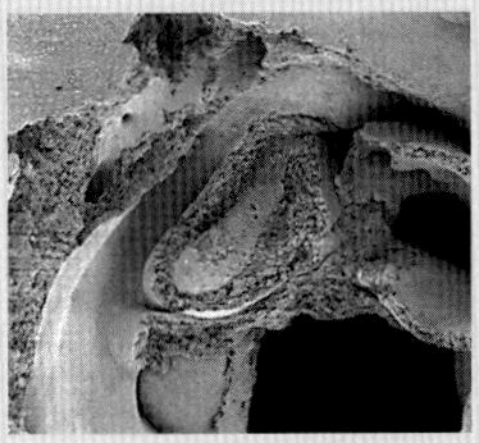

3.2.61 Lateral (right) view.
The common cardinal vein has been removed and the endoderm of the lung is exposed. The first indication of the fold separating the pleural from the peritoneal cavity is visible.
Embryo at week 5.

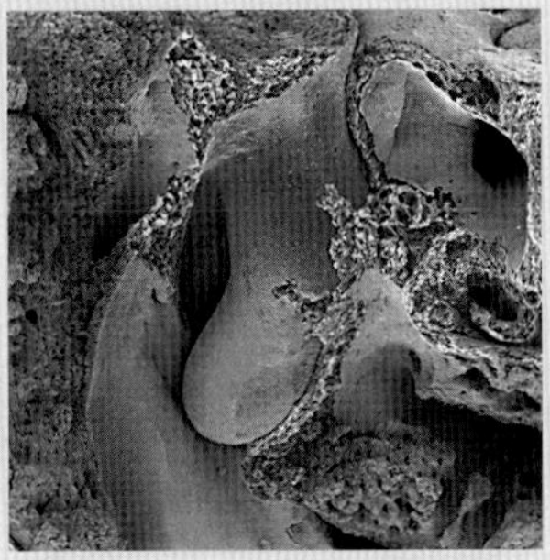

3.2.62 Ventral-right view.
After removal of the right common cardinal vein the pleural surface of the entire lung is visible.
Embryo at week 6.

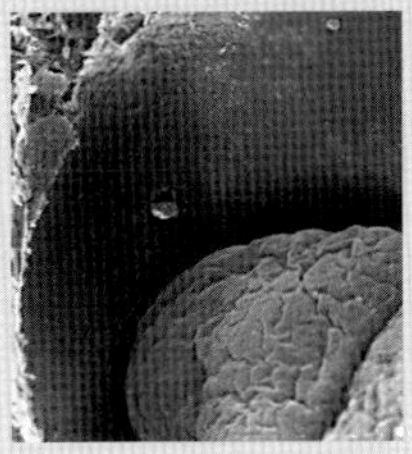

3.2.63 Cranial view of the floor of the left pleural cavity (diaphragm). A tiny communication between pleural and peritoneal cavities is still present.
Embryo at week 8.

4
Organs in the Abdomen

4.1 The Development of the Small Intestine

The early embryonic endoderm is formed by that part of the endoderm of the umbilical vesicle that abuts the embryo proper. This endodermal plate is folded in from the endoderm of the umbilical vesicle and, in contact with the embryo, forms a sort of valley, the intestinal groove (fig. 4.1.01). Due to the descent of the superior intestinal port and the ascent of the inferior intestinal port, the foregut and the hindgut are delimitated from the midgut which gives rise to the stomach, the small and partly the large intestine.

The intestinal groove deepens (fig. 4.1.02–4.1.04) and forms a tube; the initially wide communication with the umbilical vesicle thus narrows to the omphalo-enteric duct by which the intestinal lumen communicates with the lumen of the umbilical vesicle (fig. 4.1.03–4.1.05). The endoderm of the midgut is covered by the peritoneal epithelium (fig. 4.1.06) except in its dorsal circumference where the peritoneum switches over to the dorsal wall of the peritoneal cavity, the embryonic coelom, thus forming the dorsal mesentery (fig. 4.1.07, 4.1.12, 4.1.17).

In week 5, the gut begins to elongate and bends consecutively in left-right and ventral-dorsal directions (fig. 4.1.07a–4.1.09). The portion of the gut that grows considerably ventrad forms the primary intestinal loop and reaches a position in the coelom of the umbilical cord (fig. 4.1.10, 4.1.16). The omphalo-enteric duct opens at the apex of the loop (fig. 4.1.10).

Due to increasing elongation, the small intestine achieves its tangled appearance (fig. 4.1.17–4.1.19). The relocation of the small intestine into the intraembryonic coelom is realized approximately during week 9 or 10.

4.1

Abbreviations

il	intestinal loop
lv	left ventricle
mn	mesonephros
ra	right atrium
rv	right ventricle
s	stomach
2	liver, left lobe
3	liver, right lobe
4	bare area of the liver
5	lung
6	bulge of the kidney
7	genital tubercle

Week 4

4.1.01–4.1.19
Developmental stages of the gut.

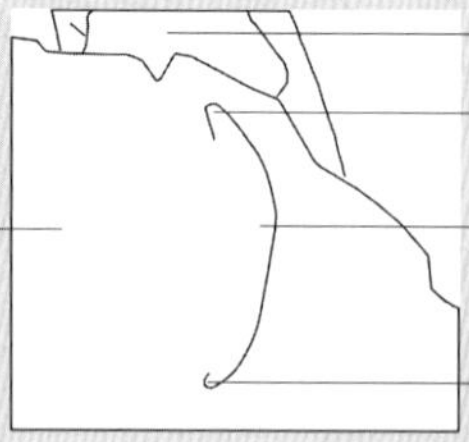

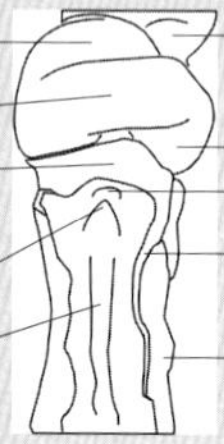

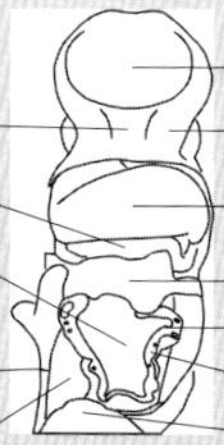

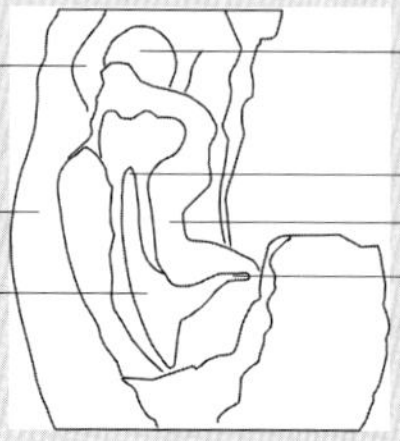

4.1.01

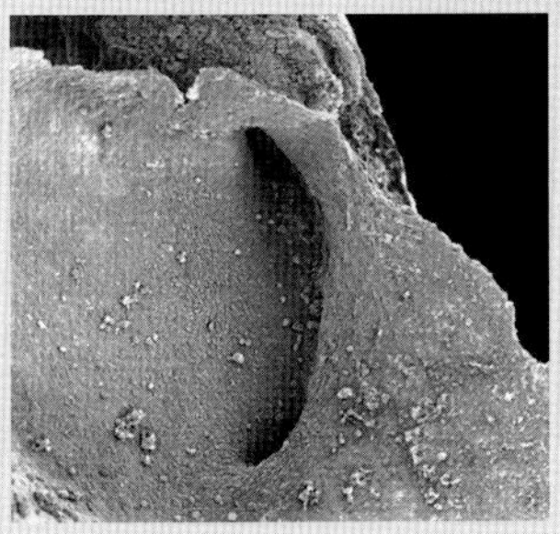

Closure of the endodermal gut plate forms the intestinal tube. Different developmental stages in embryos at week 4. Ventral view.

4.1.02

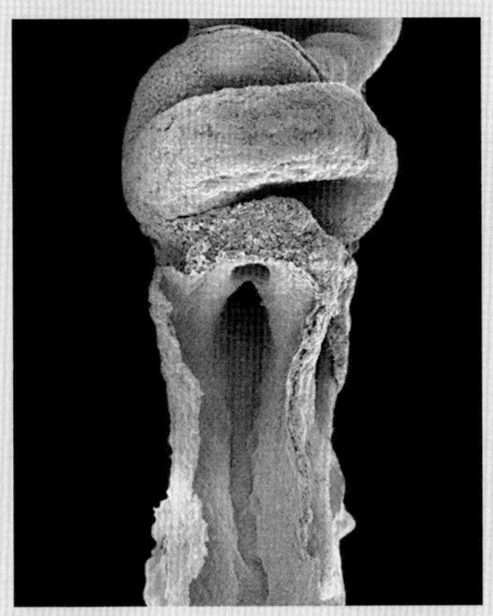

4.1.03

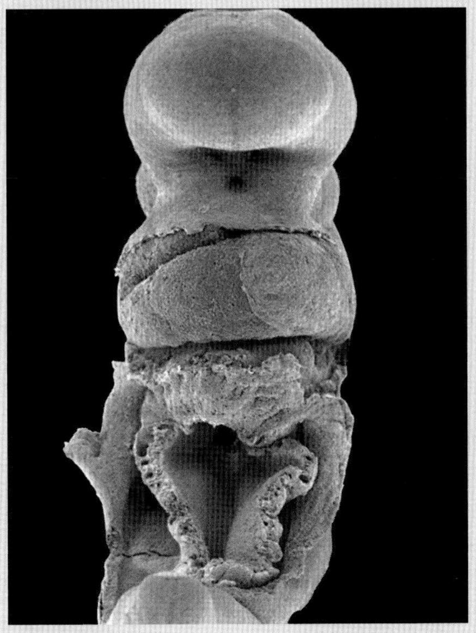

4.1.04

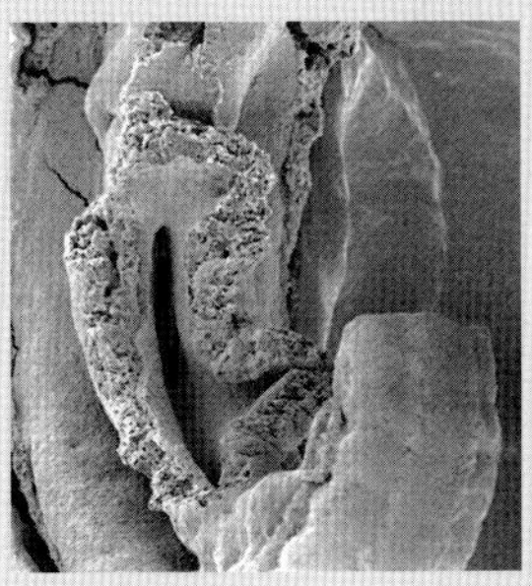

4.1.01–4.1.19
Developmental stages of the gut.

Week 4

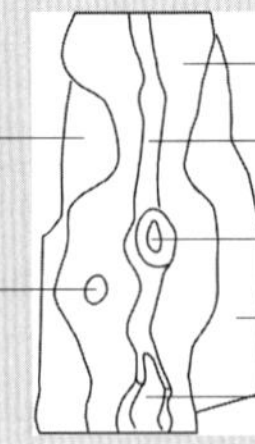

Week 5

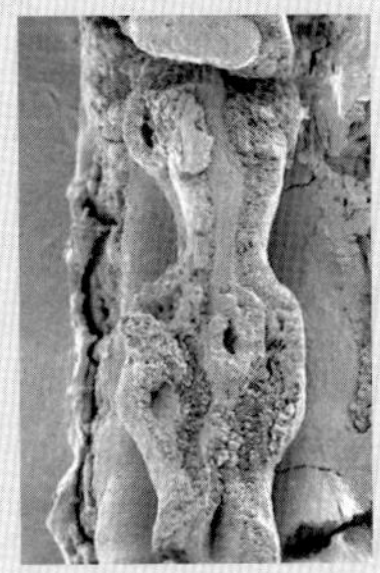

4.1.05 Closure of the endodermal gut plate forms the intestinal tube. Different developmental stages in embryo at week 4. Ventral view.

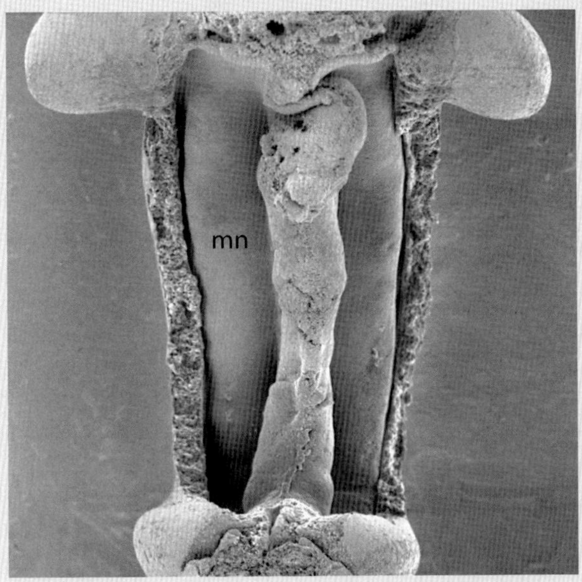

4.1.06 Ventral view of the peritoneal cavity. The liver has been removed. Embryo at week 5.

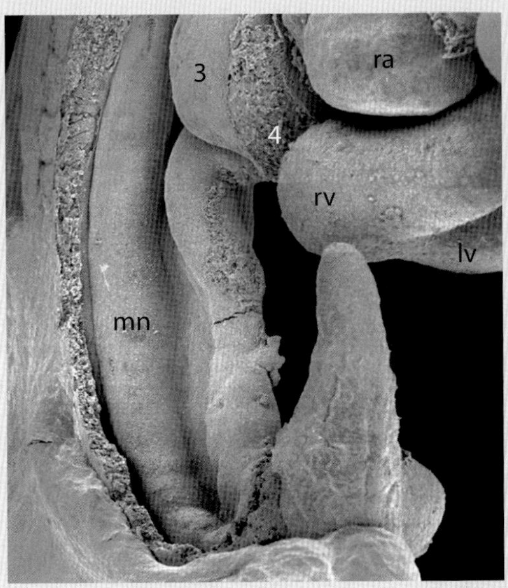

4.1.07 Ventral right view of the first bend of the intestinal tube. Embryo at week 5.

Week 5

4.1.01–4.1.19
Developmental stages of the gut.

Week 5 / 6

4.1.07a

Same embryo as in figure 4.1.07.

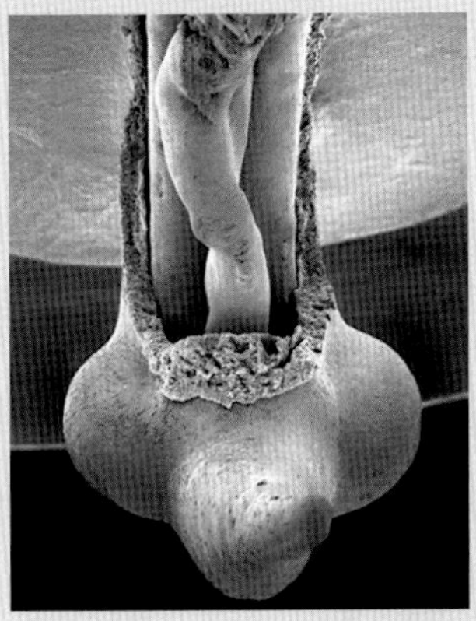

4.1.08

Ventral and ventral-right views of an embryo at the end of week 5.

4.1.09

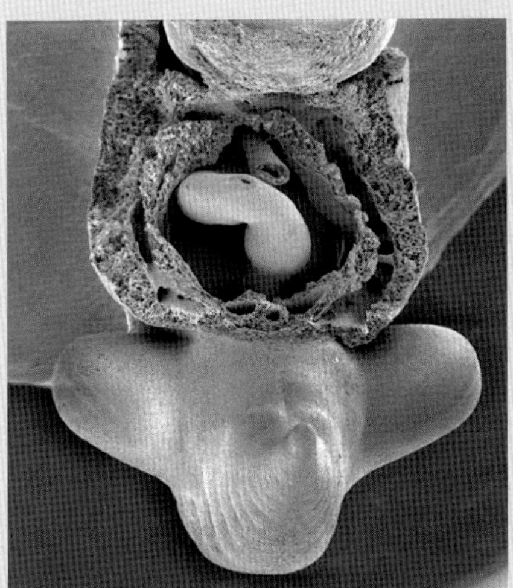

4.1.10

Ventral view of the intestinal loop. Embryo at the end of week 5/ beginning of week 6.

4.1.01–4.1.19
Developmental stages of the gut.

20:1

Week 5 / 6

Week 6

Week 7

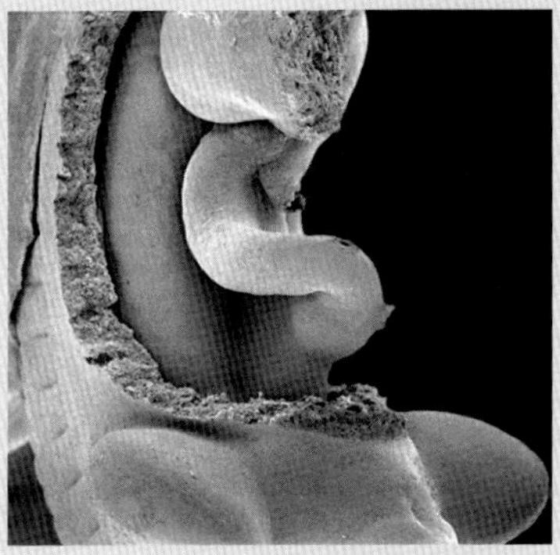

4.1.11 Same embryo as in figure 4.1.10. Ventral-right view. The body wall has been removed.

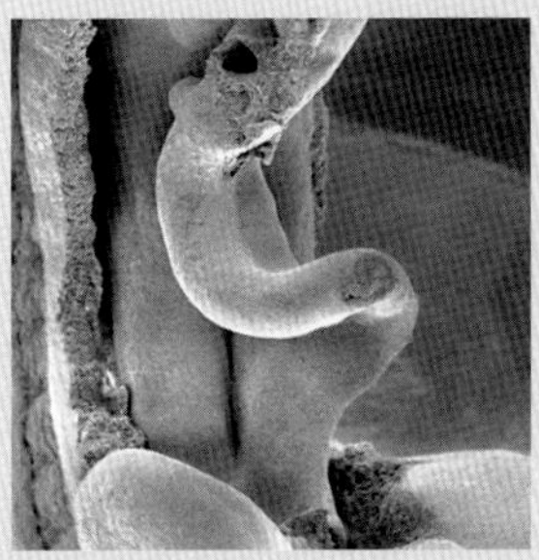

4.1.12 Ventral-right view of the intestinal loop. The liver has been removed. Embryo at week 6.

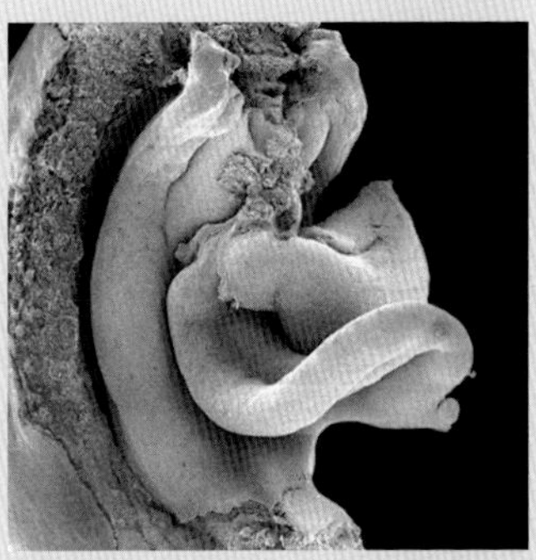

4.1.13 Ventral-right view of the intestinal loop and the stomach. The liver has been removed. Embryo at the end of week 6.

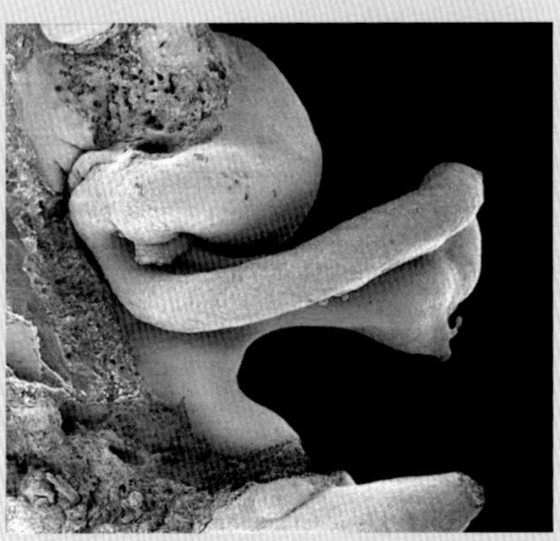

4.1.14 Ventral-right view of the intestinal loop and the stomach. The liver has been removed. Embryo at week 7.

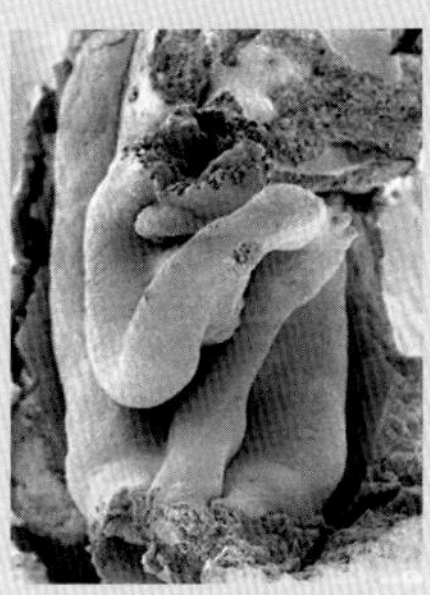

4.1.15 Ventral view of the intestinal loop. Embryo at week 7.

Week 7

4.1.01–4.1.19
Developmental stages of the gut.

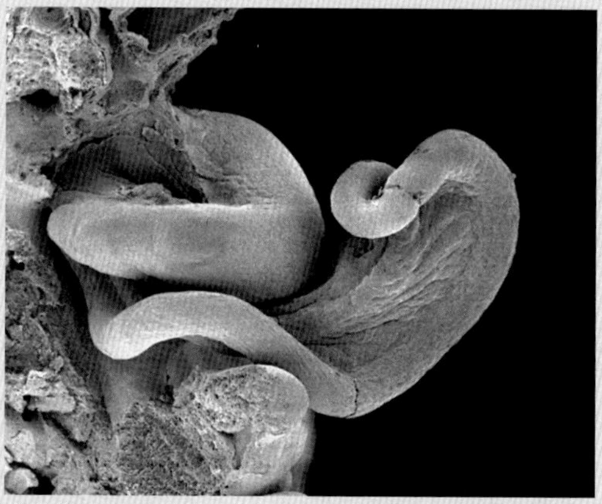

Week 9

4.1.16

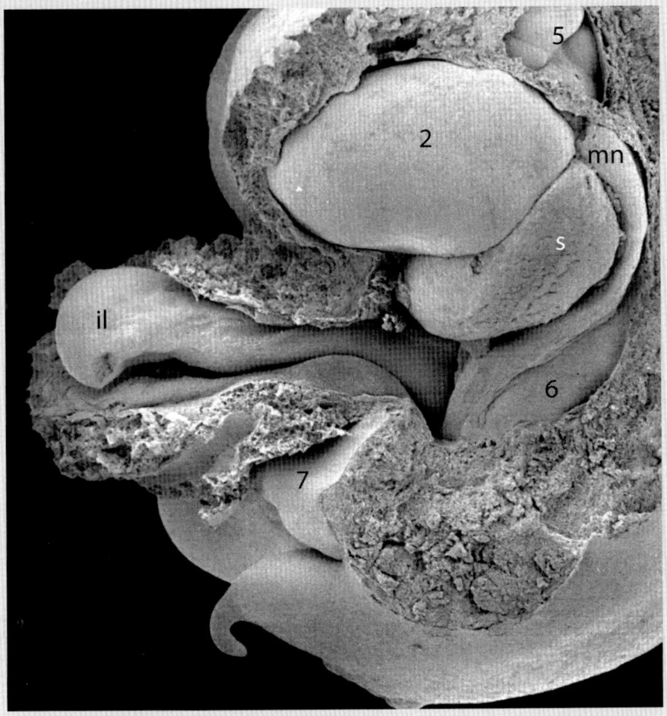

Lateral-left view of the intestinal loop, liver, stomach, and mesonephros. Embryo at week 7.

4.1.17 / 18

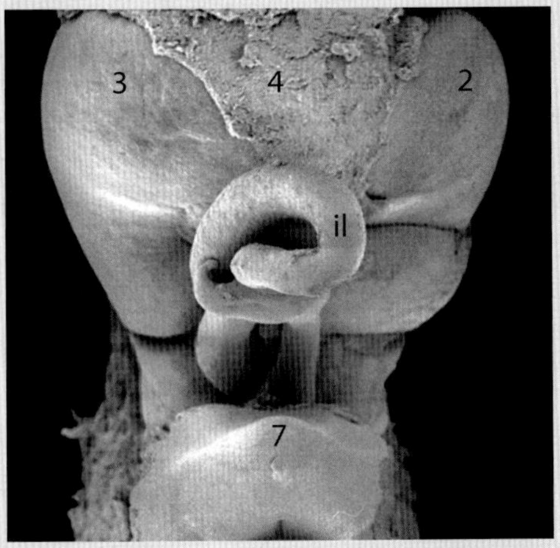

4.1.17
Ventral-right view of the intestinal loop, duodenum, and stomach. The liver has been removed. Embryo at week 7.

4.1.18
Ventral view of the intestinal loop and the liver. Embryo at week 7.

4.1.19

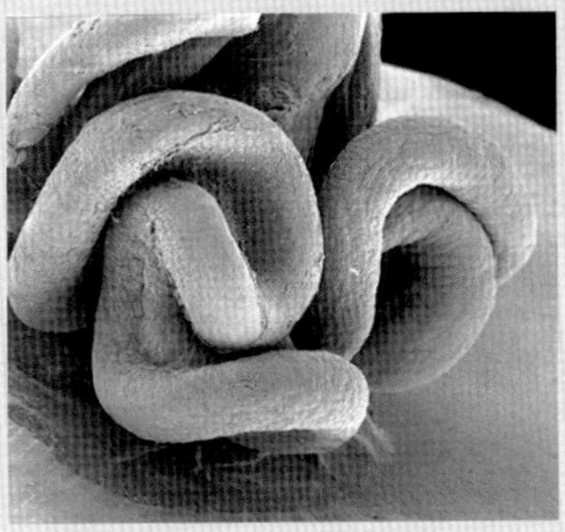

Ventral view of the small intestine. Embryo at week 9.

4.2 The Development of the Liver

The first anlage of the liver arises as an endodermal invagination into the mesenchyme of the septum transversum just above the anterior intestinal port (fig. 4.2.01–4.2.03). The endoderm of this hepatic diverticulum forms cellular cords invading the mesenchyme of the septum transversum. The parenchymal cords together with the mesenchyme and the covering coelomic epithelium form two bulges extending into the coelomic cavity (fig. 4.2.05). These primarily very small bulges increase rapidly in size and appear as the right and the left lobes of the liver (fig. 4.2.06–4.2.11).

The liver receives blood mainly from the left umbilical vein. Primarily the left umbilical vein drains into the left common cardinal vein (fig. 4.2.12), but during week 5 the left umbilical vein connects directly to the liver and the suprahepatic portion of the left umbilical vein atrophies (fig. 4.2.13, 4.2.14). Later on, the liver additionally receives blood from the portal vein (fig. 4.3.11–4.3.13).

Within the liver, between the cellular cords, the blood-conducting sinusoids appear (fig. 4.2.20–4.2.22) which are supplied by the afferent vessels mentioned above. Two types of greater veins are formed by confluent sinusoids: superficial veins at the periphery of the lobes (fig. 4.2.15) which eventually will form the hepatic veins and their tributaries; in the centre of the lobes the central veins develop (fig. 4.2.18) which receive their blood via the sinusoids from the umbilical and the portal vein. In short, the blood from the portal vein and partly from the left umbilical vein flows via their branches and the sinusoids and the central veins into the hepatic veins which initially empty into the right atrium of the heart (fig. 4.2.16, 4.2.17). Later on, the confluence of the hepatic veins is incorporated into the newly formed inferior caval vein (fig. 4.2.25, 4.2.26).

From the orifice of the left umbilical vein, a large vein develops which drains directly into the converging hepatic veins (fig. 4.2.16) and into the inferior caval vein, respectively (fig. 4.2.26). However, this ductus venosus gives rise to a multitude of small veins (fig. 4.2.23) which drain into sinusoids and eventually form branches of the portal system. The ductus venosus atrophies after birth.

4.2

Abbreviations

bw	cut edge of the body wall
il	intestinal loop
l	lung
la	left atrium
lv	left ventricle
mn	mesonephros
ra	right atrium
rv	right ventricle
s	stomach
1	liver, right lobe
2	bare area of the liver
3	liver, left lobe
4	umbilical vein, orifice
5	portal vein
6	subcardinal-hepatic anastomosis
11	duodenum
12	genital tubercle
13	hepatic vein/hepatic portion of inferior caval vein
14	venous duct

Week 4

4.2.01–4.2.05
Early stages of liver development.

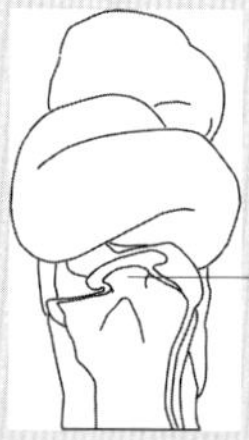

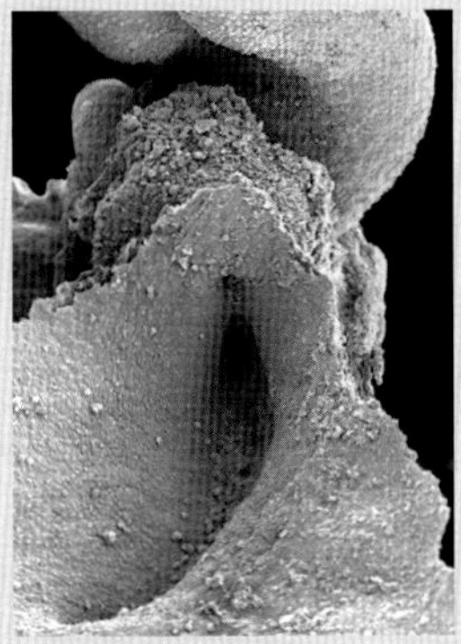

4.2.01 Ventral view of the midgut plate and the orifice of the hepatic diverticulum. Embryo early in week 4.

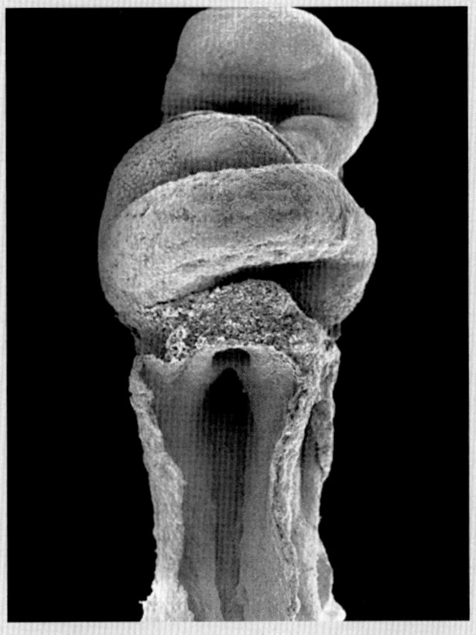

4.2.02 Ventral view of the more folded midgut plate and the orifice of the hepatic diverticulum. Embryo at week 4.

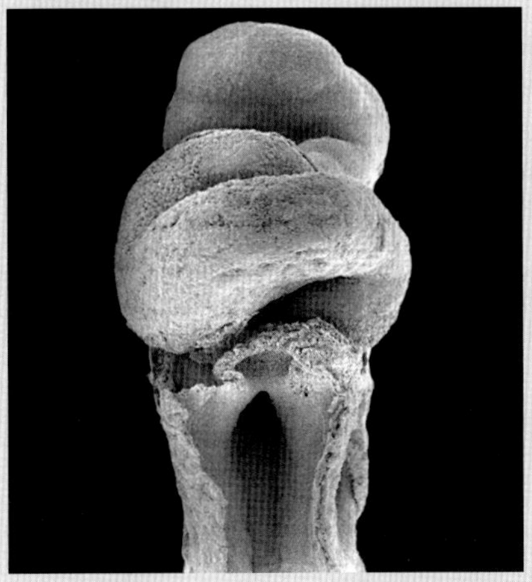

4.2.03 Same embryo as in figure 3.2.01. The hepatic diverticulum has been opened.

Week 5

Week 4

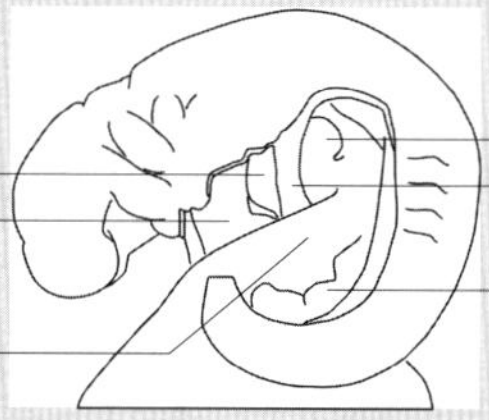

4.2.01–4.2.05
Early stages of liver development.

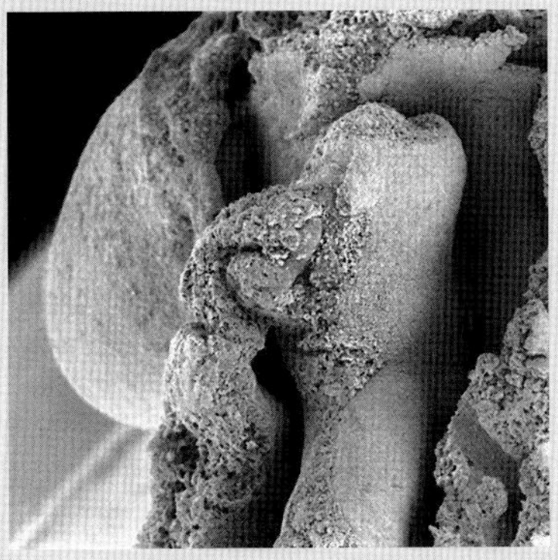

4.2.04 Ventral view of the endodermal hepatic bud. The adjacent mesenchyme of the transverse septum has been removed. Embryo at week 5.

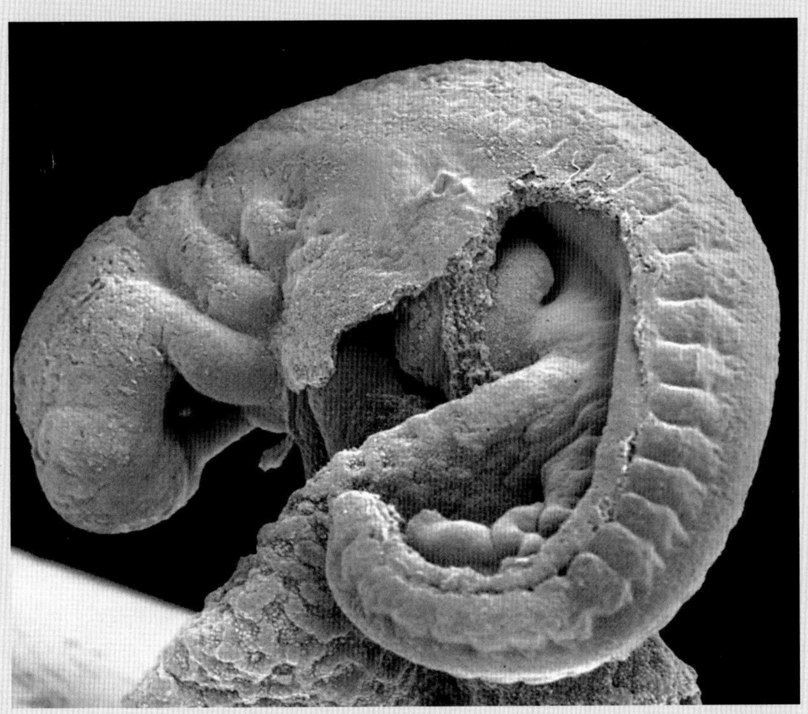

4.2.05 Left and caudal view of the peritoneal cavity showing the external appearance of the left lobe of the liver. Embryo at the end of week 4.

25:1

4.2.5a–4.2.11
Development of the external form of the liver. Ventral view.

Week 4

Week 5

Week 6

Week 7

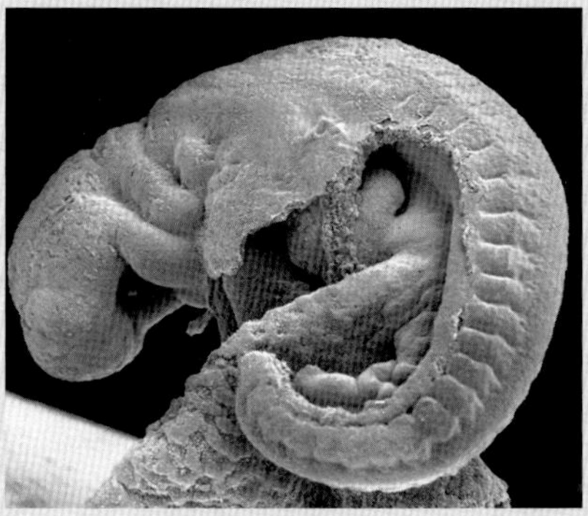

4.2.05a Same embryo as in figure 4.2.05.

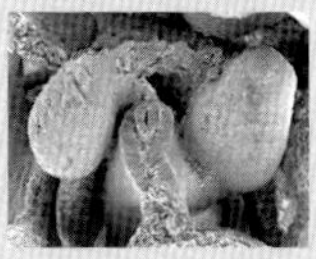

4.2.06 Right and left lobes of the liver. Embryo at week 5.

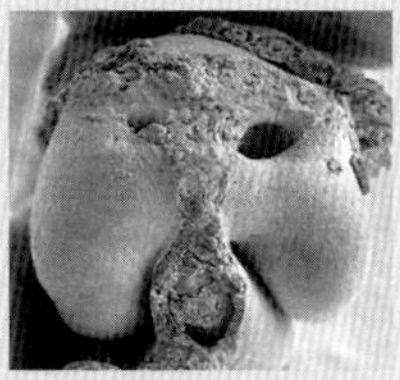

4.2.07 Right and left lobes of the liver. Embryo at the end of week 5.

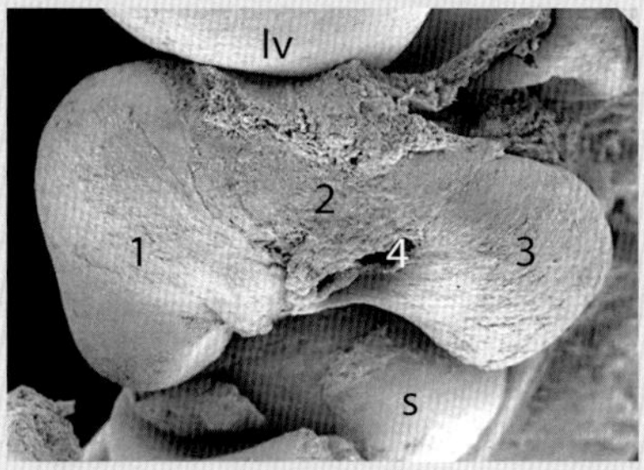

4.2.08 The liver now appears as an unpaired organ. Embryo at week 6.

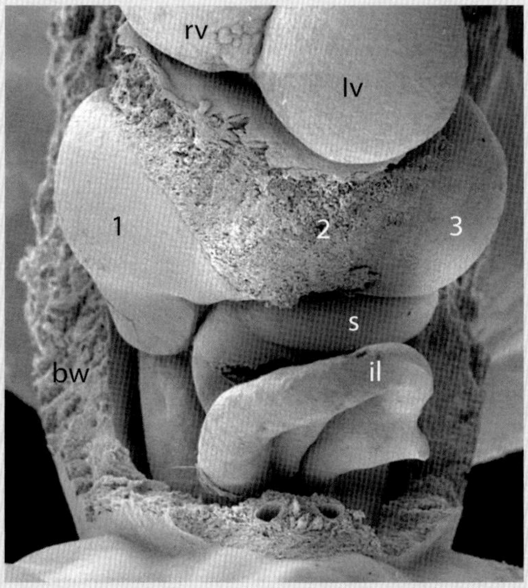

4.2.09 Embryo at week 7.

Week 8

4.2.5a–4.2.11

Development of the external form of the liver. Ventral view.

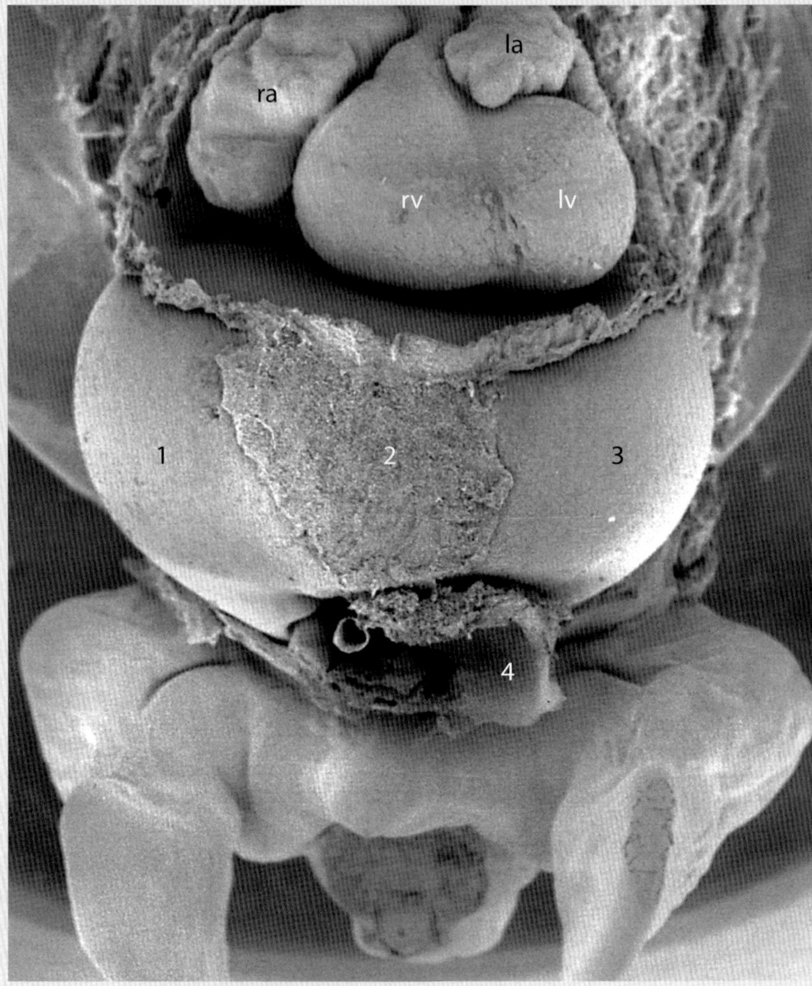

4.2.10 / 11

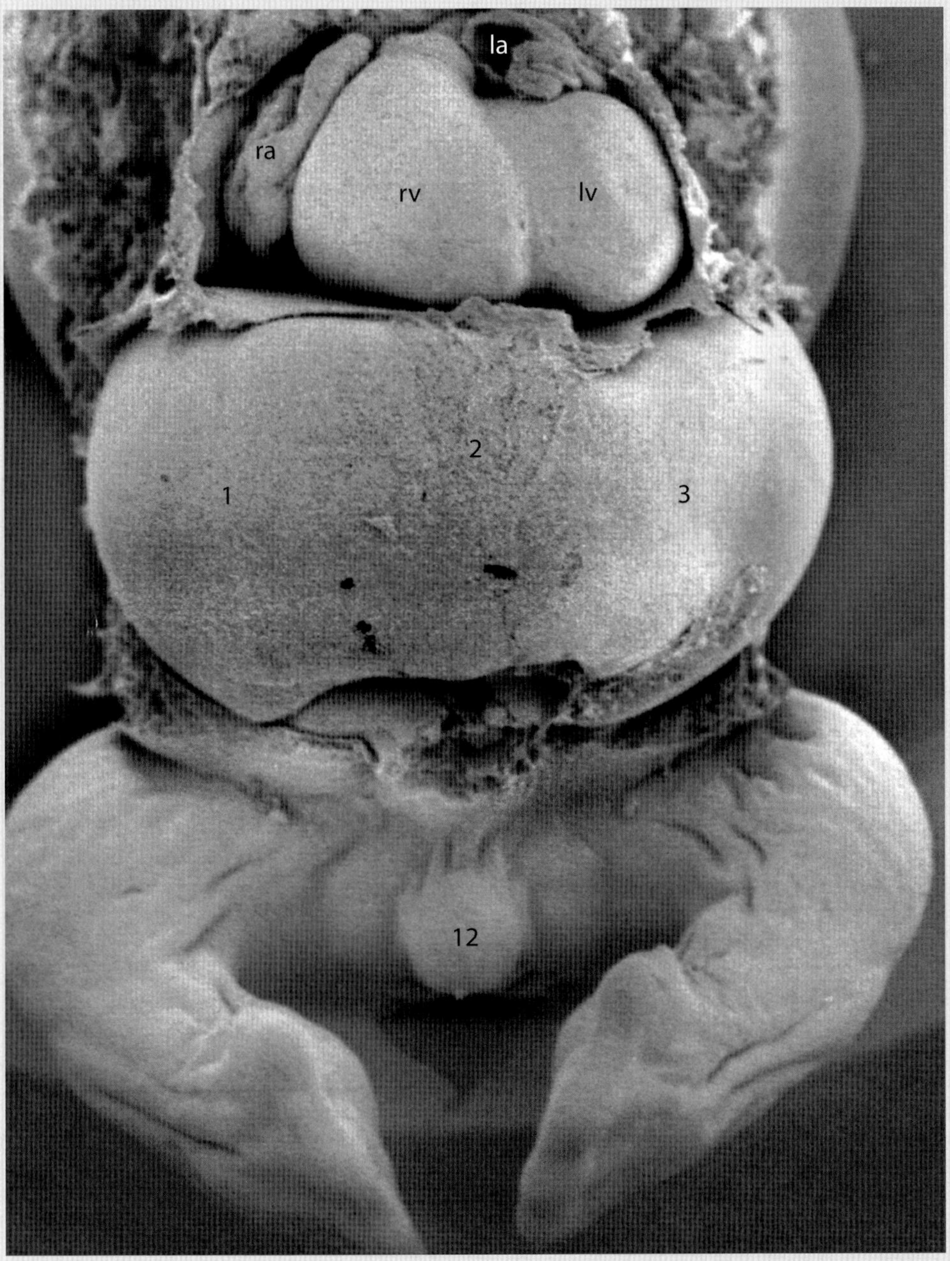

4.2.10
Embryo at week 8.

4.2.11
Embryo late in week 8.

4.2.12–4.2.14
The left umbilical vein. Lateral/ventral-left view.

Week 5

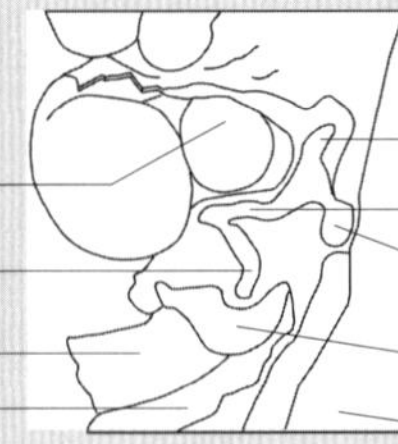

Week 6

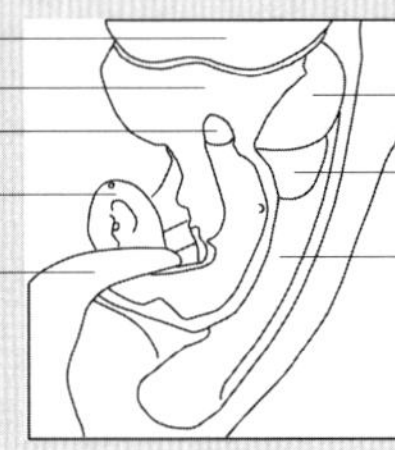

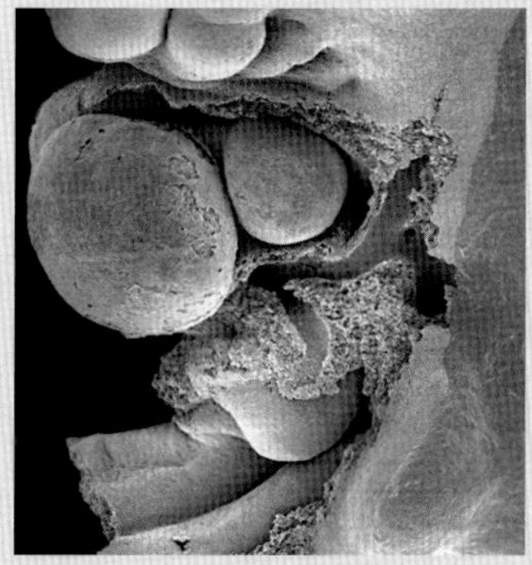

4.2.12 Embryo at week 5. The left umbilical vein drains into the sinus venosus via the left common cardinal vein. Ventral-left view.

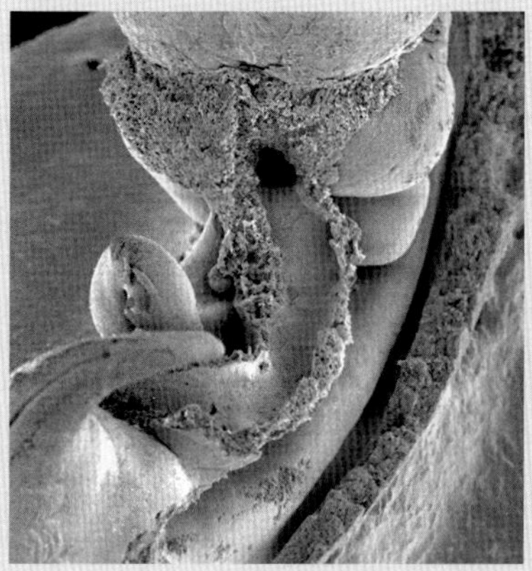

4.2.13 Embryo at week 6. The left umbilical vein has lost its connection to the sinus venosus and drains into the liver. Ventral-left view.

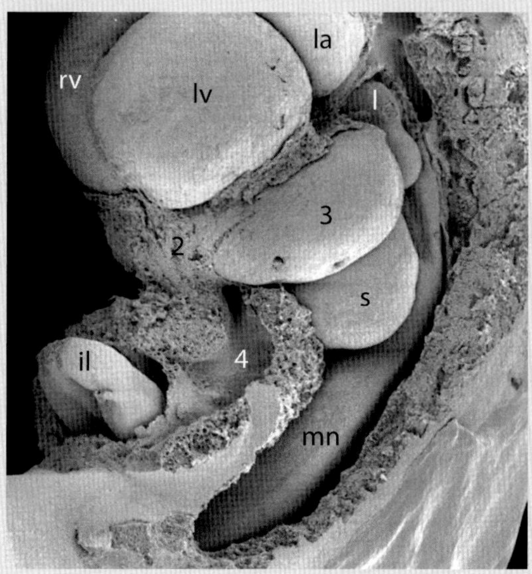

4.2.14 Embryo at the end of week 6.

4.2.15–4.2.19
**The intrahepatic veins.
Ventral-right view.**

Week 6

Week 7

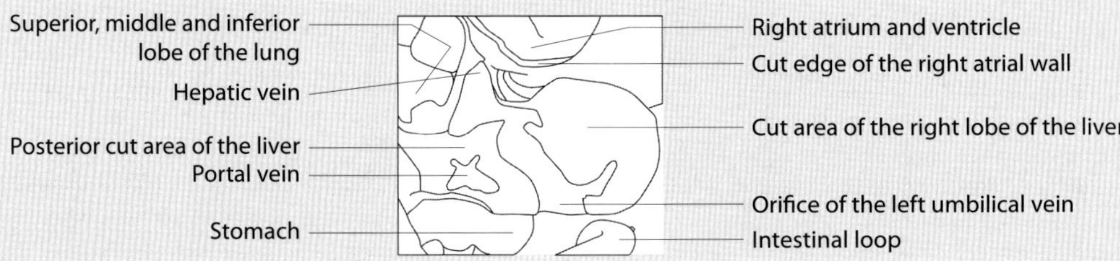

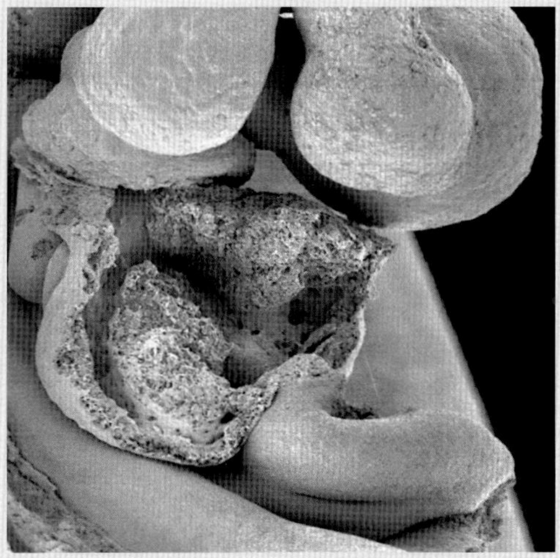

4.2.15 Embryo at the end of week 6.

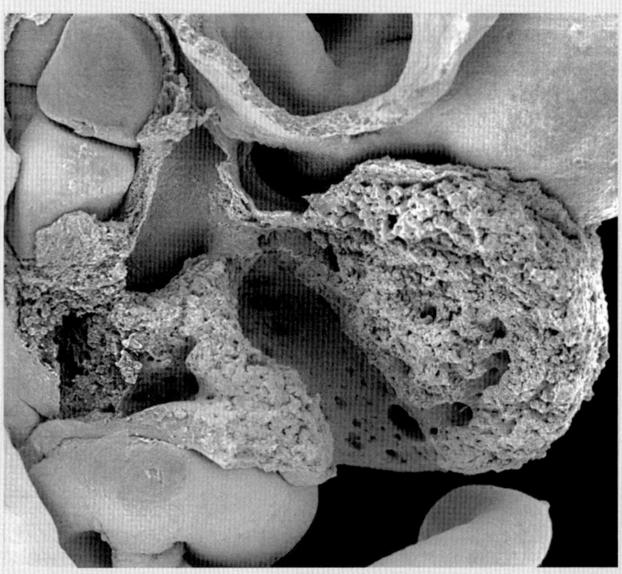

4.2.16 Embryo at week 7.

4.
Organs in the Abdomen

4.2
The Development of the Liver

4.2.15–4.2.19
The intrahepatic veins. Ventral-right view.

40:1

Week 8

25:1

Week 7

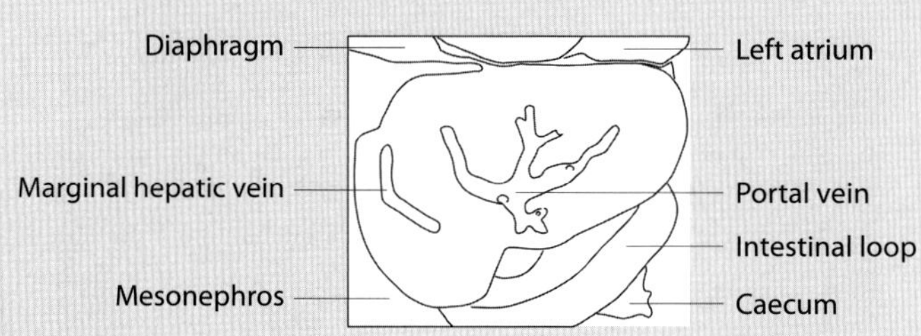

Week 5

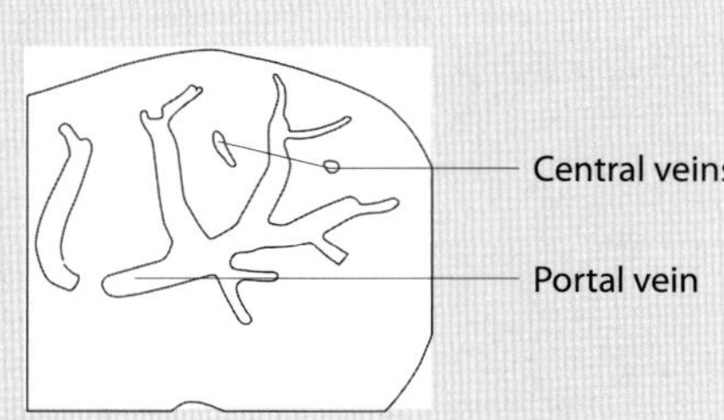

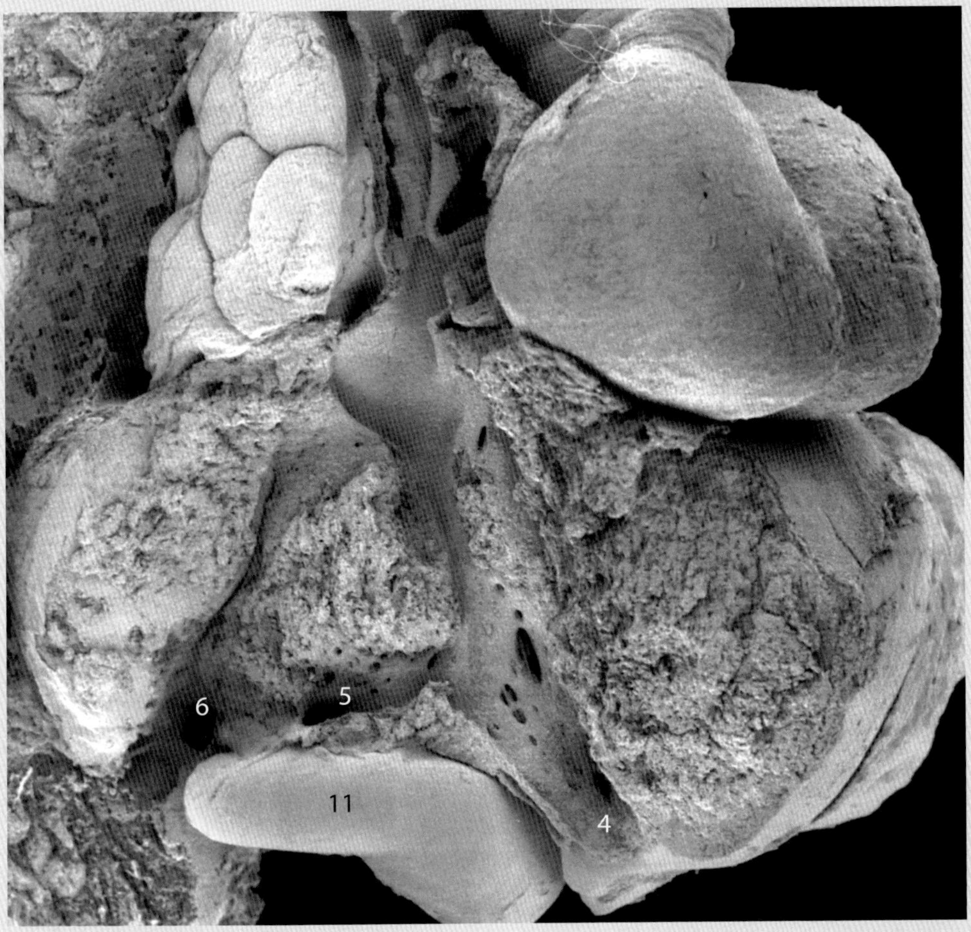

4.2.17 Embryo at week 8.

4.2.18 Ramifications of the hepatic portal vein are shown. Embryo at week 7.

4.2.19 Early arrangements of the hepatic lobules with the branches of the hepatic portal vein and the central veins. Embryo late in week 5.

Week 6

Week 7

4.2.20–4.2.23
**Structure of the liver.
Ventral view.**

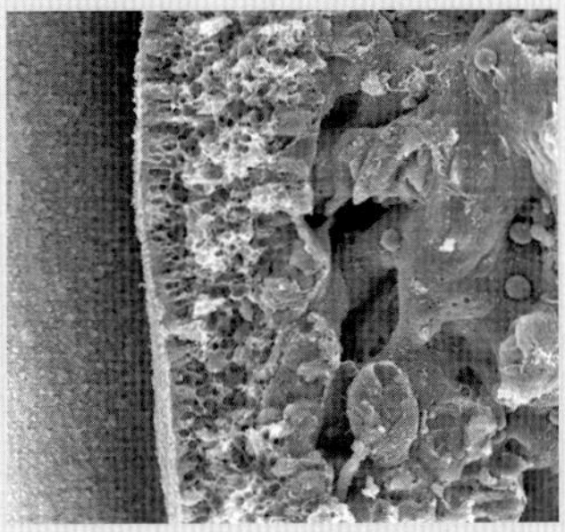

4.2.20

Surface area of the right lobe of the liver: hepatic peritoneum and parenchyme of the liver. Embryo at week 6.

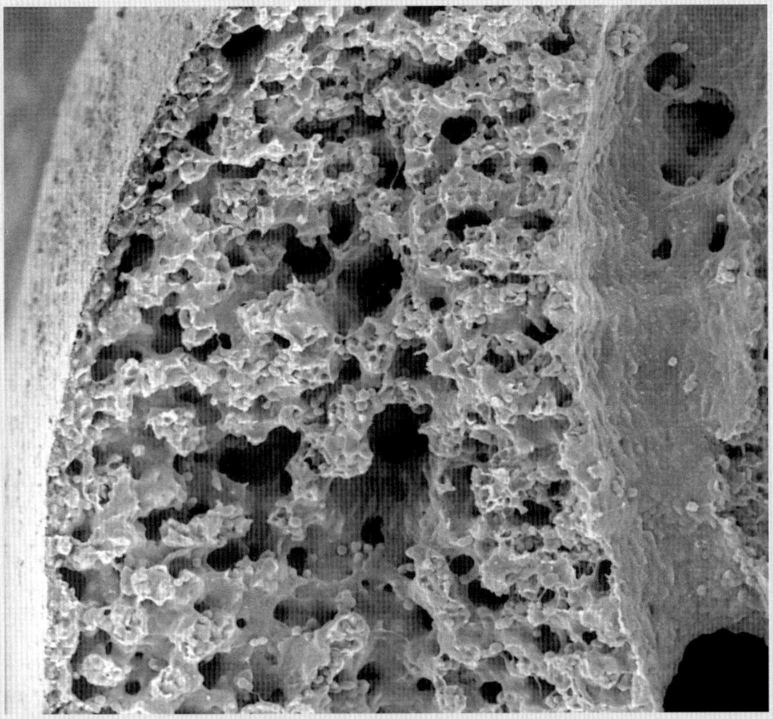

4.2.21 / 22

4.2.21
Surface area of the right lobe of the liver: hepatic peritoneum and parenchyme and right marginal branch of the portal vein. Embryo at week 7.

4.2.22
Region corresponding to that in figure 4.2.21. Embryo at the end of week 7.

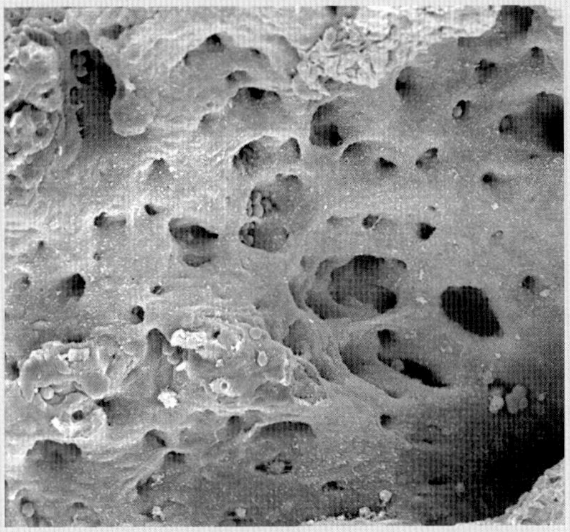

4.2.23

Surface of the venous duct showing the orifices of the venae advehentes. Embryo at week 7.

4.2.24–4.2.26
The inferior caval vein in the liver. Lateral-right view.

Week 5

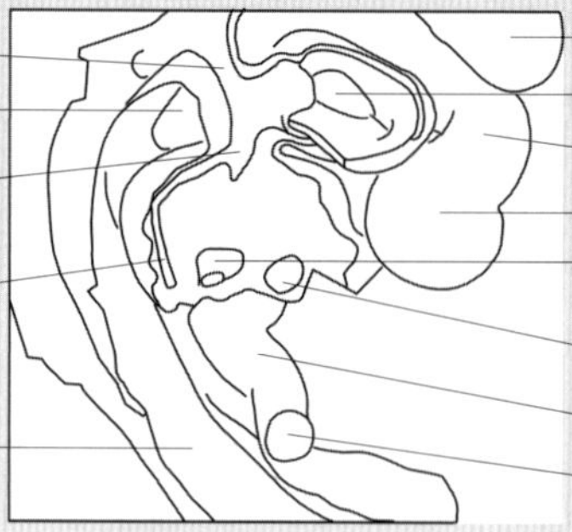

Week 6

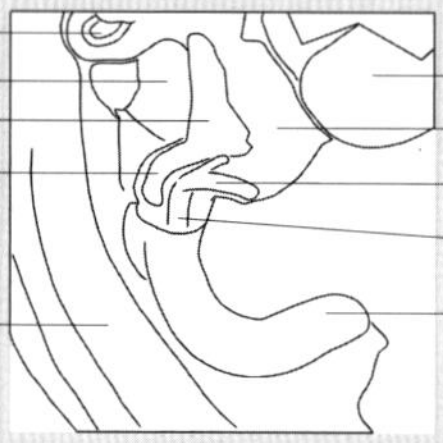

Week 8

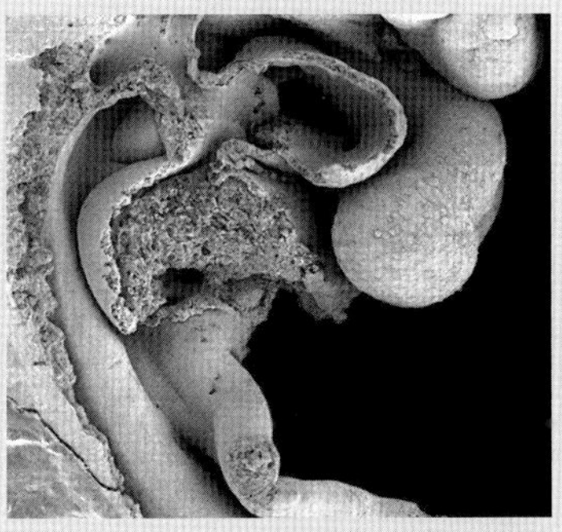

4.2.24 Embryo at week 5. Lateral-right view. The right lobe of the liver has been partly removed to show the hepatic portal vein and its tributary from the subcardinal vein (hepatic portion of the subcardinal vein).

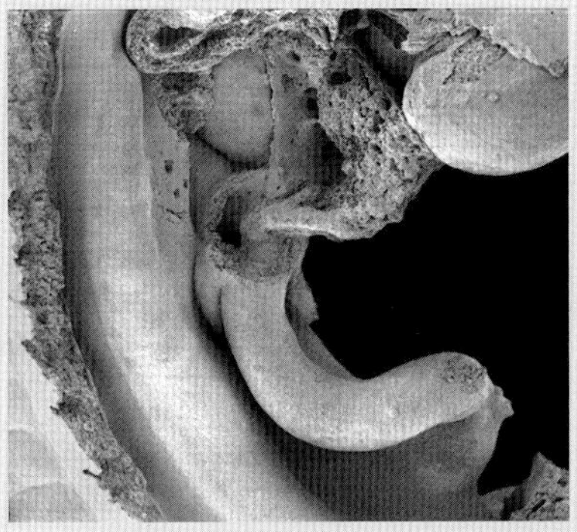

4.2.25 Embryo at week 6. Lateral-right view. The right lobe of the liver has been removed to show the subcardinal-hepatic anastomosis (communication of the hepatic portion of the subcardinal vein with the hepatic portal vein).

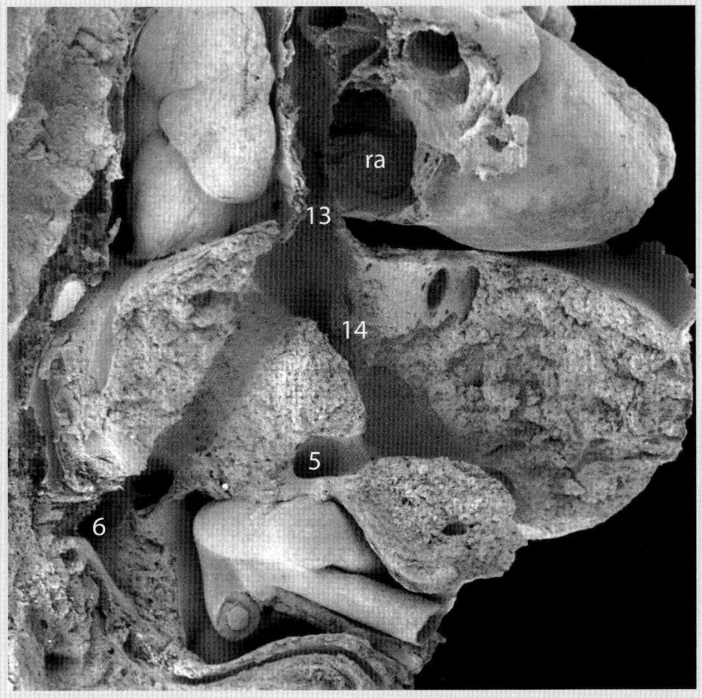

4.2.26 Embryo at week 8. The hepatic portion of the inferior caval vein has been established.

4.3

The Great Vessels of the Trunk

The umbilical veins pass from the umbilical cord into the embryonic lateral body wall (fig. 4.3.03 and 4.1.10) and primarily empty into the sinus venosus (fig. 4.3.01, 4.3.02). The right umbilical vein disappears (fig. 4.3.03) and the left one is then the only vessel that carries oxygenated blood to the embryo. Following the origin of anastomoses of the left umbilical vein with hepatic sinusoids the suprahepatic portion of the umbilical vein disappears and all the blood from the placenta reaches the embryo via the liver (fig. 4.3.05, 4.3.08, 4.3.09).

The two umbilical arteries are branches of the paired descending aorta and pass from the dorsal body wall into the superior part of the insertion of the umbilical cord (fig. 4.3.03 and 4.3.04, 4.3.06 and 4.3.07).

The inferior caval vein arises from the complex of subcardinal veins. The most cranial portion of the definitive inferior caval vein is formed by the confluence of the hepatic veins (fig. 4.3.11–4.3.13).

The superior caval veins develop from the superior (precardinal) veins (fig. 4.3.15, 4.3.16). Whereas the right superior caval vein drains via the right common cardinal vein directly into the right atrium during embryonic life span, in the fetus and the adult the proximal portion of the left superior caval vein has disappeared and empties via a transverse anastomosis cranial to the heart directly into the right superior caval vein.

The inferior (postcardinal) veins primarily drain the dorsal body wall and empty into the common cardinal veins (fig. 4.3.15–4.3.17). Eventually the inferior cardinal veins disappear except for their most proximal portions which are integrated into the azygos vein and possibly into the left superior intercostal vein.

4.3.01–4.3.18
The main blood vessels of the trunk.

4.3.01, 4.3.02
The early umbilical veins drain into the sinus venosus.

60:1

Week 5

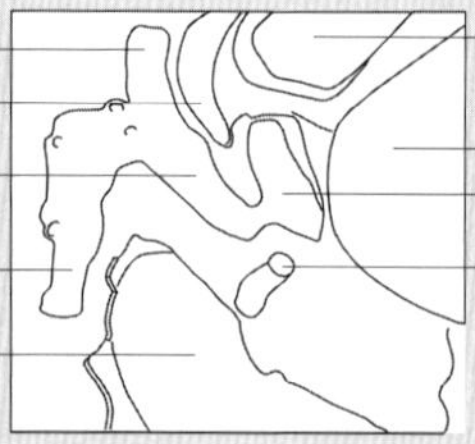

30:1

60:1

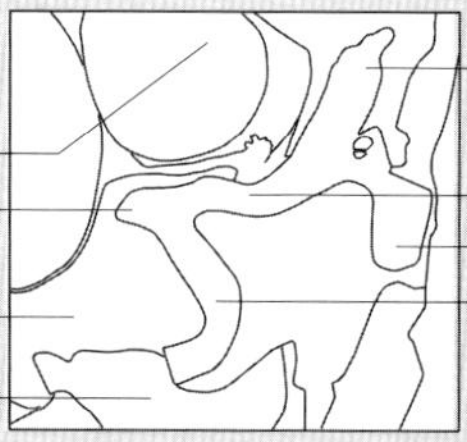

30:1

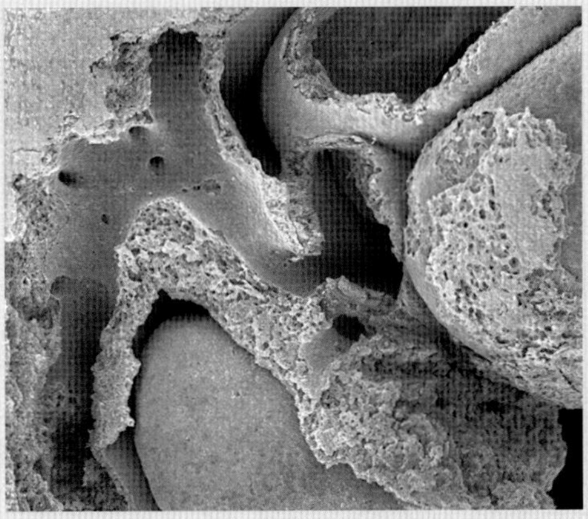

4.3.01

Right view of the right superior, inferior, common cardinal, and right umbilical vein. Embryo at week 5.

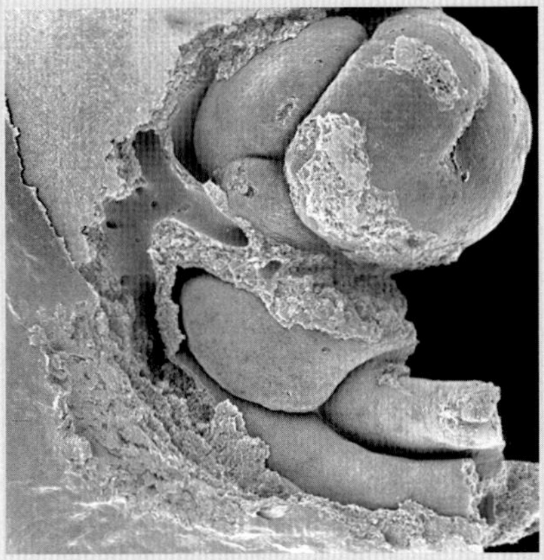

4.3.01a

Same embryo, same view. The sinus venosus has been opened.

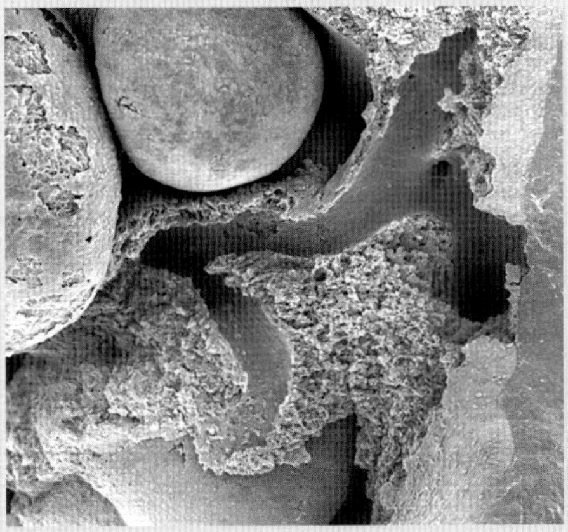

4.3.02

Same embryo, left view of the left superior, inferior, common cardinal, and the left umbilical vein.

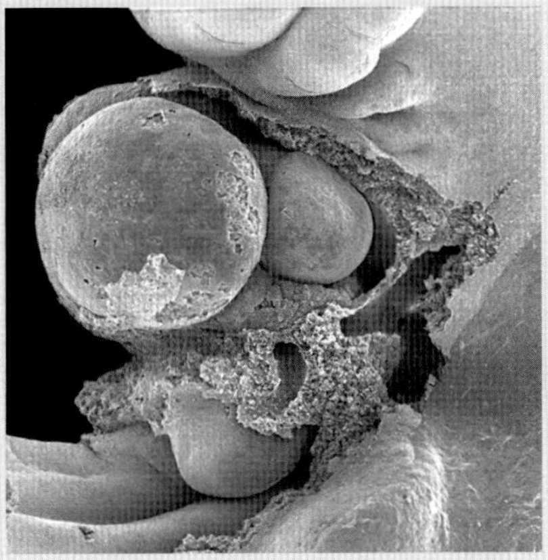

4.3.02a

Same embryo, same view. The sinus venosus has been opened.

4.
Organs in the Abdomen

4.3
The Great Vessels of the Trunk

4.3.01–4.3.18
The main blood vessels of the trunk.

4.3.03–4.3.10
The position of the umbilical vessels in the body wall.

Week 5

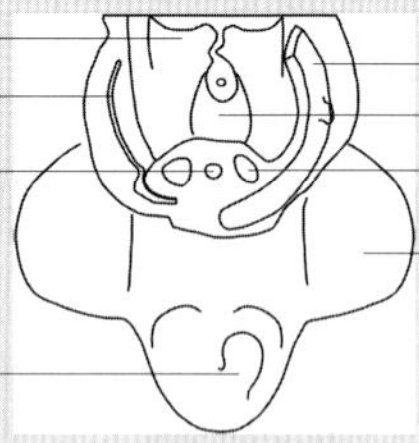

Week 5 / 6

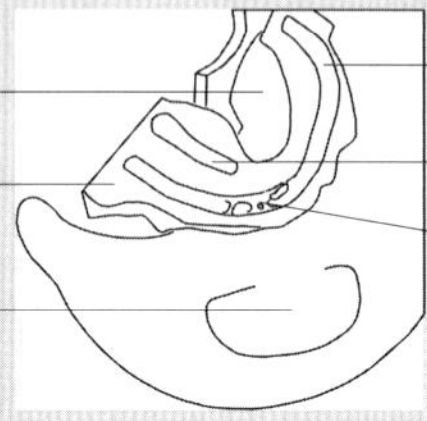

Week 6

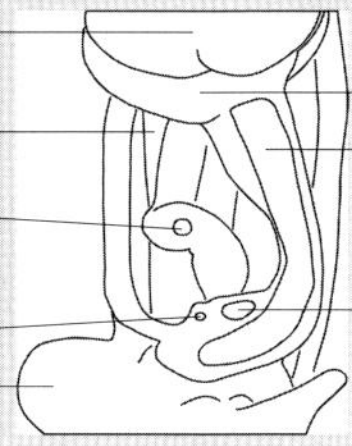

Week 5 / 6

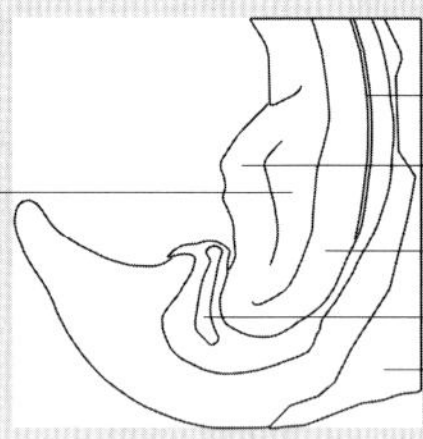

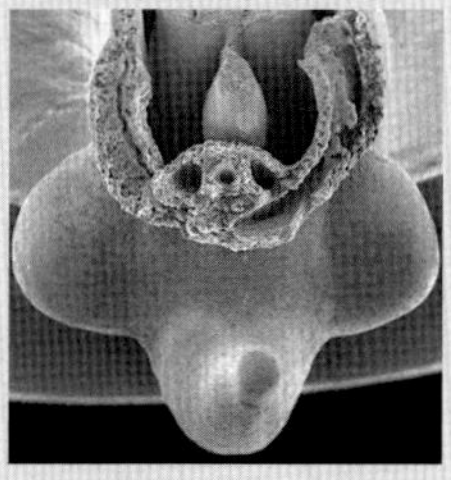

4.3.03 Ventral view of the umbilical arteries and veins. Embryo at week 5.

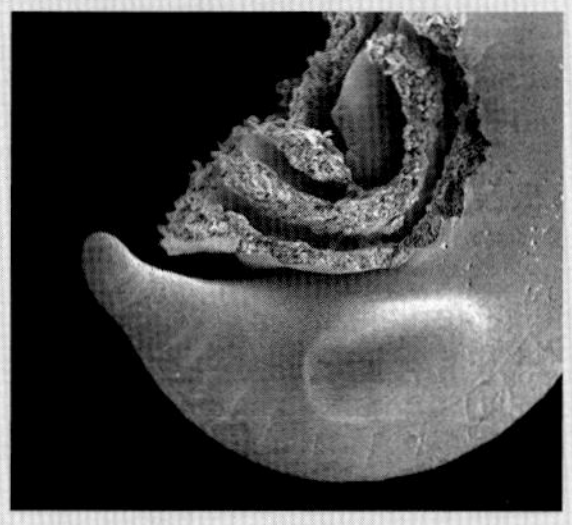

4.3.04 Lateral-left view of the left umbilical artery and vein. Embryo at the end of week 5/early in week 6.

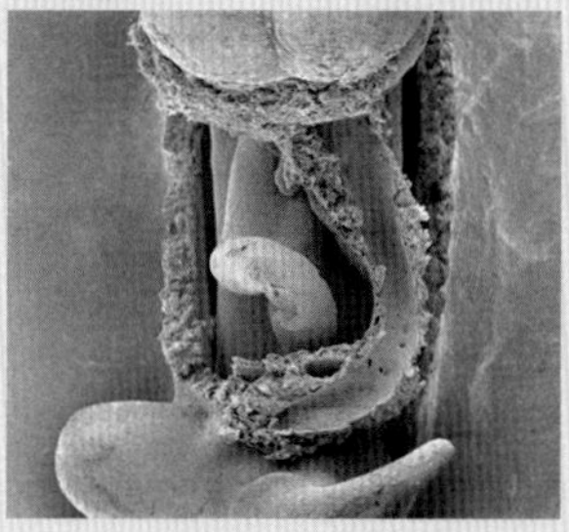

4.3.05 Ventral view of the left umbilical vein draining into the liver. Embryo at week 6.

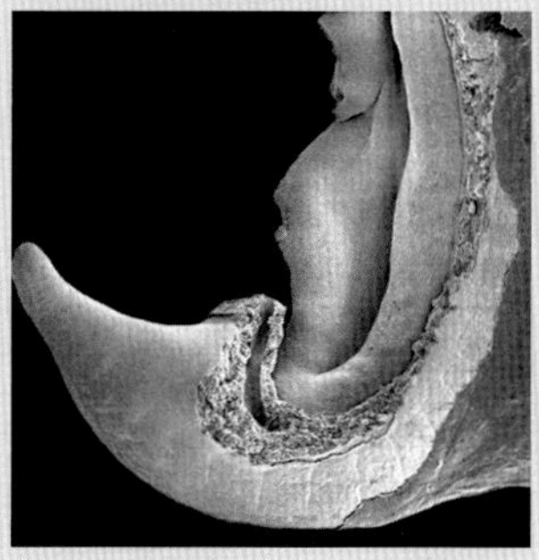

4.3.06 Lateral-left view of the left umbilical artery as it reaches the dorsal body wall. Embryo at the end of week 5/early in week 6.

Week 7

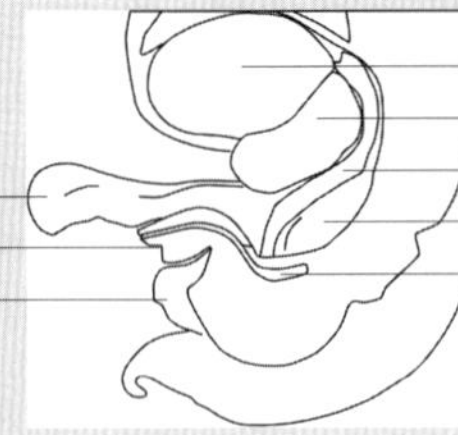

4.3.01–4.3.18

The main blood vessels of the trunk.

4.3.03–4.3.10

The position of the umbilical vessels in the body wall.

Week 5 / 6

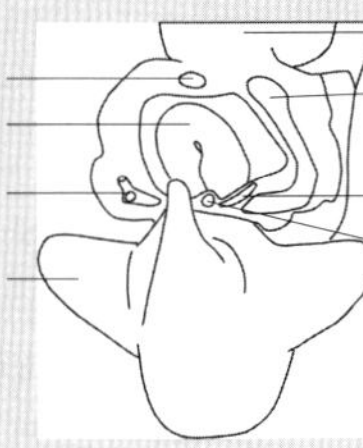

Week 6

Week 7

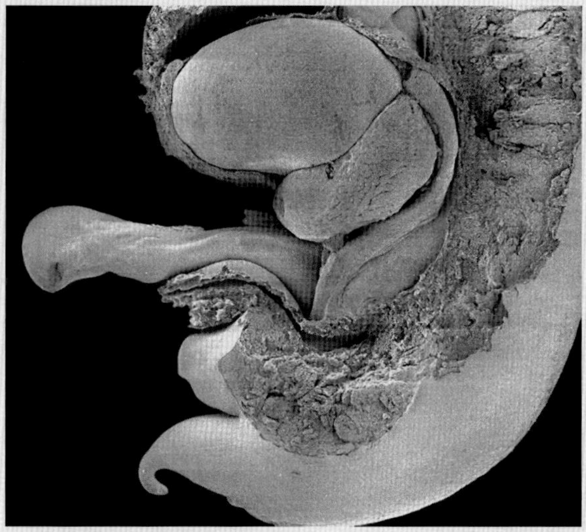

4.3.07 Lateral-left view of left umbilical artery as it reaches the dorsal body wall. Embryo at week 7.

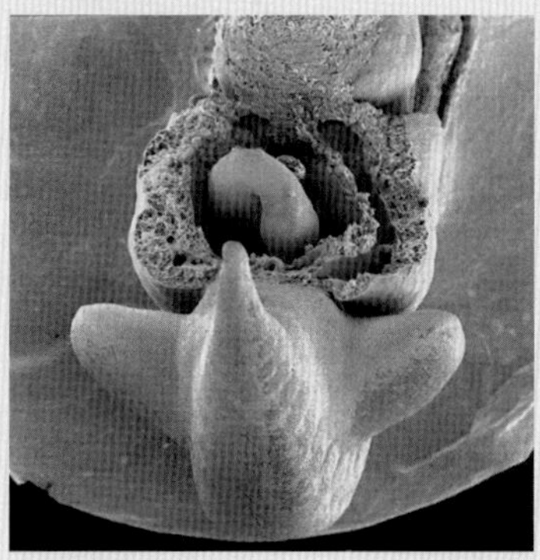

4.3.08 Ventral view of the left umbilical vein in the body wall. Embryo at the end of week 5/early in week 6.

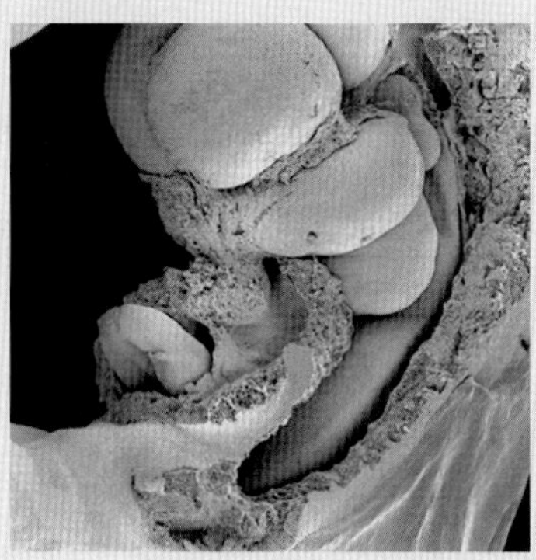

4.3.09 Ventral-left view of the left umbilical vein. Embryo late in week 6.

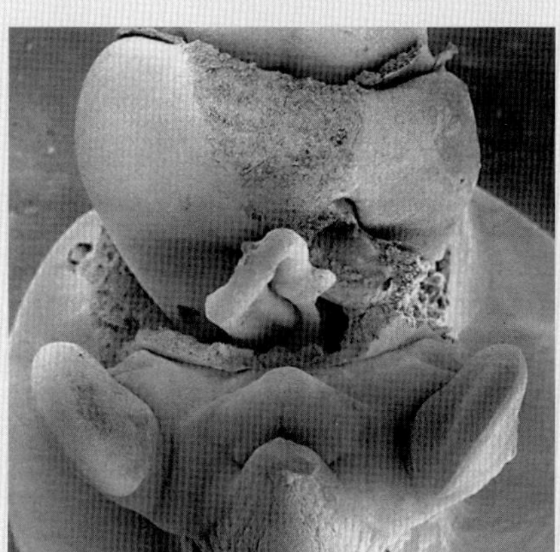

4.3.10 Ventral view of the left umbilical vein draining into the liver. Embryo at week 7.

4.3.01–4.3.18
The main blood vessels of the trunk.

4.3.11–4.3.13
Development of the hepatic portion of the inferior caval vein. Lateral-right view.

Week 6

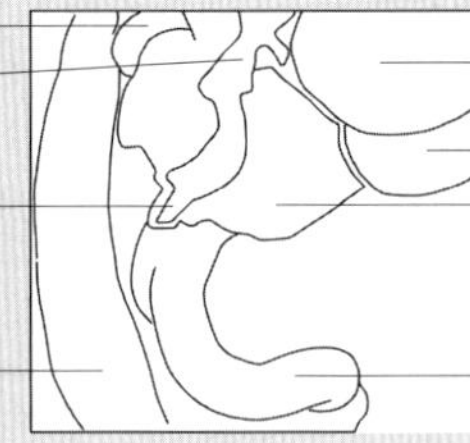

Week 8

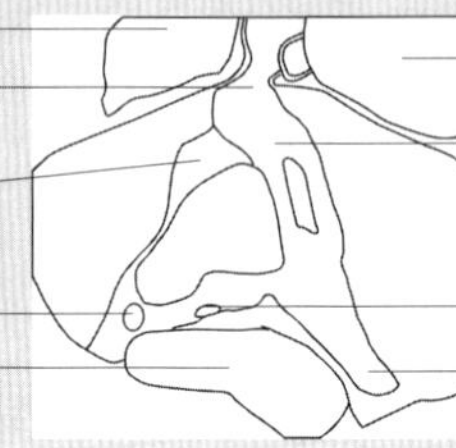

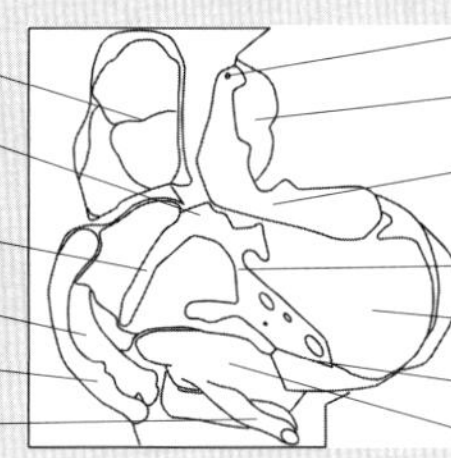

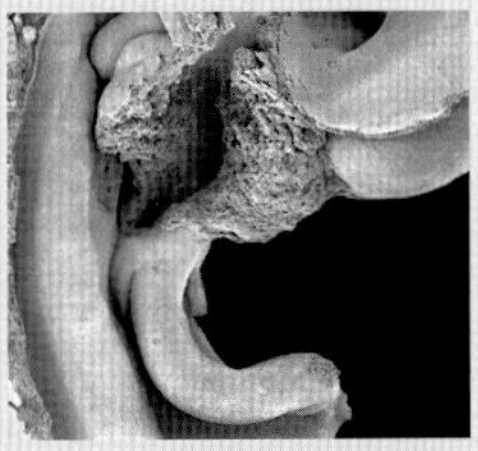

4.3.11 Embryo at week 6. Lateral-right view. The right lobe of the liver has been removed to show the communication of the hepatic portion of the subcardinal vein with the hepatic portal vein.

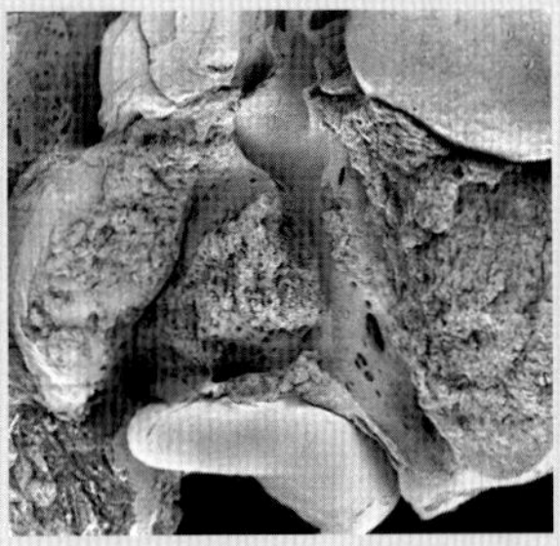

4.3.12 Embryo at week 8. The superficial portions of the right lobe of the liver have been removed to show the anastomosis of the subcardinal vein with the hepatic veins.

4.3.13 Embryo at week 8. The superficial portions of the right lobe of the liver have been removed. The hepatic portion of the inferior caval vein has reached an advanced stage.

4.
Organs in the Abdomen

4.3
The Great Vessels of the Trunk

4.3.01–4.3.18
The main blood vessels of the trunk.

4.3.14–4.3.18
The superior and inferior cardinal veins draining into the sinus venosus via the common cardinal veins.

Week 4

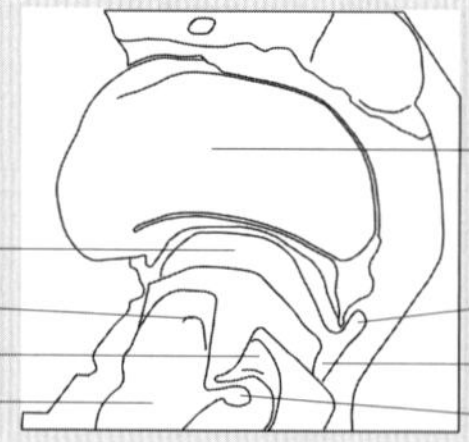

Week 5

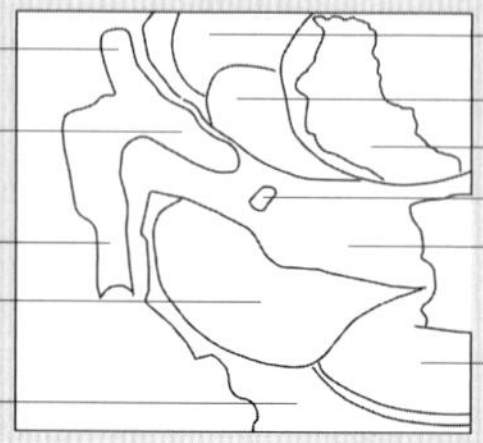

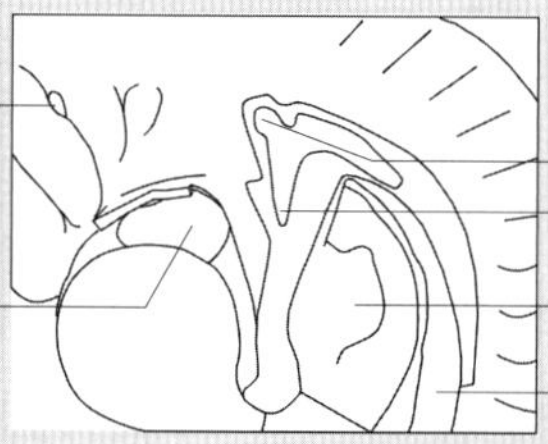

Week 6

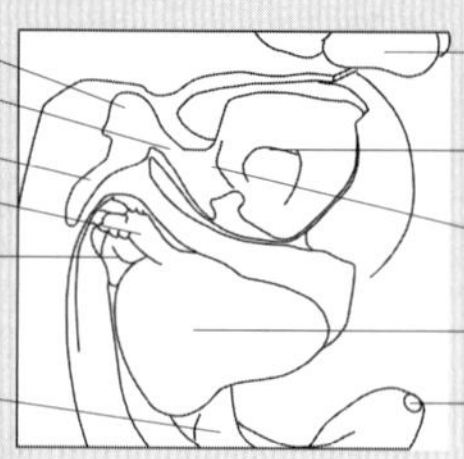

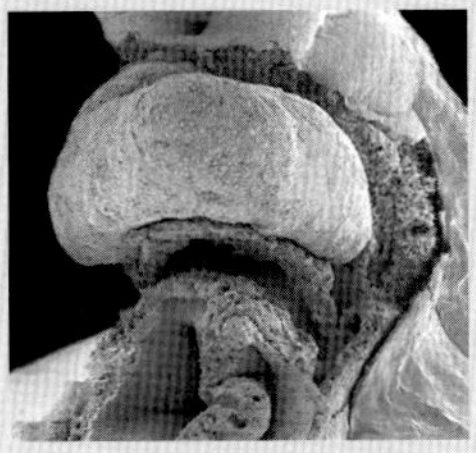

4.3.14 Embryo at the end of week 4. Caudal ventral and left view. The sinus venosus has been opened.

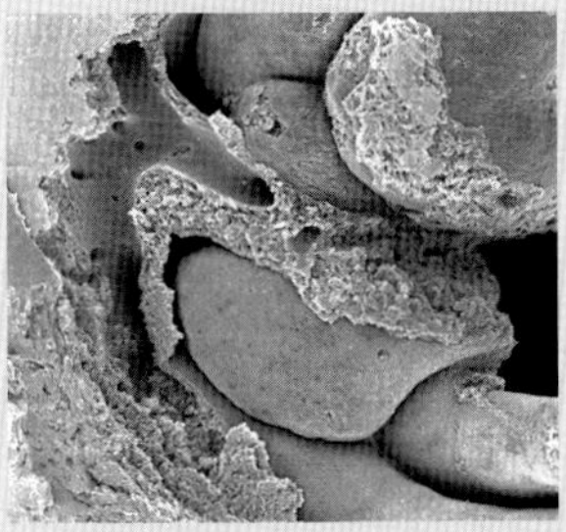

4.3.15 Embryo at week 5. Lateral-right view.

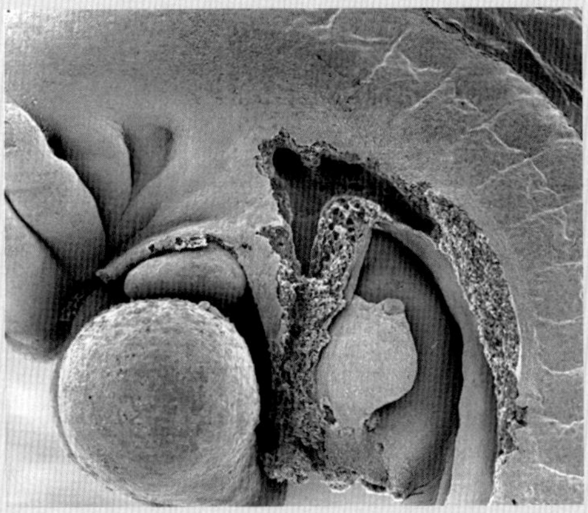

4.3.16 Embryo at week 5. Lateral-left view.

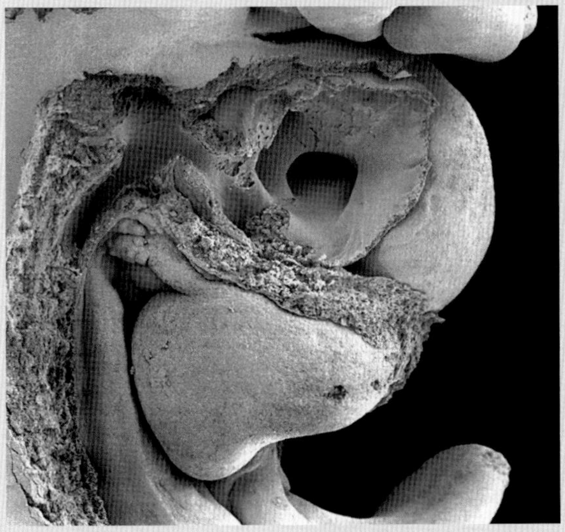

4.3.17 Embryo at the end of week 6. Lateral-right view.

Week 7

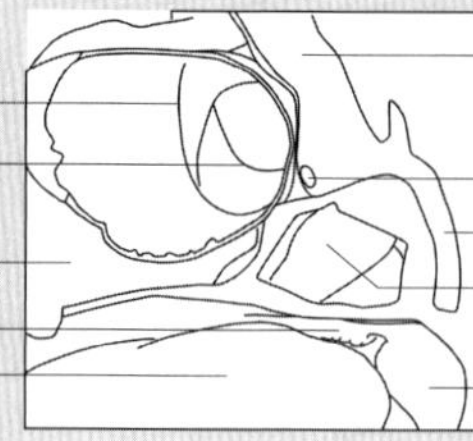

4.3.01–4.3.18
The main blood vessels of the trunk.

4.3.14–4.3.18
The superior and inferior cardinal veins draining into the sinus venosus via the common cardinal veins.

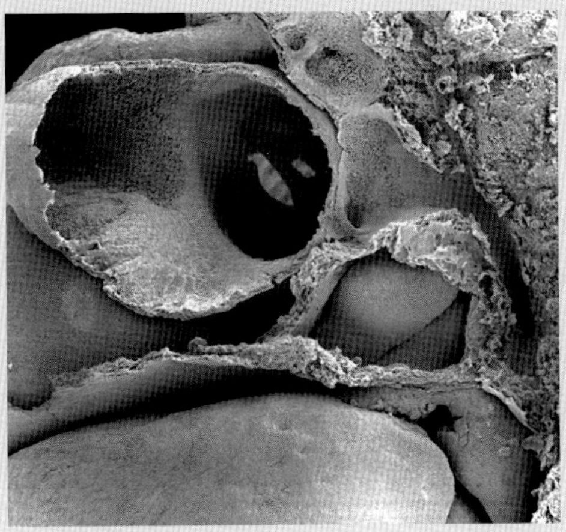

4.3.18 Embryo at week 7. Lateral-left view. The left common cardinal vein drains into the left sinus horn.

4.4
The Development of the Stomach

The stomach arises from the most cranial portion of the midgut (fig. 4.4.01). Its development is characterized by a very early formation of left-convex and corresponding right-concave bendings, i.e. the greater and the lesser curvatures (fig. 4.4.02, 4.4.03, 4.4.09–4.4.13, 4.4.15–4.4.19).

Due to this shaping process and due to the huge growth of the liver, the stomach achieves its typical position (fig. 4.4.04–4.4.08).

The coelomic cavity extends by a recess into the broad dorsal mesogastrium of the stomach, the omental bursa (fig. 4.4.20, 4.4.21), thus forming the lesser omentum (fig. 4.4.13, 4.4.14). The greater omentum arises as a recess of the omental bursa along the greater curvature (fig. 4.4.14).

4.4

Abbreviations

il	intestinal loop
la	left atrium
lv	left ventricle
mn	mesonephros
1	stomach, greater curvature
2	endoderm of the stomach
3	greater omentum
4	lesser omentum
5	spleen
6	peritoneum of the dorsal wall of the omental bursa
7	liver, left lobe
8	bare area of the liver
9	liver, right lobe
10	lung, inferior lobe
11	lung, superior lobe
12	oesophagus
13	gonad
14	genital tubercle
15	duodenum

4.4.01–4.4.27
Developmental stages of the stomach.

4.4.01–4.4.08
Development of the position of the stomach. Ventral-left/left view.

Week 4

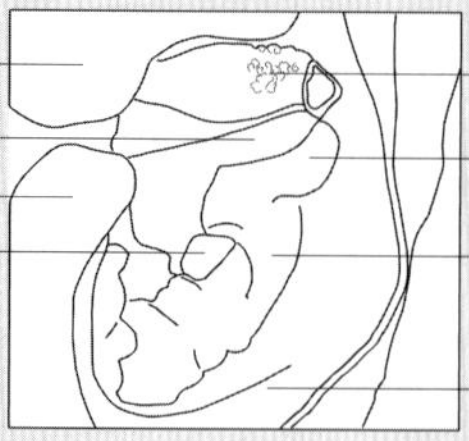

Week 5

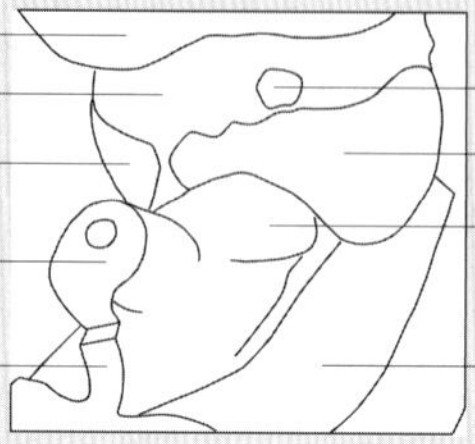

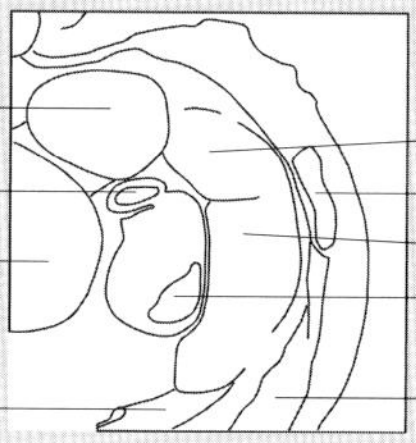

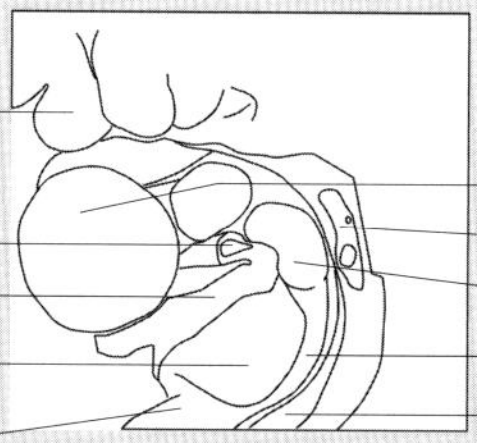

Week 6

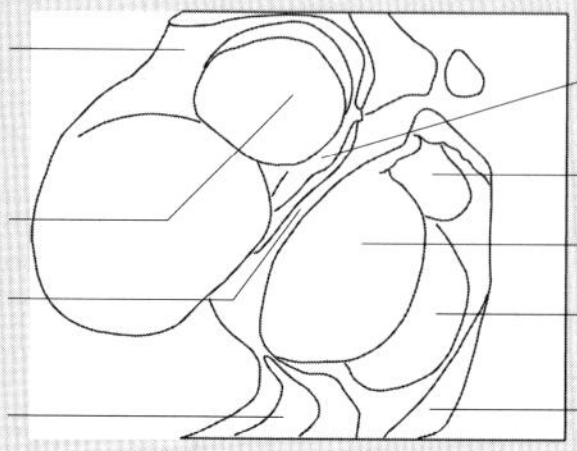

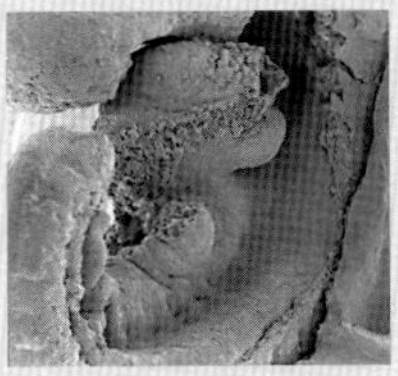

4.4.01 No indication of the bulging of the stomach. Embryo at the end of week 4.

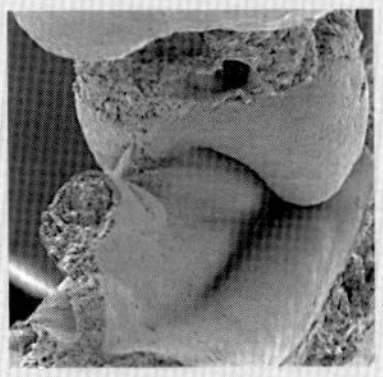

4.4.02 The first indication of the great curvature is visible. Embryo at week 5.

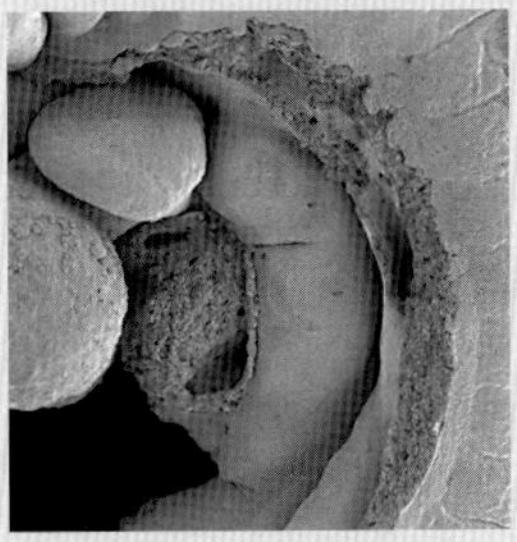

4.4.03 The left lobe of the liver has been removed. Embryo at week 5.

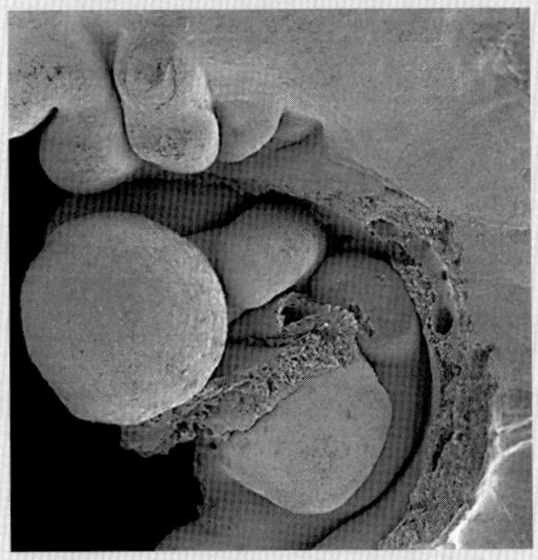

4.4.04 The liver almost completely covers the stomach. Embryo at week 5.

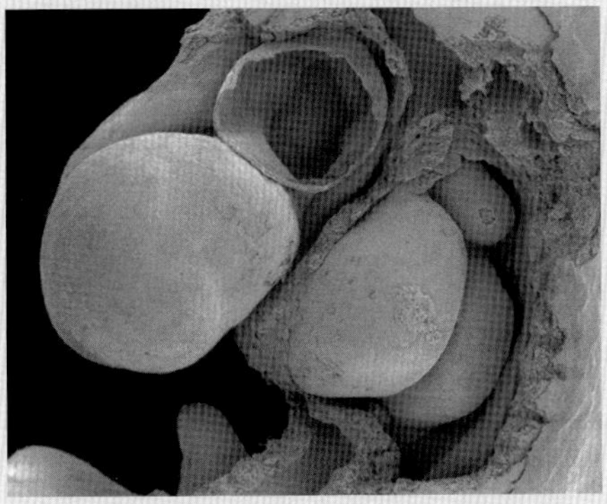

4.4.05 Embryo at week 6.

Week 6

Week 7 / 8

4.4.01–4.4.27
Developmental stages of the stomach.

4.4.01–4.4.08
Development of the position of the stomach. Ventral-left/left view.

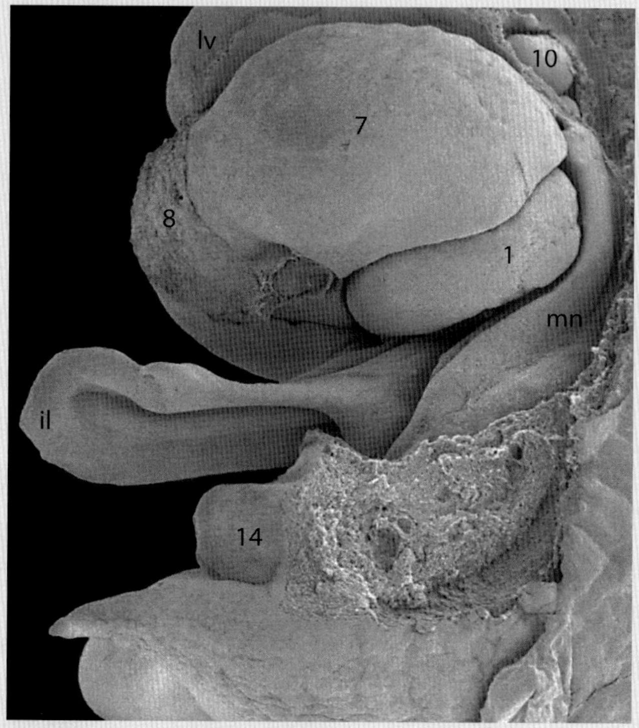

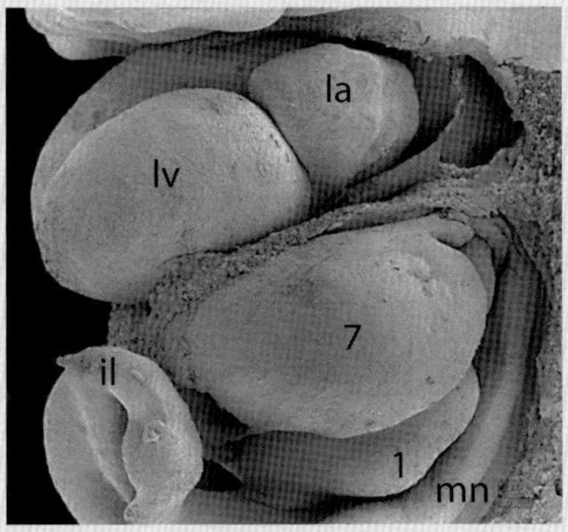

4.4.06

Embryo at week 6.

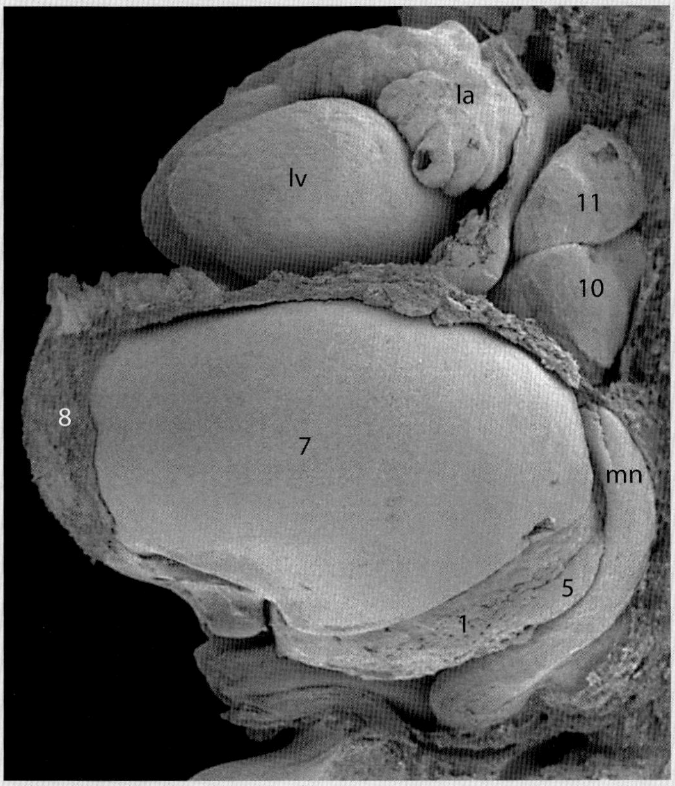

4.4.07 / 08

4.4.07
Embryo at week 7.

4.4.08
Embryo at week 8.

4.4.01–4.4.27
Developmental stages of the stomach.

4.4.09–4.4.14
Development of the external form of the stomach. The liver has been removed. Ventral-left view.

Week 5

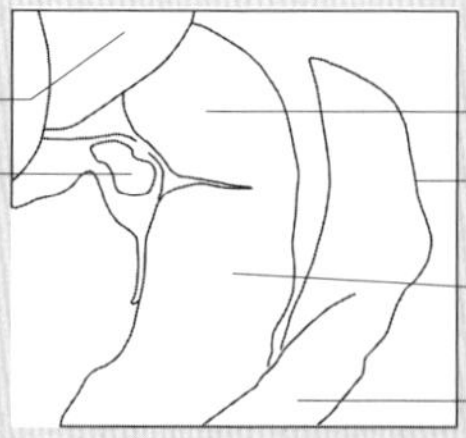

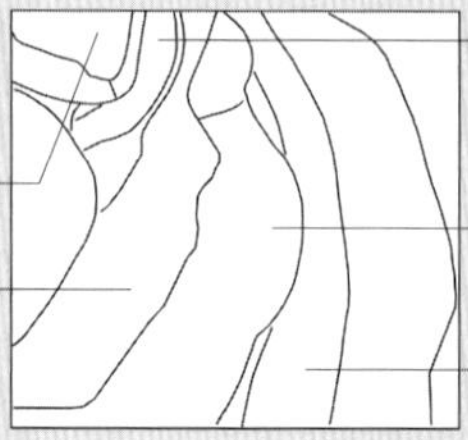

Week 5 / 6

Week 7 / 9

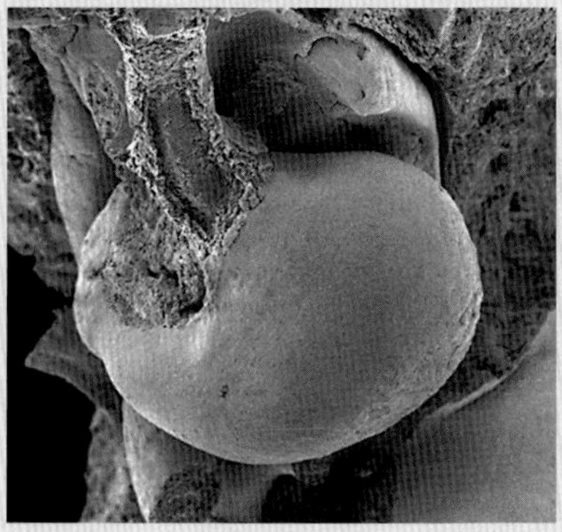

4.4.09

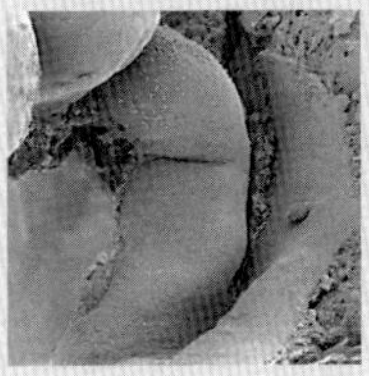

Embryos at week 5.

4.4.10

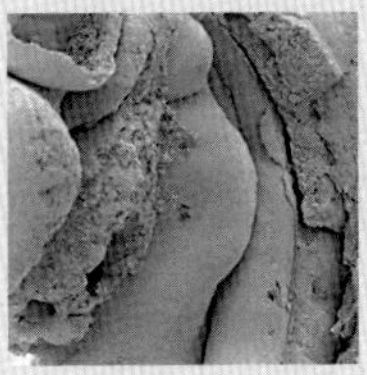

4.4.11 / 12

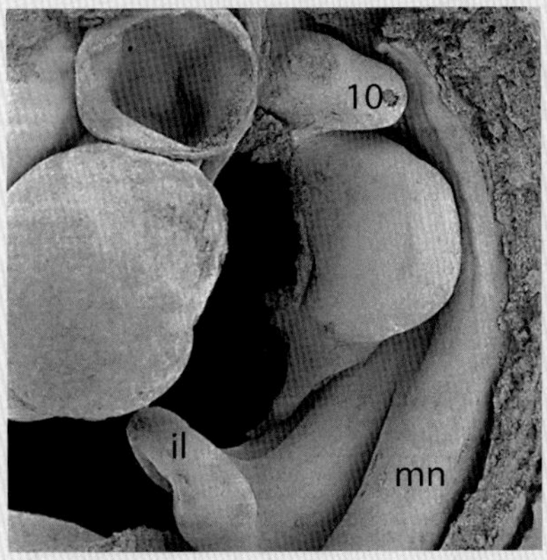

4.4.11
Embryo at week 5.

4.4.12
Embryo at week 6.

4.4.13 / 14

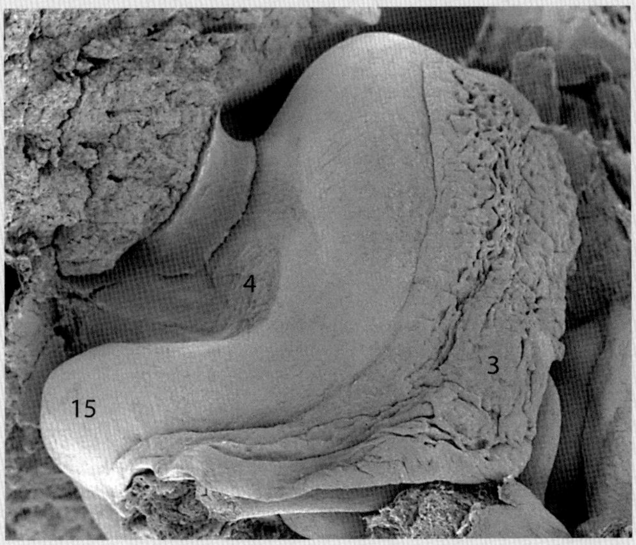

4.4.13
Embryo at week 7.

4.4.14
Embryo at week 9.

25:1

Week 5

Week 6

4.4.01–4.4.27
Developmental stages of the stomach.

4.4.15–4.4.19
Development of the internal form of the stomach. Ventral-left view.

Week 7

Week 9

4.4.15 The peritoneum and the mesenchyme covering the anterior wall of the stomach have been removed. Ventral view of the basal side of the endoderm of the stomach. Embryo at week 5.

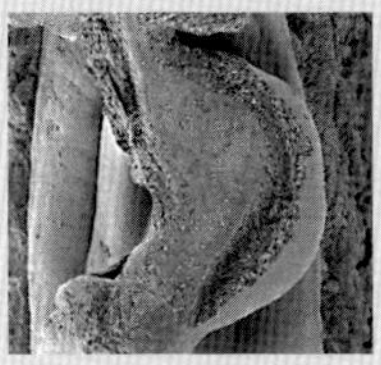

4.4.16 The peritoneum and the mesenchyme covering the anterior wall of the stomach have been removed. Ventral view of the basal side of the endoderm of the stomach. Embryo at week 6.

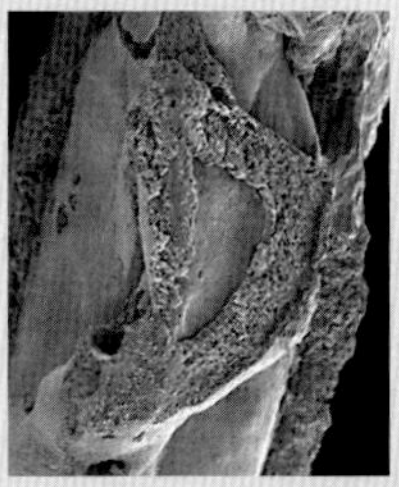

4.4.17 The ventral wall of the stomach has been removed. Embryo at week 6.

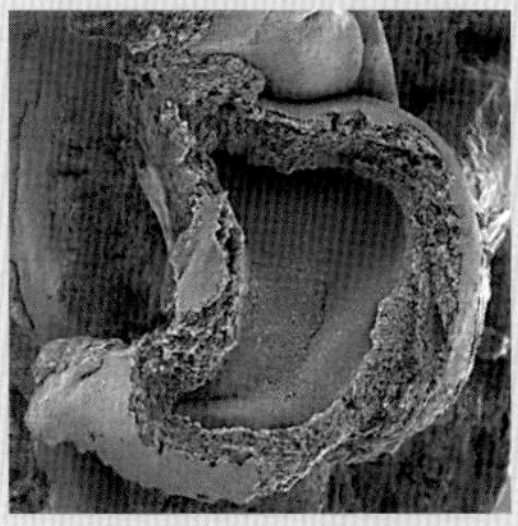

4.4.18 The ventral wall of the stomach has been removed. Embryo at week 7.

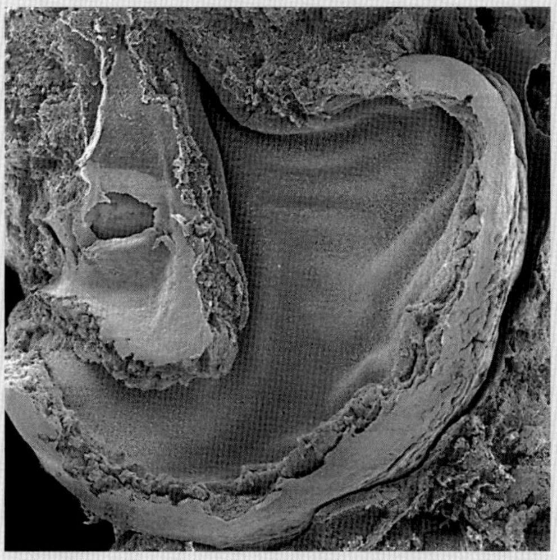

4.4.19 The ventral wall of the stomach has been removed. Embryo at week 9.

4.4.01–4.4.27
Developmental stages of the stomach.

Week 5

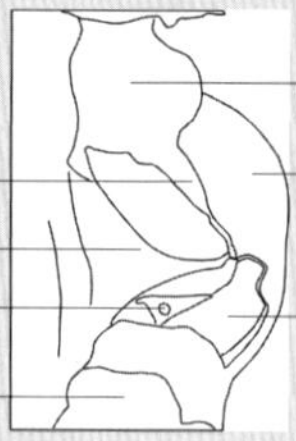

Week 6

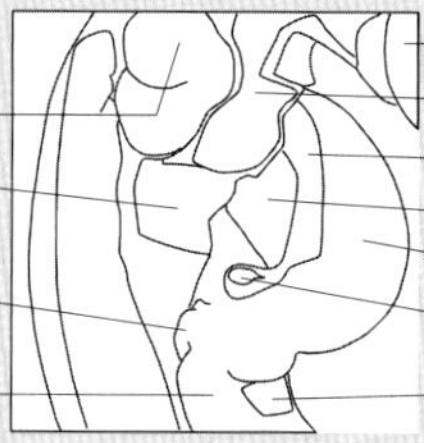

4.4.20

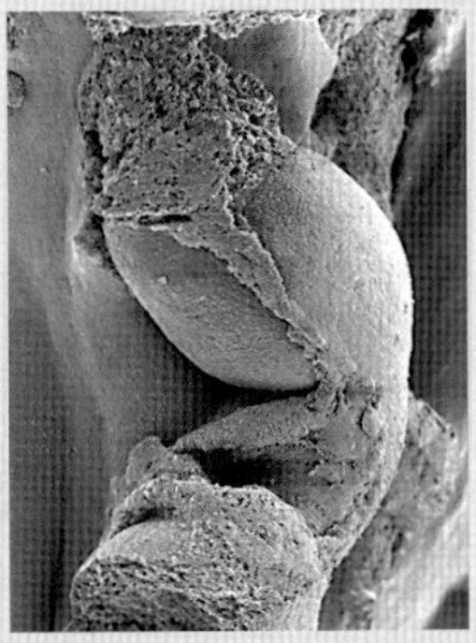

The origin of the lesser omentum and the omental bursa (lesser sac). Ventral-right view. Embryos at weeks 5 and 6.

4.4.21

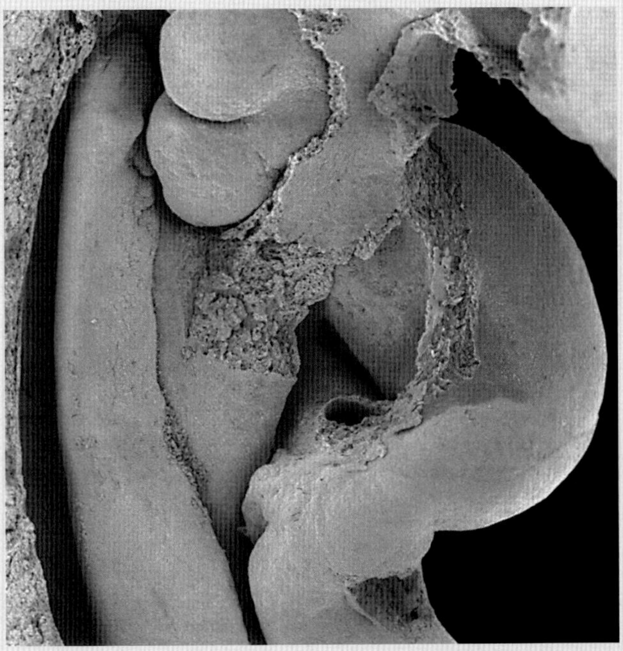

4.4.01–4.4.27
Developmental stages of the stomach.

4.4.22–4.4.27
The relationship of stomach and omental bursa.

Week 6

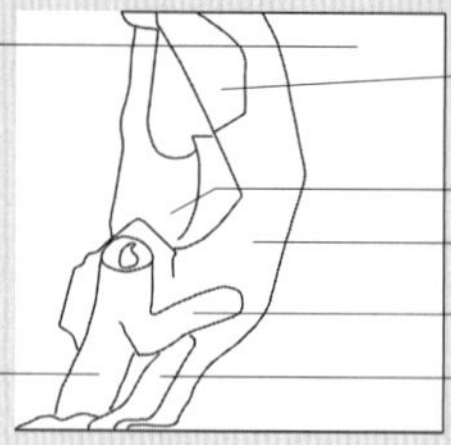

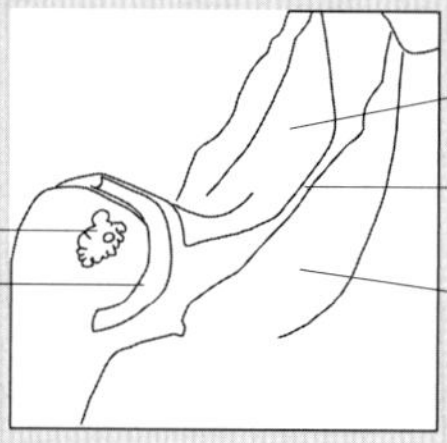

Week 7

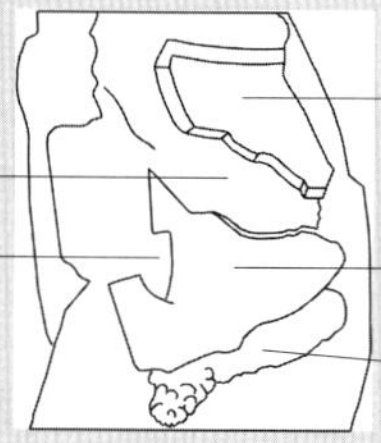

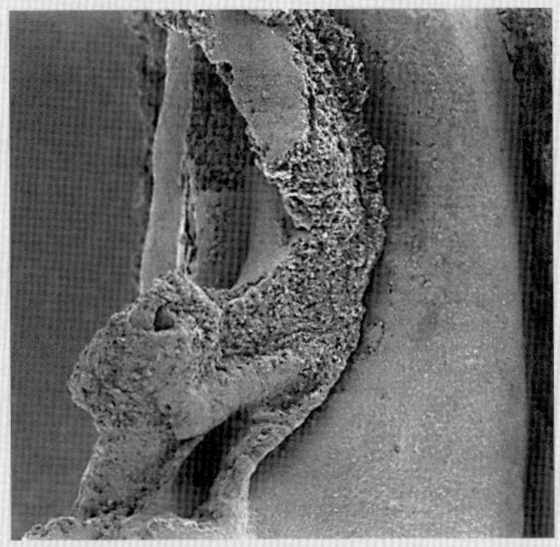

4.4.22 Ventral-left view of the omental bursa and the pancreas. Except for part of the fundus, the stomach has been removed. Embryo at week 6.

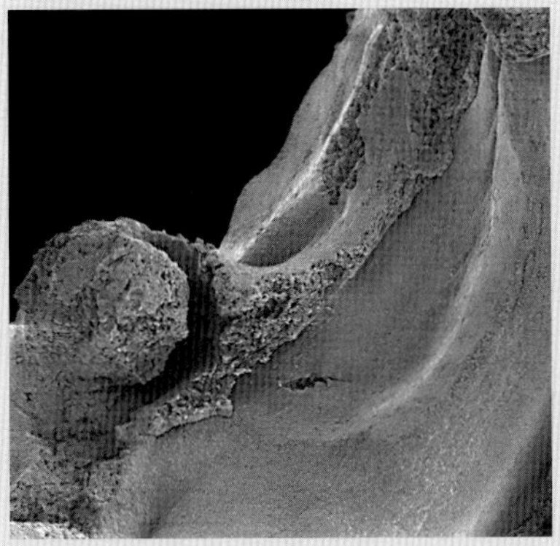

4.4.23 Lateral-left view of the omental bursa and the dorsal mesogastrium. The stomach has been removed. Embryo at week 6.

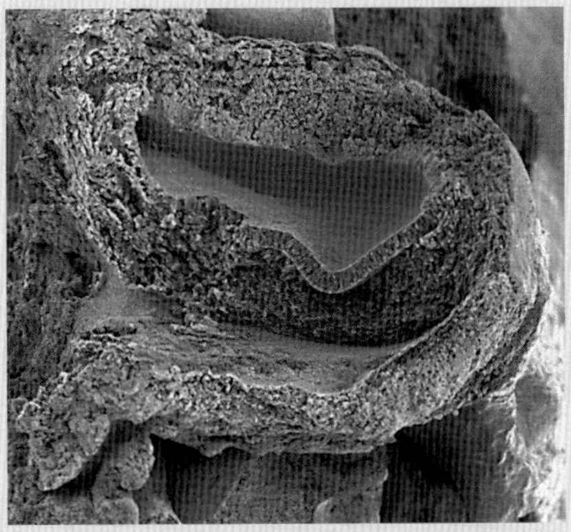

4.4.24 Caudal view of the stomach and the omental bursa. The corpus and the pylorus of the stomach have been removed. Embryo at week 7.

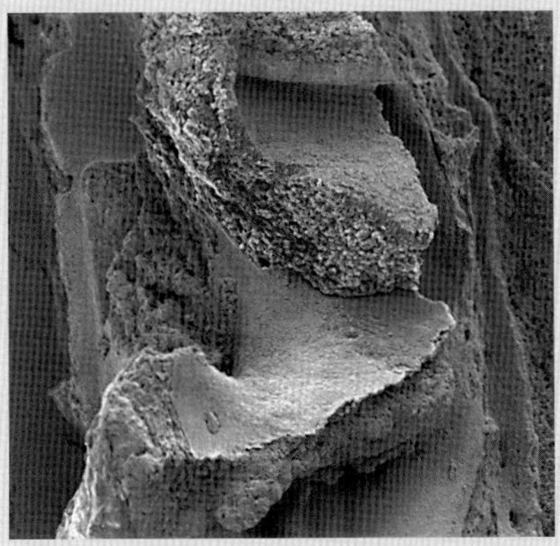

4.4.25 Ventral and caudal view of the stomach, the omental bursa and the pancreas. The corpus of the stomach has been partly and the pylorus completely removed. Embryo at week 7.

Week 7

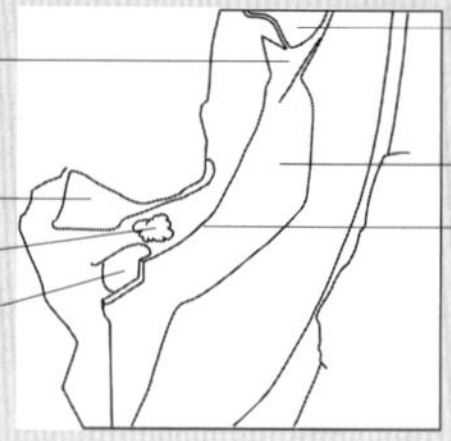

4.4.01–4.4.27
Developmental stages of the stomach.

4.4.22–4.4.27
The relationship of stomach and omental bursa.

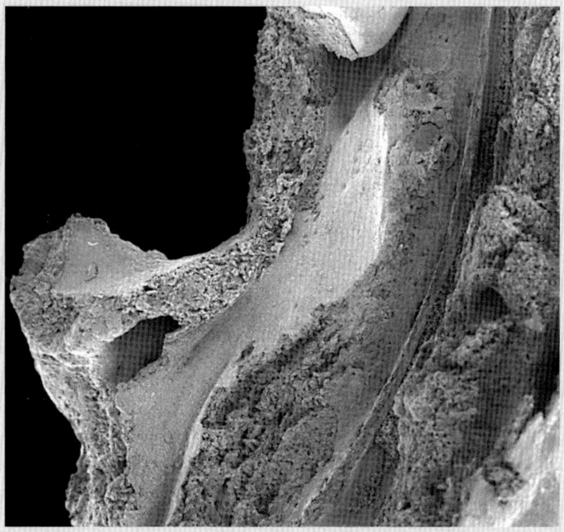

4.4.26 Lateral-left view of the omental bursa and the dorsal mesogastrium. The stomach has been removed. Embryo at week 7.

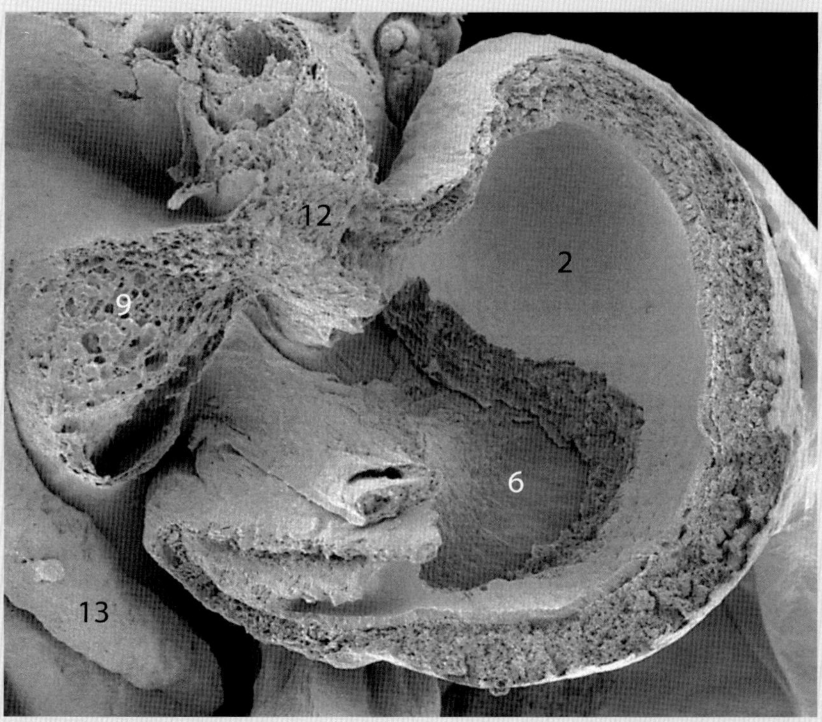

4.4.27 Ventral and caudal view of the stomach and the omental bursa. The stomach has been partly removed. Embryo at the end of week 7.

4.5
The Origin of the Spleen

The spleen arises in the upper portion of the greater curvature of the stomach (fig. 4.5.01–4.5.03). By the end of week 7 the spleen is demarcated from the wall of the stomach by deepening furrows (fig. 4.5.04, 4.5.06–4.5.10). During the fetal period these furrows will develop into the gastrosplenic and the splenophrenic ligaments.

The tail of the pancreas, located in the floor of the omental bursa, approaches the spleen when the recess of the omental bursa is maximally extended towards the left (fig. 4.5.05, 4.7.07, 4.7.08).

	4.5 **Abbreviations**
1	spleen
2	stomach, greater curvature
3	greater omentum
4	lesser omentum
5	liver, left lobe
6	gonad

4.5.01–4.5.10
Developmental stages of the spleen.

4.5.01–4.5.06
Development of the position of the spleen. Lateral/ventral-left view.

Week 7

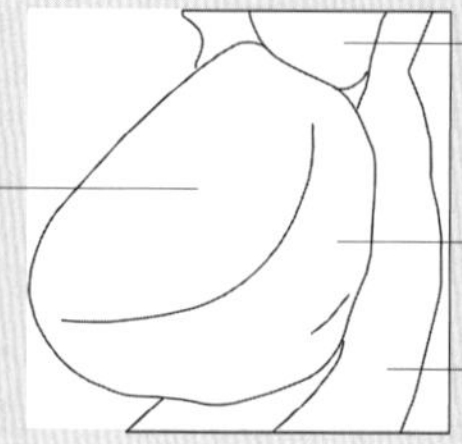

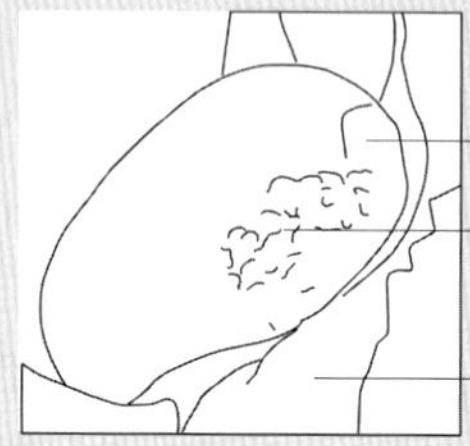

Week 9

4.5.01 The demarcation of the spleen from the upper pole of the major curvature is not yet visible. Embryo at week 7.

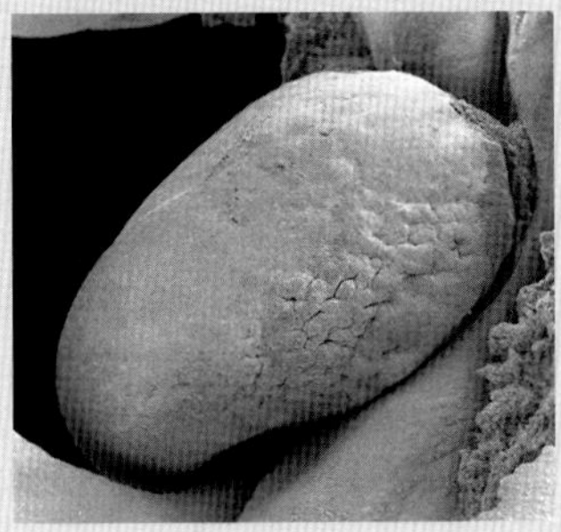

4.5.02 The first demarcations of the spleen from the upper pole of the major curvature become visible. Embryo at week 7.

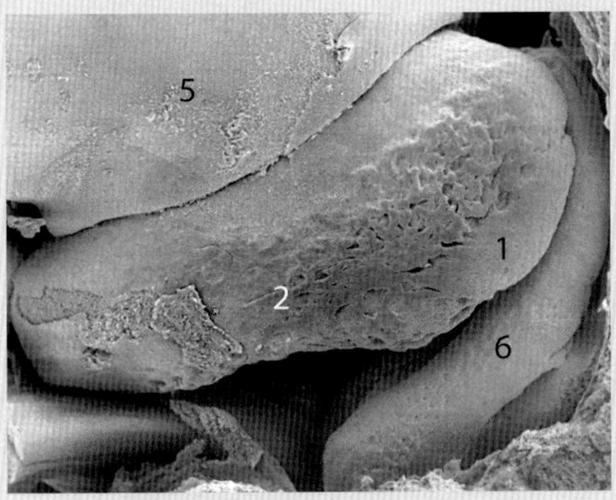

4.5.03 The demarcated spleen is still completely integrated into the wall of the stomach. Embryo at the end of week 7.

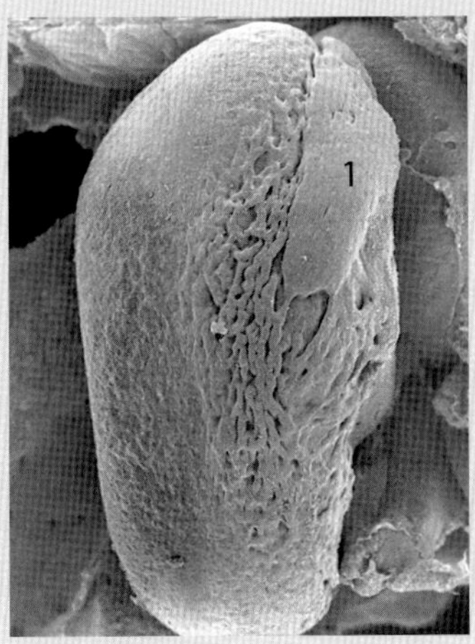

4.5.04 The demarcation of the spleen is more advanced. Embryo at week 9.

Week 9

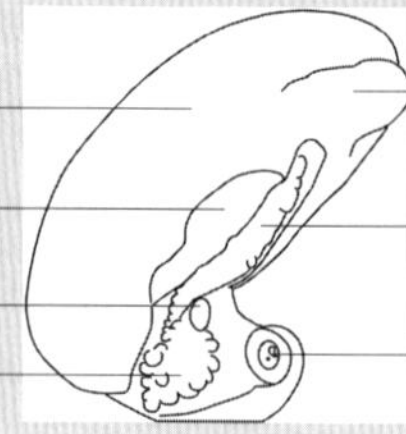

4.5.01–4.5.10
Developmental stages of the spleen.

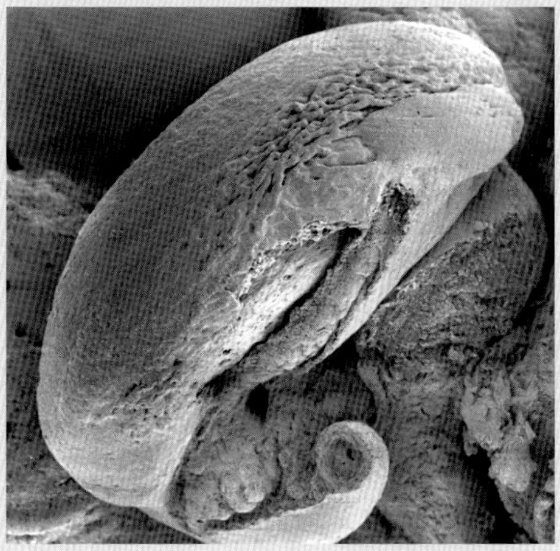

4.5.05

Same embryo as in figure 4.5.04. The omental bursa has been opened and the pancreas exposed to show the topographical relationship of the pancreas to the hilus of the spleen.

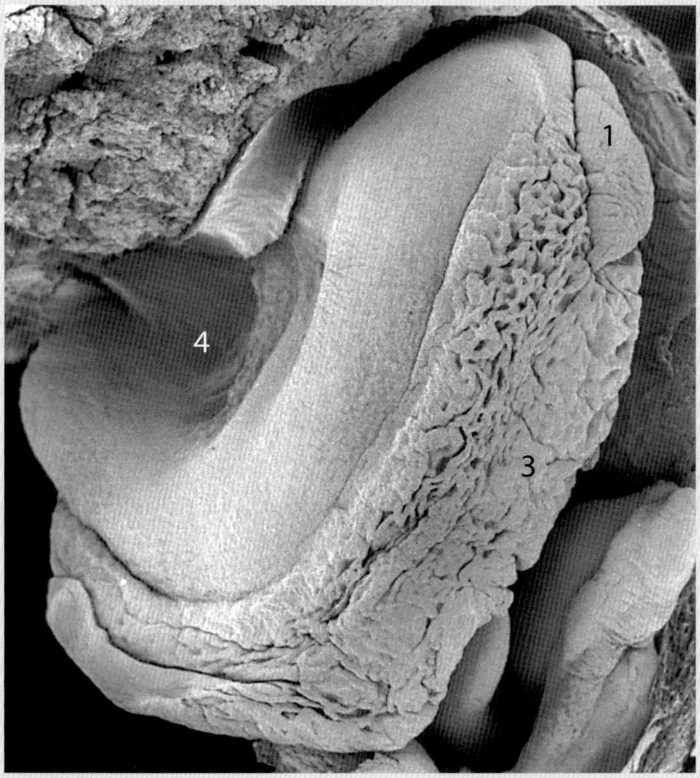

4.5.06

Ventral-left view of the spleen, the stomach and the greater omentum. Embryo at week 9.

4.5.01–4.5.10
Developmental stages of the spleen.

4.5.07–4.5.10
Detailed view of the demarcation of the spleen from the wall of the stomach.

60:1

Week 7

Week 8

Week 9

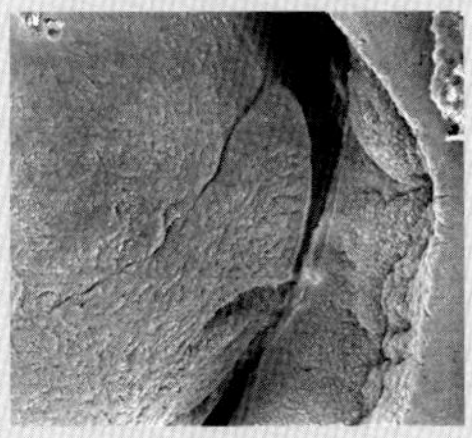

4.5.07 Embryo at week 7.

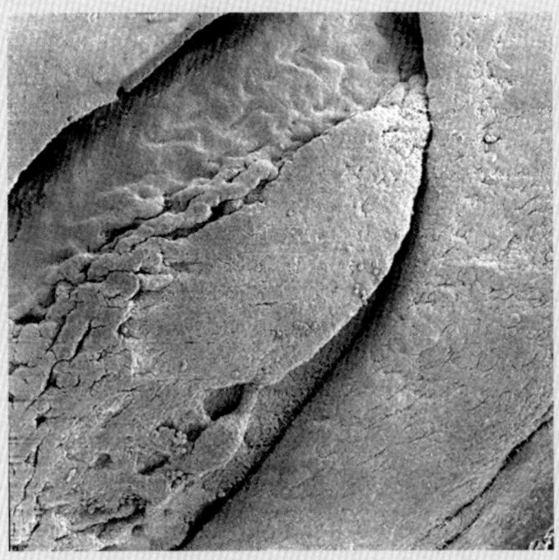

4.5.08 Embryo at week 8.

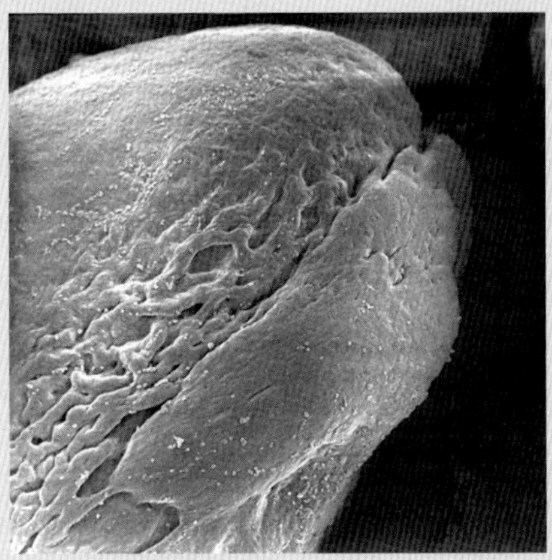

4.5.09 Embryos at week 9.

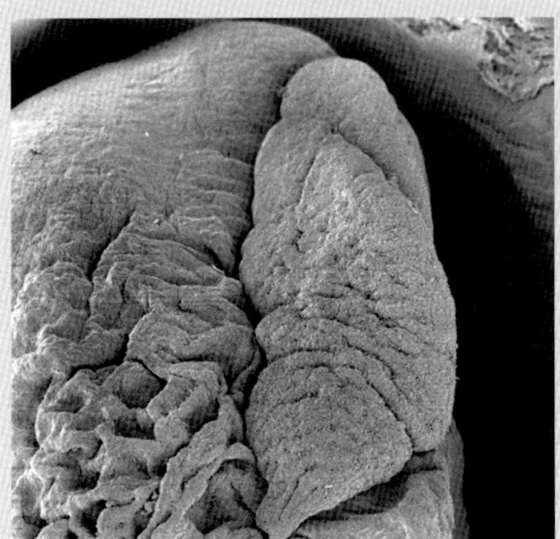

4.5.10

4.6
The Development of the Pancreas

The pancreas arises from two separate anlagen. The part called the dorsal pancreas is a diverticulum of the duodenum (fig. 4.6.02–4.6.06), the part called the ventral pancreas branches off from the duodenum near the origin of the bile duct (fig. 4.6.01, 4.6.02, 4.6.04). The two anlagen fuse and form a single organ.

The ventral pancreas remains near its origin in the concavity of the duodenum (fig. 4.6.11), forming the posterior portion of the pancreatic head. The dorsal pancreas forms the ventral portion of the head (fig. 4.6.10), the body and the tail (fig. 4.6.12). The body and the tail grow out to the left into the wall of the stomach along the inferior circumference of the greater curvature (fig. 4.6.13). Due to the development of the omental bursa, body and tail are separated from the mesenchyme of the stomach by the omental bursa and become covered by the peritoneum (fig. 4.6.07, 4.6.08). The end of the tail reaches the spleen (fig. 4.6.13).

Thus, the body and the tail of the pancreas reach their final position at the posterior wall of the omental bursa. The incorporation of the pancreas into the dorsal body wall is achieved by a reduction of the retropancreatic recess of the peritoneal cavity (fig. 4.6.08, 4.6.09).

4.6.01–4.6.16
Developmental stages of the pancreas.

50:1

Week 5

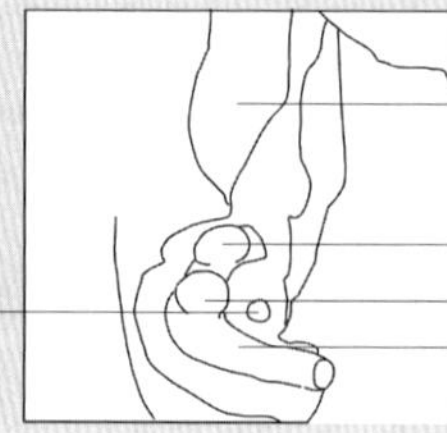

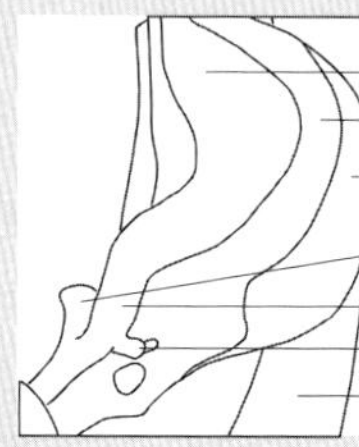

100:1

50:1

Week 6

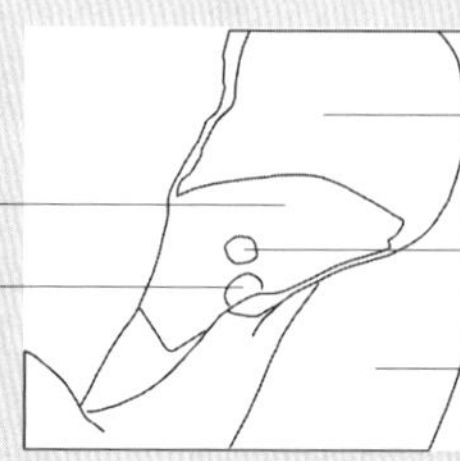

4.6.01 Ventral-caudal view of the duodenum and the buds of the bile duct and the ventral pancreas. The peritoneum of the duodenum has been removed. The buds are partly covered by the mesenchyme. Embryo at week 5.

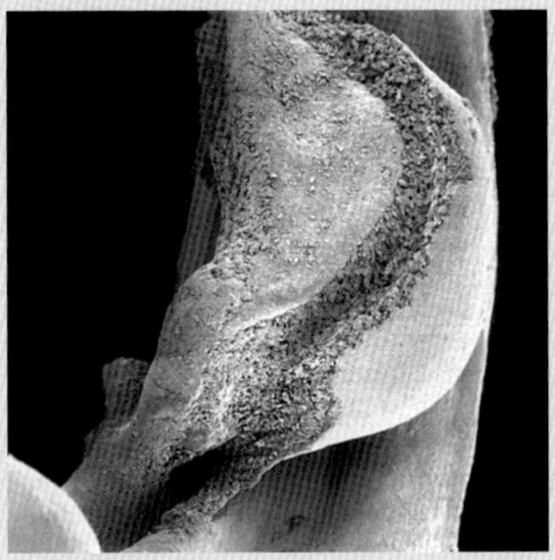

4.6.02 Ventral-caudal view of the duodenum and the bud of the dorsal pancreas. The peritoneum of the duodenum has been removed. Embryo at week 5.

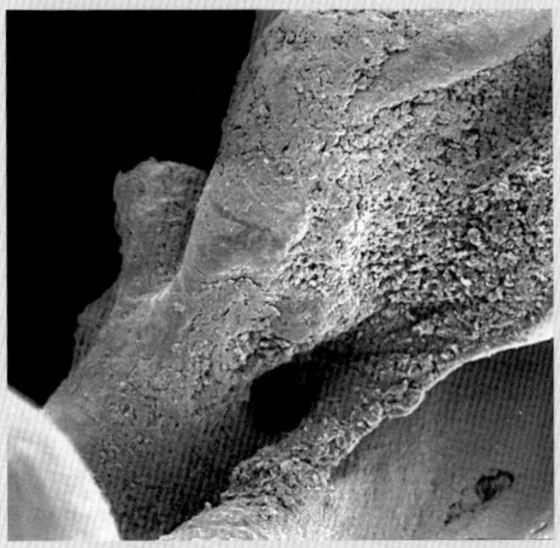

4.6.02a Same embryo as in figure 4.6.02.

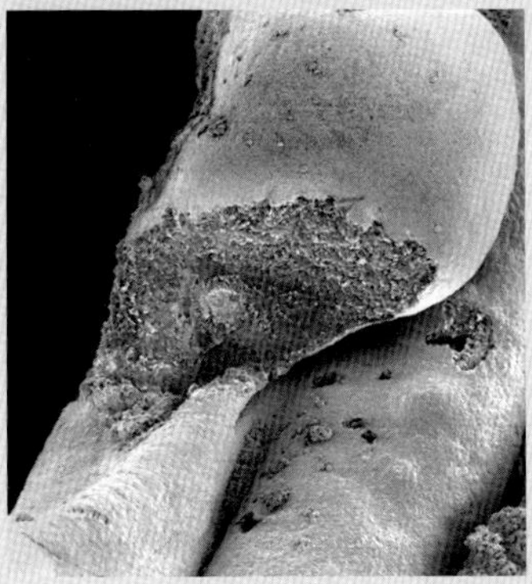

4.6.03 Ventral-caudal view of the pancreas within the mesenchyme of the stomach. Embryo at week 6.

4.6.01–4.6.16
Developmental stages of the pancreas.

Week 6

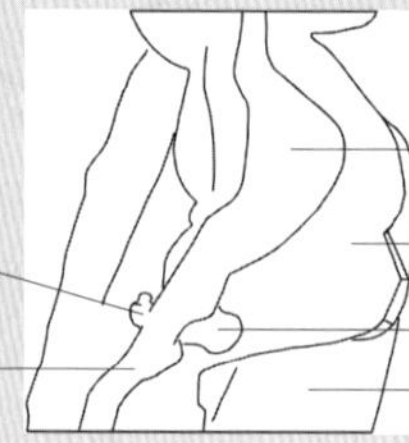

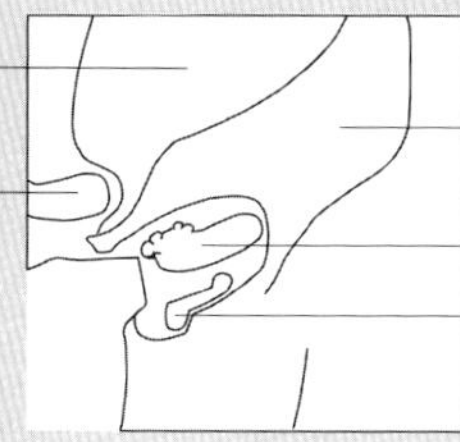

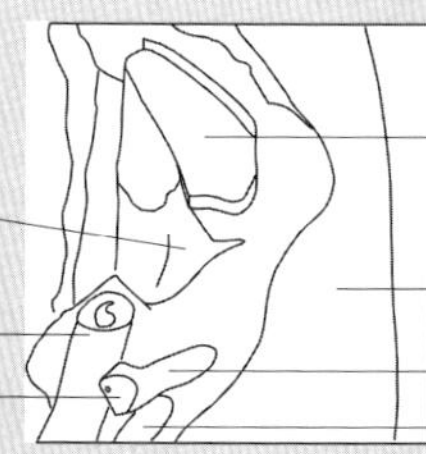

Week 7

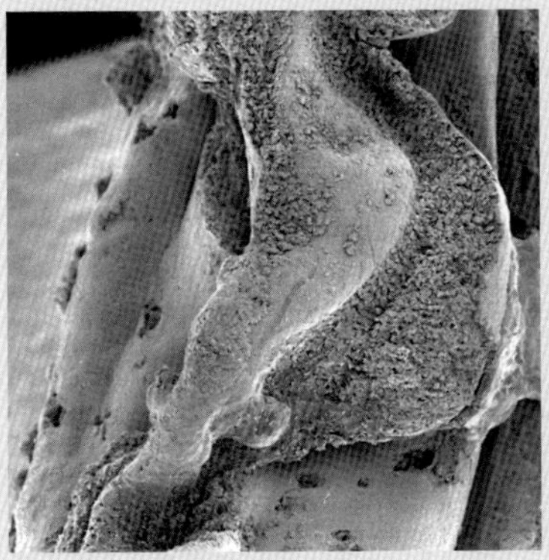

4.6.04 Ventral view of the basal endoderm of the stomach, the duodenum, and the buds of the bile duct and the dorsal pancreas. The ventral mesenchyme of the stomach has been removed to show the position of the dorsal pancreas. Embryo at week 6.

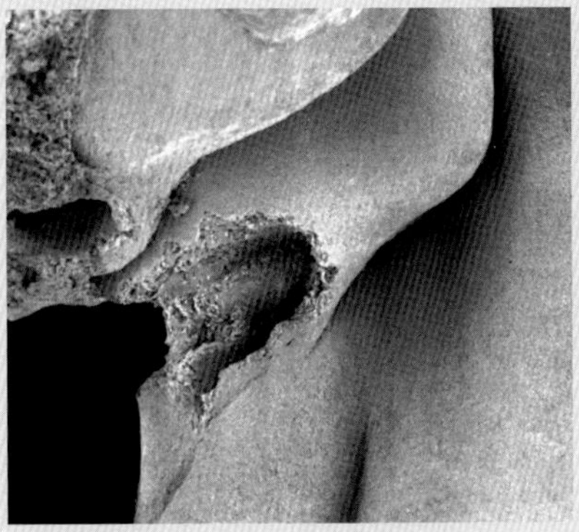

4.6.05 Ventral-caudal view of the pancreas within the mesenchyme of the stomach. Embryo at the end of week 6.

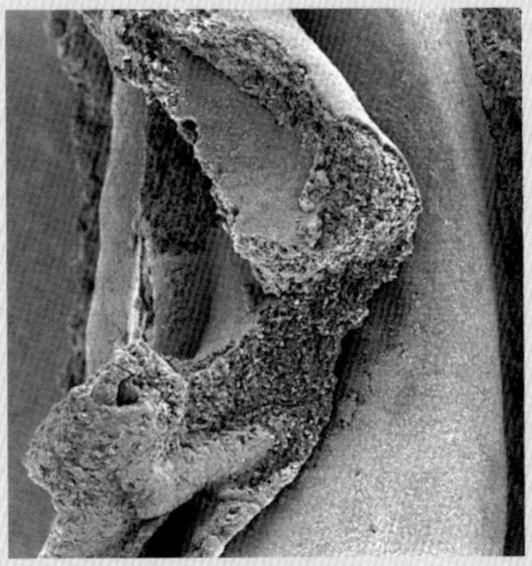

4.6.06 The caudal portion of the opened stomach has been removed to show the omental bursa. The dorsal pancreas within the mesenchyme of the stomach has been exposed. Ventral-caudal view. Embryo at the end of week 6.

4.6.07 The caudal portion of the opened stomach has been removed to show the omental bursa and its position with respect to the body of the pancreas. The pancreas is now directly beneath the epithelium of the omental bursa. The mesenchyme of the greater curvature of the stomach has been removed. Embryo at week 7.

4.6.01–4.6.16
Developmental stages of the pancreas.

Week 7

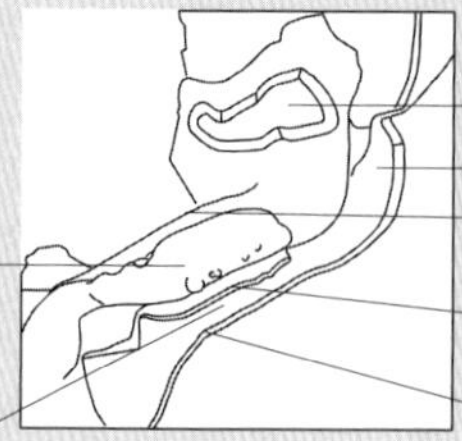

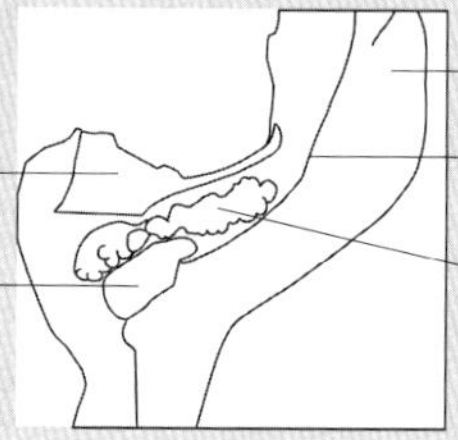

Week 8

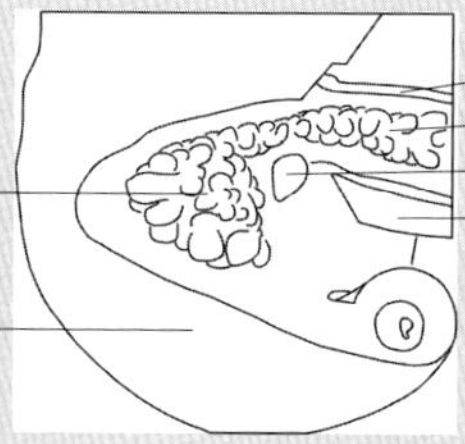

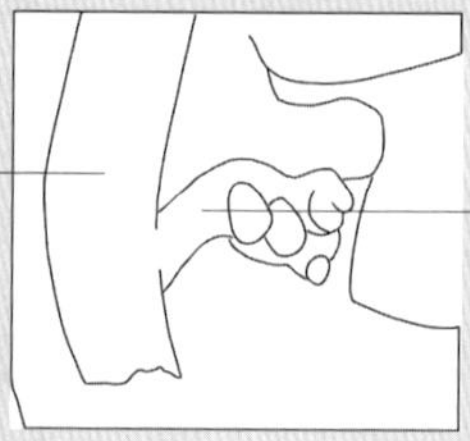

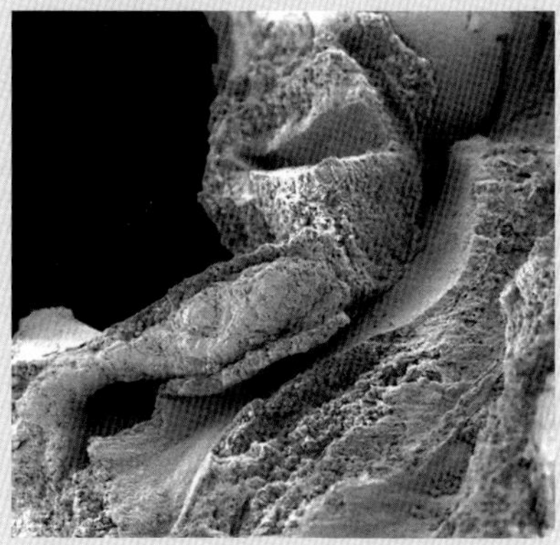

4.6.08 Caudal view of the pancreas to show its relation to the omental bursa and the retropancreatic peritoneal recess. Same embryo as in figure 4.6.07.

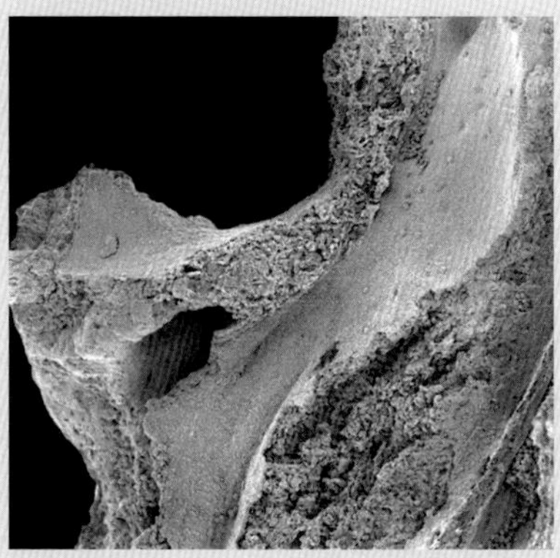

4.6.09 Caudal-left view of the primordial radix of the mesentery of the stomach, the cranial border of the retropancreatic peritoneal recess. The stomach has been removed. Embryo at week 7.

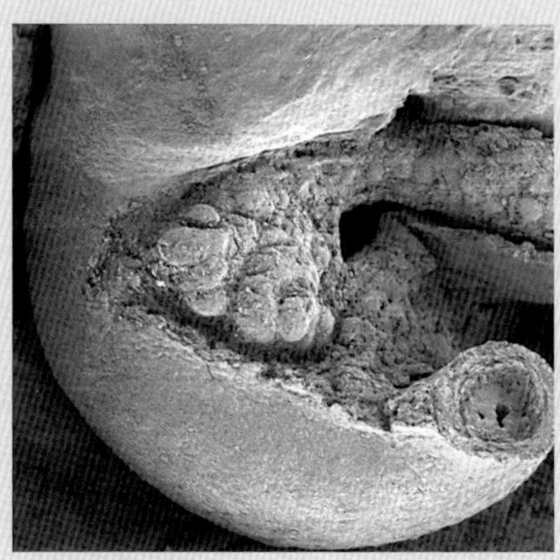

4.6.10 Ventral view of the exposed head of the pancreas. Embryo at week 8.

4.6.11 The peritoneum and the mesenchyme of the duodenum and the ventral portion of the head of the pancreas have been removed to show the origin of the bile duct and the part of the pancreatic head that has developed from the ventral pancreatic bud. Embryo at week 8.

4.6.01–4.6.16
Developmental stages of the pancreas.

25:1

Week 9

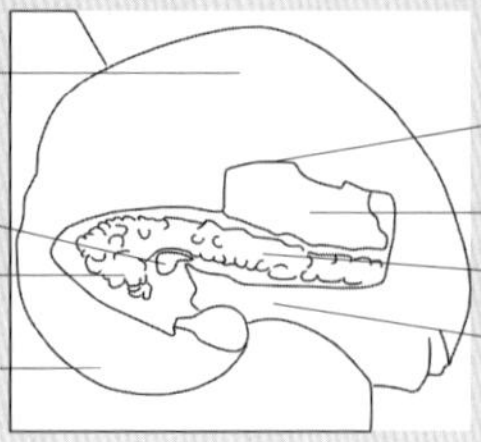

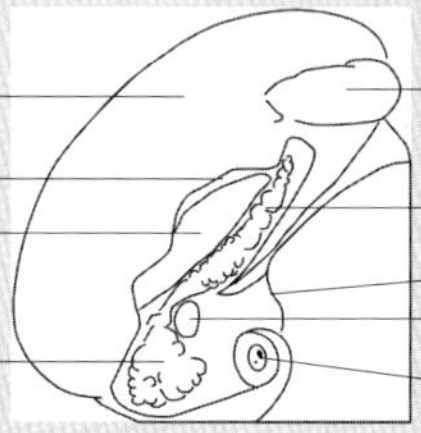

50:1

Week 8

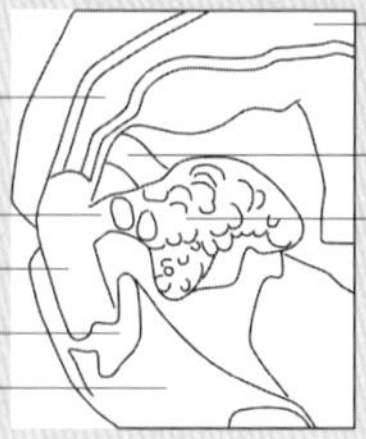

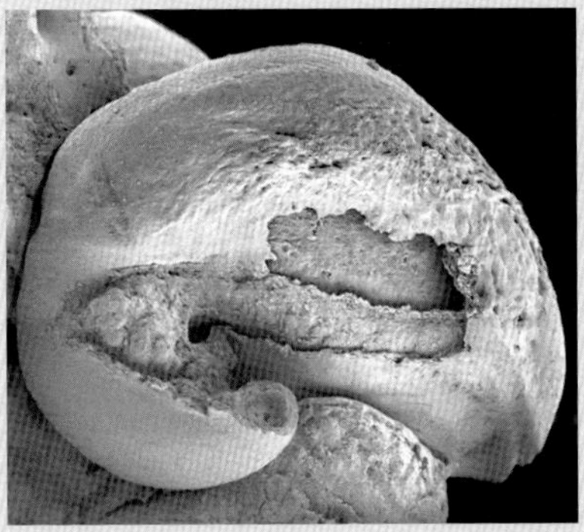

4.6.12 Caudal view of the stomach and the exposed pancreas. The omental bursa is partly opened. Embryo at week 9.

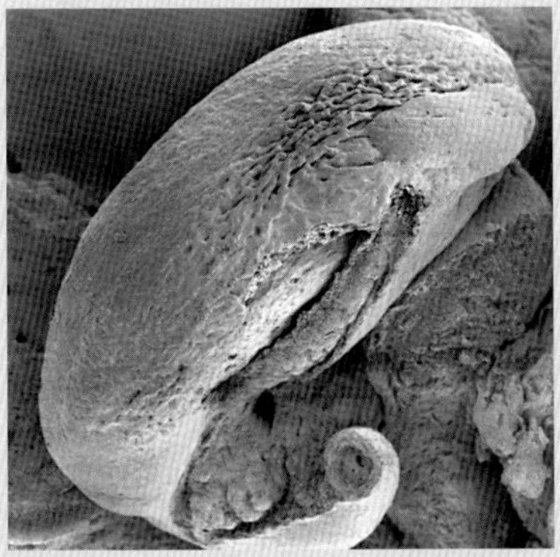

4.6.13 Caudal-left view of the stomach, the spleen, and the exposed pancreas to show the positional relation of the tail of the pancreas to the spleen. The omental bursa is partly opened. Embryo at week 9.

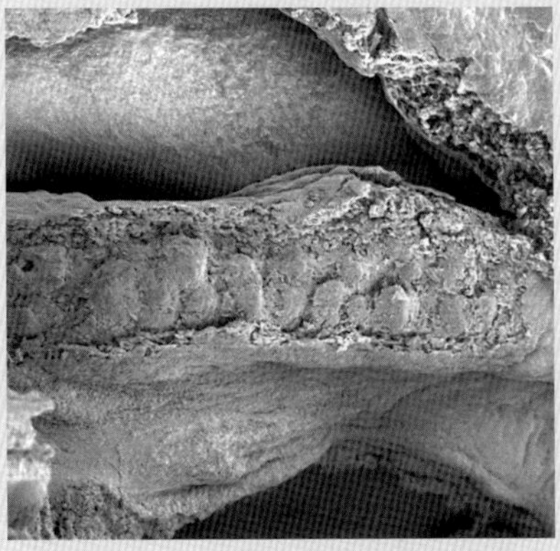

4.6.14 Details of the secreting lobules of the body of the pancreas. Embryo at week 9.

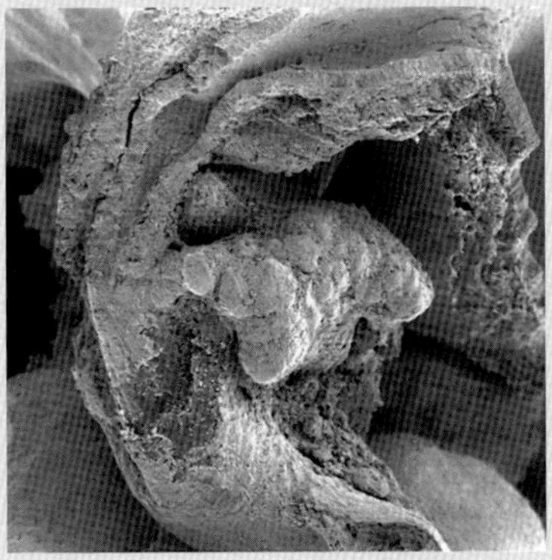

4.6.15 Ventral view of the head of the pancreas and the bile duct. The duodenum has been opened. Embryo at week 8.

Week 8

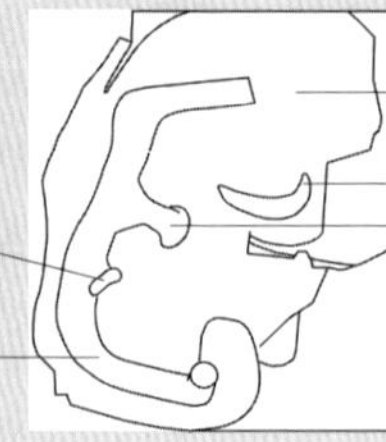

4.6.01–4.6.16
Developmental stages of the pancreas.

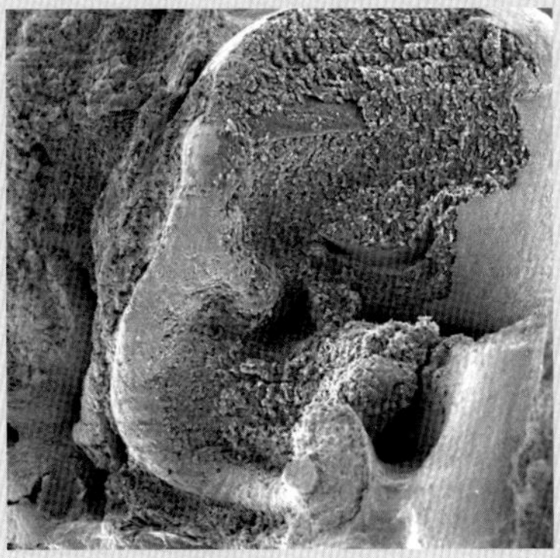

4.6.16 Ventral view of the exposed endoderm of the duodenum and the bile duct. The head of the pancreas is partly removed. Embryo at week 8.

5
Urogenital Organs

5.1–5.4
The Development of the Urinary Organs

The urinary organs arise in two main steps. Initially, beginning in week 4, the mesonephros develops as the excretory organ of the embryo; thereafter, starting at the end of week 5, the metanephros, the excretory organ of the adult, begins to develop. Both organs are connected by the mesonephric duct to the urogenital sinus, into which the mesonephric duct empties together with the gut. The urinary bladder is separated from the rectum by the urorectal septum in week 6 or 7.

The caudal part of the mesonephric duct is involved in the origin of the ureter, while the urogenital sinus gives rise to the urinary bladder and the urethra. The ductal system of the mesonephros, the mesonephric and the paramesonephric ducts (in the classical nomenclature the wolffian and the müllerian ducts) play an important role in the origin of the male and female genital ductal organs, the ductal system of the testis and the ovary.

The first anlage of the mesonephros is a bulging in the coelomic epithelium of the dorsal body wall (fig. 5.1.01–5.1.04). In its mesenchymal core, condensations of mesenchymal cells occur which form the epithelia of the mesonephric duct, the mesonephric vesicles, the glomeruli, and the mesonephric tubules (fig. 5.1.10–5.1.12). The embryonic urine is produced in the glomeruli (fig. 5.1.10, 5.1.11, 5.1.13–5.1.18) and, via the tubules and the mesonephric duct, reaches the region of the still undivided urogenital sinus (fig. 5.2.02–5.2.11), which by the end of week 7 or 8 is divided into the bladder and the rectum (fig. 5.2.03–5.2.08, 5.2.11–5.2.14).

The metanephros, the kidney of the fetus and of the adult, arises as an epithelial outgrowth of the mesonephric duct near the urinary bladder, the ureteric bud (fig. 5.3.01–5.3.05). The urine-conducting system, including the pelvis, the calyces, and the collecting ducts, arises from the epithelial ureteric bud (fig. 5.3.17, 5.3.18). The urine-producing system originates from the metanephric (metanephrogenic) blastema which develops from the mesenchyme that abuts the epithelium of the ureteric bud (fig. 5.3.04–5.3.11).

The ureteric bud elongates and divides consecutively into up to about 16 branches, which form the pelvis of the kidney, the calyces, and the collecting ducts. Finally, the expanded ends of the collecting ducts (fig. 5.3.19) form anchor-like ramifications (fig. 5.3.20–5.3.24). The mesenchymal cells in contact with them become arranged as epithelia (fig. 5.3.20, 5.3.22) and develop into the urine-producing glomeruli and the urine-concentrating tubular system of the nephron (fig. 5.3.23, 5.3.24).

The ureter develops from the common outlet of the ureteric bud and the mesonephric duct (fig. 5.3.05–5.3.08). Due to the positional development of the kidney, the ascent to the lumbar level, the ureter elongates considerably (fig. 5.3.13–5.3.16).

The urinary bladder arises in the region where the allantoic diverticulum branches off from the end-gut, the urogenital sinus (in the older literature often called cloaca) (fig. 5.2.02).

The urogenital sinus is separated into urinary bladder and rectum by the urorectal septum (fig. 5.2.03–5.2.09). The free edge of this septum folds up into the urogenital sinus, and in week 6 the separation of the bladder from the rectum is completed (fig. 5.2.11–5.2.14). The most caudal portion of the gut, the postanal gut (fig. 5.2.09), remains small and is eventually absorbed.

The common orifice of the mesonephric duct and the ureter (fig. 5.4.03–5.4.05) becomes duplicated due to the growth of the spur between the mesonephric duct and the ureter (fig. 5.4.03, 5.4.05). This spur reaches the lumen of the bladder, thus creating two isolated orifices. The exit of the ureter is situated more lateral, whereas the orifice of the mesonephric duct is situated more medial (fig. 5.4.06–5.4.09). Due to the growth of the bladder, the exit of the mesonephric duct reaches its position in the proximal portion of the urethra (fig. 5.4.09, 5.4.10).

5.1

Abbreviations

d	diaphragm
il	intestinal loop
mn	mesonephros
ra	right atrium
rv	right ventricle
1	mesonephric duct
2	mesonephric vesicle
3	mesonephric glomerulus
4	secretory tubule, orifice
5	secretory tubule
6	bare area of the liver
7	right lobe of the liver
8	region of the developing gonad/gonad
9	lung
10	suprarenal gland
11	paramesonephric duct

5.2

Abbreviations

1	urethra, pelvic part
2	urethra, phallic part
3	urinary bladder
4	anal groove
5	rectum, endoderm
6	peritoneum of the rectum
7	anus
8	müllerian tubercle

5.3

Abbreviations

mn	mesonephros
1	metanephros
2	ureter
3	mesonephric duct
4	common outlet for meso- and metanephros
5	urinary bladder
6	urethra, pelvic part
7	urethra, phallic part
8	endoderm of the rectum
9	peritoneum of the rectum

Week 4

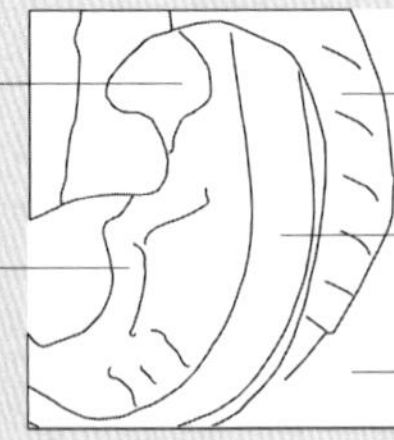

Week 5

5.1.01–5.1.18
Developmental stages of the mesonephros.

5.1.01–5.1.04
The urogenital fold.

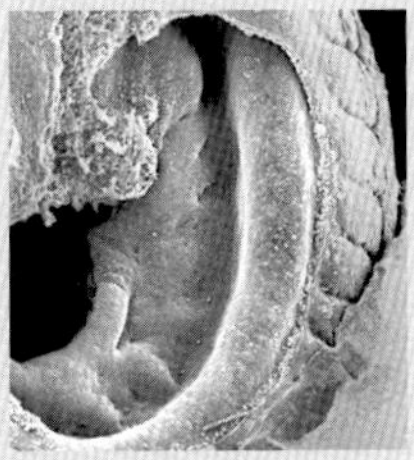

5.1.01 Embryo at the end of week 4. Ventral-left view.

5.1.02 individual variations of the urogenital fold. Ventral-right view. Embryo at week 5.

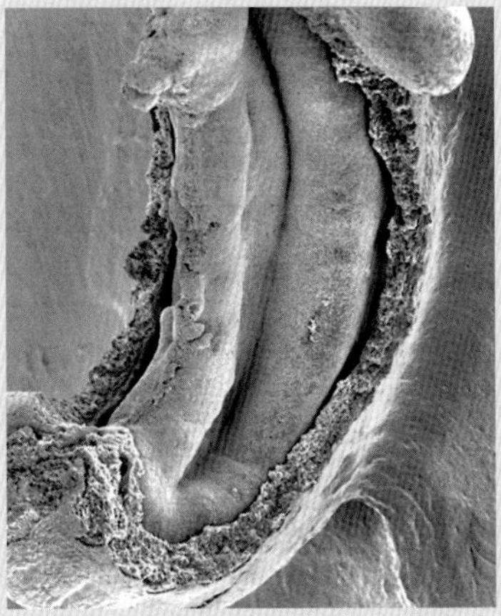

5.1.03 Individual variations of the urogenital fold. Ventral-left views. Embryos at week 5.

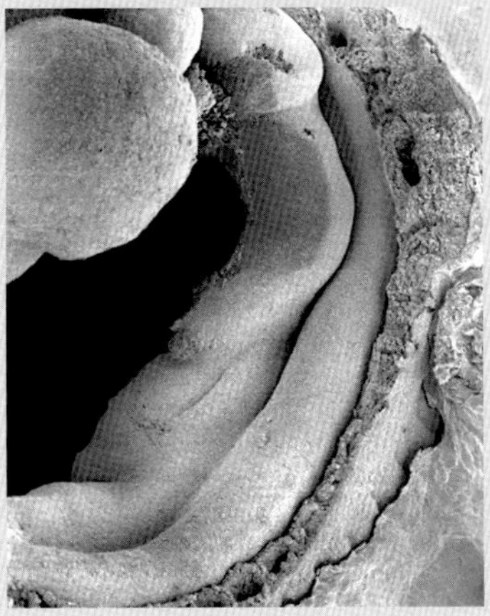

5.1.04

5.1.01–5.1.18
Developmental stages of the mesonephros.

5.1.05–5.1.09
The subdivision of the urogenital fold into the mesonephric and the gonadal fold.

Week 6

Week 7

Week 8

5.1.05

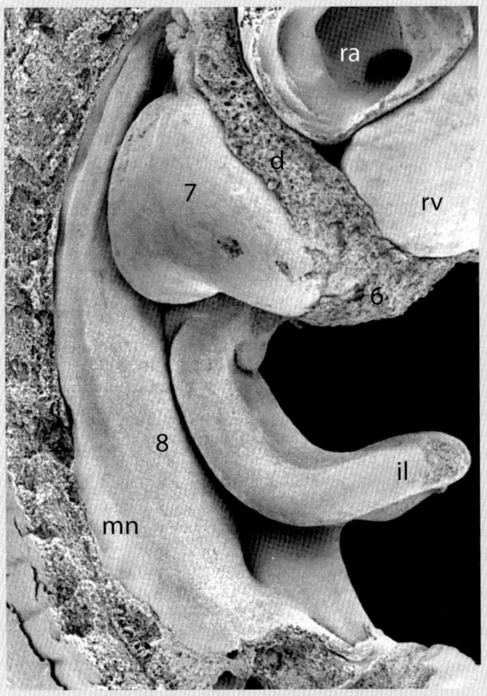

Slight indication of the slit-like demarcation between the mesonephros and the gonad. Lateral-right view. Embryo at the end of week 6.

5.1.06 / 07

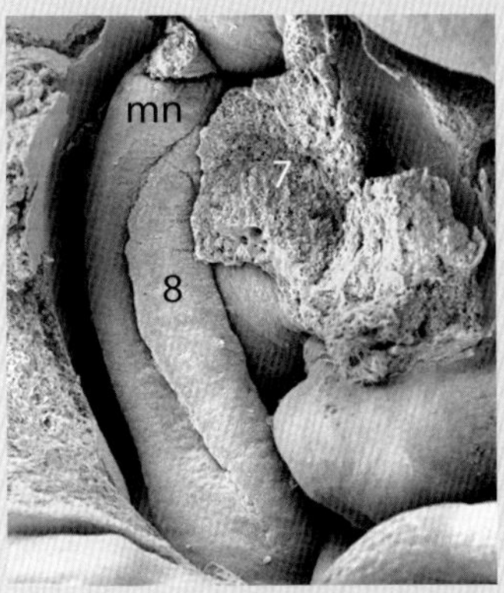

Due to the recess-like deepening between mesonephros and gonad and the developing recess on the medial side, the gonad achieves a greater free peritoneal surface. The right lobe of the liver has been removed. Ventral views. Embryos at week 7 (5.1.06, 5.1.07) and 8 (5.1.08).

5.1.08

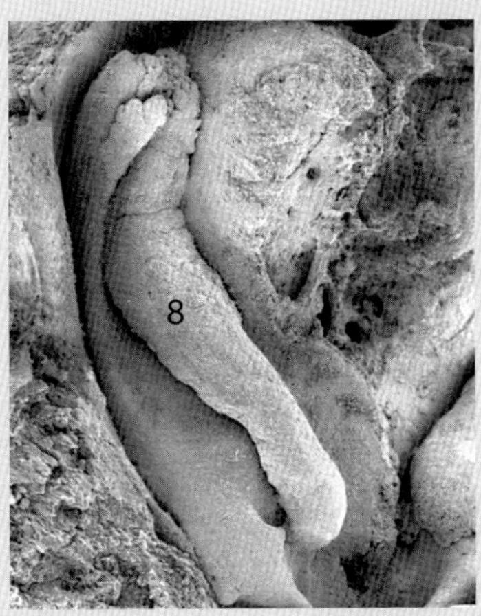

Week 9

5.
Urogenital Organs

5.1–5.4
The Development of
the Urinary Organs

5.1.01–5.1.18
Developmental stages of the mesonephros.

5.1.05–5.1.09
The subdivision of the urogenital fold into the mesonephric and the gonadal fold.

5.1.10–5.1.18
The secretory organs of the mesonephros.

150:1

Week 5 / 7

Week 9

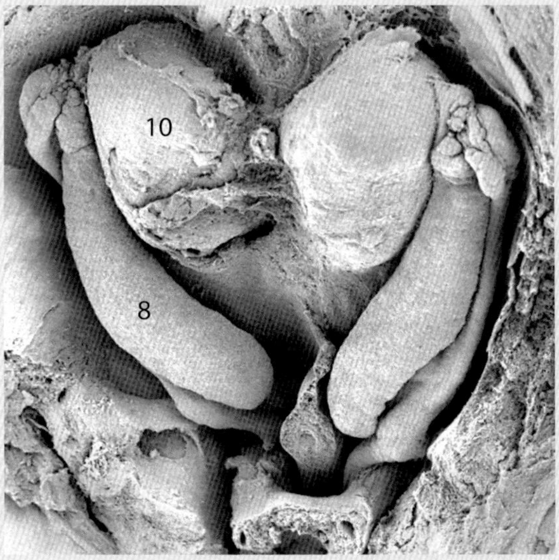

5.1.09

Embryo at week 9. Ventral view. Due to their increase in size the gonads partly overlap the mesonephos.

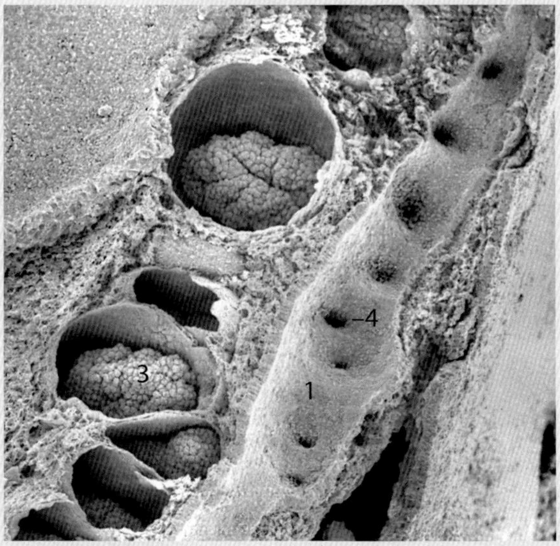

5.1.10 / 11

5.1.10
Ventral view of the left mesonephros. The mesonephric vesicles and the mesonephric (wolffian) duct have been exposed. Embryo at week 5.

5.1.11
Ventral view of the left mesonephros. Some of the mesonephric vesicles and the mesonephric (wolffian) duct have been opened. Embryo at week 7.

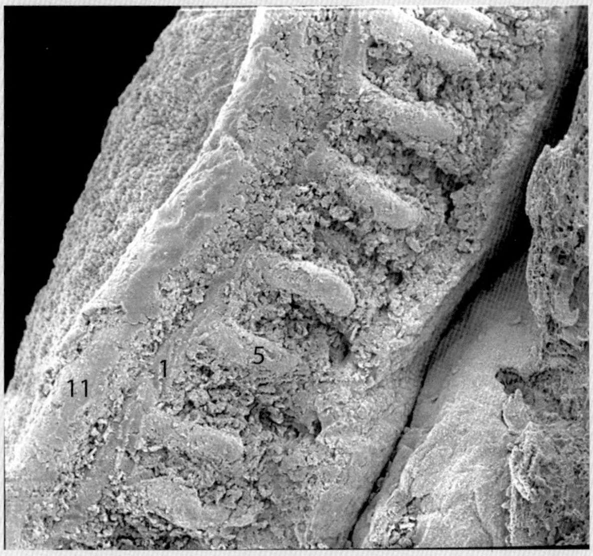

5.1.12

Lateral left view of the left mesonephros. The mesonephric (wolffian) duct and the collecting tubules have been exposed. The parameso-nephric duct is partly covered by peritoneum. The glomeruli remain hidden. Embryo at week 9.

Week 6 / 7

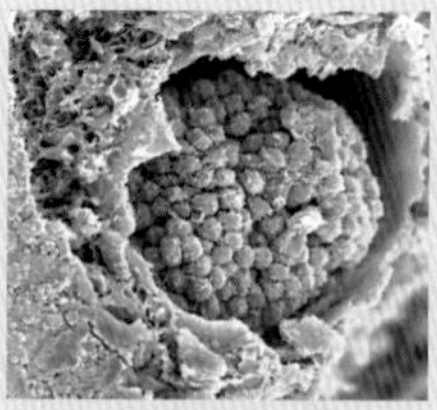

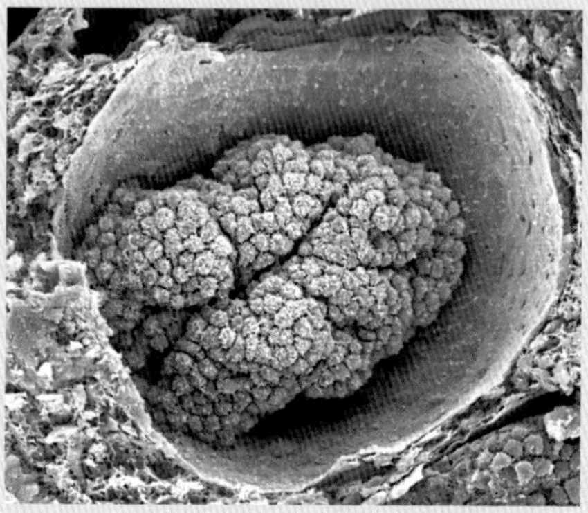

5.1.01–5.1.18
Developmental stages of the mesonephros.

5.1.10–5.1.18
The secretory organs of the mesonephros.

Week 5

Week 6

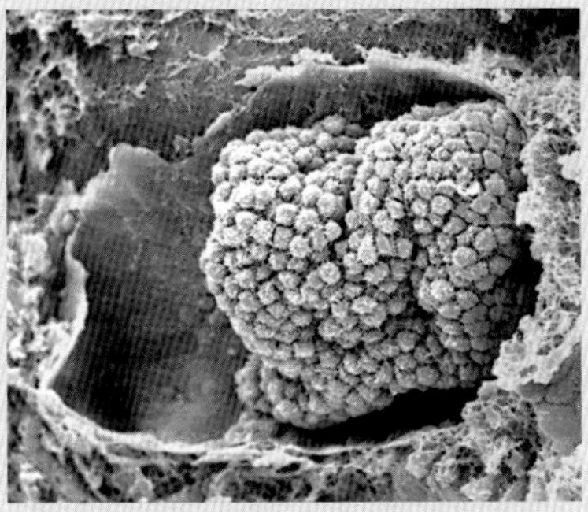

5.1.13 / 14

5.1.15 / 16

In the opened mesonephric vesicles, the capillary loop of the glomeruli covered by the inner layer of Bowman's capsule is visible. Due to the advancing development of clefts, the originally smooth surface becomes re-arranged into an increasing number of capillary loops. Embryos at week 6 and 7.

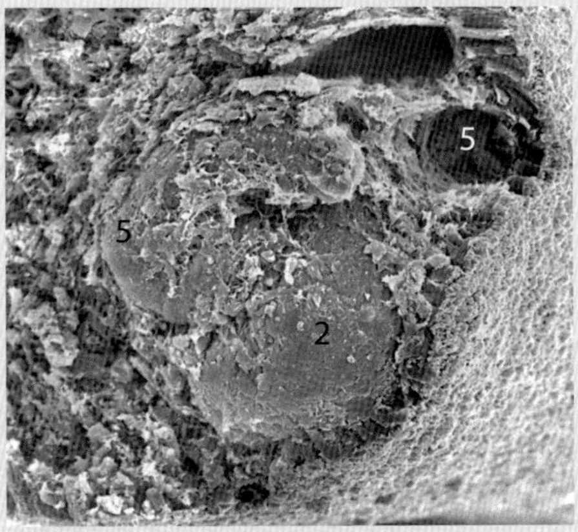

5.1.17

Surface of a mesonephric vesicle with the outlet of the secretory tubule. Embryo at week 5.

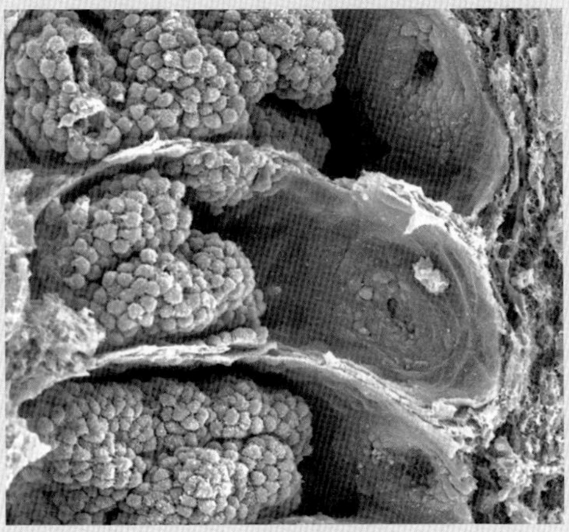

5.1.18

An opened mesonephric vesicle shows the capillary loop of the glomerulus and the outlet of the mesonephric tubule. Embryo at the end of week 6.

5.2.01–5.2.14

Development of the urogenital sinus and the urinary bladder.

Week 5

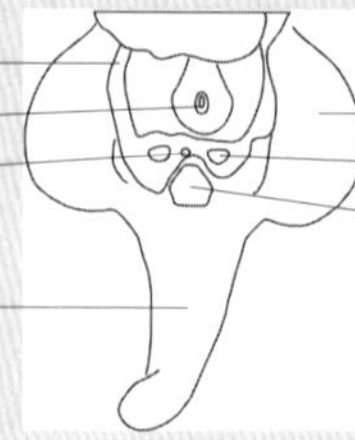

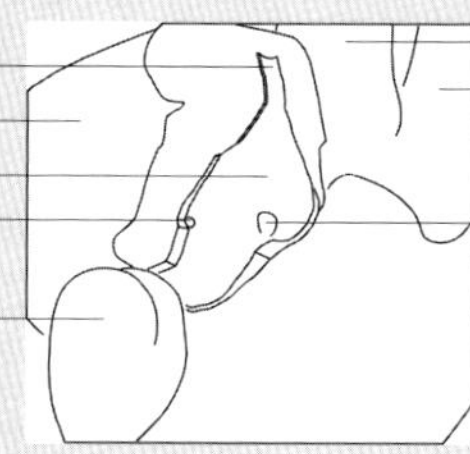

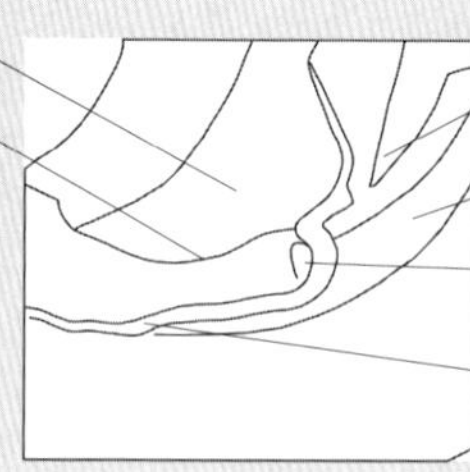

Week 6

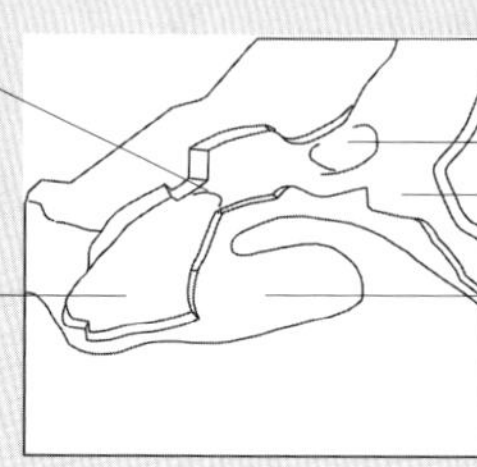

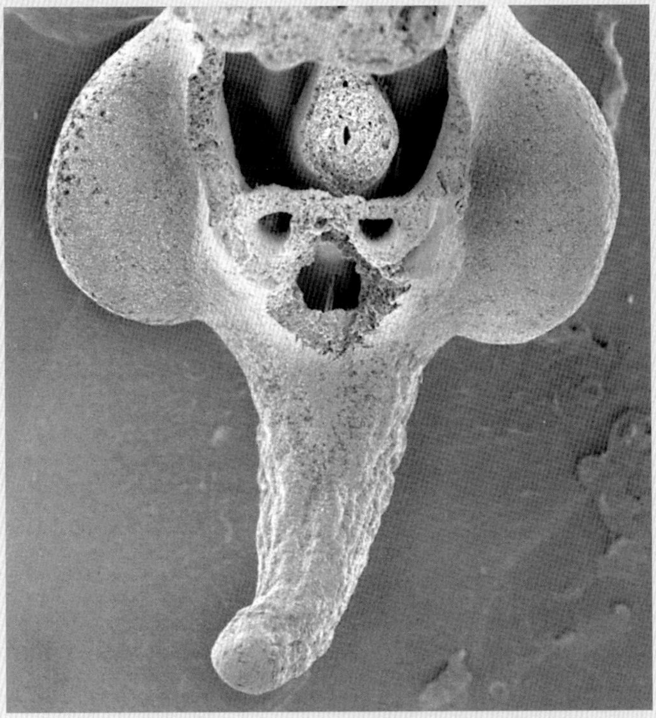

5.2.01 Overall view of the position of the urogenital sinus and the urinary bladder. The peritoneal cavity and the urogenital sinus are opened. Ventral and cranial view. Embryo at week 5.

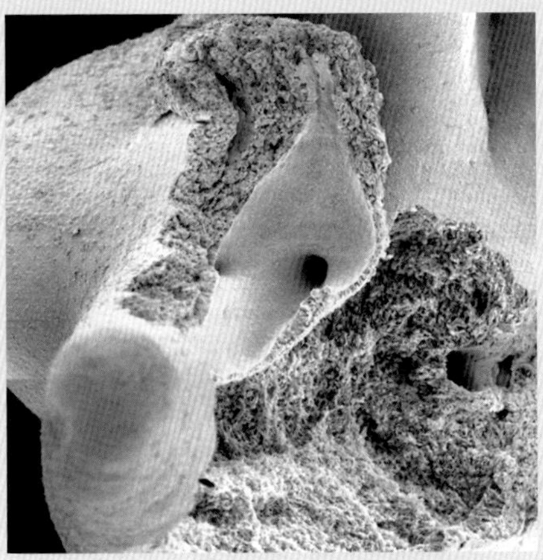

5.2.02 Ventral and left view of the exposed and opened urogenital sinus. Embryo at week 5.

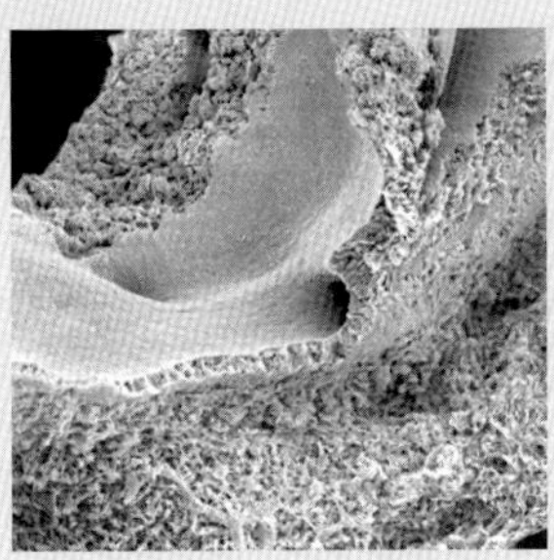

5.2.03 Left lateral view of the exposed and opened urogenital sinus shows the first indication of the urorectal septum. Embryo at week 5.

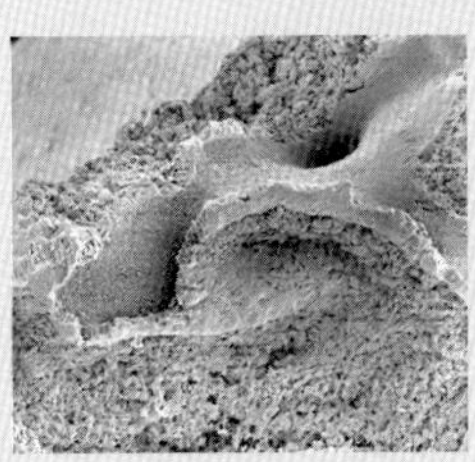

5.2.04 Ventral and left view of the exposed and opened urogenital sinus and the urinary bladder. The urorectal septum has developed further. Embryo at week 6.

Week 6

5.2.01–5.2.14

Development of the urogenital sinus and the urinary bladder.

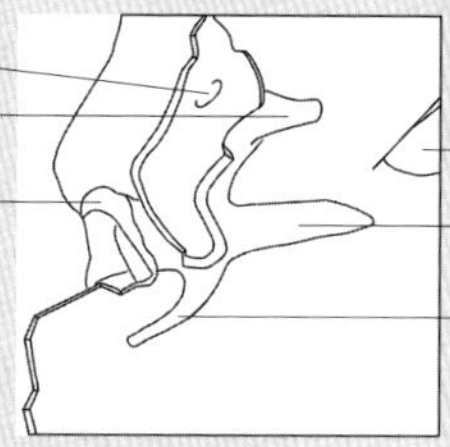

Week 7

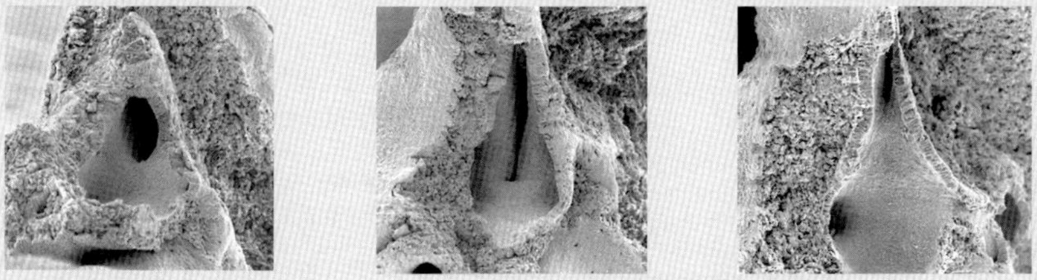

5.2.05 / 06 / 07

Cranial view of the opened urinary bladder shows different individual developmental stages of the urorectal septum. Embryos at week 6.

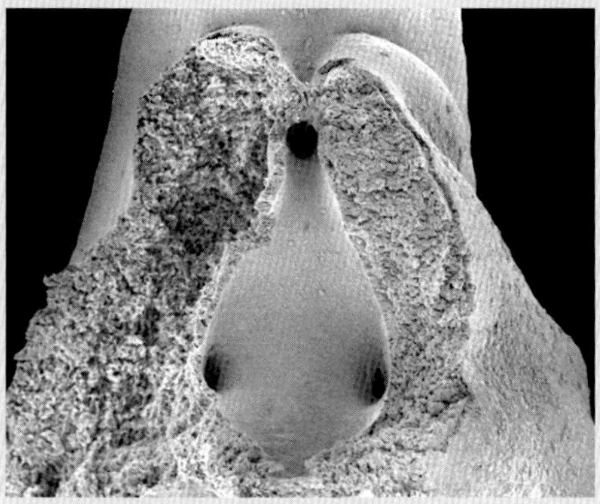

5.2.08

Cranial view of the opened urinary bladder. The separation of the anal canal from the urinary bladder is almost complete. Embryo at the end of week 6.

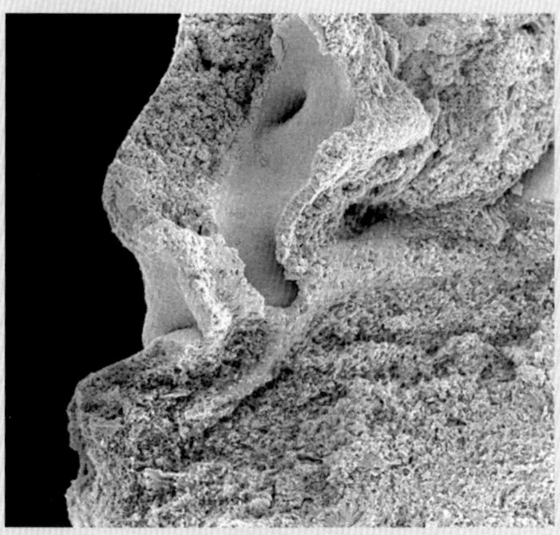

5.2.09

Lateral-left view of the exposed and opened urogenital sinus, the urinary bladder and the postanal rectum. Embryo at week 6.

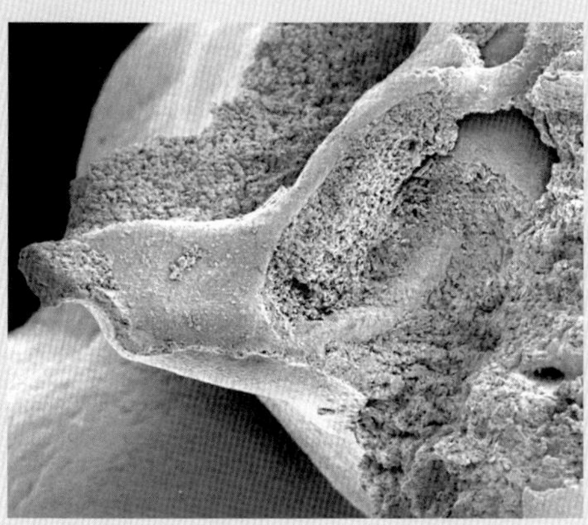

5.2.10

Lateral-left view of the external form of the urinary bladder, the urethra and the urogenital sinus. Embryo at week 7.

Week 7

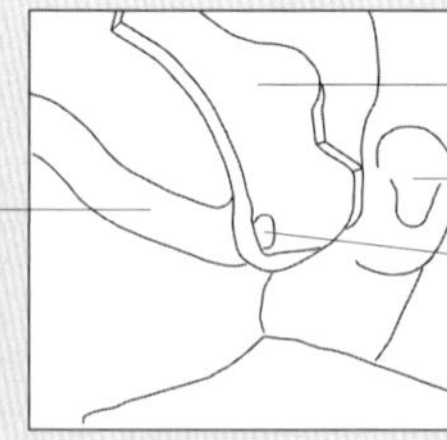

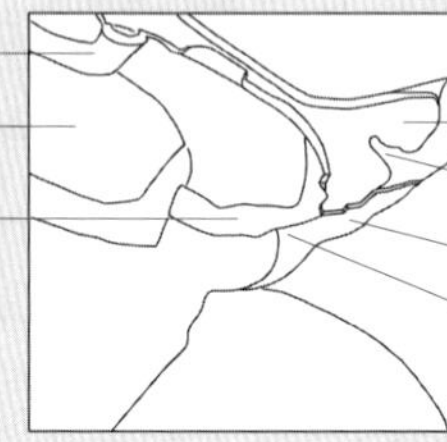

5.2.01–5.2.14

Development of the urogenital sinus and the urinary bladder.

5.2.11–5.2.14

Stages of the development of the perineum. The urinary bladder and the urethra have been opened.

Week 8 / 9

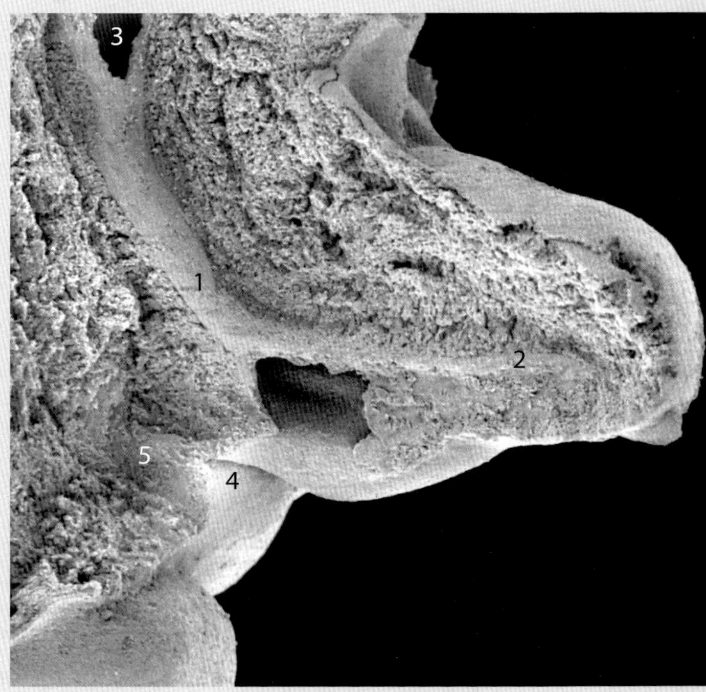

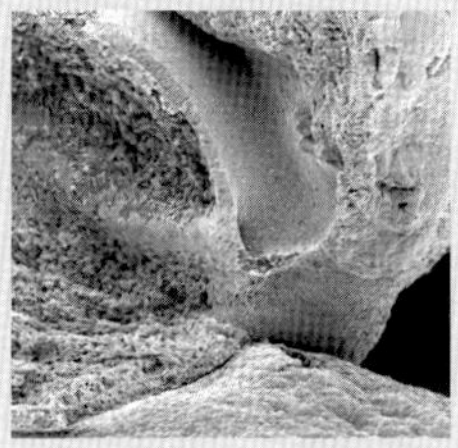

5.2.11

The anal canal still opens into the urogenital sinus. Lateral-right view. Embryo at week 7.

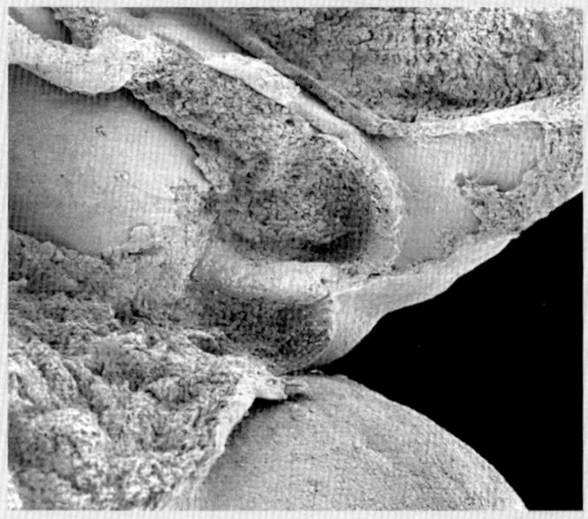

5.2.12

The separation of the anal canal from the urinary bladder is advanced. Lateral-right view. Embryo at the end of week 7.

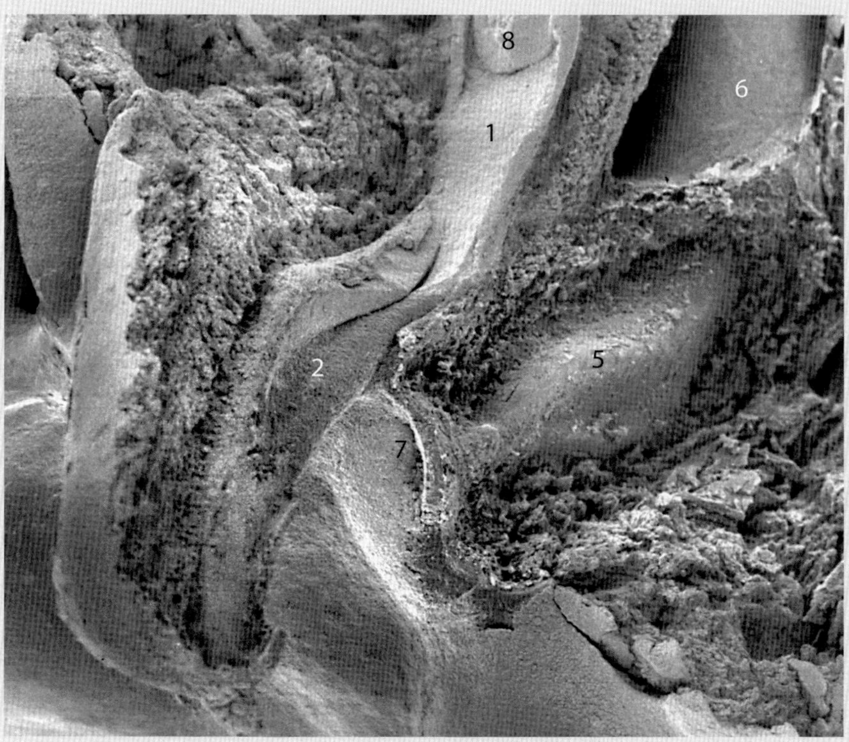

5.2.13 / 14

5.2.13
The separation of the anal canal from the urinary bladder is complete. Lateral-right view. Embryo at week 8.

5.2.14
Embryo at week 9. Lateral-left view.

Week 5

5.3.01–5.3.24
Developmental stages of the metanephros (kidney).

5.3.01–5.3.11
Development of the left ureteric bud. Lateral-left view.

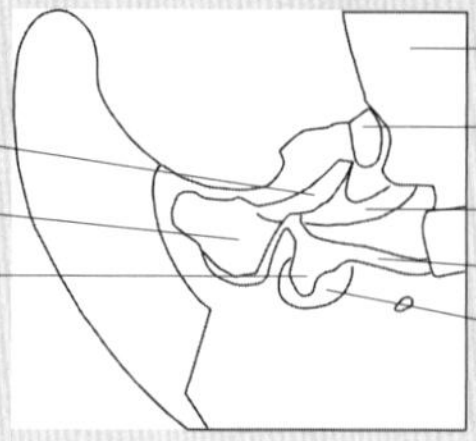

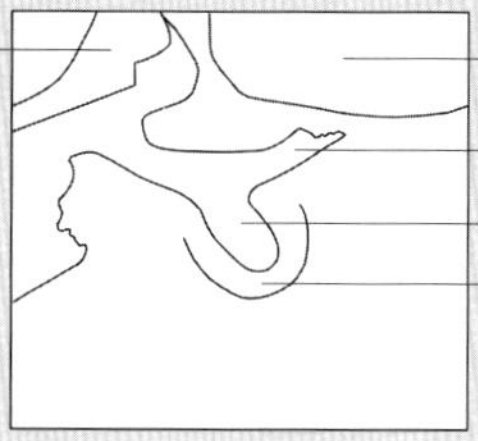

200:1

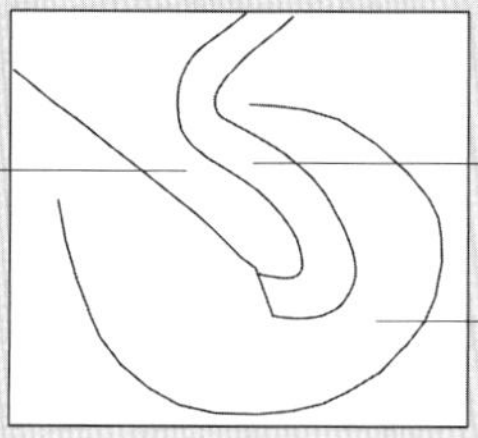

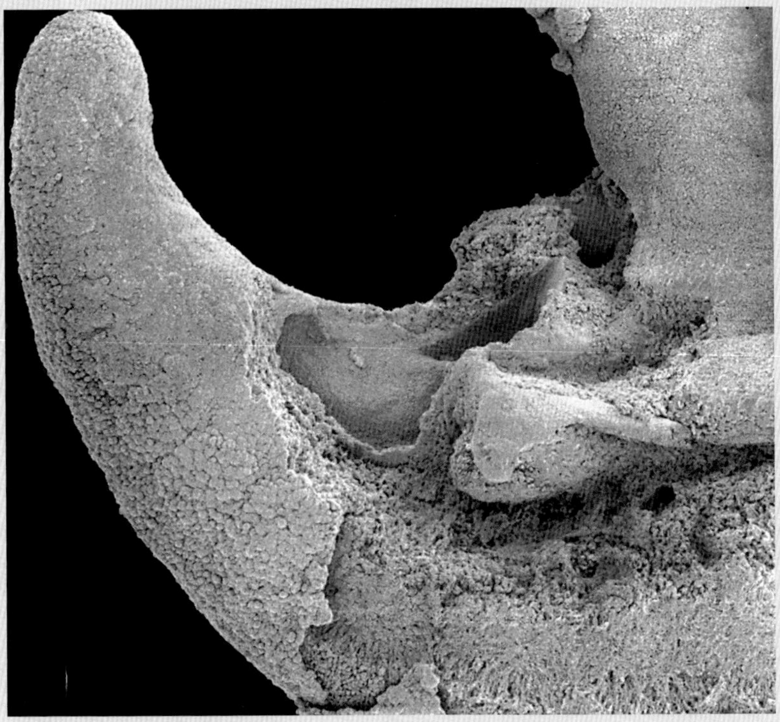

5.3.01 Overall left view of the position of the ureteric bud, the mesonephric (wolffian) duct, the urogenital sinus, and the urinary bladder. Embryo at week 5.

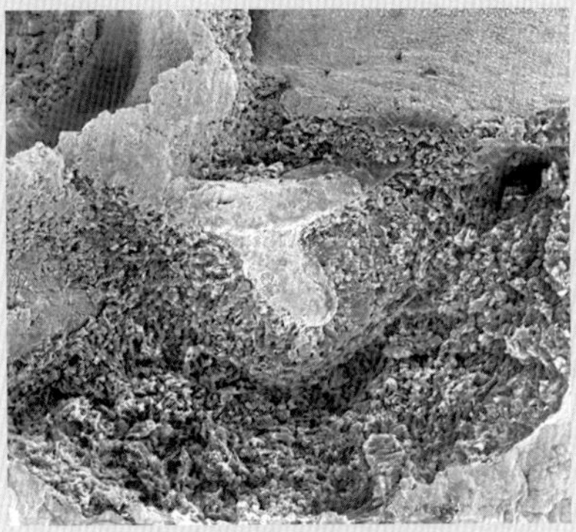

5.3.02 The epithelium of the ureteric bud, embedded in the metanephric blastema, has been exposed. Embryo at week 5.

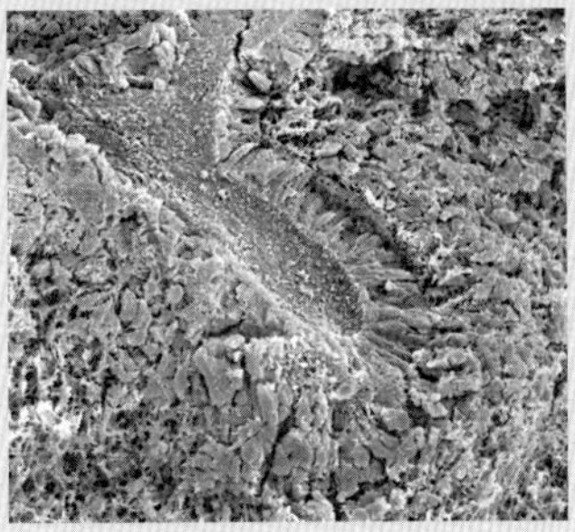

5.3.03 The ureteric bud has been opened. Embryo at week 5.

5.3.01–5.3.24
Developmental stages of the metanephros (kidney).

5.3.01–5.3.11
Development of the left ureteric bud. Lateral-left view.

Week 6

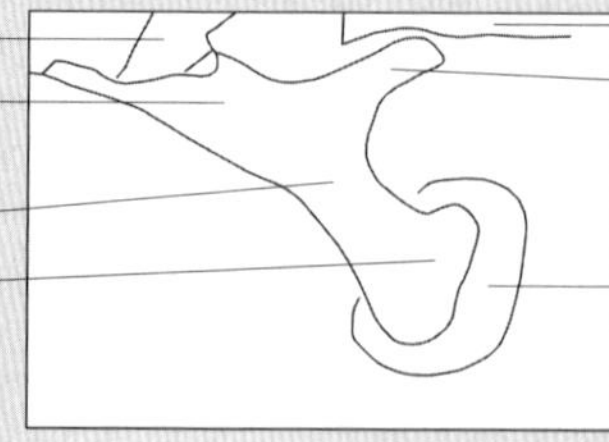

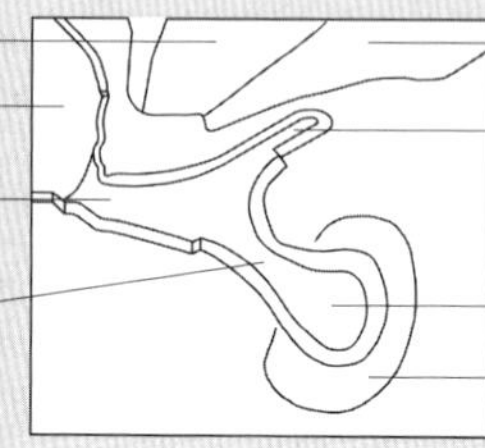

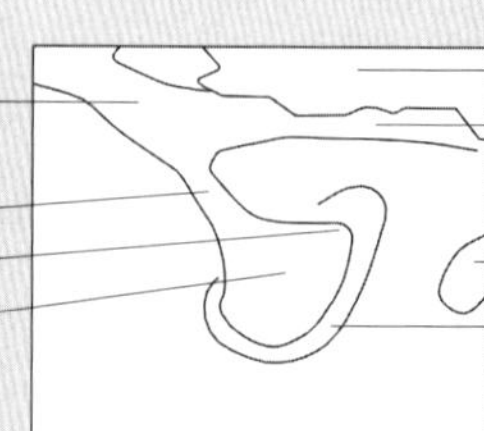

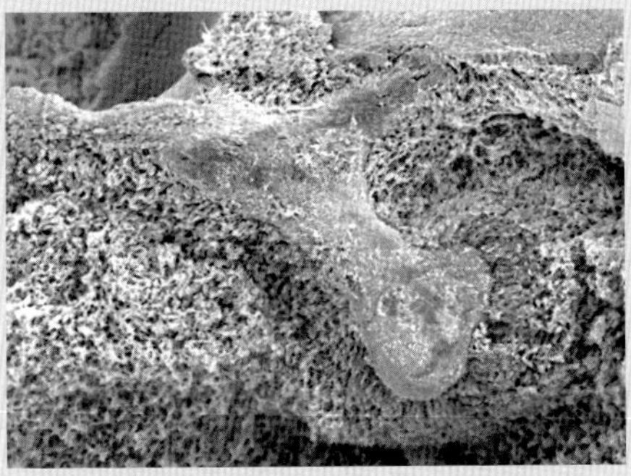

5.3.04 The epithelium of the ureteric bud has been exposed. The ureteric bud, embedded in the metanephric blastema, moves away from the mesonephric duct, thus giving rise to the ureter. Embryo at week 6.

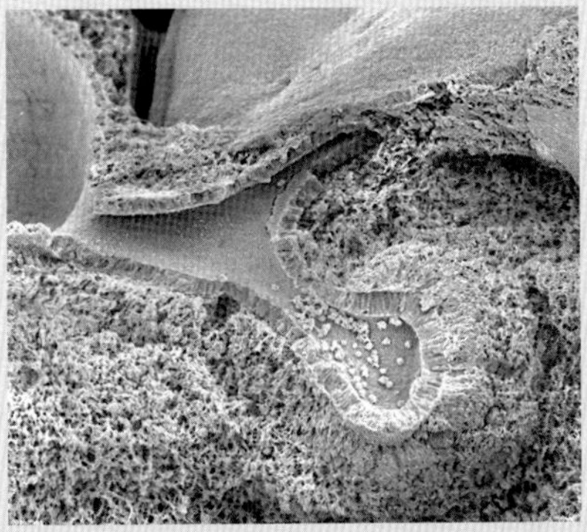

5.3.05 Same embryo as in figure 5.3.04. The ureteric bud, the ureter, and the mesonephric duct have been opened.

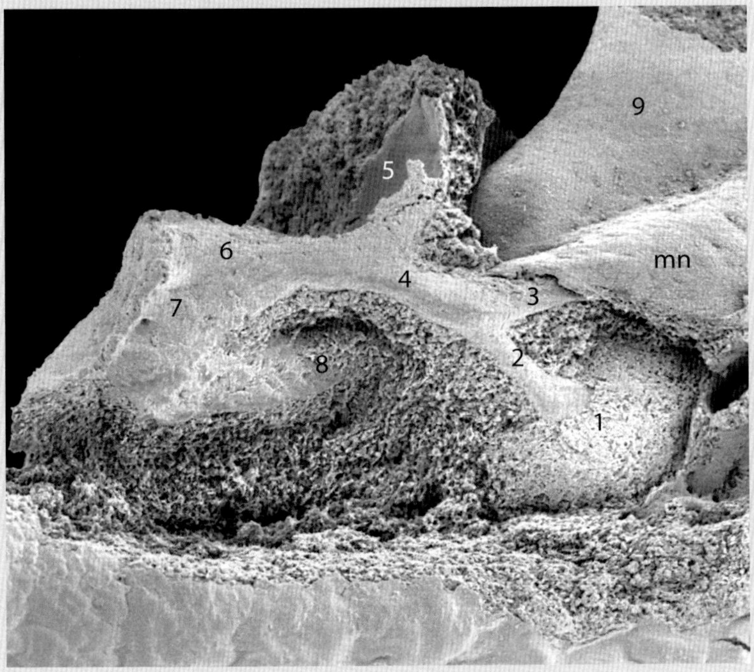

5.3.06 Lateral-left view of the metanephric blastema enveloping the ureteric bud. The epithelium of the ureter, the mesonephric duct, and the urogenital sinus have been exposed. Embryo at week 6.

5.3.07 Same embryo as in figure 5.3.06. The epithelium of the ureteric bud and the mesonephric duct have been exposed.

Week 6

5.3.01–5.3.24
Developmental stages of the metanephros (kidney).

5.3.01–5.3.11
Development of the left ureteric bud. Lateral-left view.

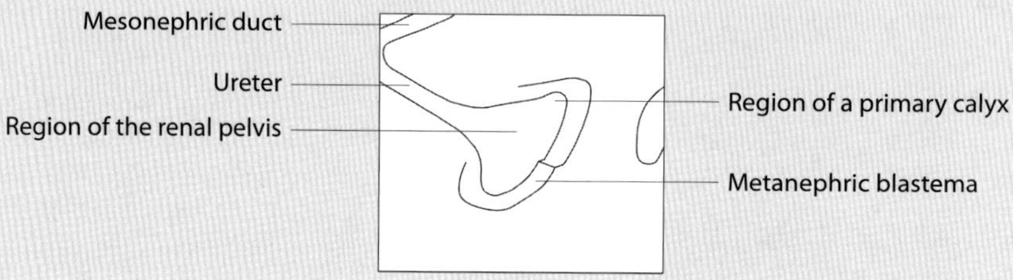

500:1

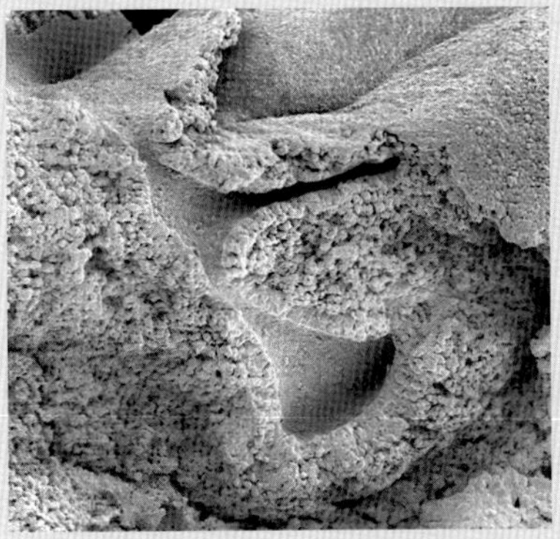

5.3.08 Lateral-left view of the opened ureteric bud, the ureter and the mesonephric duct. Due to the elongation in the cranio-caudal direction, the ureteric bud becomes kidney-shaped and could be called the pelvis of the future kidney. Embryo at the end of week 6.

5.3.09 Lateral-left view of the exposed epithelium of the pelvis and the ureter. Embryo at the end of week 6.

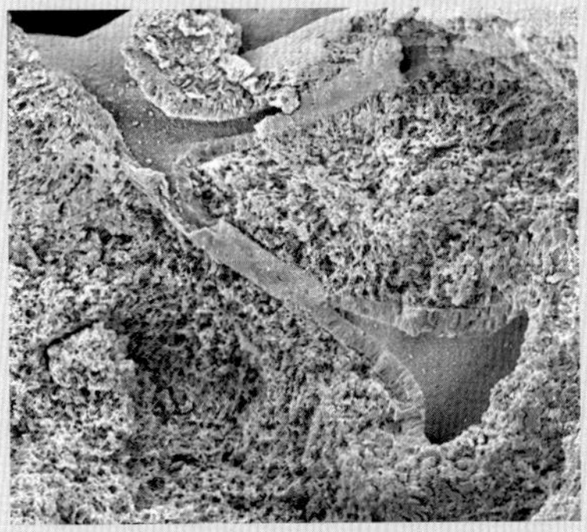

5.3.10 Same embryo as in figure 5.3.09. The pelvis has been opened. The ureter is lengthened.

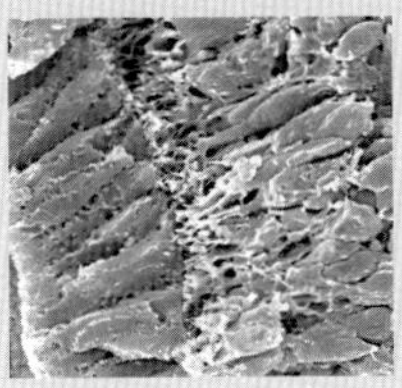

5.3.11 Detailed view of the contact zone of the pelvis with the metanephric blastema.

5.
Urogenital Organs

5.1–5.4
The Development of the Urinary Organs

5.3.01–5.3.24
Developmental stages of the metanephros (kidney).

5.3.12–5.3.16
Development of the position of the left kidney. Lateral-left view.

Week 6

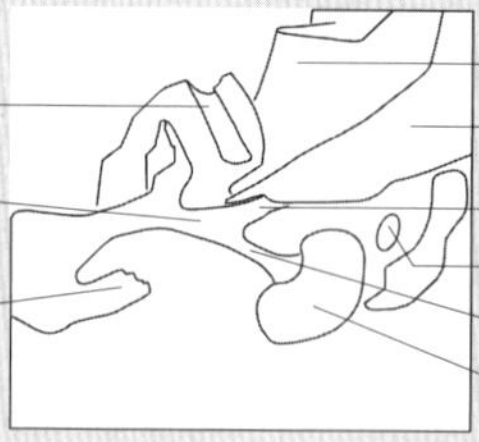

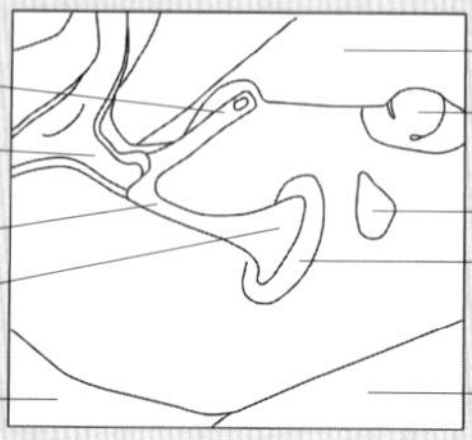

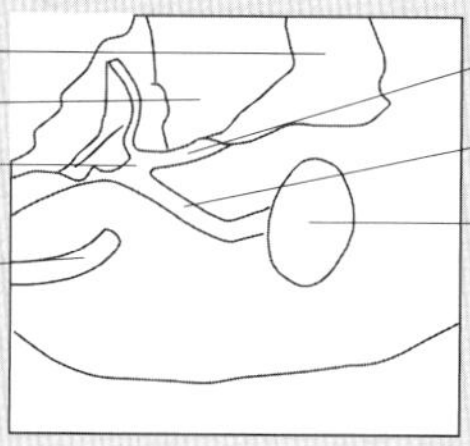

Week 7

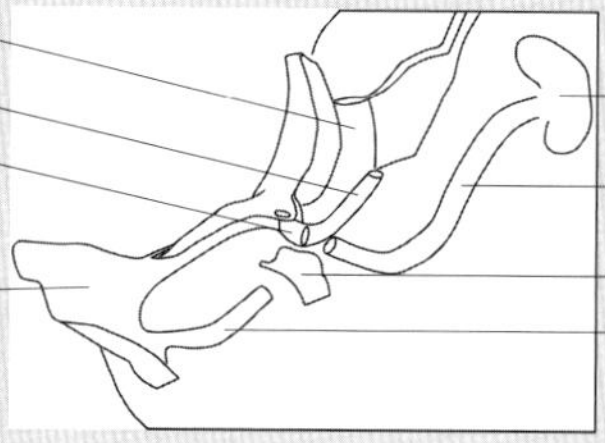

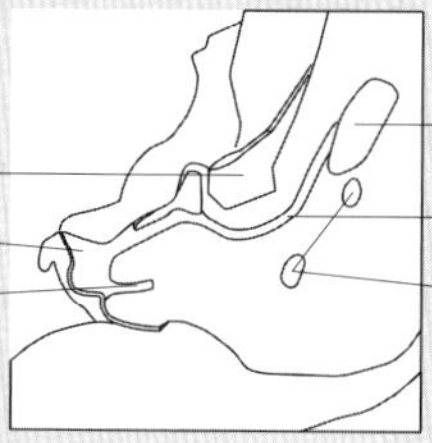

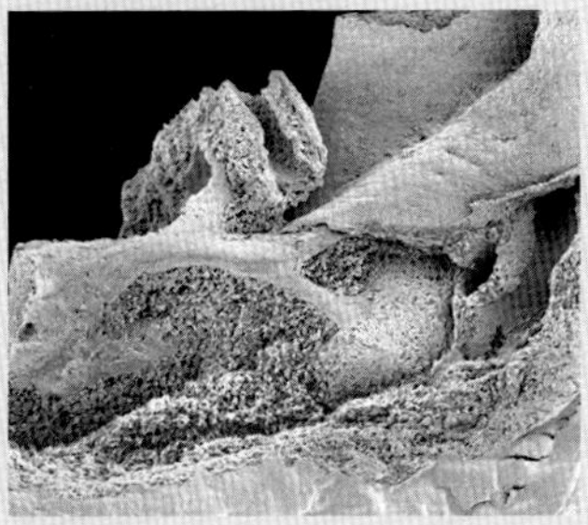

5.3.12 Embryo at week 6.

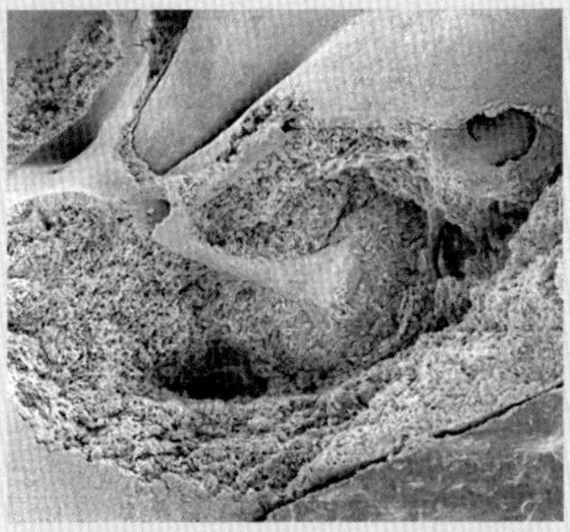

5.3.13 Embryo at the end of week 6.

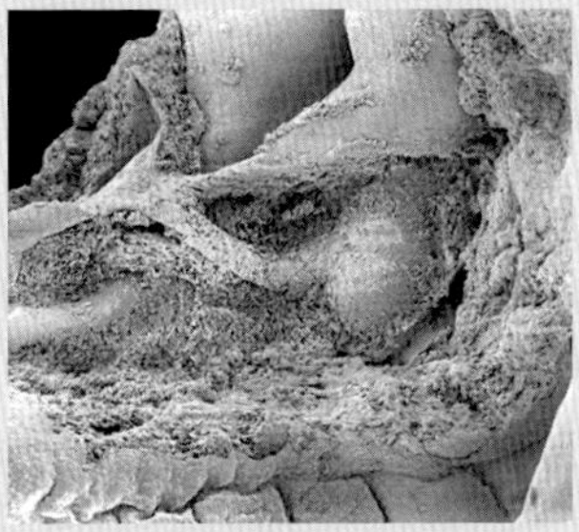

5.3.14 Embryo at the end of week 6. More advanced stage.

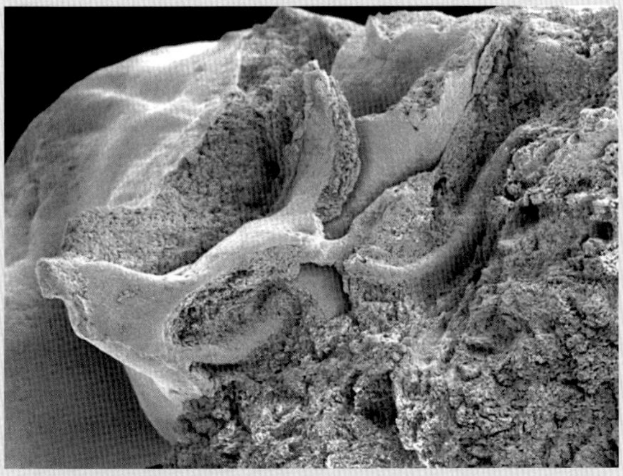

5.3.15 Embryo at week 7.

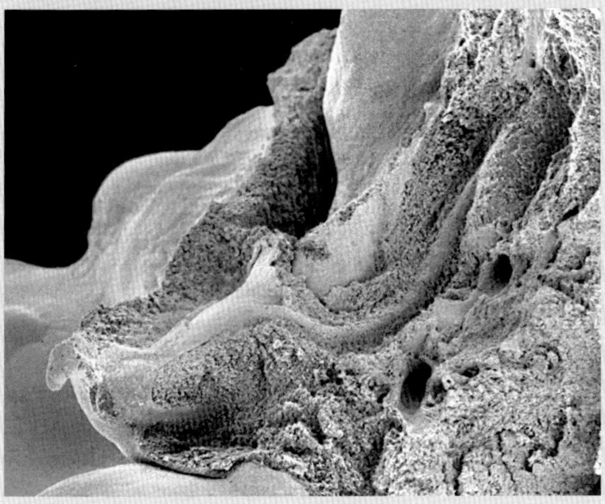

5.3.16 Embryo in the middle of week 7.

5.3.01–5.3.24
Developmental stages of the metanephros (kidney).

5.3.19–5.3.24
Early development of the nephron.

40:1

Week 7 / 9

500:1

Week 7

250:1

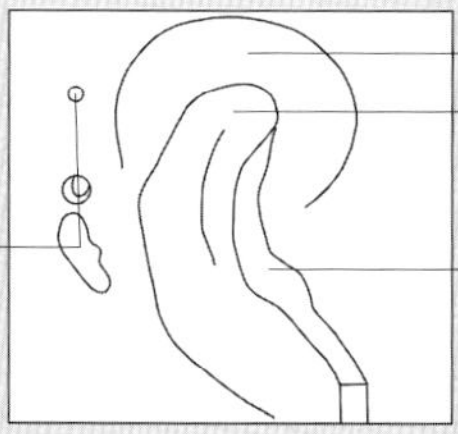

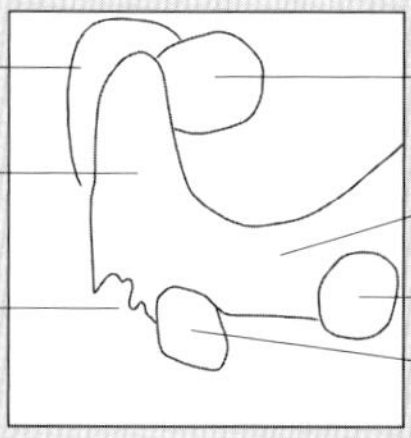

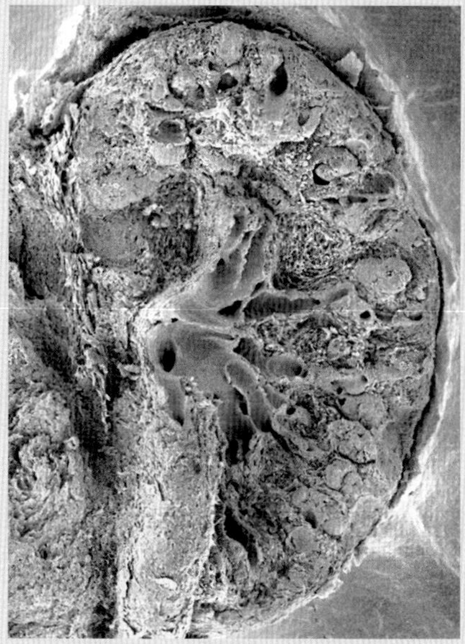

5.3.17 / 18 Ventral overall view of the kidney. Embryos at weeks 7 and 9.

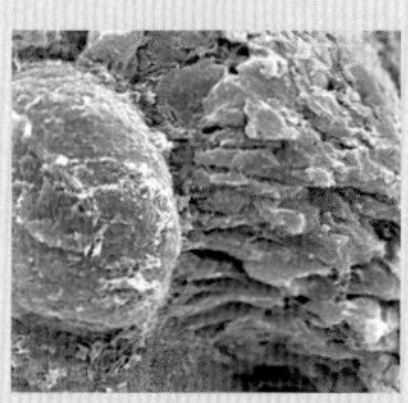

5.3.19 The contact zone of a collecting duct with the metanephric blastema is shown. Embryo at week 7.

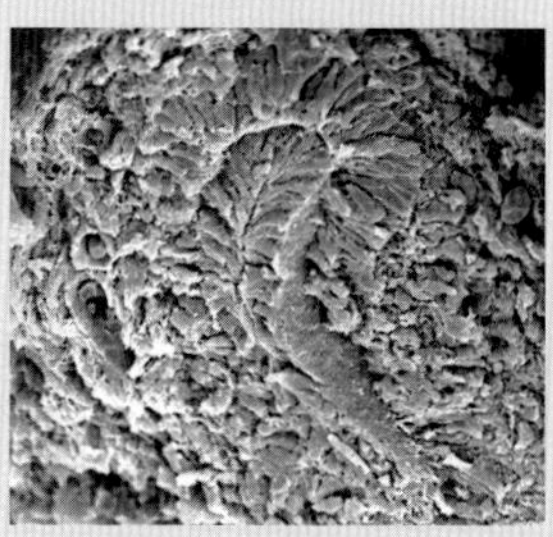

5.3.20 A fracture through the contact zone of a branch of a collecting duct with the metanephric blastema is shown. Embryo at week 7.

5.3.21 The epithelium of the two branches of a collecting duct. Embryo at week 7.

5.3.01–5.3.24
Developmental stages of the metanephros (kidney).

5.3.19–5.3.24
Early development of the nephron.

500:1

Week 8

250:1

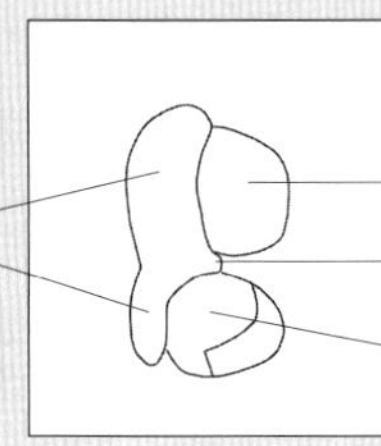

Week 9

5.3.22 The contact zone of a branch of a collecting duct with the metanephric blastema that will convert into the metanephric epithelium. Embryo at week 8.

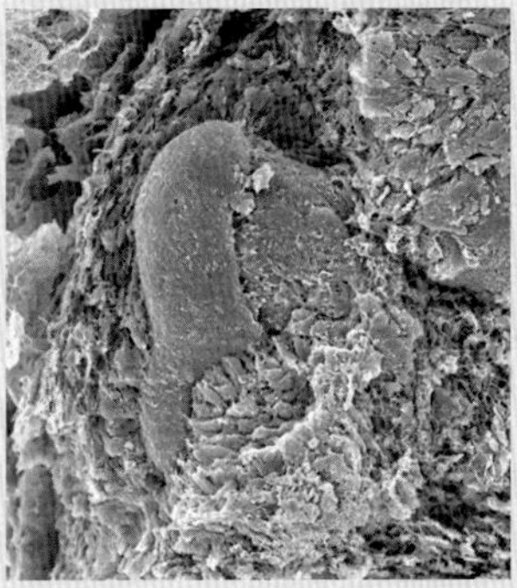

5.3.23 The metanephric blastema in contact with a branch of a collecting duct has converted into vesicles of metanephric epithelium. Embryo at week 8.

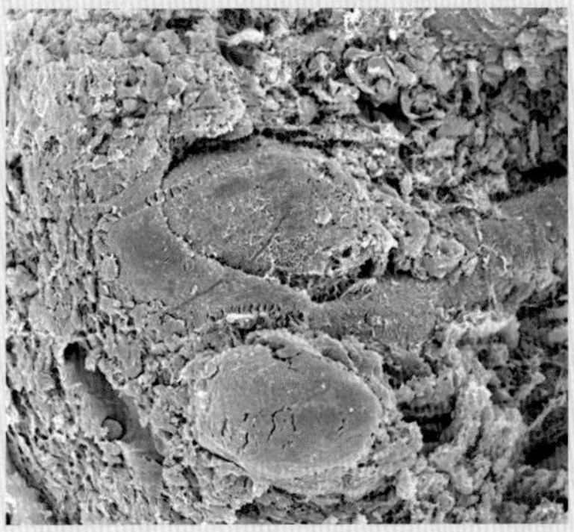

5.3.24 The branch of a collecting duct and the vesicle of the metanephric epithelium that will make up the nephron. The tubulus system is not yet developed. Embryo at week 9.

5.4.01–5.4.10
Development of the urinary pathways.

5.4.01 / 02
Position of the urinary bladder.

5.4.03–5.4.10
The orifices of the mesonephric (wolffian) duct and the ureter leading into the urinary bladder.

Week 6

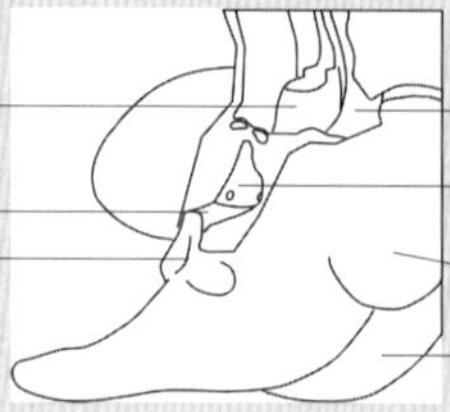

Week 7

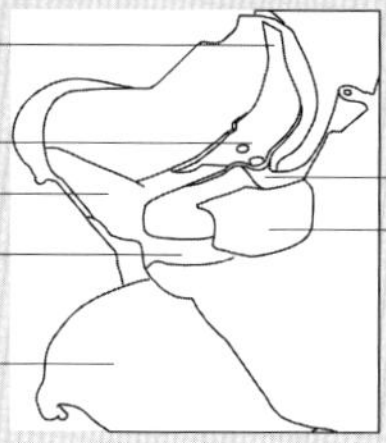

Week 6

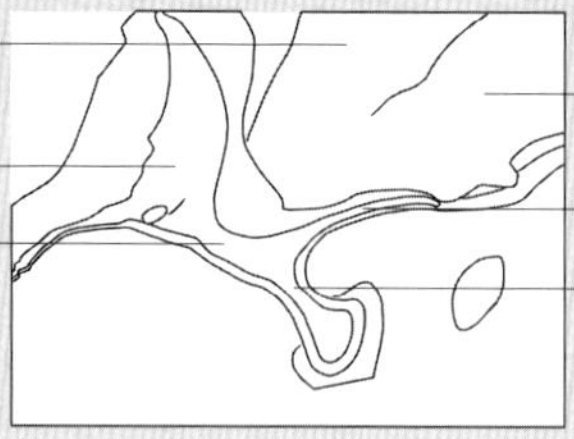

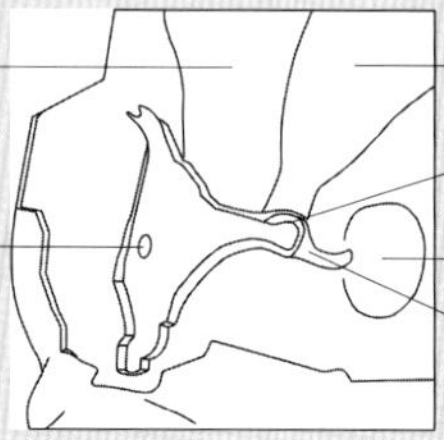

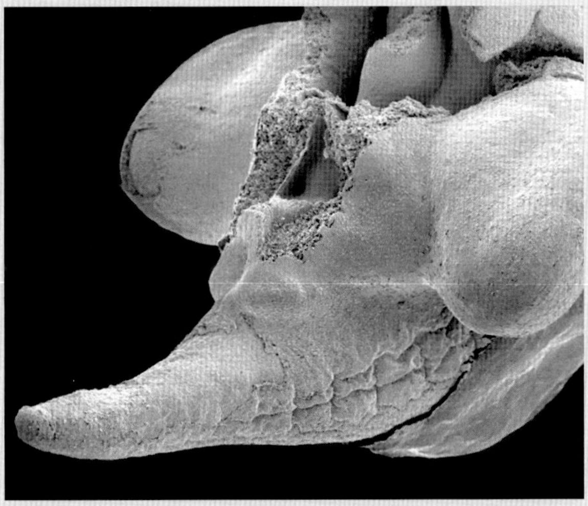

5.4.01 Ventral-left view of the opened urinary bladder and the epithelium of the urethra. Embryo at week 6.

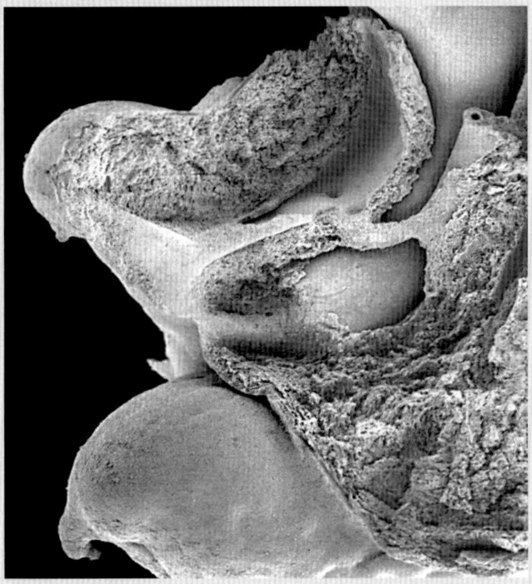

5.4.02 Lateral-left view of the exposed and opened urinary bladder and the epithelium of the urethra. Embryo at the end of week 7.

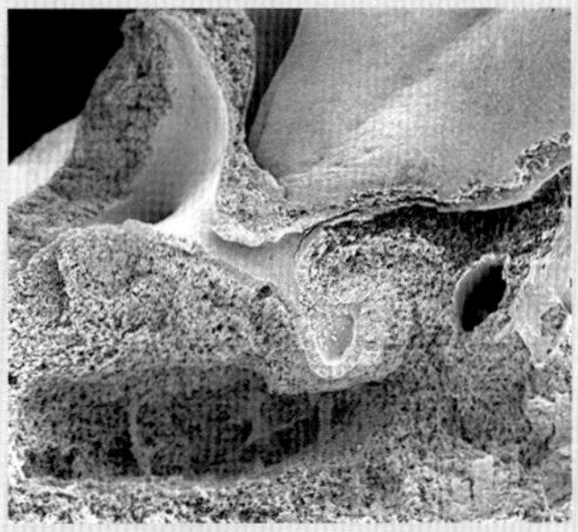

5.4.03 Ventral-left view of the left ureteric bud, the ureter, the mesonephric duct, and the bladder. Ureter and mesonephric duct have a common excretory duct leading into the bladder. Embryo at week 6.

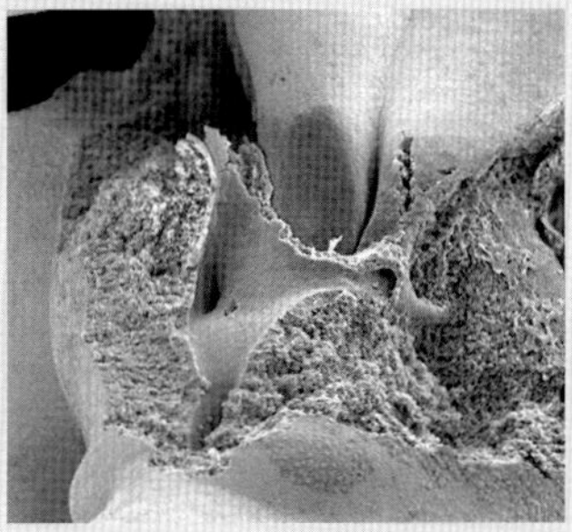

5.4.04 Lateral-left view of the orifice of the ureter and the mesonephric duct. The common excretory duct is shortened due to elongation of the separating fold between the mesonephric duct and the ureter. Embryo at the end of week 6.

5.4.01–5.4.10
Development of the urinary pathways.

5.4.03–5.4.10
The orifices of the mesonephric (wolffian) duct and the ureter leading into the urinary bladder.

40:1

Week 7

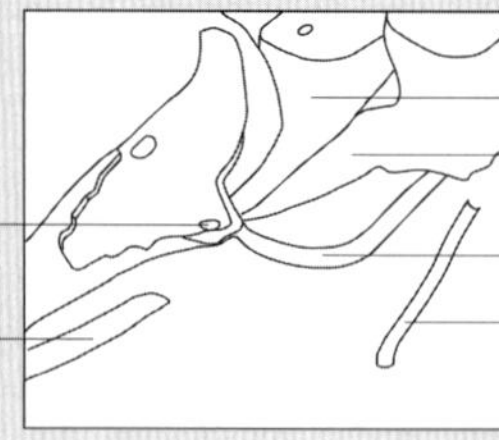

200:1

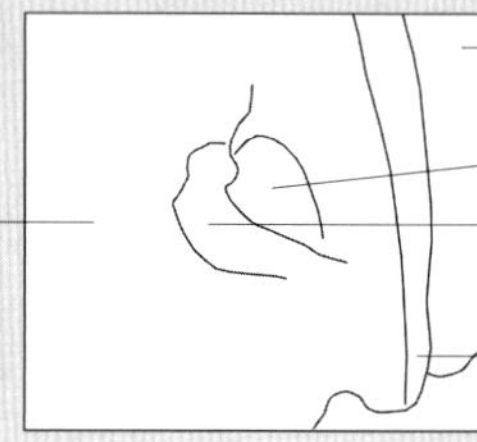

120:1

Week 8

20:1

Week 9

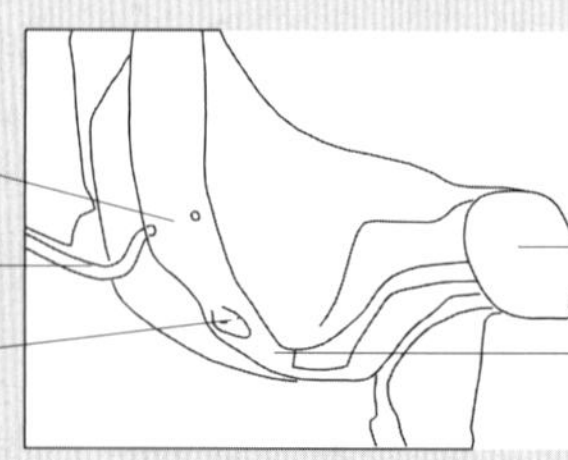

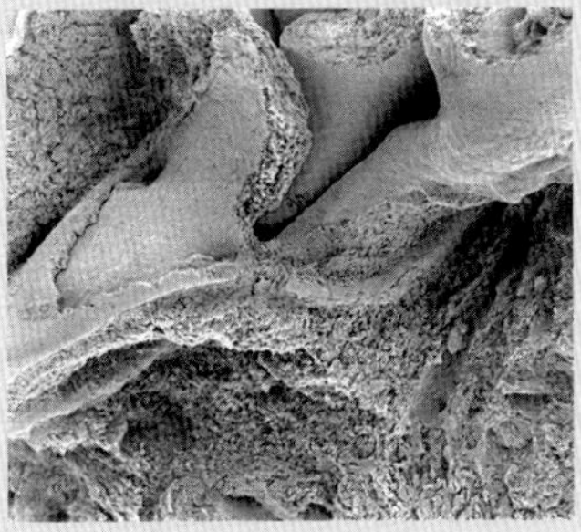

5.4.05 Ventral-left view of the bladder, the ureter, and the mesonephric duct. The orifice of the ureter has achieved a more lateral position than the opening of the mesonephric duct. Embryo at week 7.

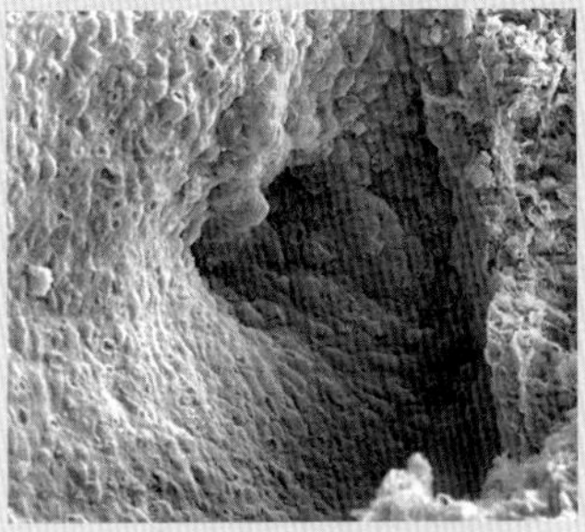

5.4.06 Detailed view of the opened urinary bladder. The left ureter and the left mesonephric duct have achieved two separate orifices. Embryo at the end of week 7.

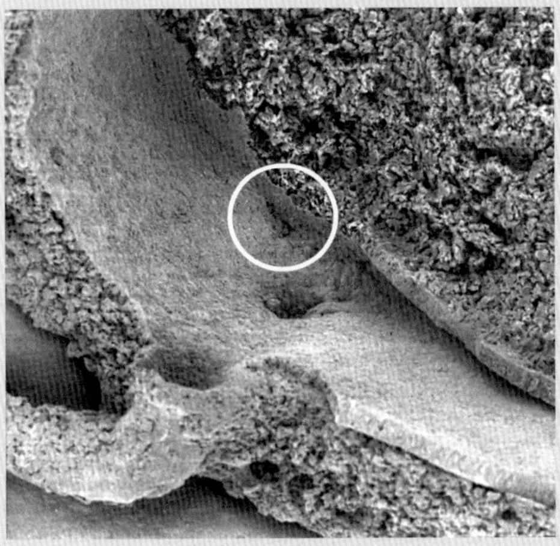

5.4.07 Ventral-right view of the bladder. Orifices of the mesonephric ducts, the orifice of the left ureter is marked by a circle. Embryo at week 8.

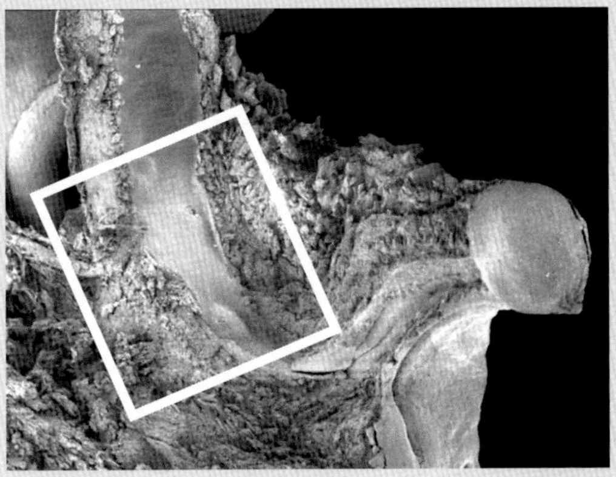

5.4.08 Ventral-right overall view of the opened bladder, the urethra, and the orifice of the right ureter. Embryo at week 9.

Week 9

5.4.01–5.4.10
Development of the urinary pathways.

5.4.03–5.4.10
The orifices of the mesonephric (wolffian) duct and the ureter leading into the urinary bladder.

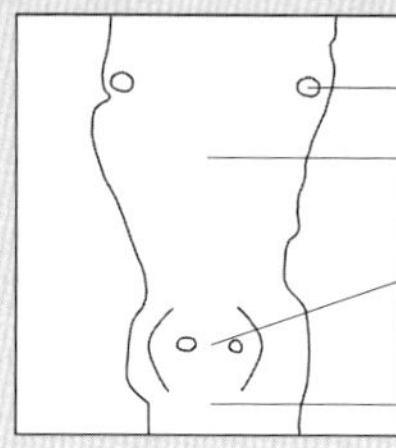

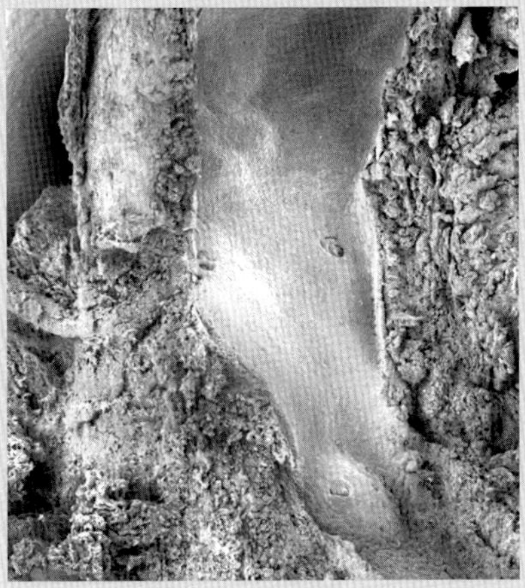

5.4.09 Detailed view of the same embryo as in figure 5.4.08. The lateral position of the orifice of the ureter and the medial position of the orifice of the mesonephric duct on the müllerian tubercle are clearly seen.

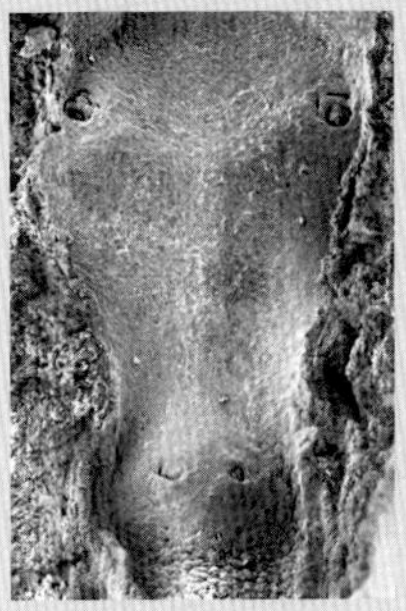

5.4.10 Detailed ventral and cranial view of the same embryo as in figure 5.4.09. The different positions of the orifices of the ureter and of the mesonephric duct are obvious.

5.
Urogenital Organs

5.5
The Development of the Gonads

The gonads develop from the medial portions of the urogenital folds (fig. 5.3.01–5.3.04). A longitudinal furrow separates the urogenital fold into the lateral mesonephros and the medial gonad (fig. 5.5.01–5.5.05).

Till week 5, the gonads are termed 'indifferent' because, histologically, a distinction between female and male gonads has remained impossible up to today. Beginning in week 6, the differences between ovary and testis become evident in internal differentiation and external shape as well (fig. 5.5.06, 5.5.07). Even the structure of the peritoneum covering the gonads shows a specific appearance from week 8 onwards (fig. 5.5.08–5.5.10).

5.5

Abbreviations

mn	mesonephros
1	ovary
2	uterine tube
3	round ligament
4	testis
5	gubernaculum testis
6	rectum
7	suprarenal gland

5.
Urogenital Organs

5.5
The Development of the Gonads

5.5.01–5.5.05
Development of the gonadal folds.

Week 6

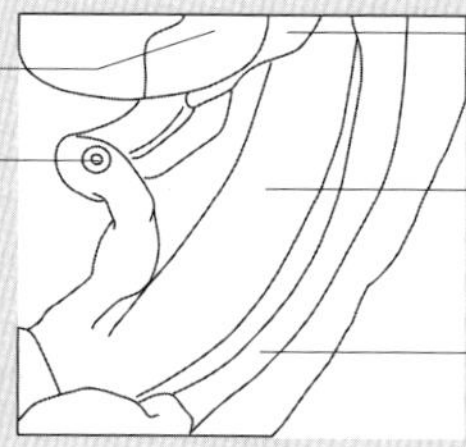

Week 7

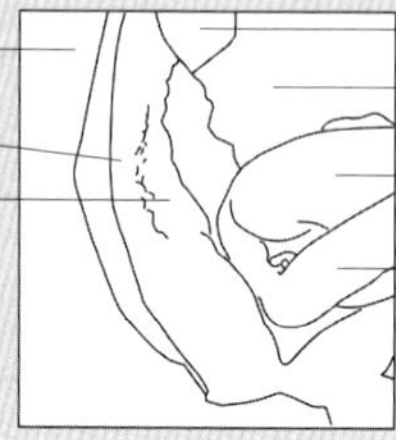

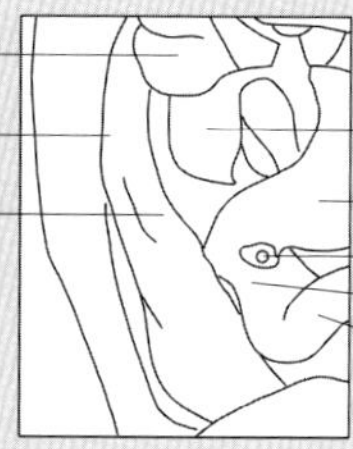

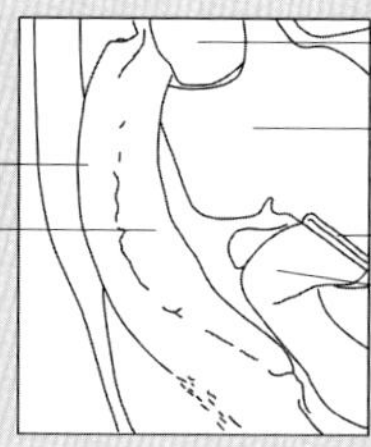

Week 8

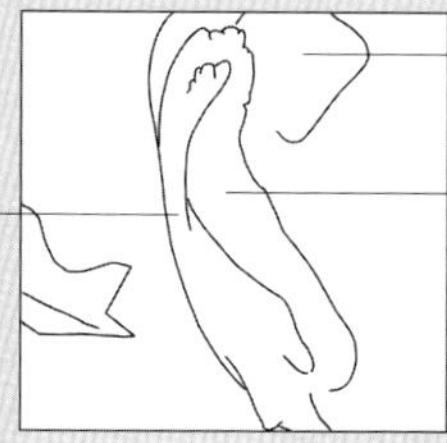

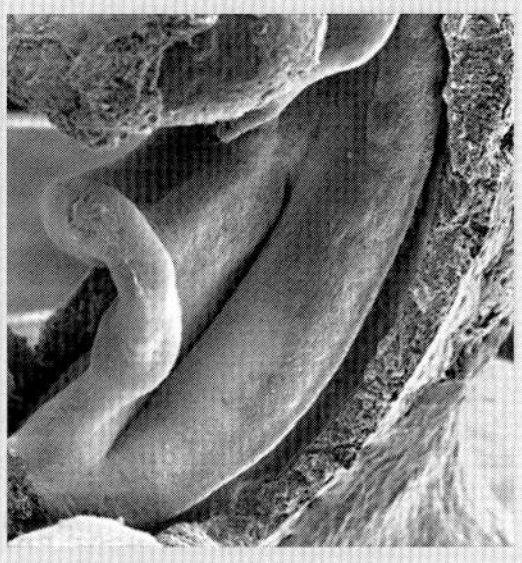

5.5.01

Ventral-left view of the left urogenital fold. Embryo at week 6.

5.5.02

Ventral-right view of the advancing demarcation of the right gonad from the mesonephros. Different developmental stages of embryos at week 7 (5.5.02–5.5.04) and 8 (5.5.05).

5.5.03

5.5.04

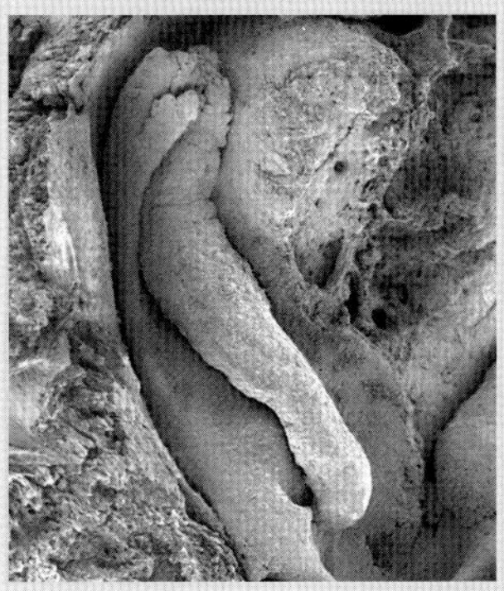

5.5.05

20:1

Week 9

5.5.06–5.5.10

First outwardly visible differences between ovary and testis. Embryos at week 9.

200:1

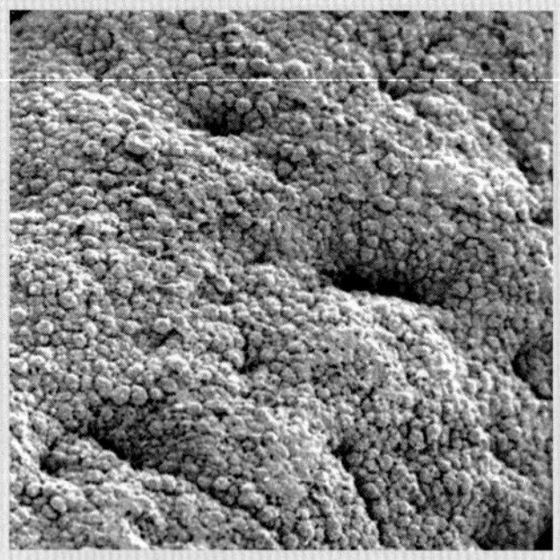

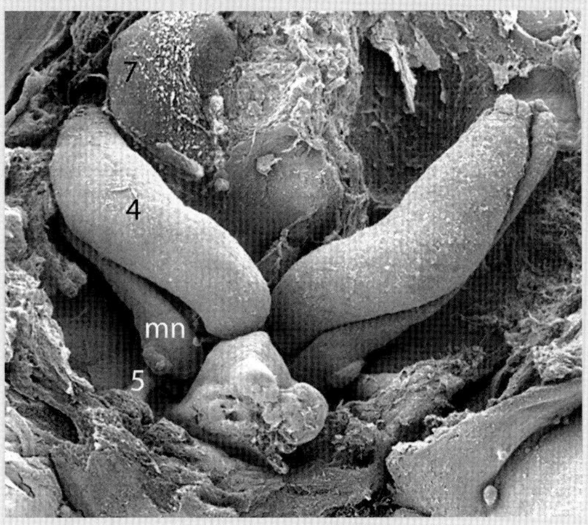

5.5.06 Ventral view of the testes.

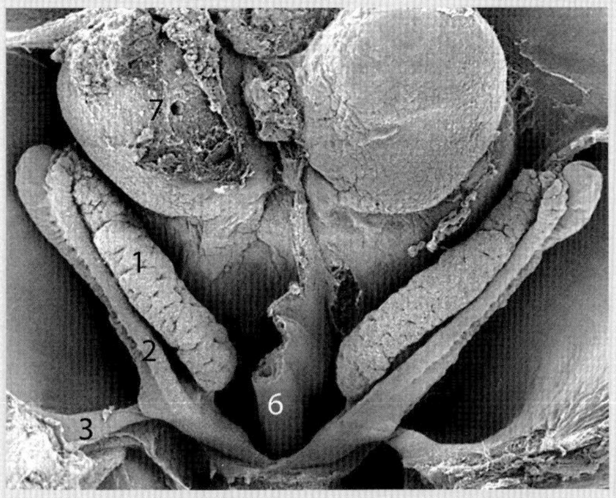

5.5.07 Ventral view of the ovaries.

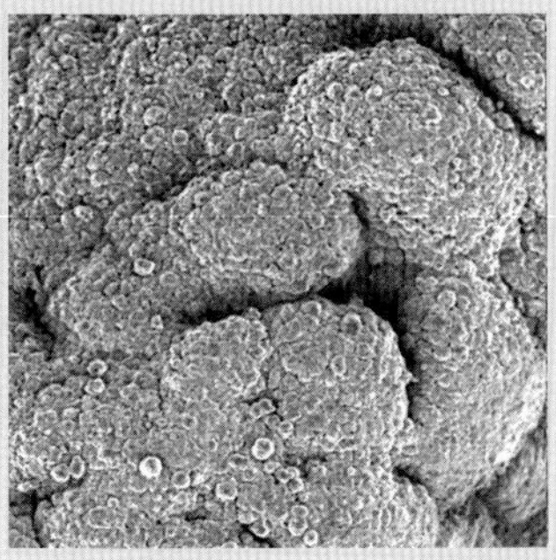

5.5.08 / 09 Individual differences in the surface arrangement of the peritoneum of the ovaries.

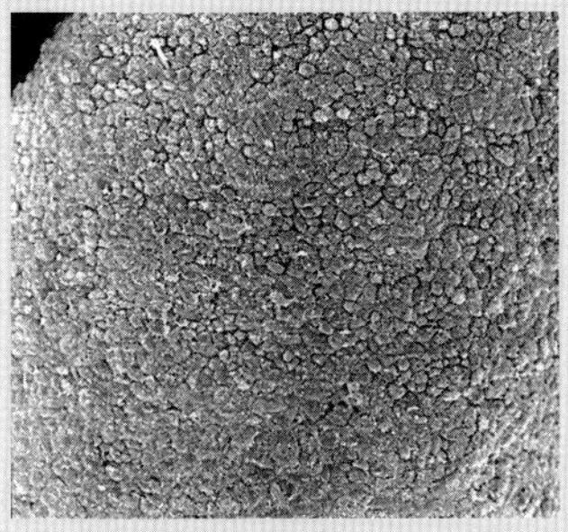

5.5.10 Unlike the ovary, the peritoneum of the testis has a soft surface.

5.6
The Reproductive Pathways

The female reproductive pathways of the uterus and the uterine tube develop from the paramesonephric ducts (müllerian ducts) which arise dorsally and laterally from the mesonephric ducts within the mesonephric folds (fig. 5.6.01). In the pelvic region the paramesonephric ducts bend in a medial direction, cross over the mesonephric ducts (fig. 5.6.02) and eventually fuse in the midline forming the uterus (fig. 5.6.04–5.6.06). The fused paramesonephric ducts end at the müllerian tubercle.

The unfused paramesonephric ducts form the uterine tubes. Their abdominal end remains open and is termed the ostium abdominale of the uterine tube (fig. 5.6.07, 5.6.08).

The male excretory duct of the testis, the ductus deferens, develops from the mesonephric (wolffian) duct (fig. 5.6.10, 5.6.11). The orifice of the ductus deferens initially located in the sinus urogenitalis (fig. 5.4.03, 5.4.04) changes its position and arrives at its final position in the urethra just beneath the urinary bladder (fig. 5.4.09, 5.4.10).

5.6.01–5.6.11
The reproductive pathways.

5.6.01–5.6.08
The female genital pathways.

80:1

Week 7

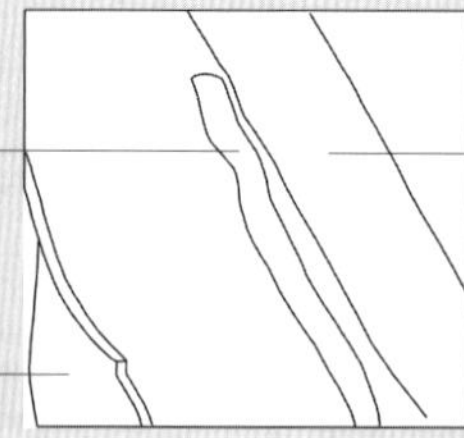

200:1

80:1

Week 8

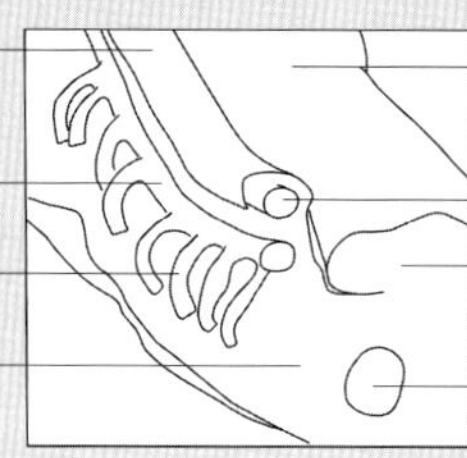

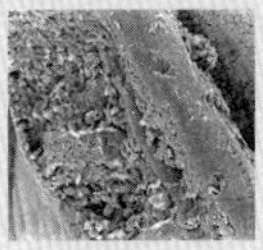

5.6.01

Ventral-right view of the median part of the right mesonephros. The epithelia of the mesonephric (wolffian) and paramesonephric (müllerian) ducts have been exposed. Embryo at week 7.

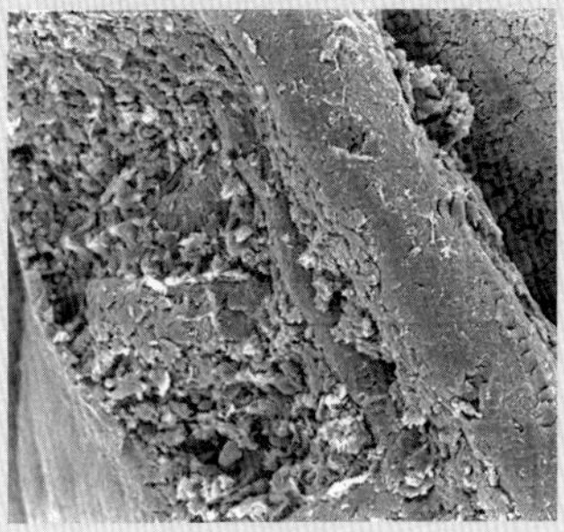

5.6.01a

Same embryo as in figure 5.6.01, same view.

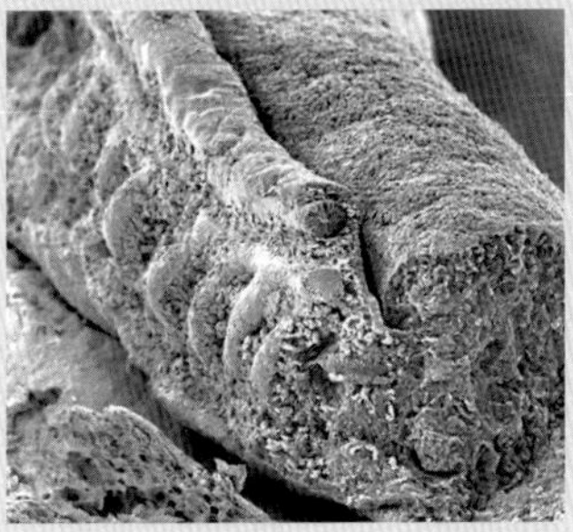

5.6.02

Ventral-right view of the caudal part of the right mesonephros. The epithelia of the mesonephric duct, the secretory tubules and the paramesonephric duct have been exposed. The mesonephric and the paramesonephric ducts have been cut. The paramesonephric duct has crossed over the mesonephric duct. Embryo at week 8.

5.6.01–5.6.11
The reproductive pathways.

5.6.01–5.6.08
The female genital pathways.

Week 8

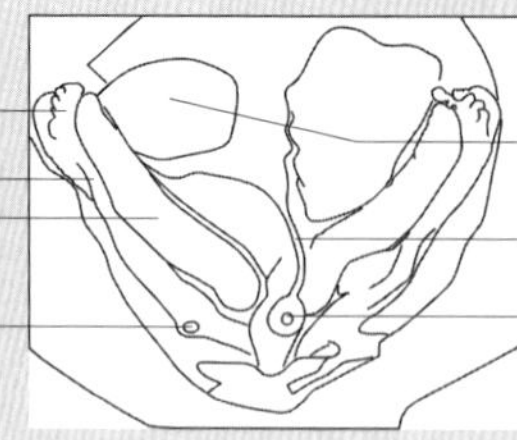

Week 9

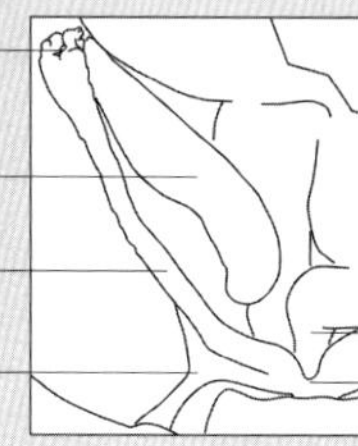

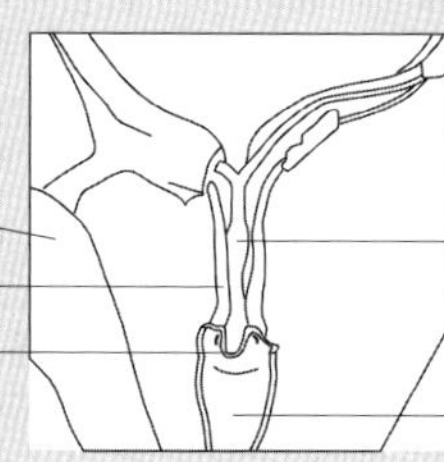

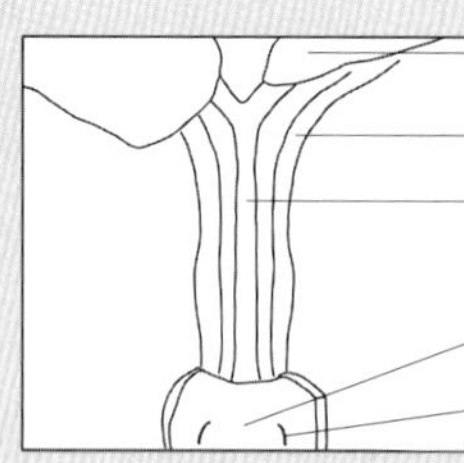

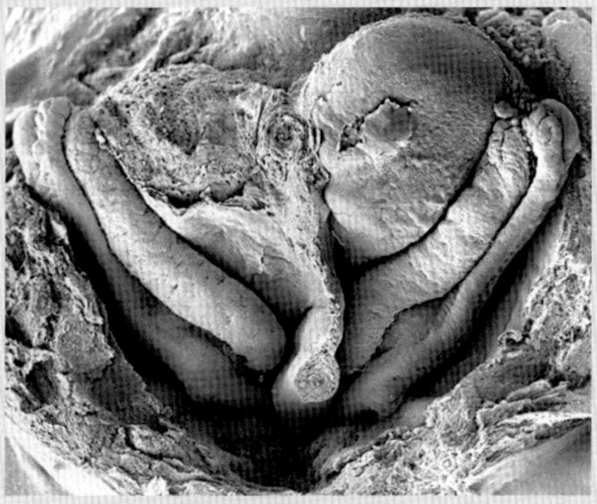

5.6.03 Ventral view of the position of the ovaries and the uterine tubes. Embryo at week 8.

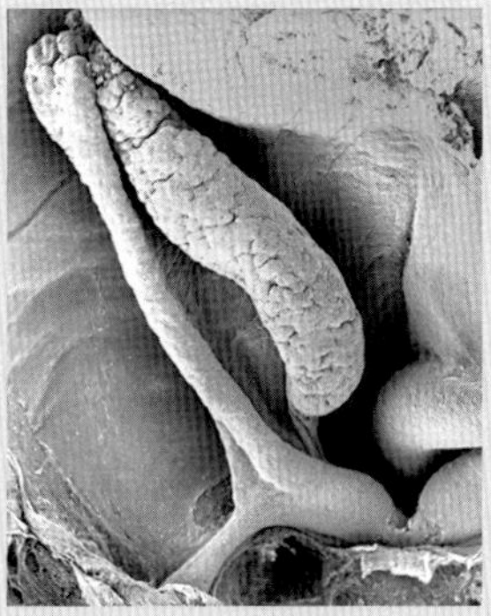

5.6.04 Ventral view of the right ovary with the mesovarium, the uterine tube and uterus, and the round ligament. Embryo at week 9.

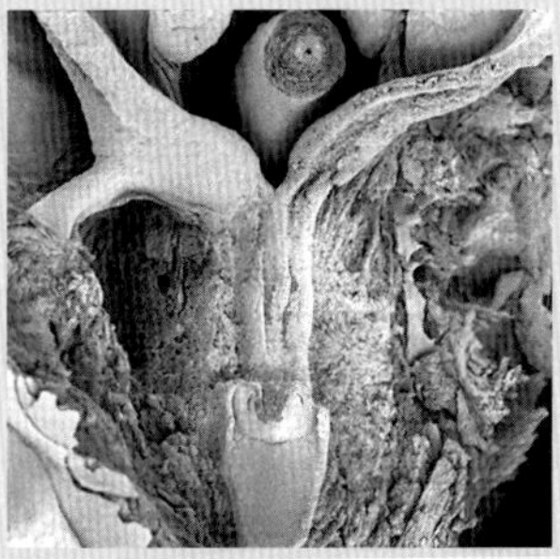

5.6.05 Ventral view of the exposed left paramesonephric (müllerian) duct crossing the mesonephric (wolffian) duct. The fused paramesonephric ducts (the uterus) reach the sinusal tubercle (müllerian tubercle). Embryo at week 9.

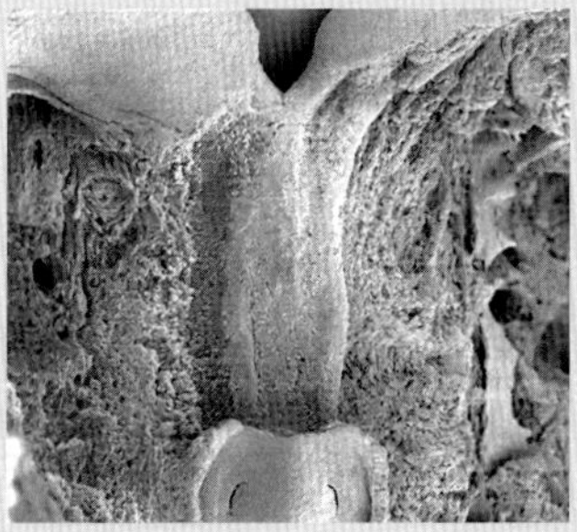

5.6.06 Detailed ventral view of the exposed uterus and the mesonephric ducts (Gartner's duct). Embryo at week 9.

80:1

Week 7

5.
Urogenital Organs

5.6
The Reproductive
Pathways

Week 9

5.6.01–5.6.11
The reproductive pathways.

5.6.01–5.6.08
The female genital pathways.

5.6.09–5.6.11
The male genital pathways.

20:1

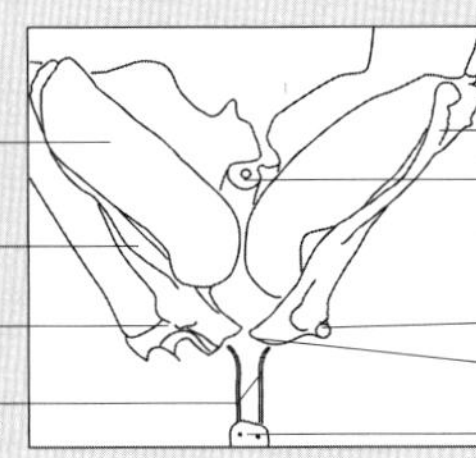

40:1

80:1

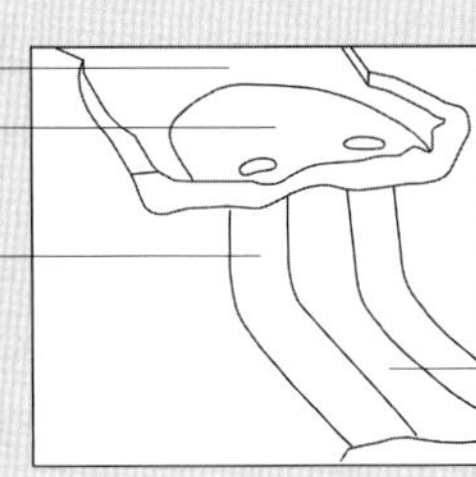

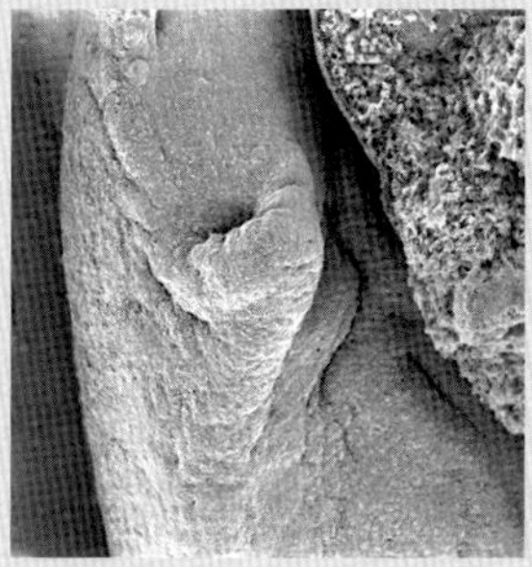

5.6.07 Cranial view of the abdominal ostium of the right paramesonephric (müllerian) duct, now termed the uterine tube. Embryo at week 7.

5.6.08 Cranial view of the abdominal ostium of the right paramesonephric (müllerian) duct. Embryo at week 9.

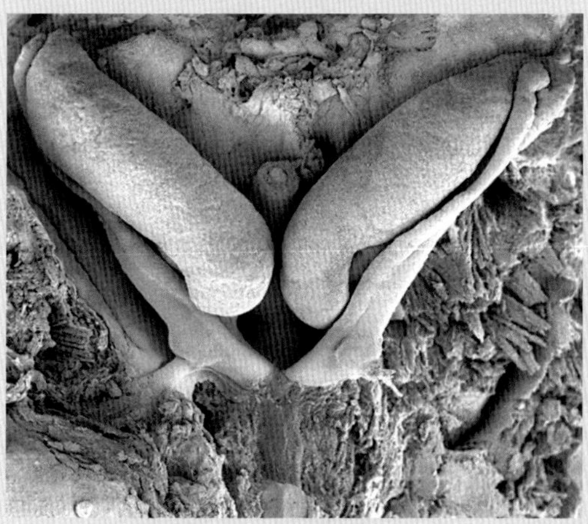

5.6.09 Ventral view of the testes and the mesonephric duct (ductus deferens). Embryo at week 9.

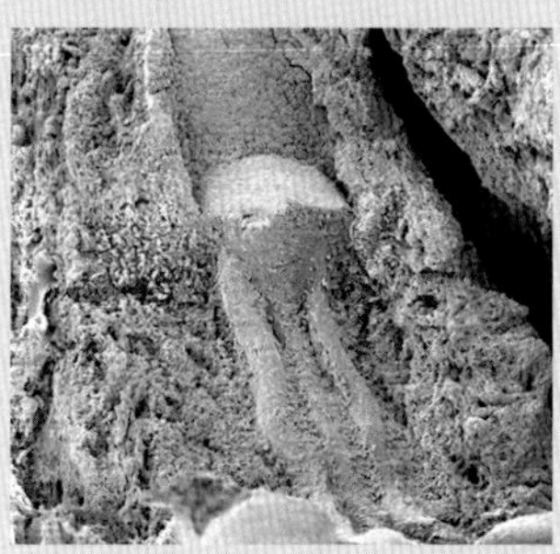

5.6.10 Same embryo as in figure 5.6.09. Dorsal and cranial view of the exposed epithelia of the mesonephric ducts (ductus deferentes) and the regressing fused paramesonephric ducts.

5.6.11 Same embryo as in figure 5.6.09. Detailed view of the orifices of the ductus deferentes.

6

Brain and Sensory Organs

6.1
The Development of the Brain

The structure of the human brain is not only externally but also internally highly complex. Therefore, a complete presentation of its development is beyond the scope of this atlas. Since the most significant processes of brain development take place during the fetal period, only a few typical stages of the embryonic period proper will be presented here.

The development of the brain proceeds from a simple club-shaped swelling of the cranial end of the neural tube. Together with the closure of the cranial neuropore (fig. 6.1.01), the optic vesicles are formed (fig. 6.1.02, 6.1.35, 6.1.36). Due to the flexures of the brain tube in a ventral direction, the main regions of the forebrain (prosencephalon), the middle brain (mesencephalon) and the hindbrain (rhombencephalon) gradually become demarcated (fig. 6.1.22–6.1.25).

The topmost region of the prosencephalon expands and gives rise to the two hemispheres of the telencephalon (fig. 6.1.04–6.1.11, 6.1.31–6.1.34), whereas the basal region of the prosencephalon develops into the diencephalon (fig. 6.1.04–6.1.10, 6.1.36 and 6.1.37, 6.1.41 and 6.1.43, 6.1.45–6.1.47).

The mesencephalon (fig. 6.1.12–6.1.26, 6.1.36 and 6.1.37, 6.1.43–6.1.46) remains relatively small and gets covered by the cerebral hemispheres (fig. 6.1.28, 6.1.29).

The rhombencephalon (fig. 6.1.23) is the transition zone between the brain and the spinal cord. The cranial portion of the rhombencephalon (metencephalon) forms the pons and the cerebellum (fig. 6.1.56, 6.1.57), and the caudal part the medulla oblongata (myelencephalon) (fig. 6.1.54–6.1.57).

The lumen of the diencephalon corresponds to the third ventricle (fig. 6.1.37, 6.1.42, 6.1.43), the lumina of the cerebral hemispheres correspond to the lateral ventricles (fig. 6.1.38 and 6.1.39 and 6.1.43 and 6.1.44). The third ventricle leads through the lumen of the mesencephalon (fig. 6.1.37, 6.1.43–6.1.46), the aqueduct, to the fourth ventricle, the lumen of the rhombencephalon (fig. 6.1.54–6.1.57) and eventually into the central canal of the spinal cord.

Inside the lateral ventricle, the basal ganglia develop at its base (fig. 6.1.39, 6.1.49–6.1.53, 6.1.65). At the superior medial wall of the lateral ventricle, the choroid plexus is folded inward (fig. 6.1.39, 6.1.61–6.1.65).

At the base of the fourth ventricle the bulges of the rhombomeres are formed (fig. 6.1.54–6.1.57).

The Origin of the Hypophysis (fig. 6.1.66–6.1.73)

The endodermal component of the hypophysis, the adenohypophysis, is an evagination of the roof of the early oral cavity (fig. 6.1.66–6.1.69). The adenohypophysial pouch is in close contact with the floor of the diencephalon (fig. 6.1.68) where the neurohypophysial diverticulum arises (fig. 6.1.73).

The origin of the adenohypophysial pouch is narrowed (fig. 6.1.70–6.1.72) and eventually loses its connection with the epithelium of the oral cavity.

6.1

	Abbreviations
di	diencephalon
I	pharyngeal arch I
m	mesencephalon
p	prosencephalon
rh	rhombencephalon
t	telencephalon
1	optic vesicle
2	optic cup
3	optic stalk
4	orifice of Rathke's pouch
5	site of the lamina terminalis
6	trigeminal ganglion
8	hypothalamus
9	basal ganglia
11	retina
12	mesencephalic flexure
13	olfactory diverticulum
15	otic vesicle
17	vagal nerve

6.1.01–6.1.34
Developmental stages of the external form of the brain.

6.1.01–6.1.11
Ventral view of the brain.

60:1

Week 4

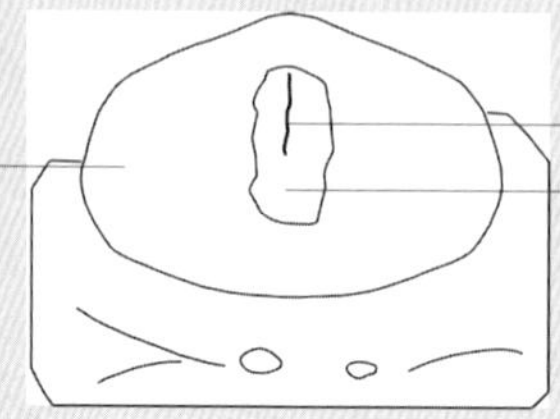

Week 5

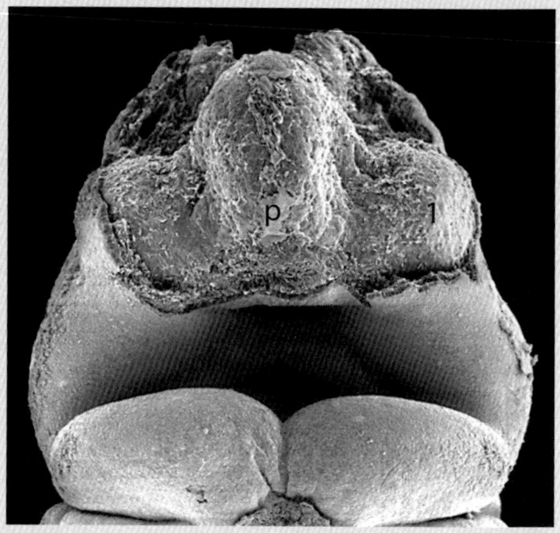

6.1.01

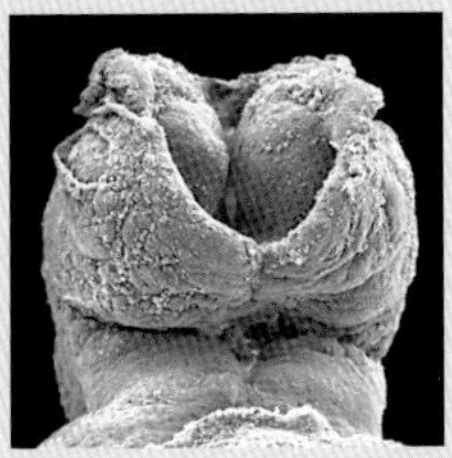

The cranial neuropore is still open. Embryo at week 4.

6.1.02

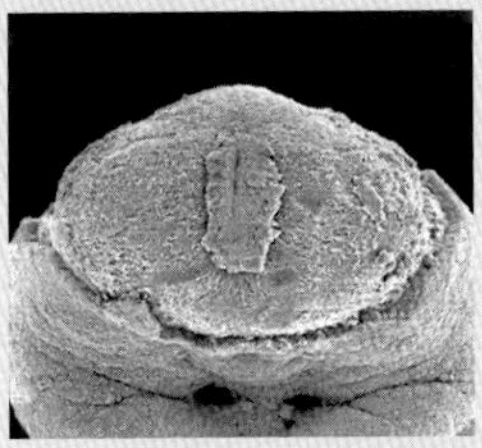

At the site of closure of the cranial neuropore, the ectoderm is still in contact with the brain. Embryo in the middle of week 4.

6.1.03

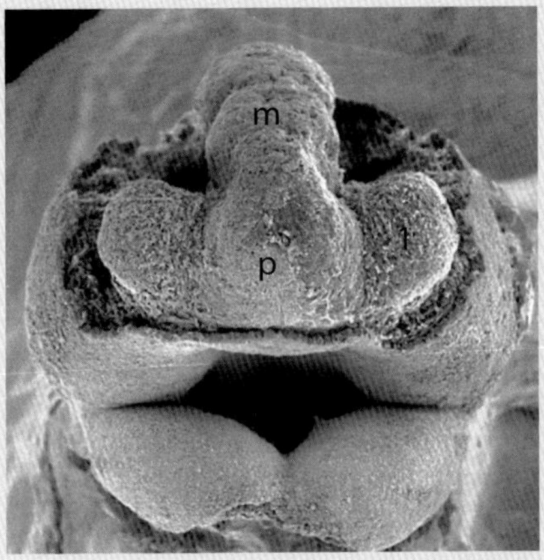

The optic vesicle has emerged. Embryo at the end of week 4.

6.1.04 / 05

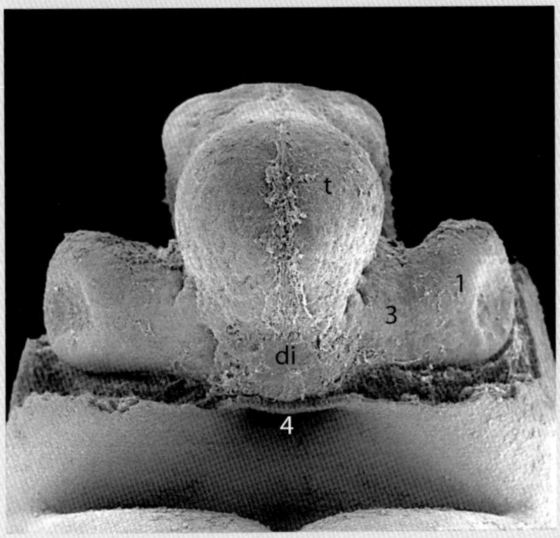

Embryos at week 5.

6.
Brain and
Sensory Organs

6.1
The Development of
the Brain

6.1.01–6.1.34
Developmental stages of the external form of the brain.

6.1.01–6.1.11
Ventral view of the brain.

Week 5

Week 6

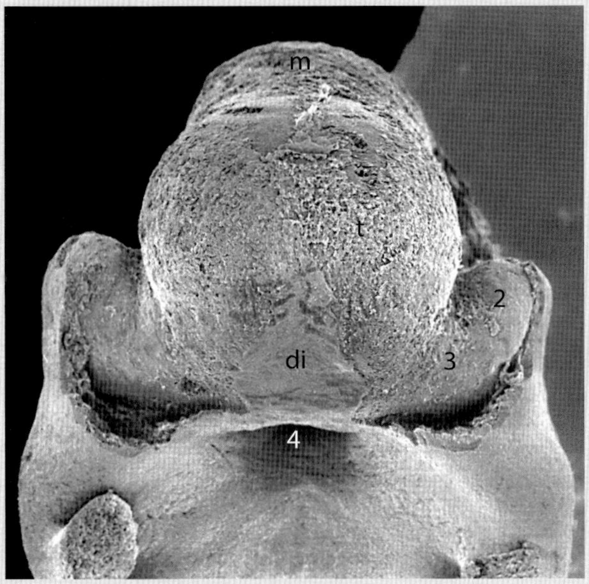

6.1.06 Embryo at week 5.

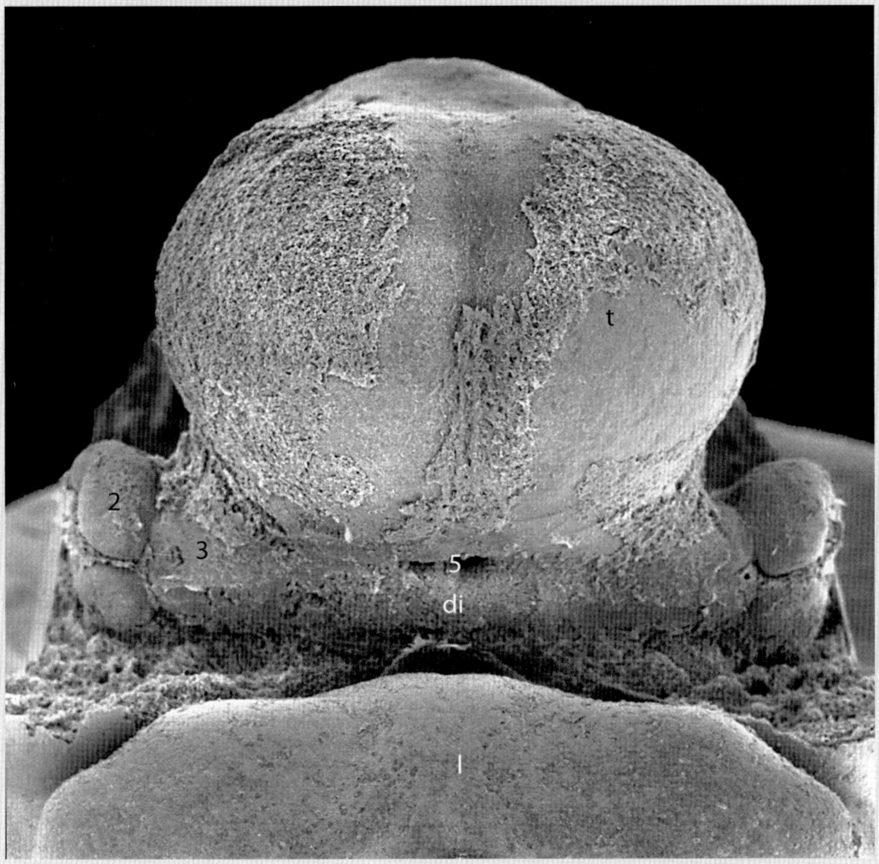

6.1.07 Embryo at week 6.

Week 6

6.1.01–6.1.34
Developmental stages of the external form of the brain.

6.1.01–6.1.11
Ventral view of the brain.

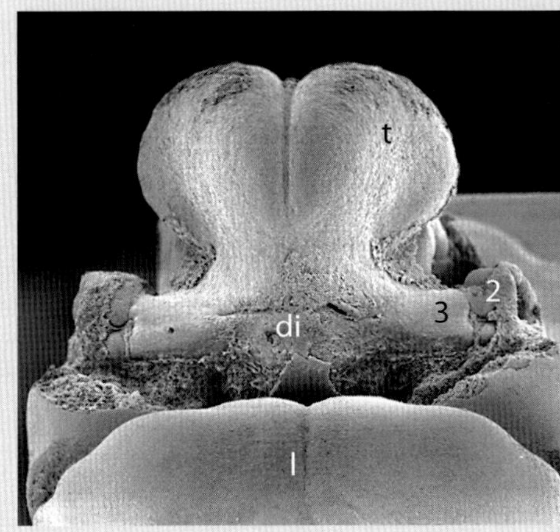

Week 7

6.1.07a

Same embryo as in figure 6.1.07.

6.1.08 / 09

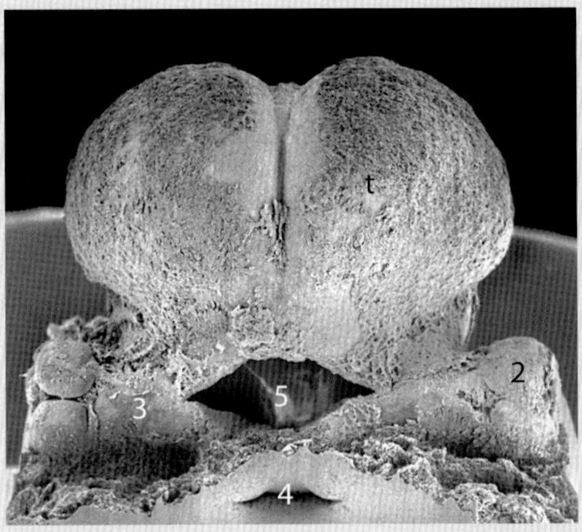

6.1.08
Embryo at week 6.

6.1.09
Embryo at the end of week 6.

6.1.10

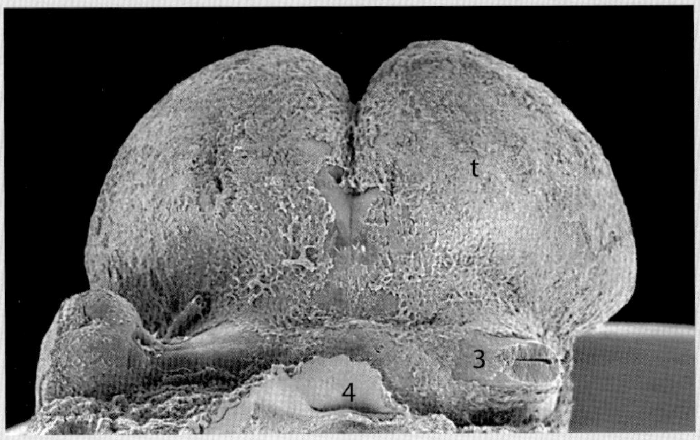

Embryo at week 7.

30:1

Week 8

6.1.01–6.1.34
Developmental stages of the external form of the brain.

6.1.01–6.1.11
Ventral view of the brain.

6.1.12–6.1.21
Cranial view of the brain.

60:1

Week 4

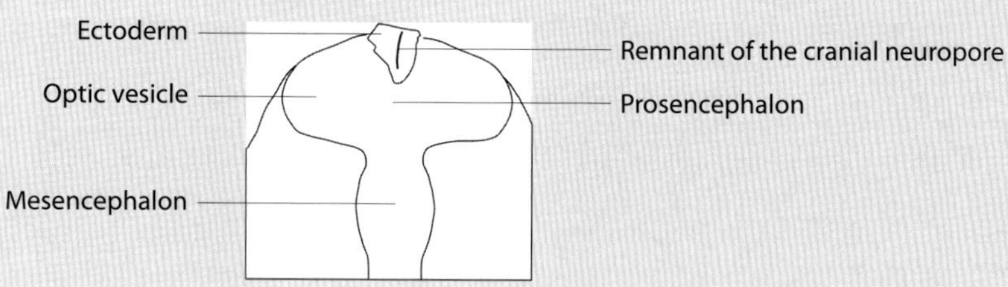

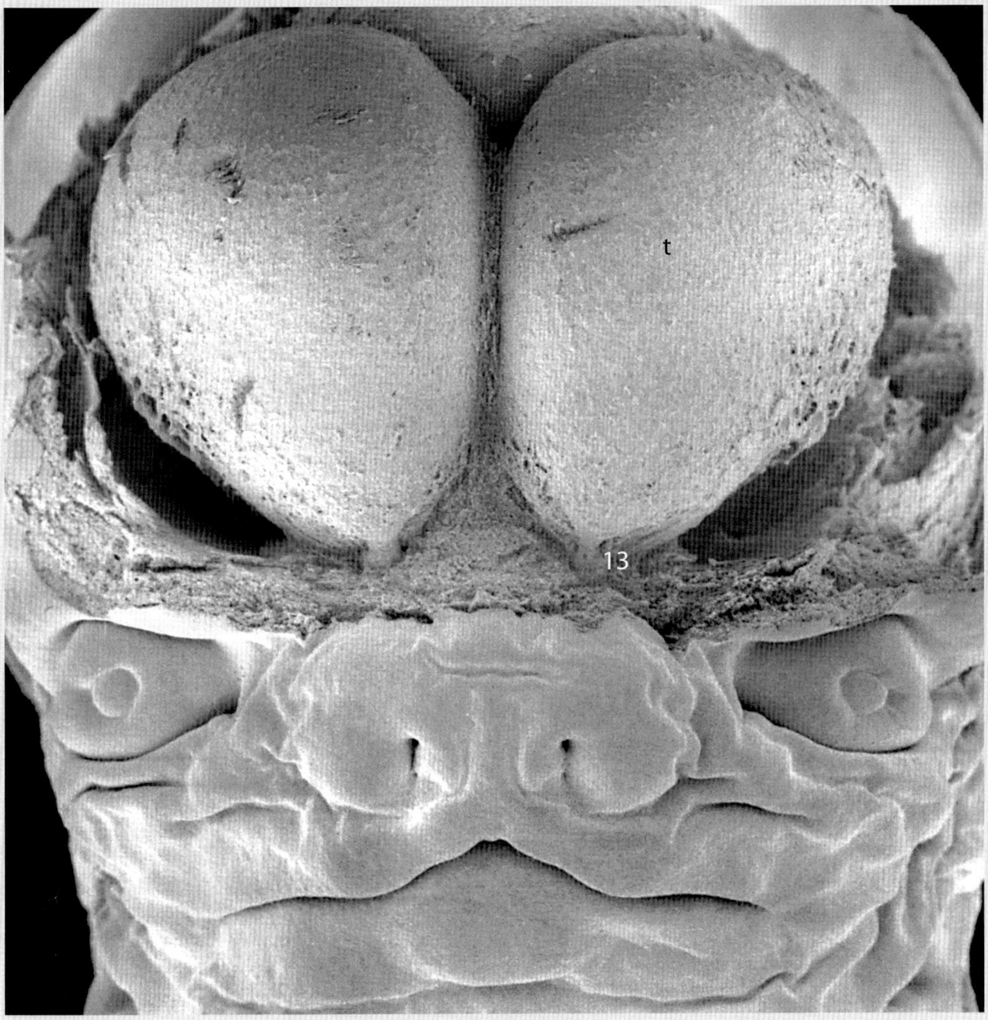

6.1.11 Embryo at week 8.

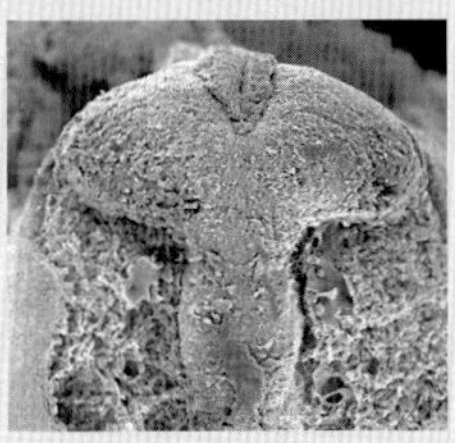

6.1.12 Embryo in the middle of week 4.

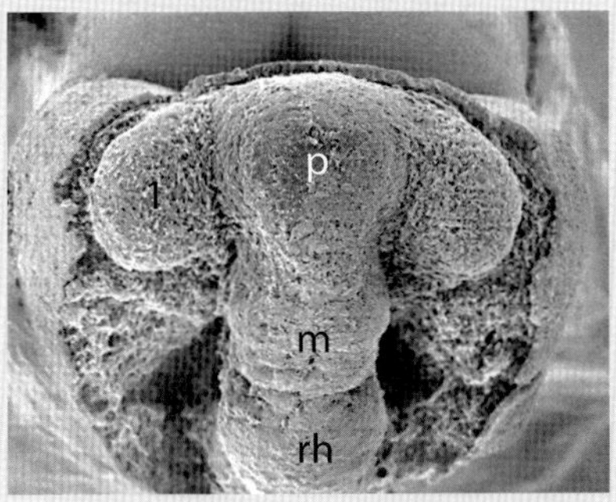

6.1.13 Embryo at the end of week 4.

6.1.01–6.1.34
Developmental stages of the external form of the brain.

6.1.12–6.1.21
Cranial view of the brain.

Week 5

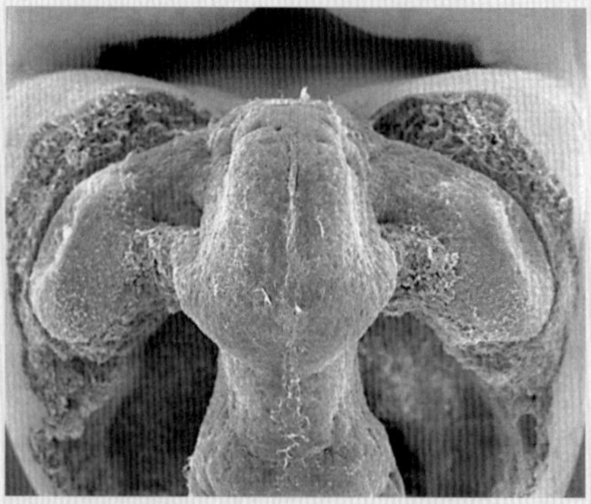

Week 5 / 6

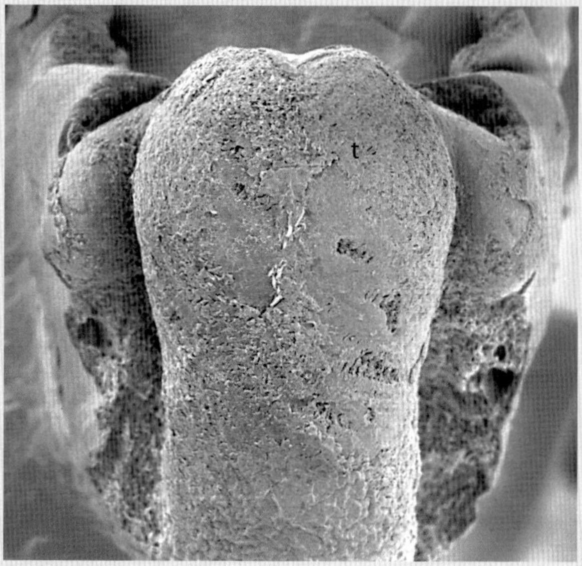

6.1.14 / 15

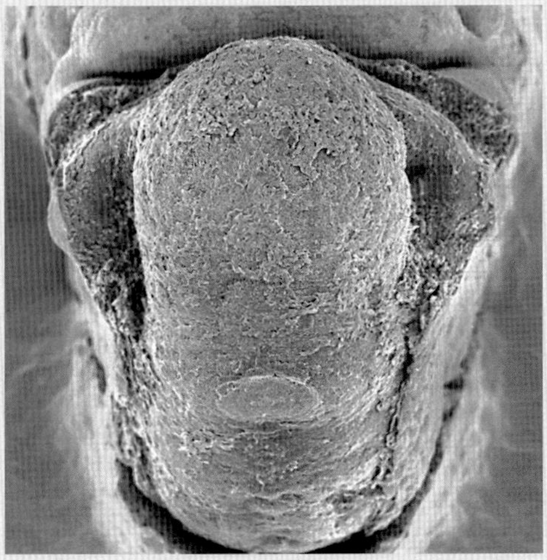

Embryos at week 5.

6.1.16 / 17

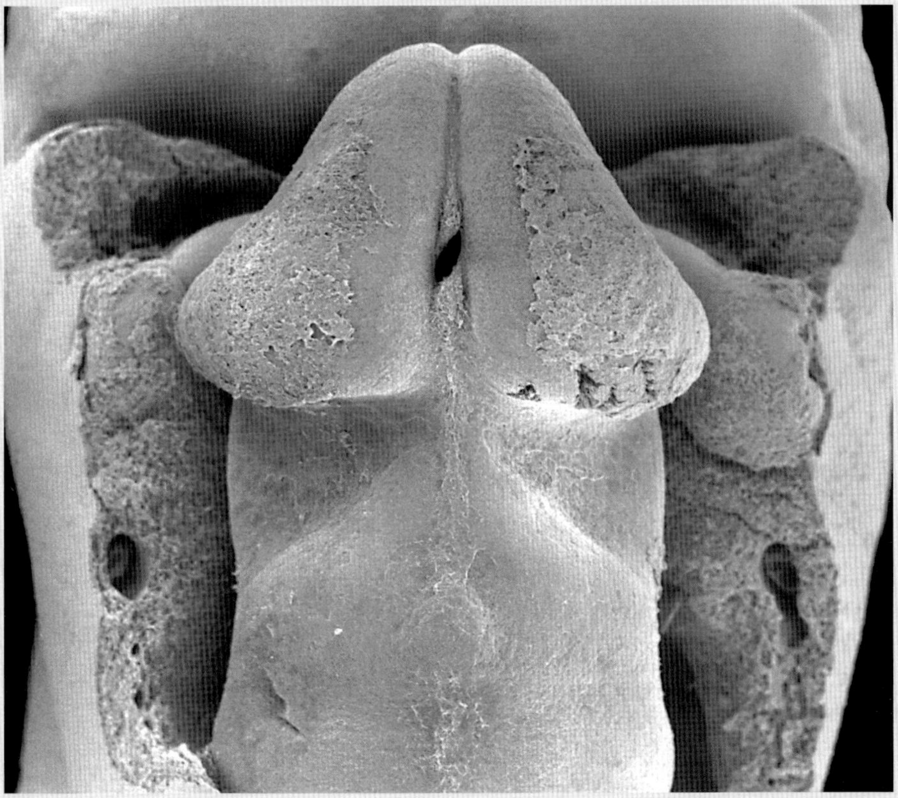

6.1.16
Embryo at week 5.

6.1.17
Embryo early in week 6.

Week 6

6.1.01–6.1.34
Developmental stages of the external form of the brain.

6.1.12–6.1.21
Cranial view of the brain.

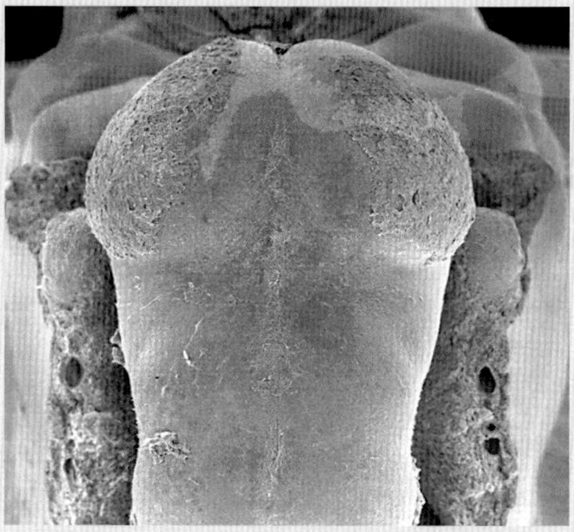

Week 7 / 8

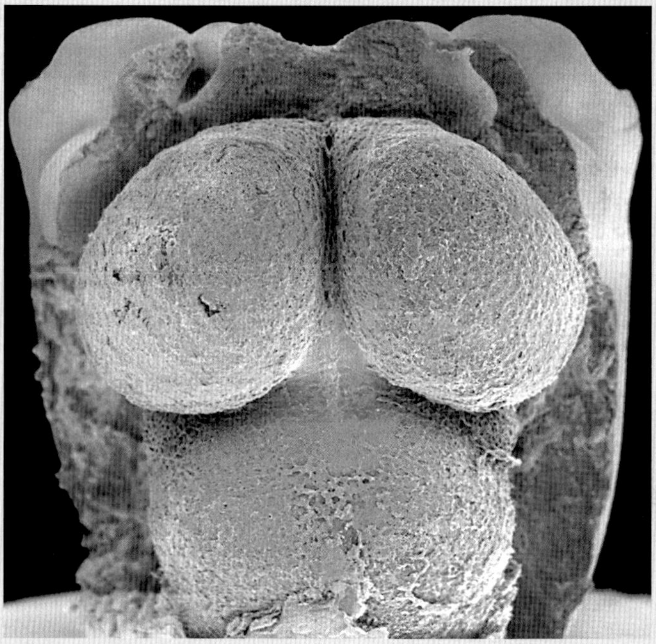

6.1.17a

Same embryo as in figure 6.1.17.

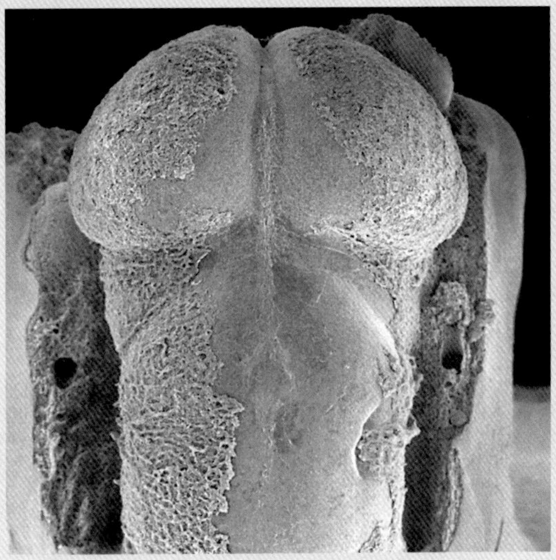

6.1.18 / 19

Embryos in the middle and at the end of week 6.

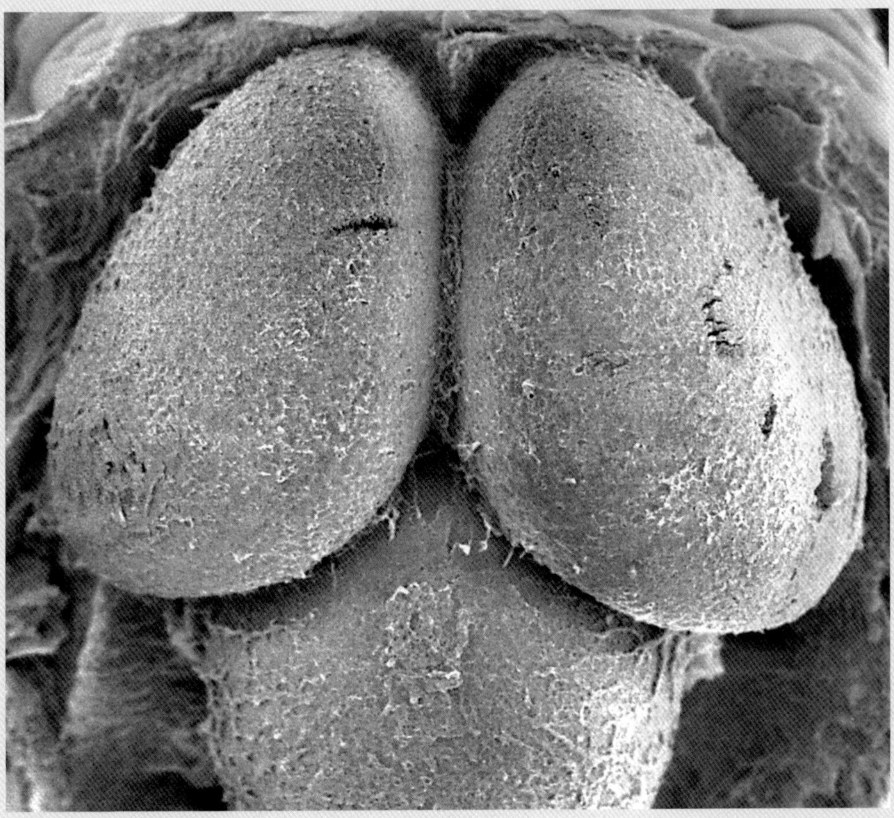

6.1.20 / 21

6.1.20
Embryo at week 7.

6.1.21
Embryo at week 8.

6.1.01–6.1.34
Developmental stages of the external form of the brain.

6.1.22–6.1.29
Lateral-right view of the brain.

Week 4

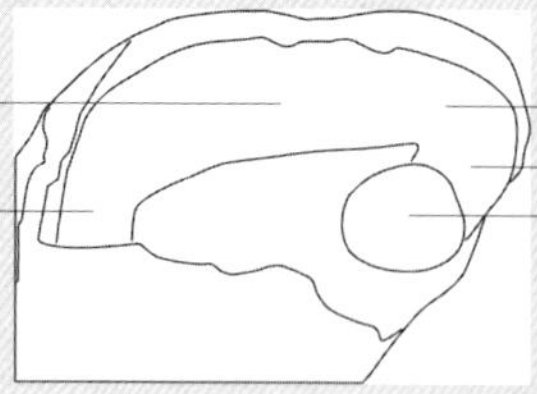

Week 5

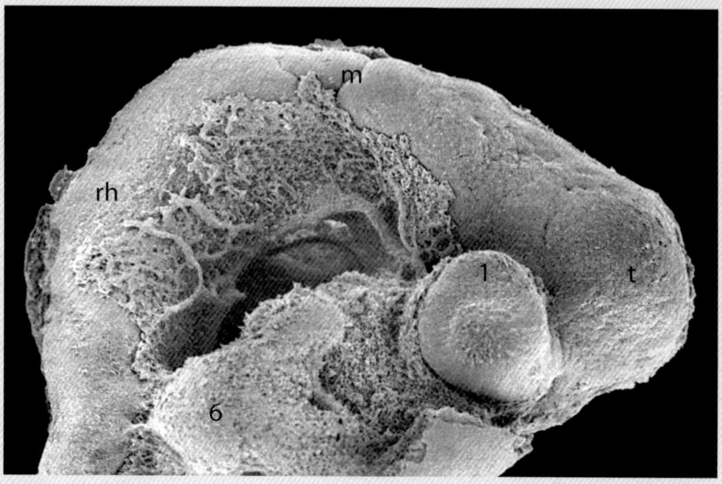

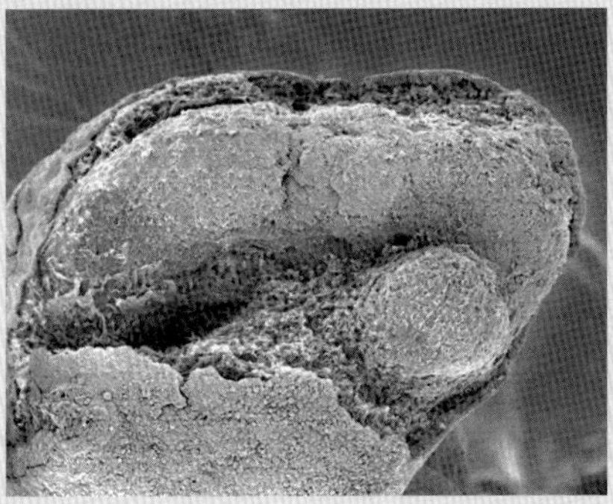

6.1.22 Embryo at the end of week 4.

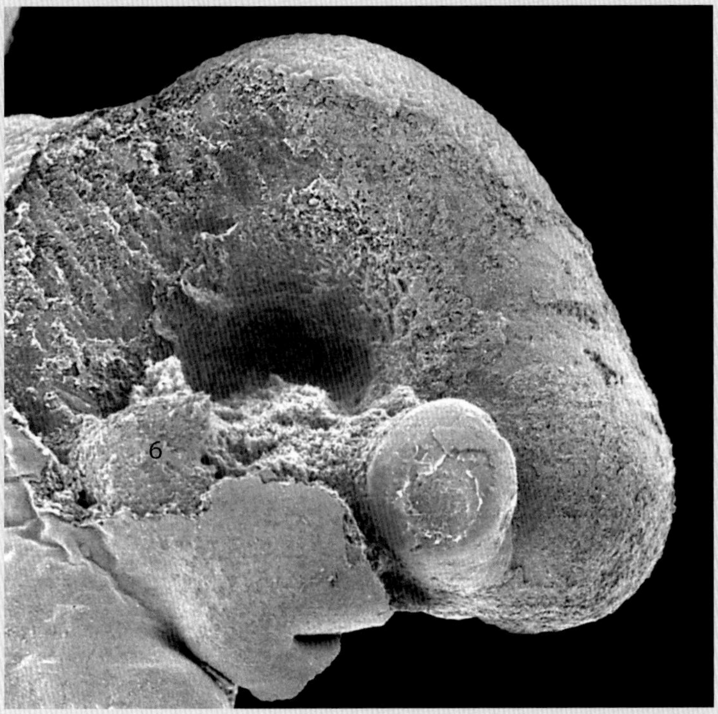

6.1.23 / 24 Embryos at week 5.

Week 5

6.1.01–6.1.34
Developmental stages of the external form of the brain.

6.1.22–6.1.29
Lateral-right view of the brain.

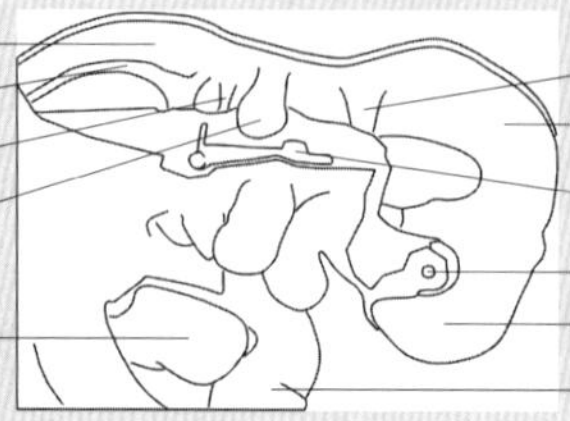

6.1.25 Embryo at week 5.

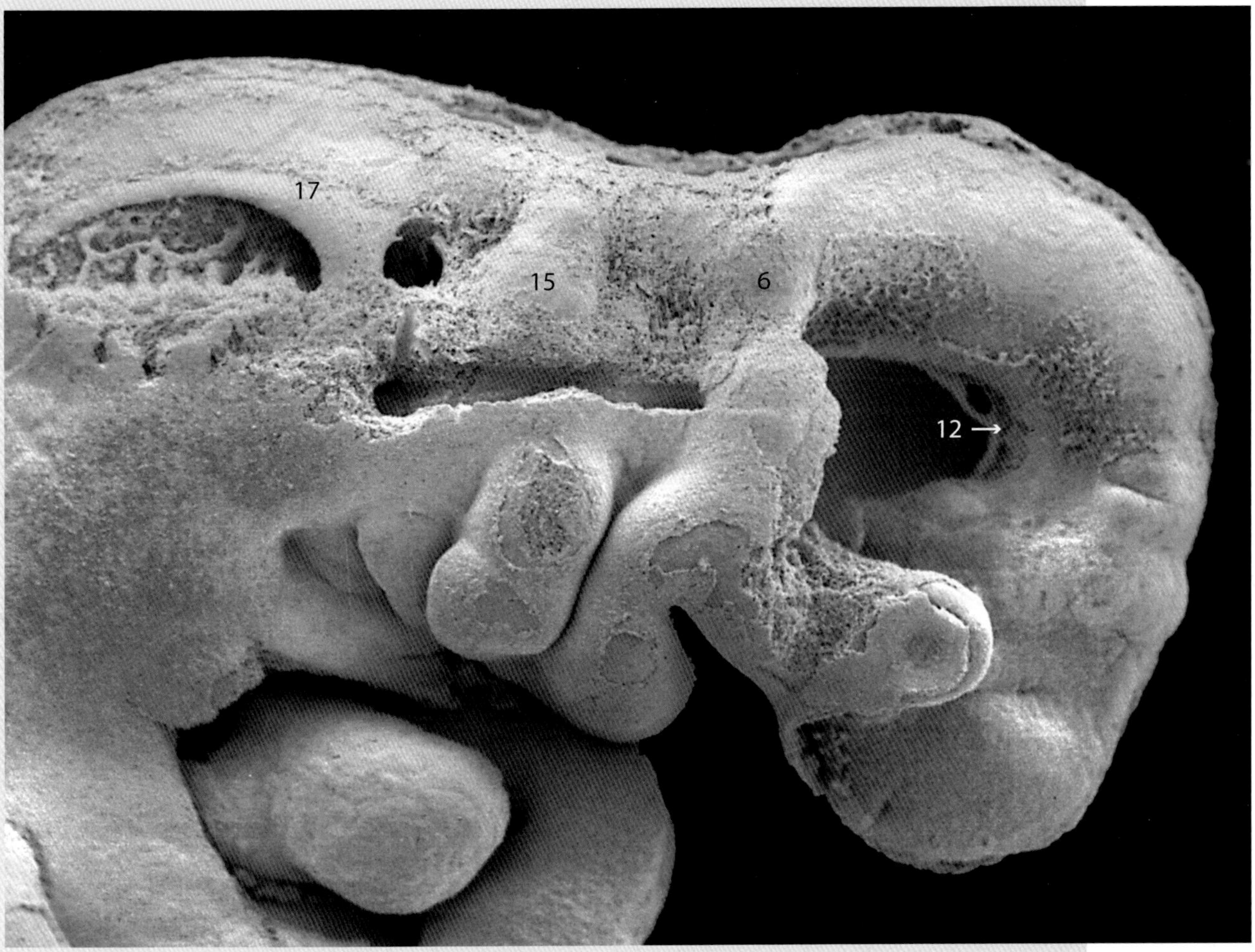

6.1.01–6.1.34
Developmental stages of the external form of the brain.

6.1.22–6.1.29
Lateral-right view of the brain.

Week 5

Week 6

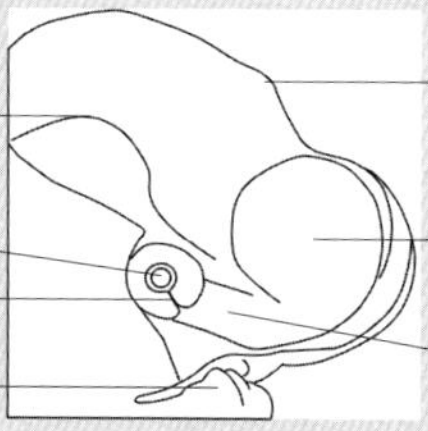

Week 7

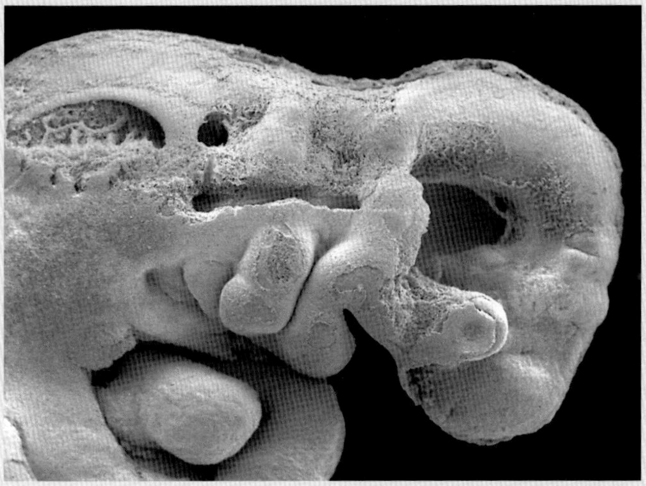

6.1.25a Same embryo as in figure 6.1.25.

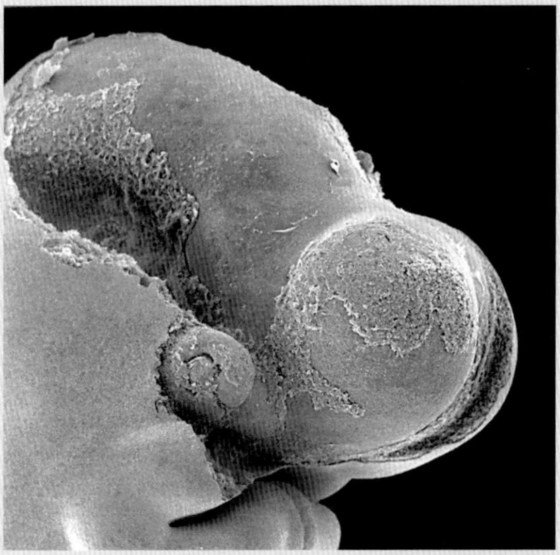

6.1.26 Embryo at week 6.

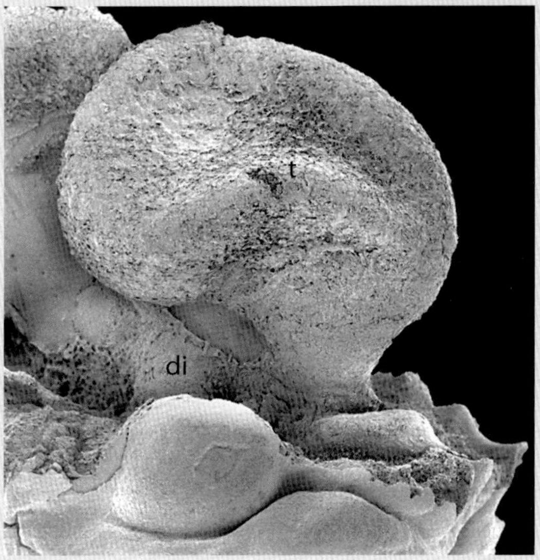

6.1.27 Embryo at week 7.

6.1.01–6.1.34
Developmental stages of the external form of the brain.

6.1.22–6.1.29
Lateral-right view of the brain.

Week 7

Week 8

Week 9

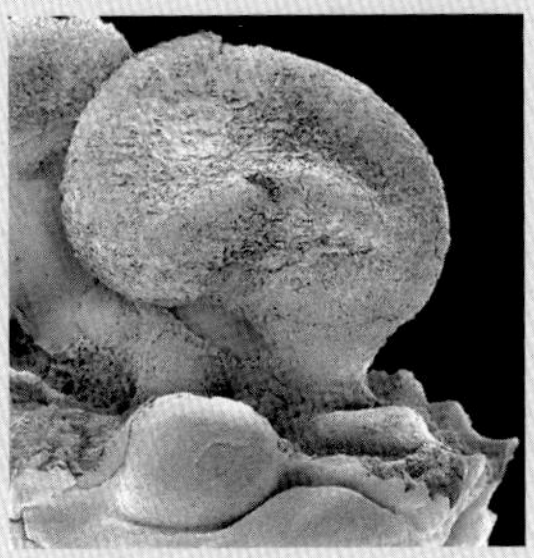

6.1.27a Same embryo as in figure 6.1.27.

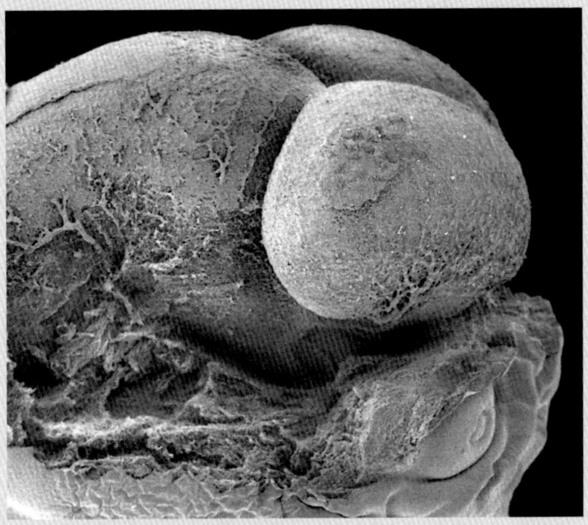

6.1.28 Embryo at week 8.

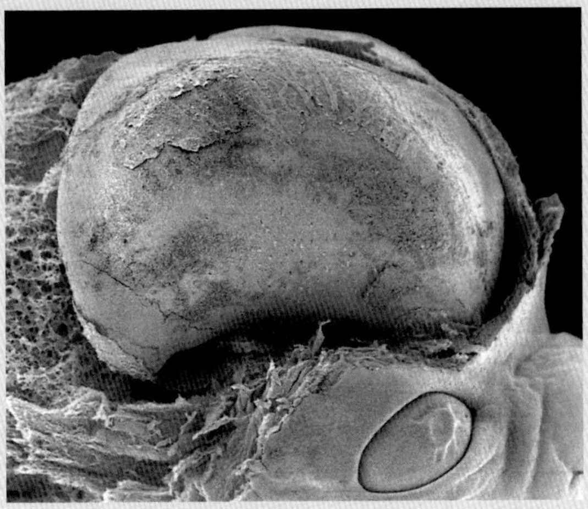

6.1.29 Embryo at week 9.

6.1.01–6.1.34
Developmental stages of the external form of the brain.

6.1.30–6.1.34
Ventral-right view of the brain.

Week 5

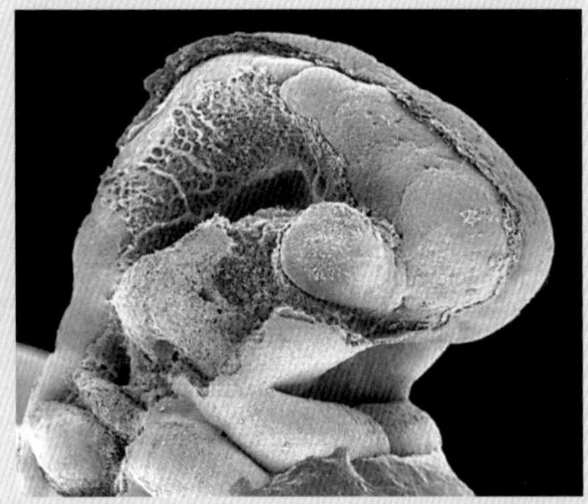

Week 6

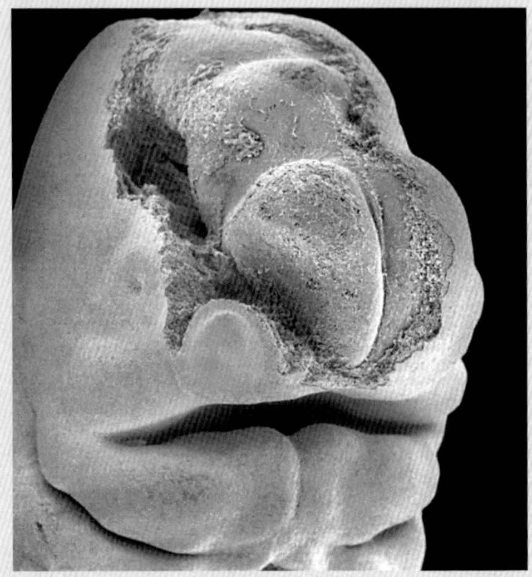

Week 7

6.1.30 / 31 Embryos at week 5.

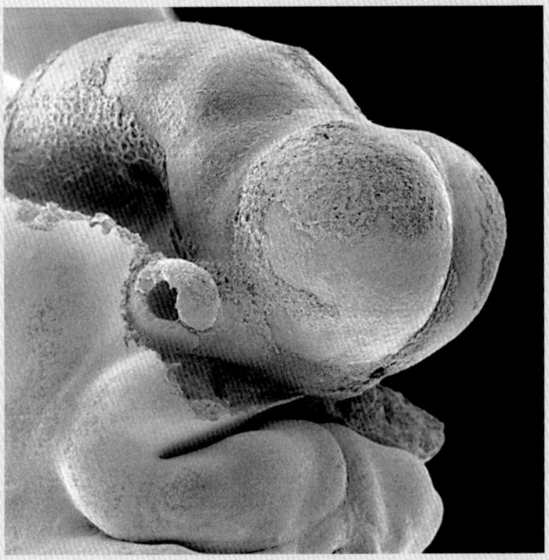

6.1.32 / 33 Embryos at week 6.

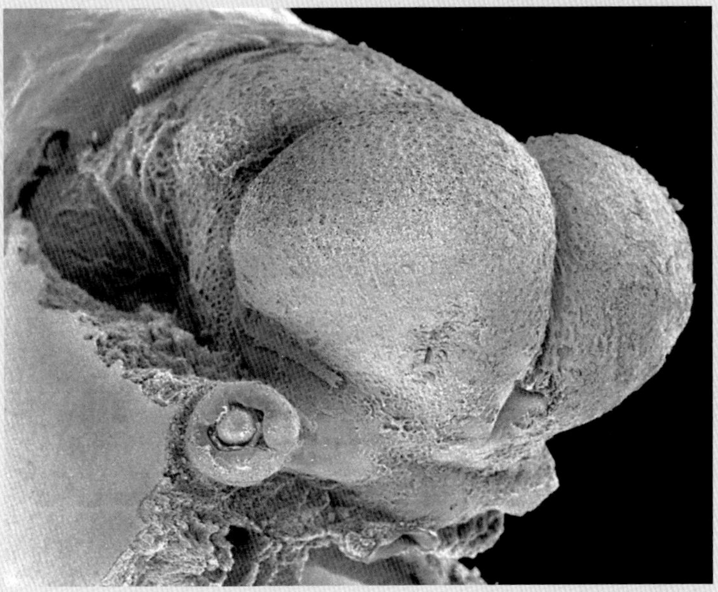

6.1.34 Embryo at week 7.

6.1.35–6.1.57
Developmental stages of the internal form of the brain.

6.1.35–6.1.39
Ventral view into the opened brain.

Week 4

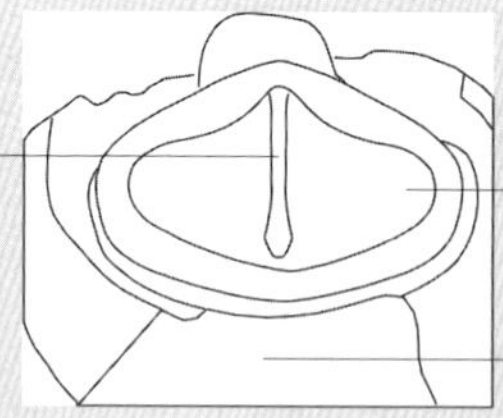

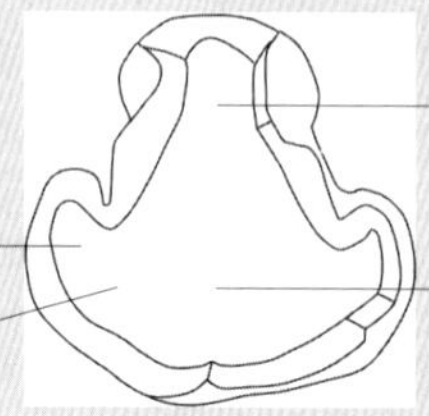

Week 5

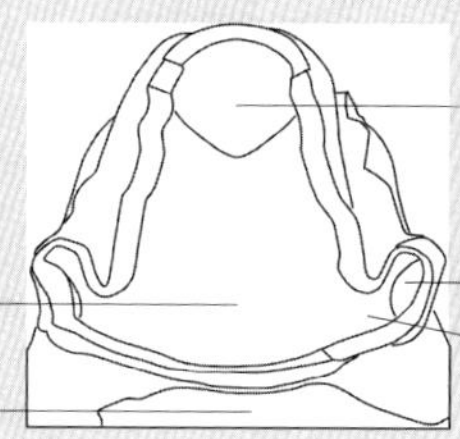

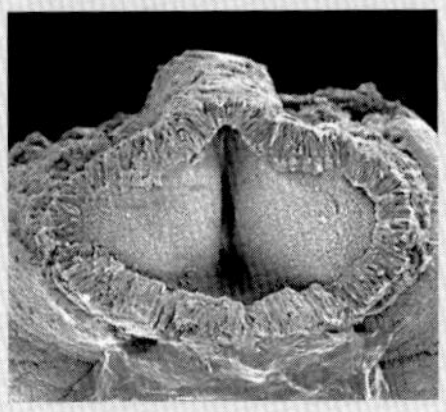

6.1.35 Ventral view into the optic vesicles and the prosencephalon. Embryo in the middle of week 4.

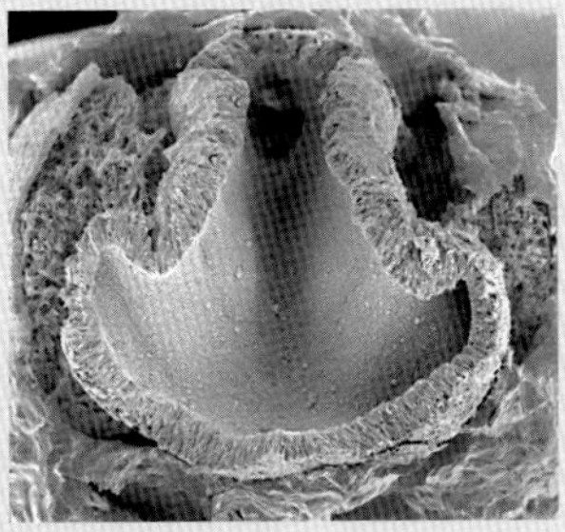

6.1.36 Ventral view into the optic vesicles and the prosencephalon. Embryo at the end of week 4.

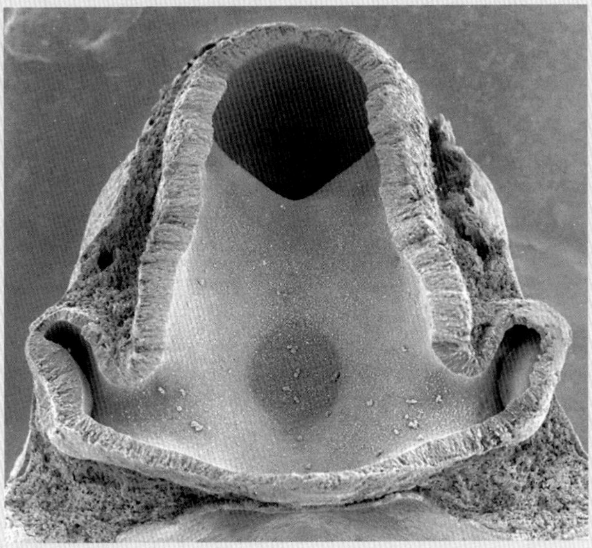

6.1.37 Ventral view into the optic vesicles, the diencephalon and the mesencephalon. Embryo at week 5.

Week 5

Week 7 / 8

6.1.35–6.1.57
Developmental stages of the internal form of the brain.

6.1.35–6.1.39
Ventral view into the opened brain.

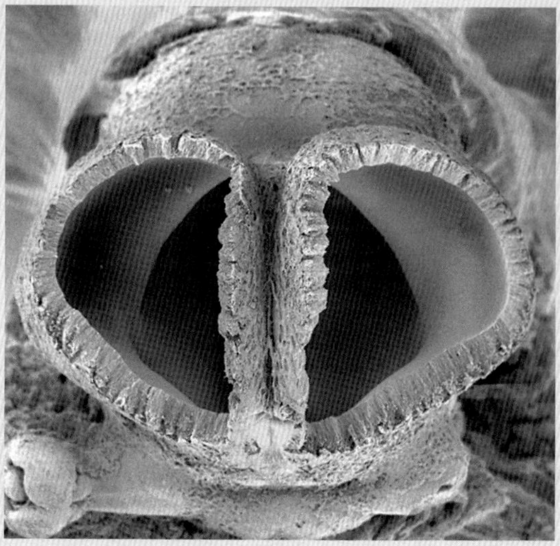

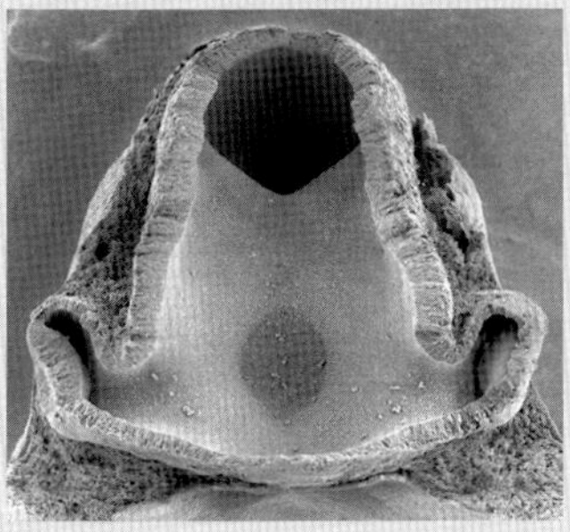

6.1.37a

Same embryo as in figure 6.1.37.

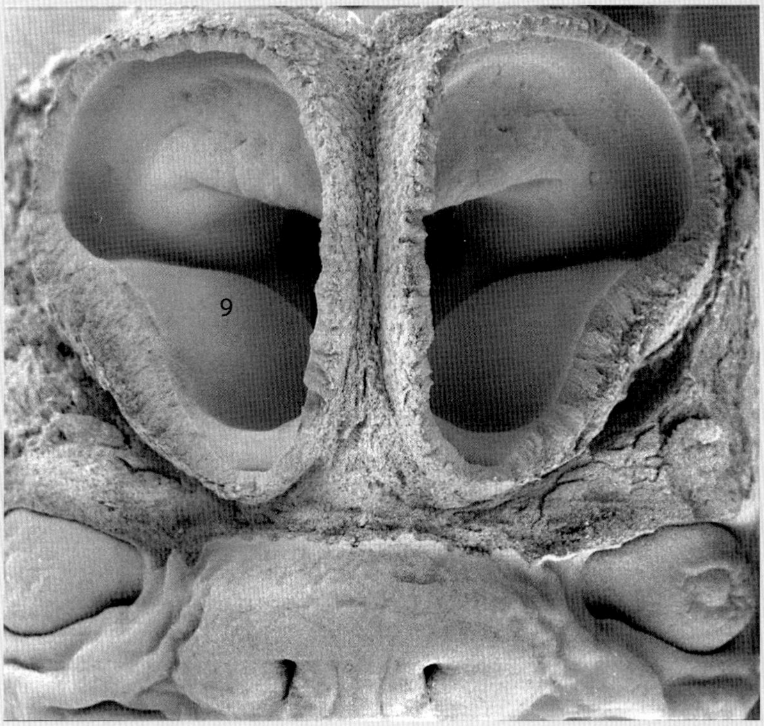

6.1.38 / 39

6.1.38
Ventral view into the cerebral hemispheres. Embryo at week 7.

6.1.39
Ventral view into the cerebral hemispheres. Embryo at week 8.

6.
Brain and
Sensory Organs

6.1
The Development of
the Brain

6.1.35–6.1.57
Developmental stages of the internal form of the brain.

6.1.40–6.1.44
Cranial view into the opened brain.

Week 4

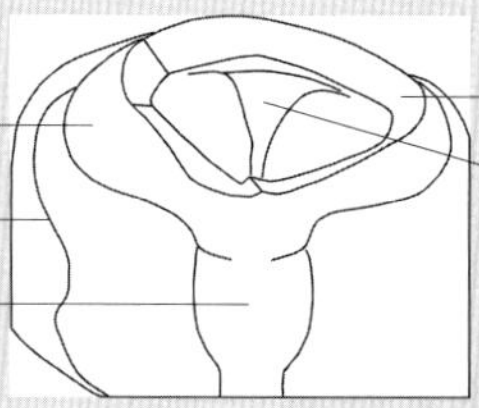

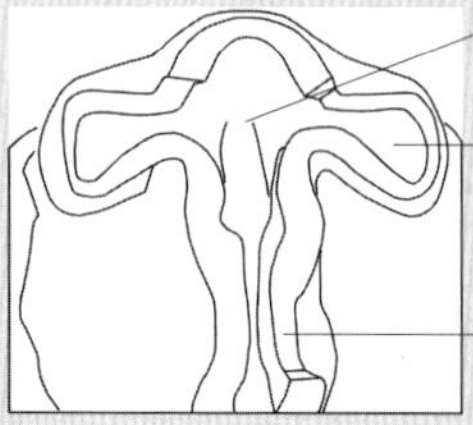

Week 6

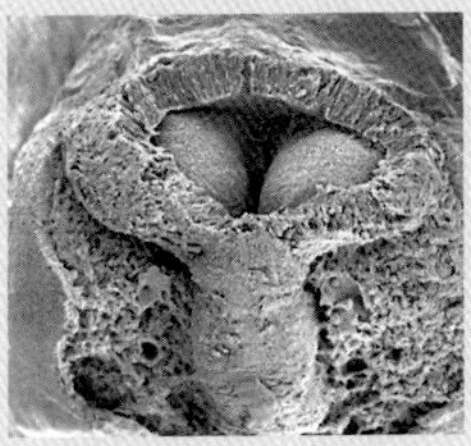

6.1.40 Cranial view into the optic vesicles and the prosencephalon. Embryo in the middle of week 4.

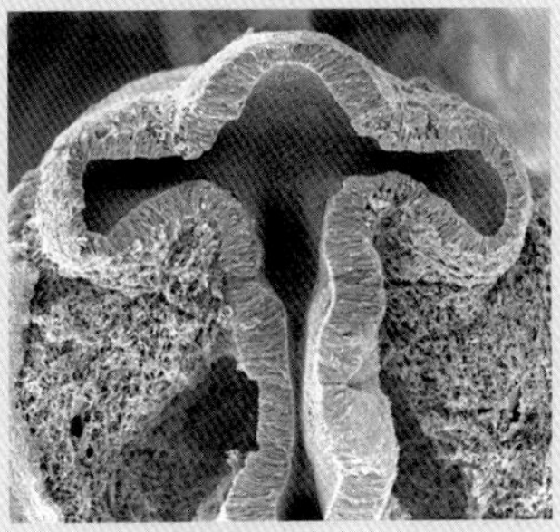

6.1.41 Cranial view into the optic vesicles and the prosencephalon. Embryo at the end of week 4.

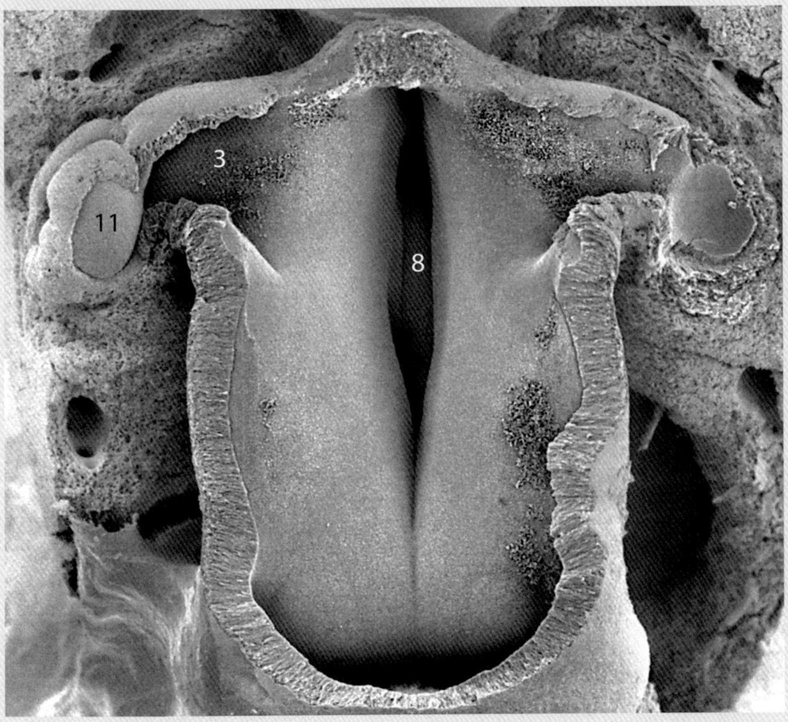

6.1.42 Cranial view into the diencephalon. Embryo at week 6.

Week 6

Week 6 / 8

6.1.35–6.1.57
Developmental stages of the internal form of the brain.

6.1.40–6.1.44
Cranial view into the opened brain.

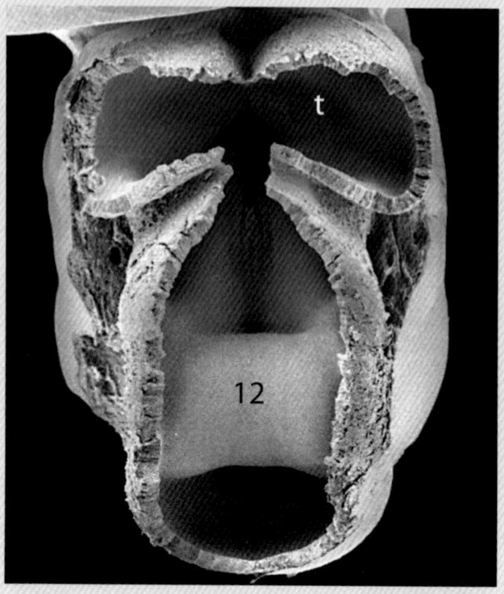

6.1.42a

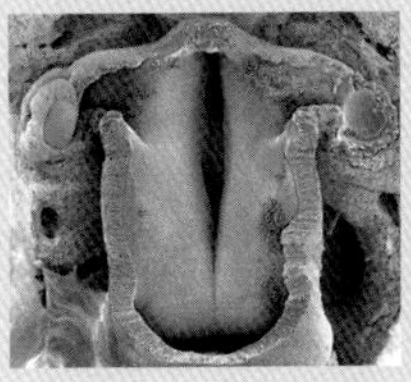

Same embryo as in figure 6.1.42.

6.1.43 / 44

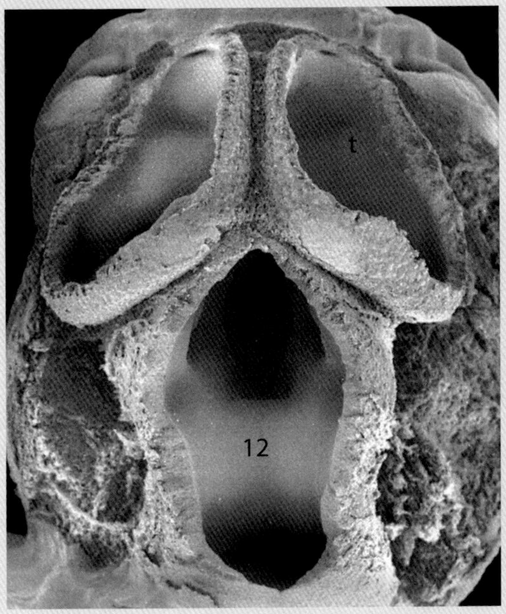

6.1.43
Cranial view into the cerebral hemispheres, the diencephalon, and the mesencephalon. Embryo at the end of week 6.

6.1.44
Cranial view into the cerebral hemispheres, the diencephalon, and the mesencephalon. Embryo at week 8.

6.1.35–6.1.57
Developmental stages of the internal form of the brain.

6.1.45–6.1.49
Lateral-right view into the opened brain.

Week 5

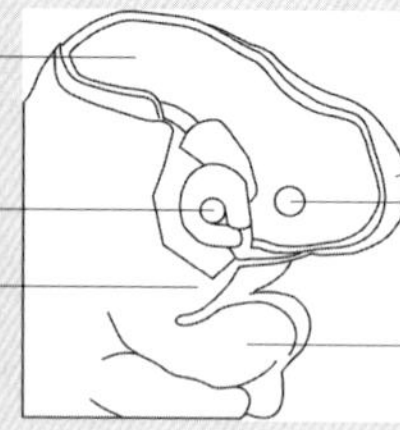

Week 6

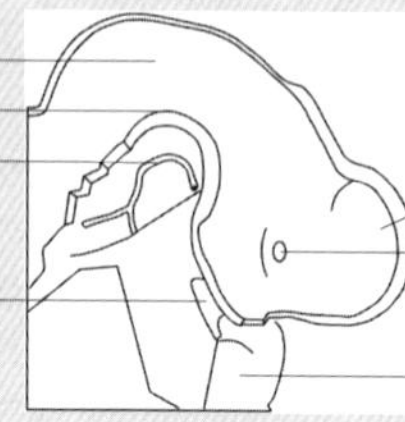

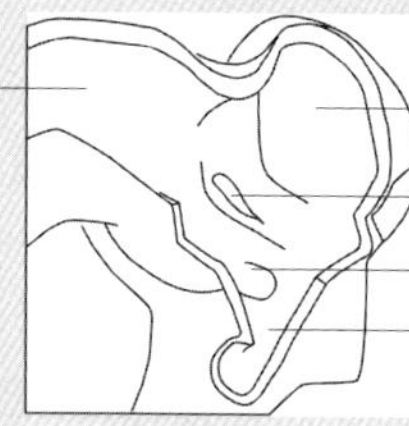

Week 7

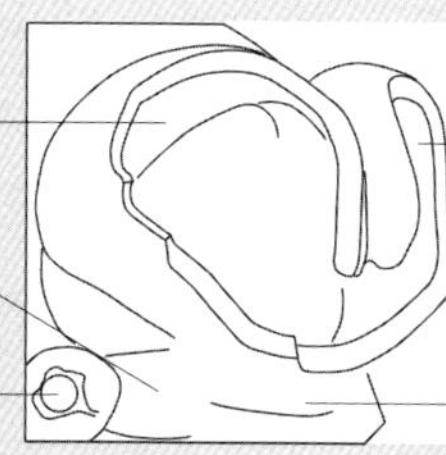

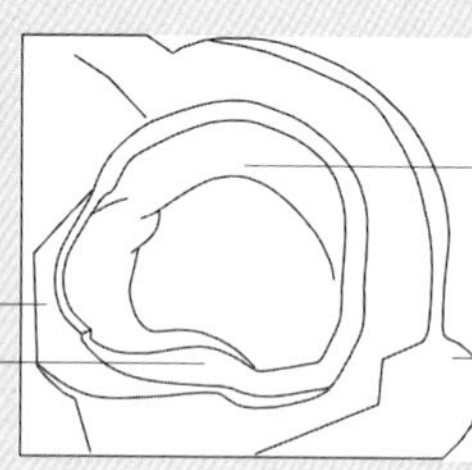

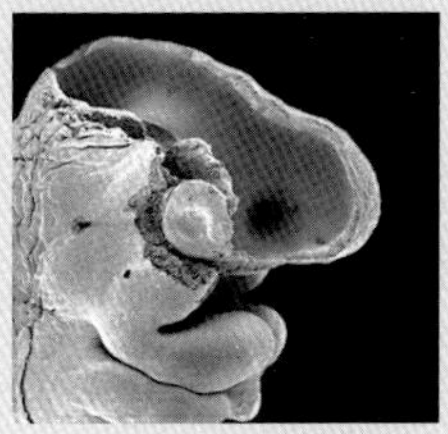

6.1.45 Lateral-right view into the left cerebral hemisphere, the diencephalon, and the mesencephalon. Embryo at week 5.

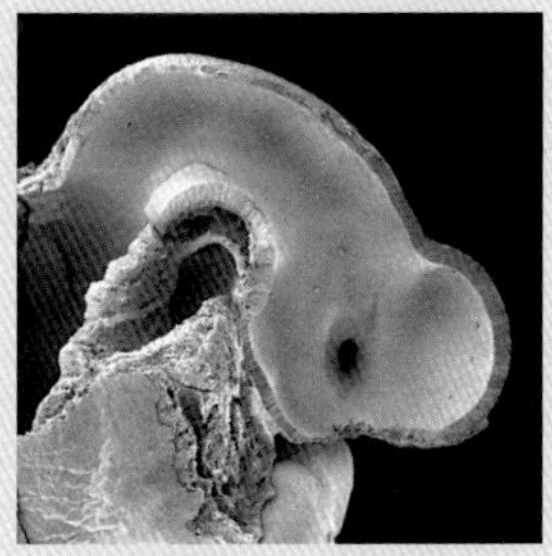

6.1.46 Lateral-right view into the left cerebral hemisphere, the diencephalon, and the mesencephalon. Embryo at week 6.

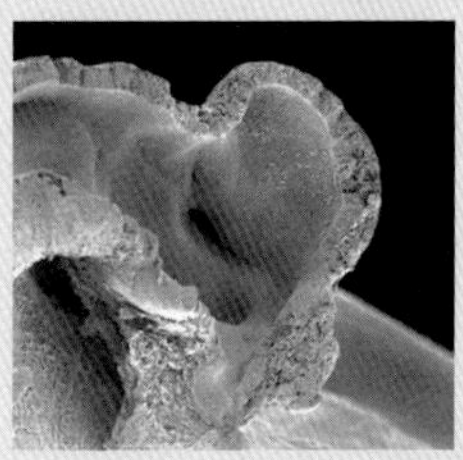

6.1.47 Lateral-right view into the left cerebral hemisphere and the diencephalon. Embryo at the end of week 6.

6.1.48 Lateral-right view into the right cerebral hemisphere. Embryo at week 7.

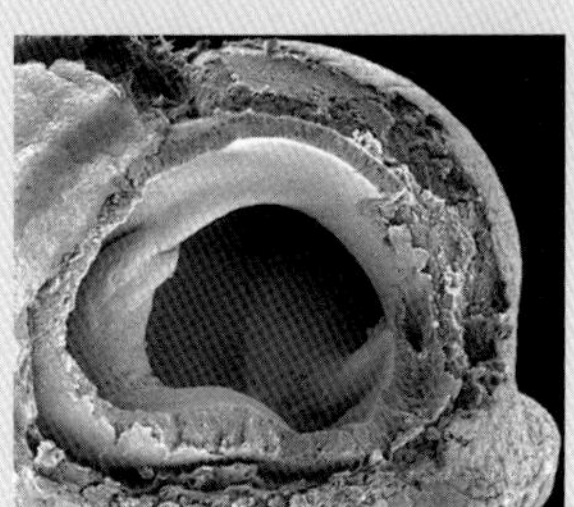

6.1.49 Lateral-right view into the right cerebral hemisphere. Embryo at week 7.

Week 8

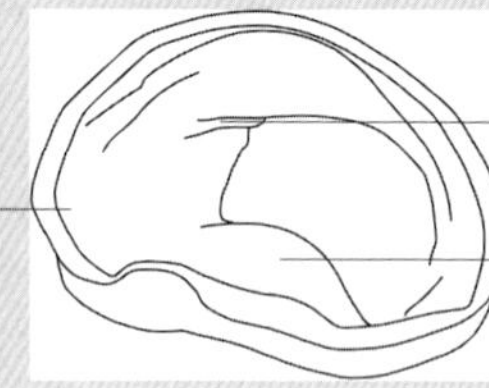

6.1.35–6.1.57
Developmental stages of the internal form of the brain.

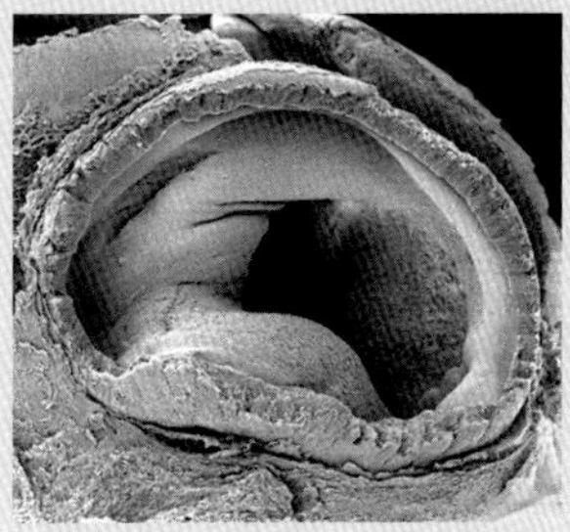

6.1.50

Lateral-right view of different developmental stages of the internal arrangement of the right cerebral hemisphere. Embryos at week 8.

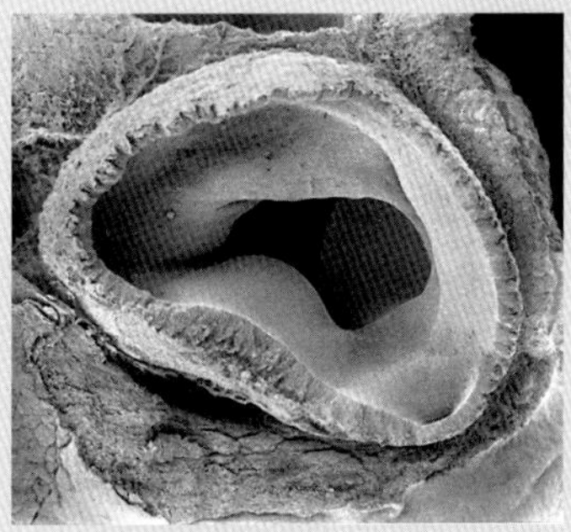

6.1.51

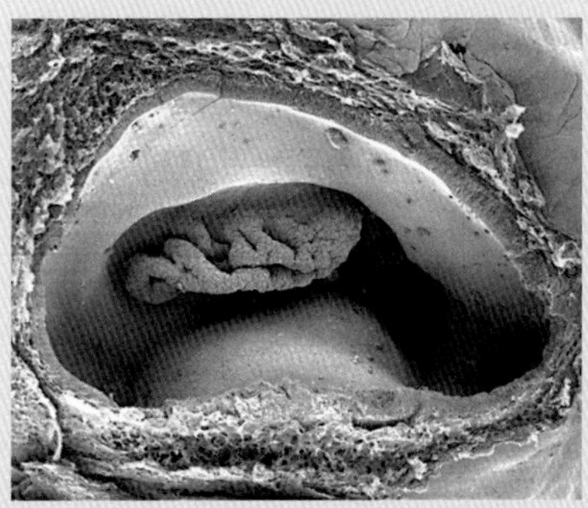

6.1.52

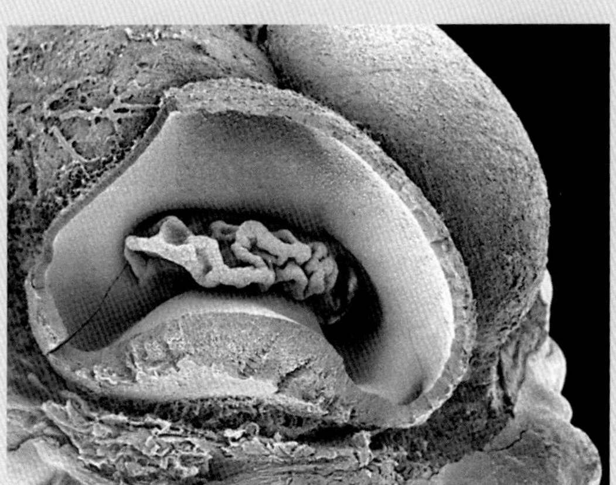

6.1.53

Week 5

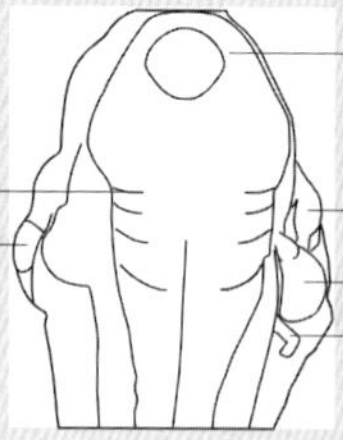

6.1.35–6.1.57

Developmental stages of the internal form of the brain.

6.1.54–6.1.57

Dorsal and cranial view into the floor of the rhombencephalon. The roof of the hindbrain has been removed to show the rhombomeres.

Week 6

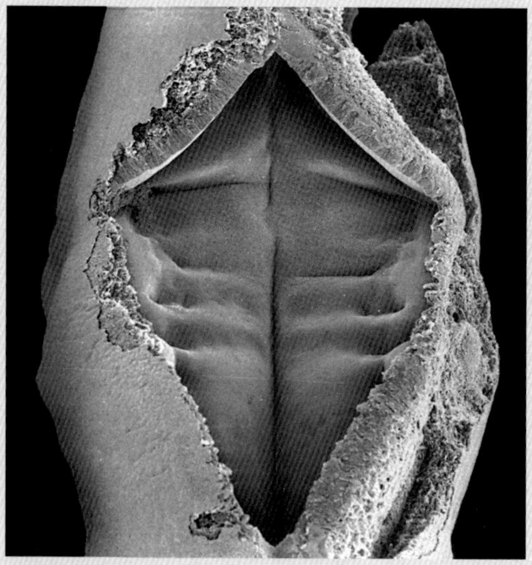

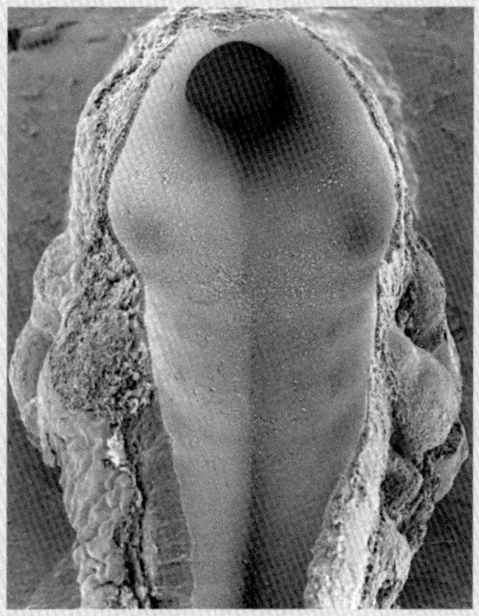

6.1.54 Embryos at week 5.

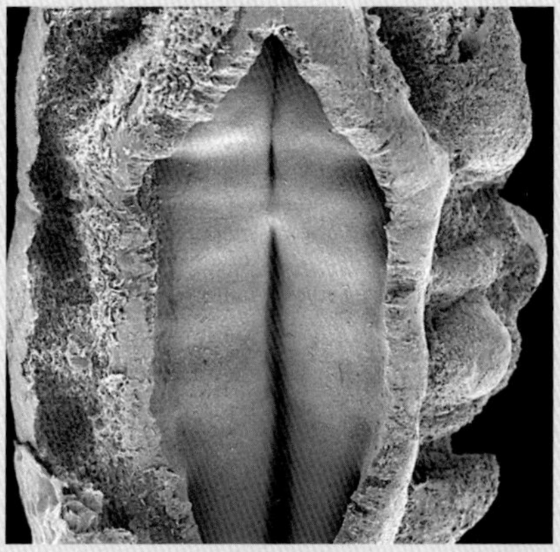

6.1.55

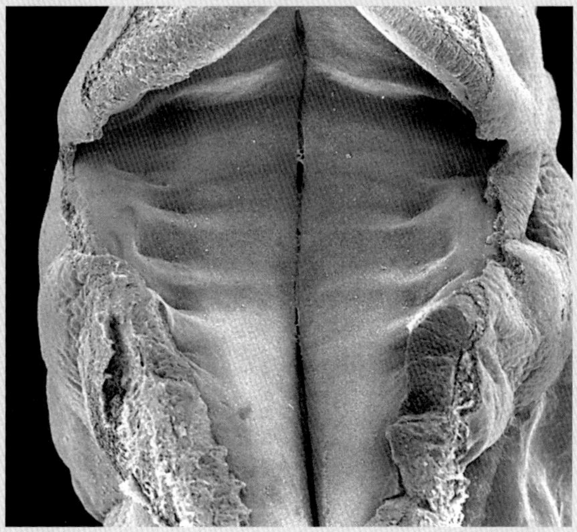

6.1.56 / 57 Embryos at week 6.

6.1.58–6.1.60

Lateral-left view of the left cranial nerves.

Week 5

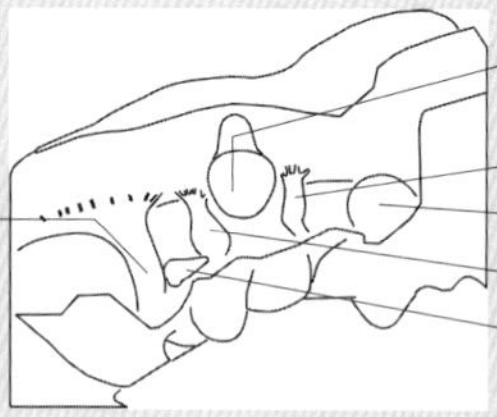

Week 6

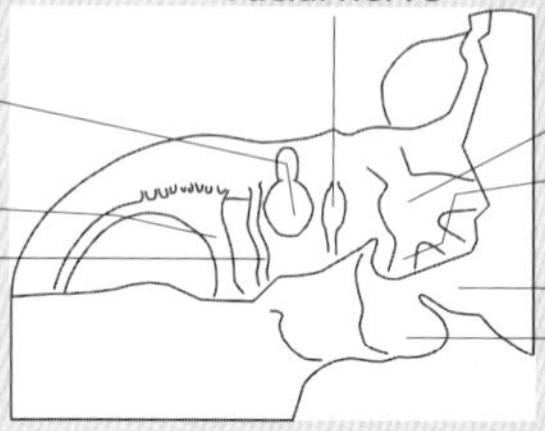

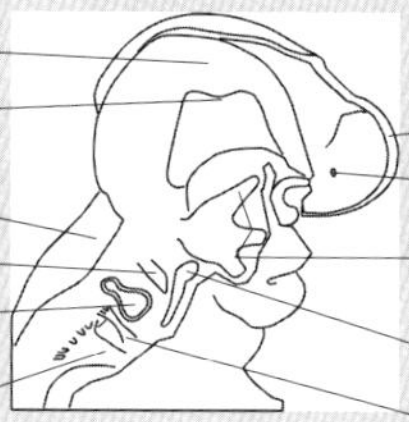

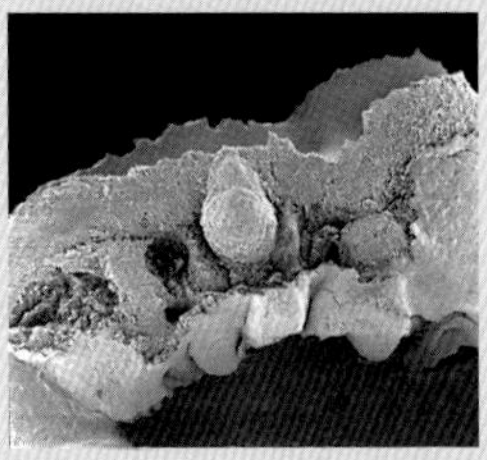

6.1.58 Embryo at week 5.

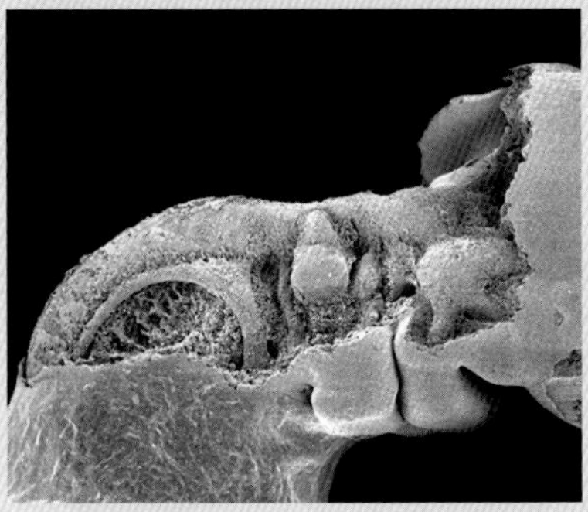

6.1.59 Embryos at week 6.

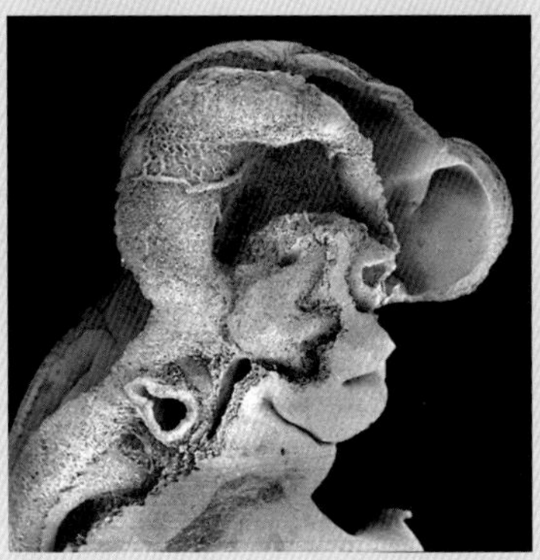

6.1.60

Week 8 / 9

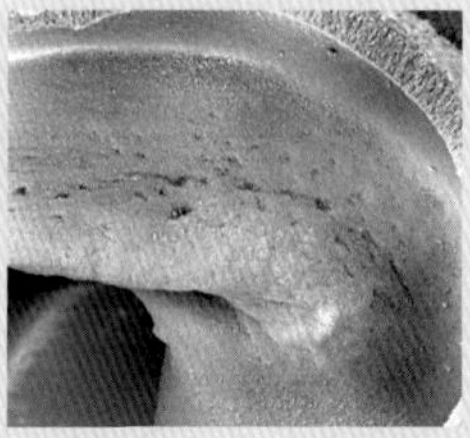

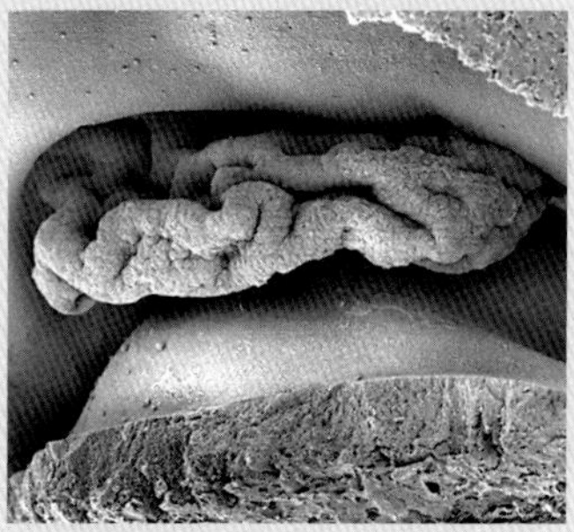

6.1.61

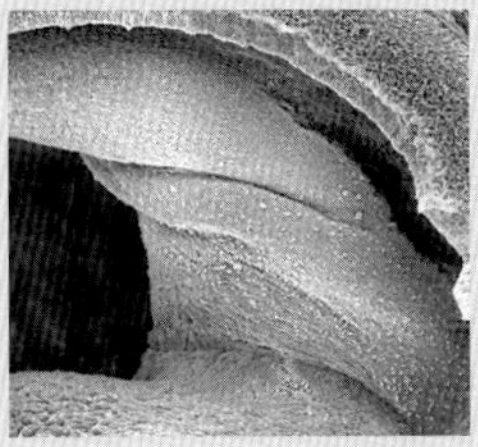

6.1.62 / 63

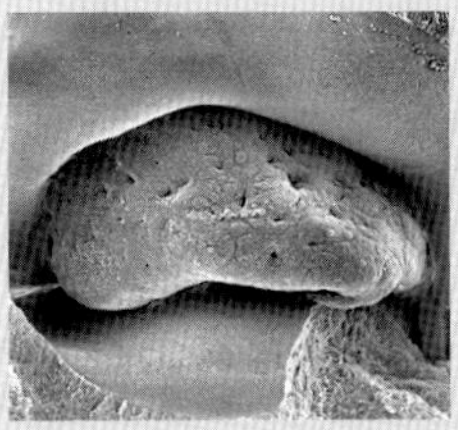

6.1.64 / 65

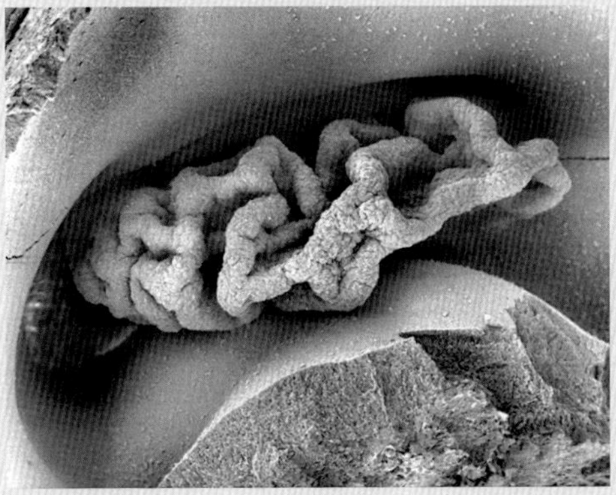

Lateral-left view of different developmental stages of the choroid plexus in the left lateral ventricle. Embryos at weeks 8 and 9.

6.1.66–6.1.73
The hypophysis.

30:1

Week 5

60:1

Week 6

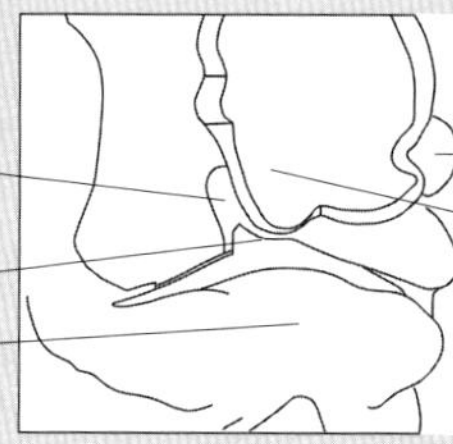

30:1

Week 7

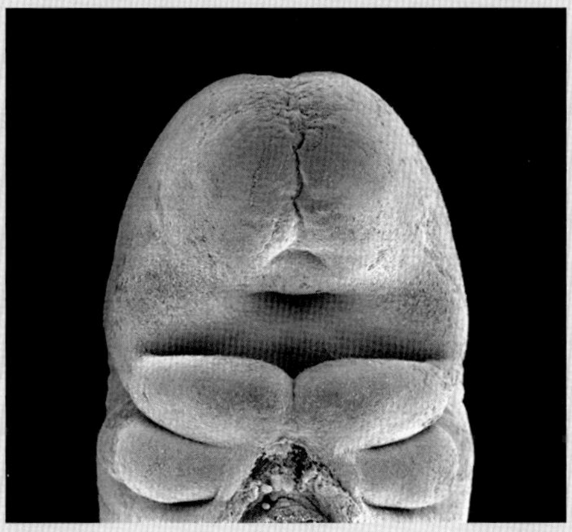

6.1.66 Ventral view of the oral cavity showing the orifice of the adenohypophysial (Rathke's) pouch. Embryo at week 5.

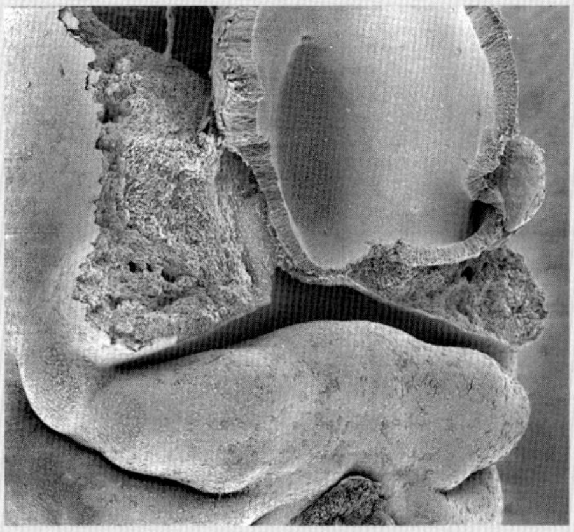

6.1.67 Ventral-right view of the exposed adenohypophysial pouch and its contact region with the anlage region of the neurohypophysis. Embryo at week 6.

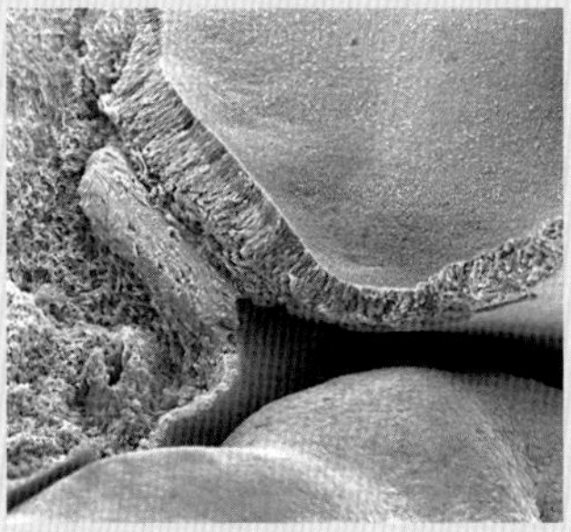

6.1.68 Overall view of the position of the adenohypophysis and neurohypophysis. The right eye with the optic stalk and the right cerebral hemisphere have been removed. Embryo at week 7.

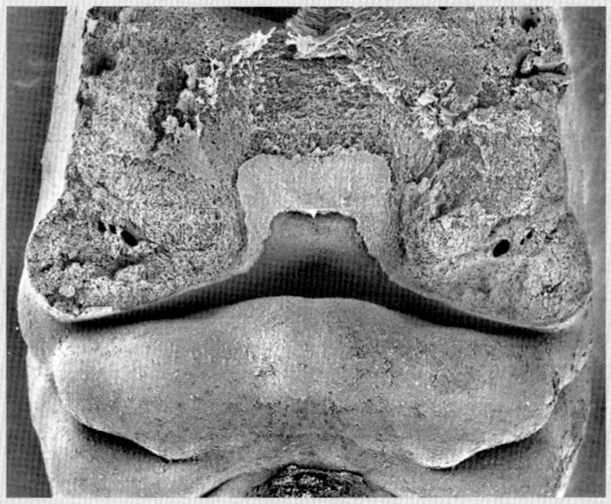

6.1.69 Same embryo as in figure 6.1.68. Ventral view of the adenohypophysial pouch. The diencephalon has been removed.

Week 7

6.1.66–6.1.73
The hypophysis.

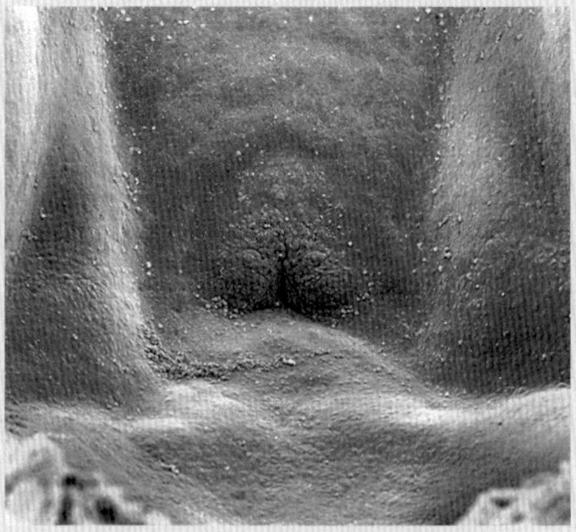

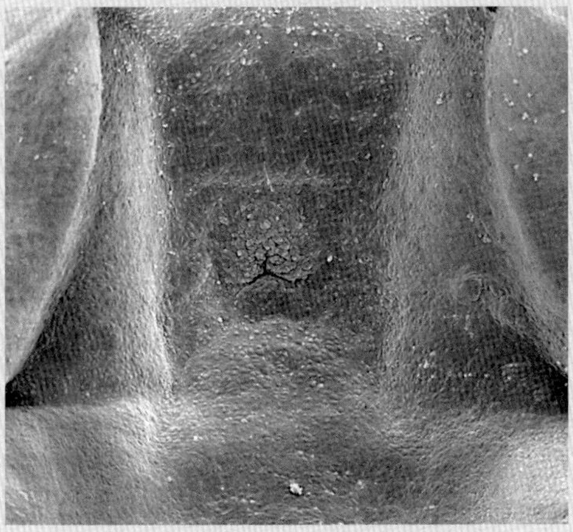

6.1.70 / 71

The caudal view of the oral cavity shows different individual examples of the phase of the orifice of Rathke's pouch losing the connection with the roof of the mouth. Embryos at week 7.

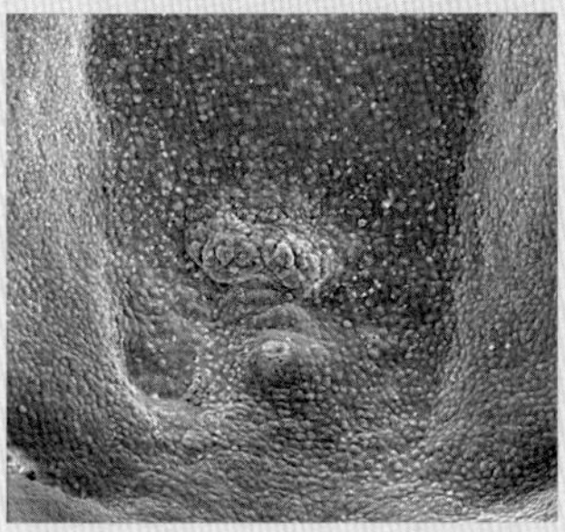

6.1.72

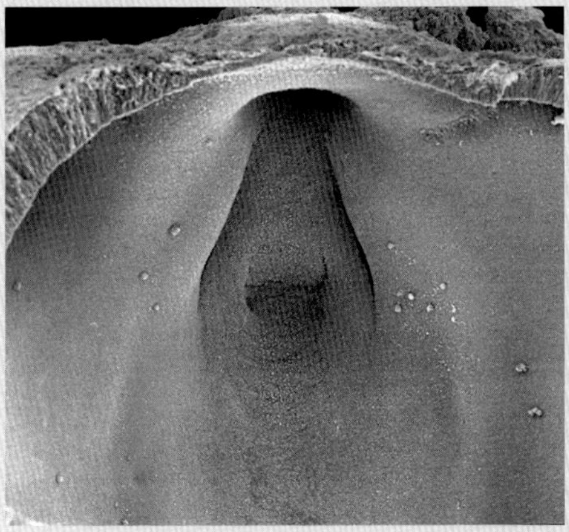

6.1.73

Cranial view of the floor of the diencephalon showing the anlage region of the neurohypophysis. Embryo at week 7.

6.2
The Development of the Eye

The first anlage of the eye becomes visible in week 3 in the region of the forebrain as a slight sulcus which deepens and evaginates as the optic vesicle (fig. 6.1.01, 6.2.02–6.2.06). At the site where the evaginated neural epithelium abuts the ectoderm, the neural epithelium is thickened, thus forming the anlage of the retina (fig. 6.2.15) which invaginates forming the double-layered optic cup (fig. 6.2.07–6.2.13), its inner layer forming the retina and its outer layer the pigmented epithelium (fig. 6.2.10, 6.2.13). Because the invagination of the retina is incomplete, at the inferior pole of the optic cup the retinal fissure remains, which is the pathway between the optic nerve and the retina and which provides access for the hyaloid artery to the lens (fig. 6.2.11, 6.2.12, 6.2.14, 6.2.30). The retinal fissure is closed approximately in week 9.

The connection between the optic cup and the brain, the optic stalk (fig. 6.2.08, 6.2.12, 6.2.13), develops into the optic nerve (fig. 6.2.32).

The ectoderm folds in into the optic cup and forms the lens pit (fig. 6.2.17–6.2.19) which closes to become the lens vesicle (fig. 6.2.20, 6.2.21). The posterior circumference of the epithelium of the lens vesicle thickens, thus narrowing the lumen of the lens vesicle (fig. 6.2.22–6.2.24) until it disappears in week 9. The elongating cells of the thickened dorsal epithelium of the lens vesicle form the lens fibres (fig. 6.2.25–6.2.28).

6.2
Abbreviations

m	mesencephalon
p	prosencephalon
1	optic vesicle
3	retina
4	pigmented epithelium
5	lens
6	optic stalk, optic nerve
7	retinal (choroid) fissure

Week 4

6.2.01–6.2.33
Developmental stages of the eye.

6.2.01–6.2.06
Stages of the optic vesicle.

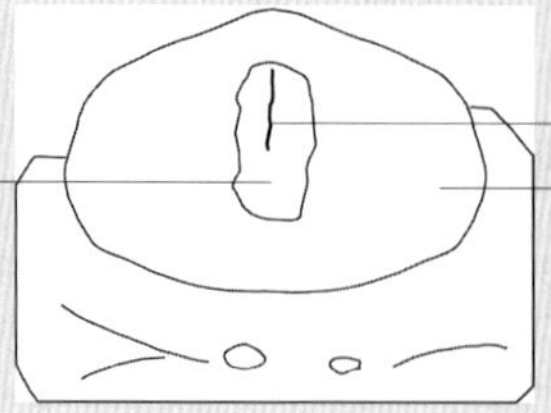

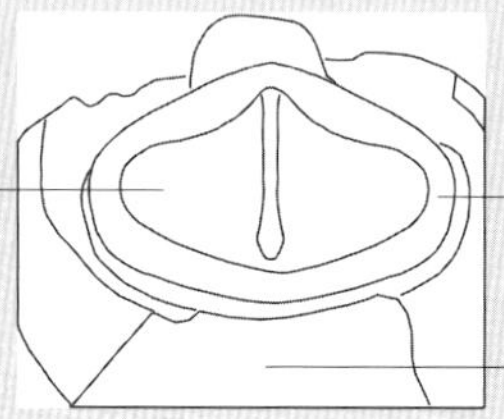

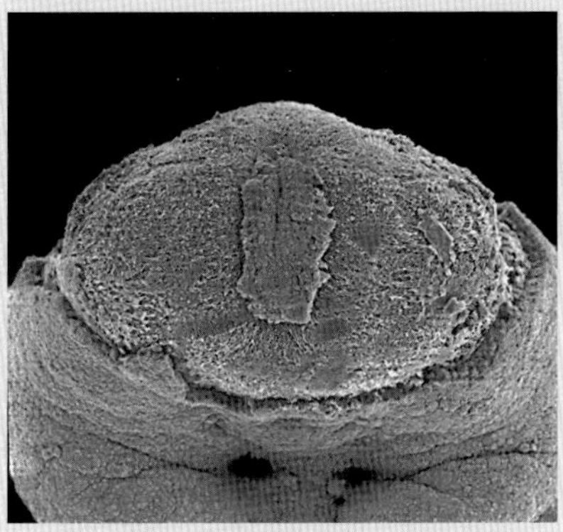

6.2.01

Ventral views of the prosencephalon and the optic vesicles, external and internal aspects. Embryo in the middle of week 4.

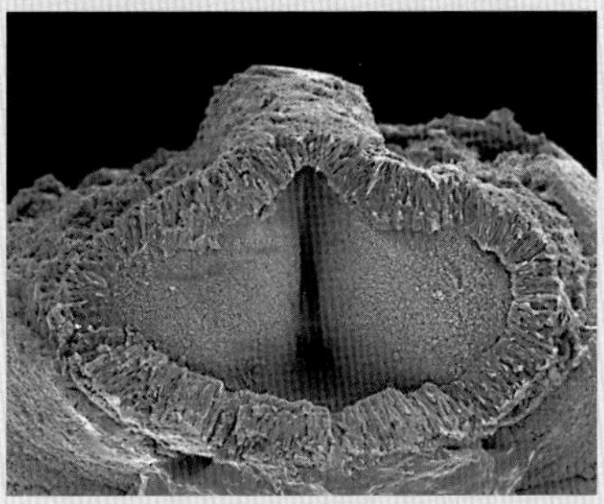

6.2.02

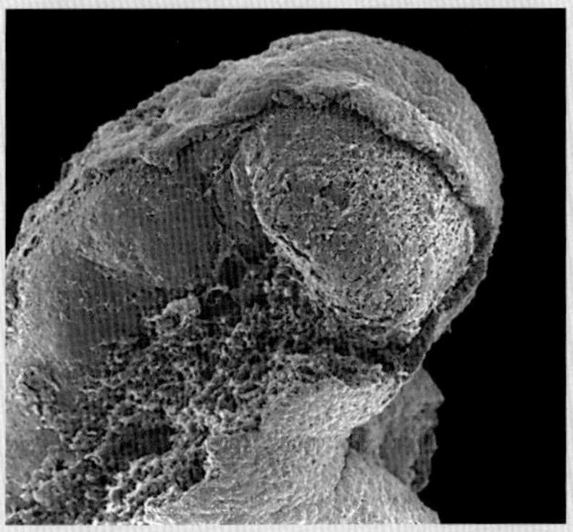

6.2.03

Lateral-right view of the right optic vesicle. Embryo in the middle of week 4.

Week 4

6.2.01–6.2.33
Developmental stages of the eye.

6.2.01–6.2.06
Stages of the optic vesicle.

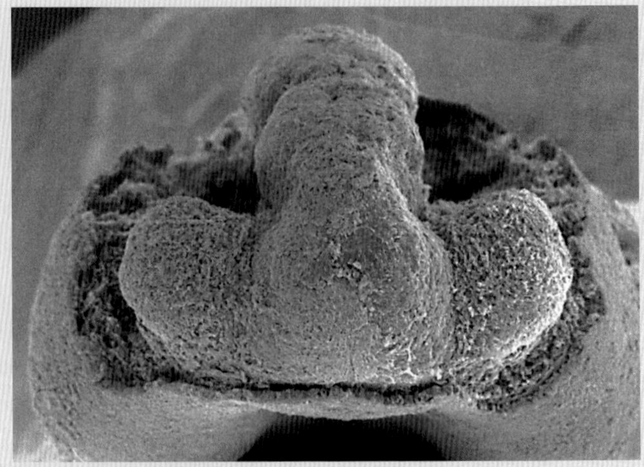

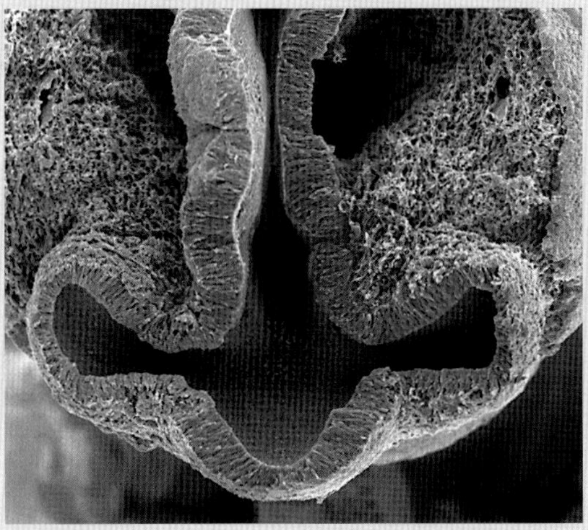

6.2.04 / 05

Ventral-cranial views of the prosencephalon and the optic vesicles, external and internal aspects. Embryo at the end of week 4.

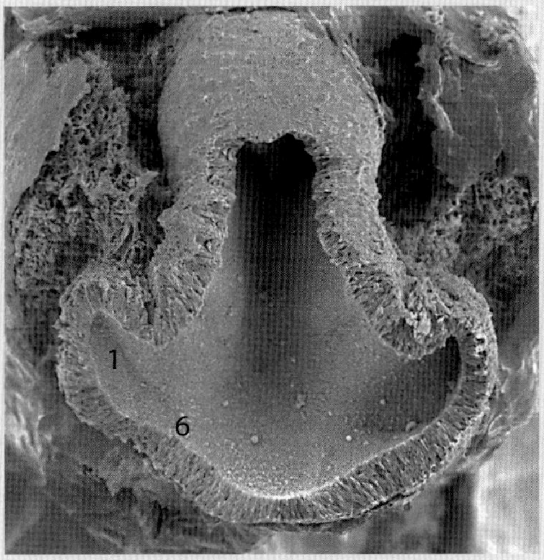

6.2.06

Ventral-cranial view of the opened prosencephalon and the optic vesicles. Early transformation of the optic vesicle into the optic cup. More advanced stage of an embryo at the end of week 4.

Week 5

6.2.01–6.2.33
Developmental stages of the eye.

6.2.07–6.2.14
Formation of the optic cup. Ventral-left views of the right eye. The overlying ectoderm with the anlage of the lens has been removed.

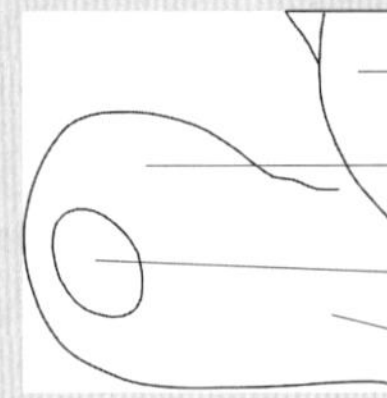

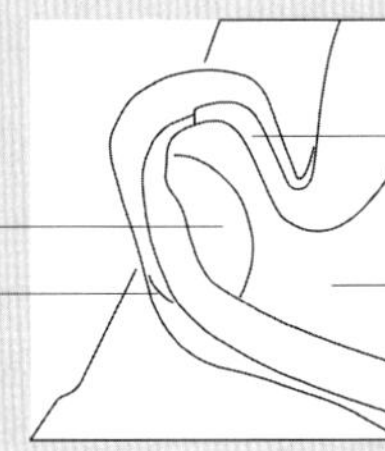

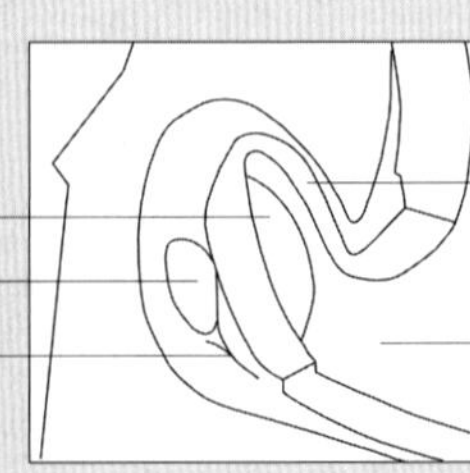

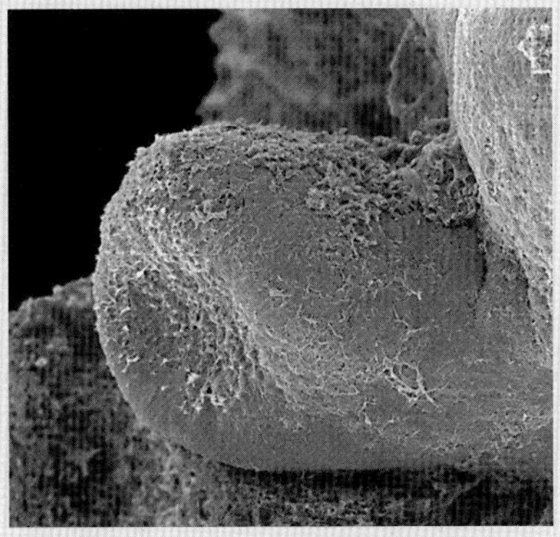

6.2.07

Early invagination of the retinal disc, external and internal views. Embryos at week 5.

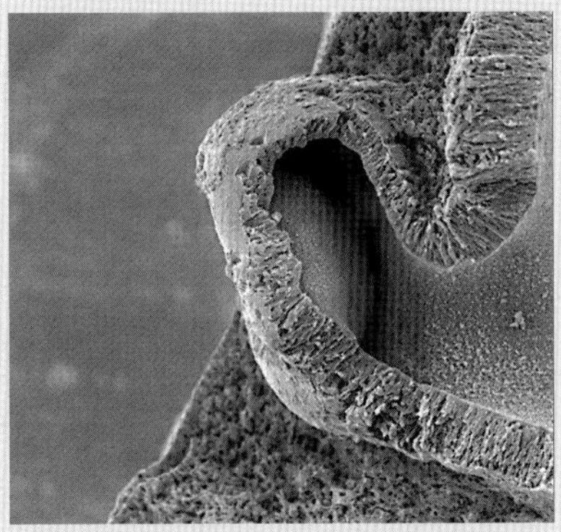

6.2.08

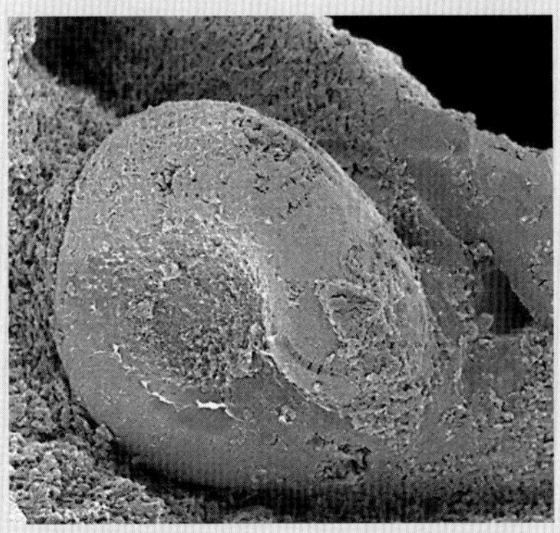

6.2.09

Advanced stage of invagination, origin of the retinal fissure. Embryo in the middle of week 5.

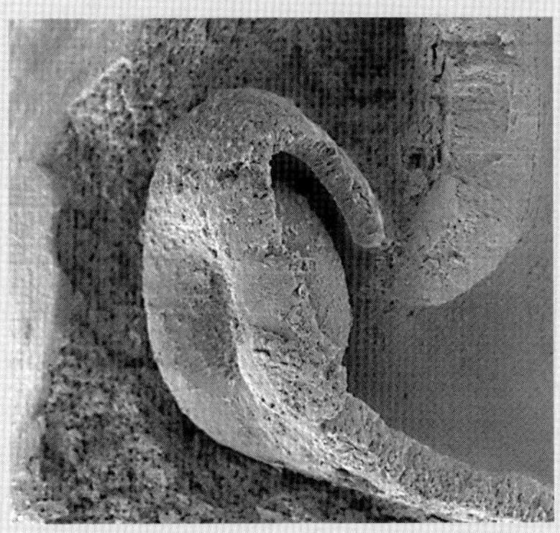

6.2.10

6.2.01–6.2.33
Developmental stages of the eye.

6.2.07–6.2.14
Formation of the optic cup. Ventral-left views of the right eye. The overlying ectoderm with the anlage of the lens has been removed.

Week 5

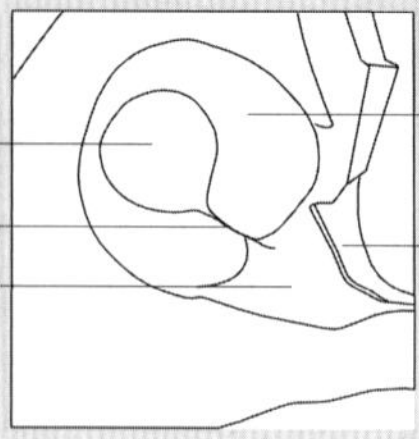

Week 6

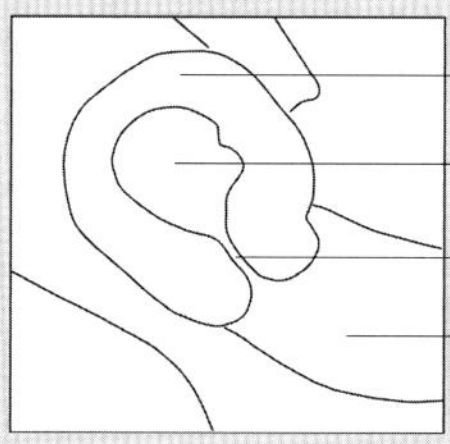

Week 5

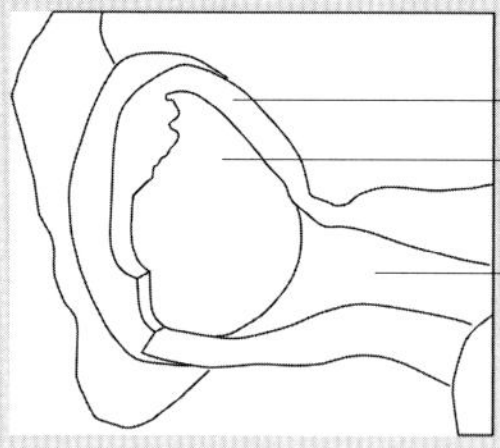

Week 6

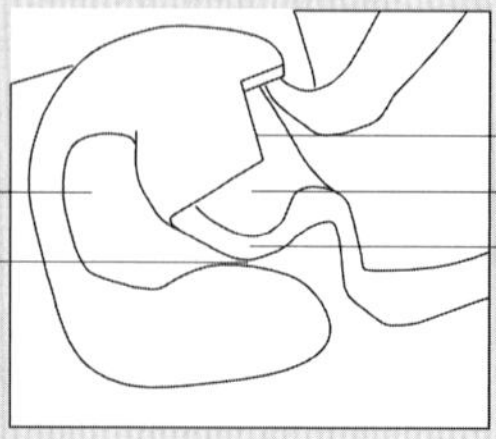

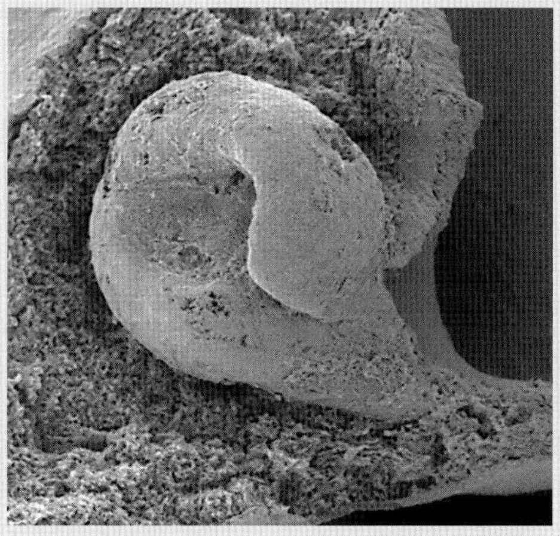

6.2.11 Embryo in the middle of week 5.

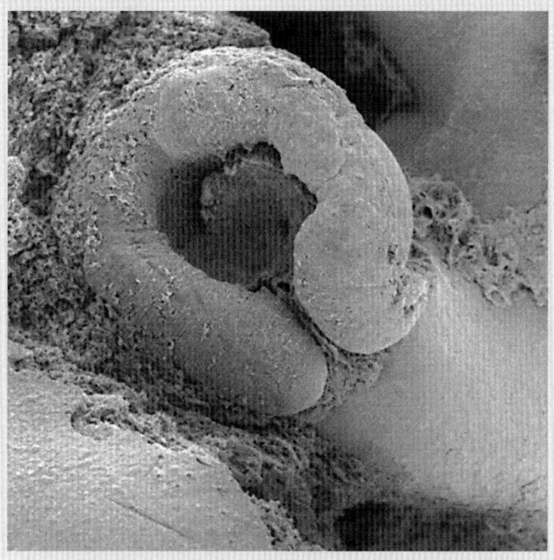

6.2.12 Embryo at week 6.

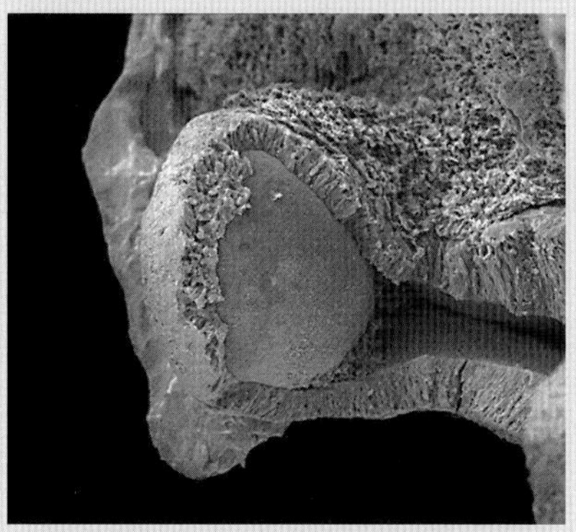

6.2.13 The pigmented layer of the optic cup has been partly removed and the optic stalk, the future optic nerve, has been opened. Embryo at week 5.

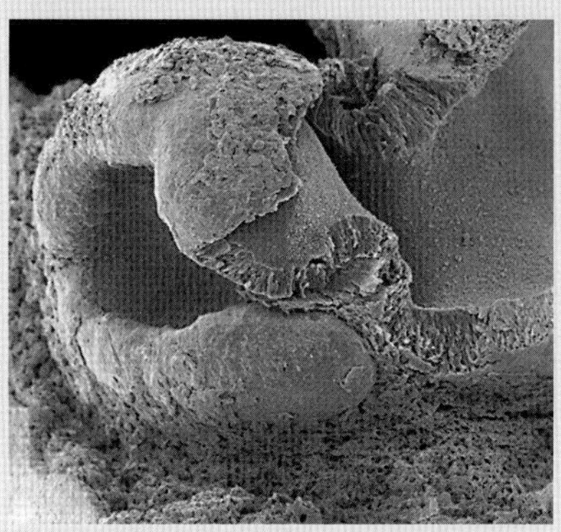

6.2.14 The pigmented layer has partly been removed to show the retinal (chorioid) fissure. Embryo at week 6.

Week 5

6.2.01–6.2.33
Developmental stages of the eye.

6.2.15–6.2.30
Development of the lens.

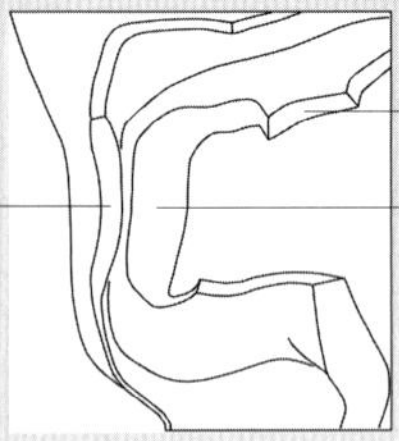

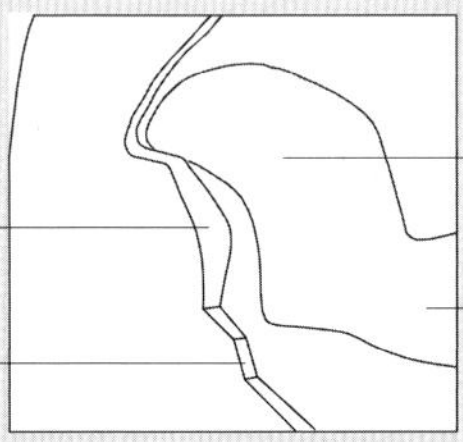

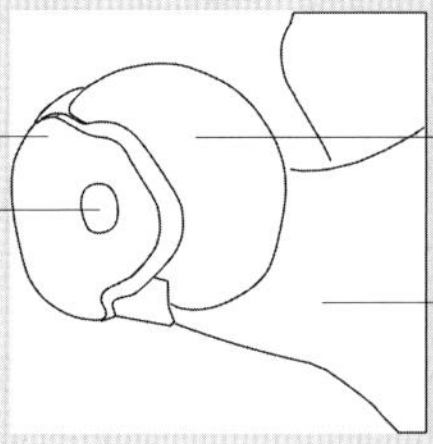

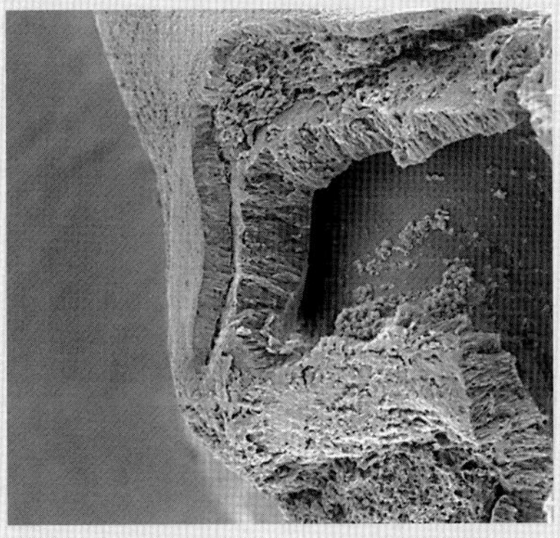

6.2.15 The optic cup has been opened to show the retinal disc (placode) in contact with the lens disc (placode). Embryo at week 5.

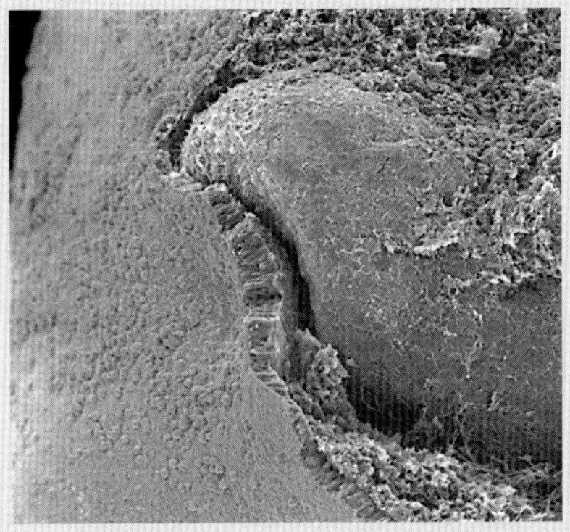

6.2.16 Early invagination of the lens disc to form the lens vesicle. Embryo at week 5.

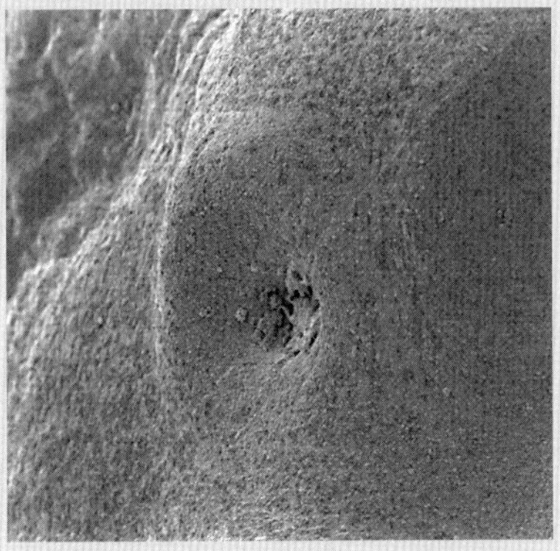

6.2.17 External view of the lens pit. Embryo in the middle of week 5.

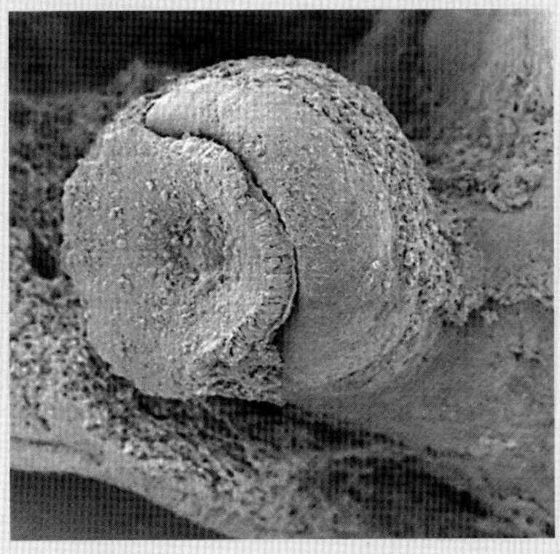

6.2.18 The optic cup has been exposed to show the position of the lens ectoderm. Embryo at week 5.

100:1

Week 5

Week 6

6.2.01–6.2.33
Developmental stages of the eye.

6.2.15–6.2.30
Development of the lens.

200:1

Week 7 / 8

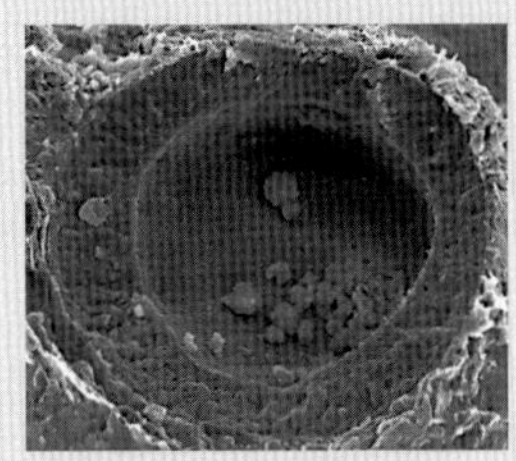

Week 8

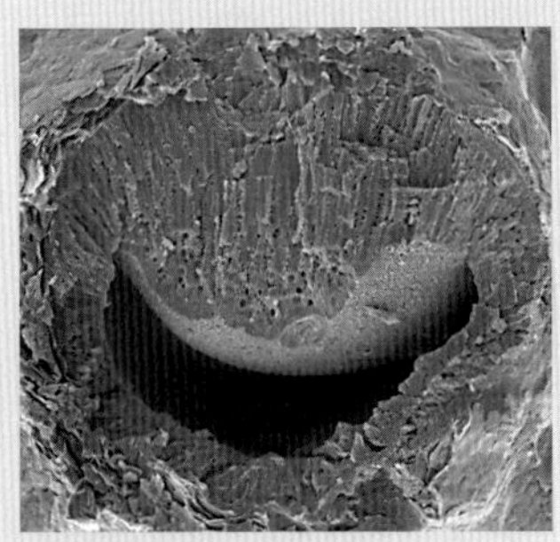

6.2.19

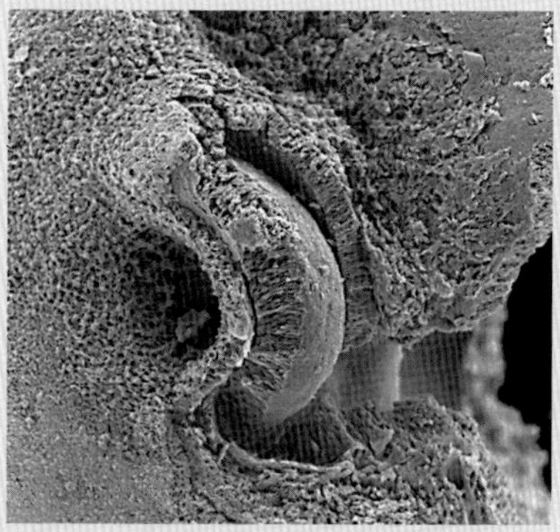

The pigmented layer has been removed. Advanced stage of invagination of the lens. Ventral-left view of the right eye. Embryo at week 5.

6.2.20

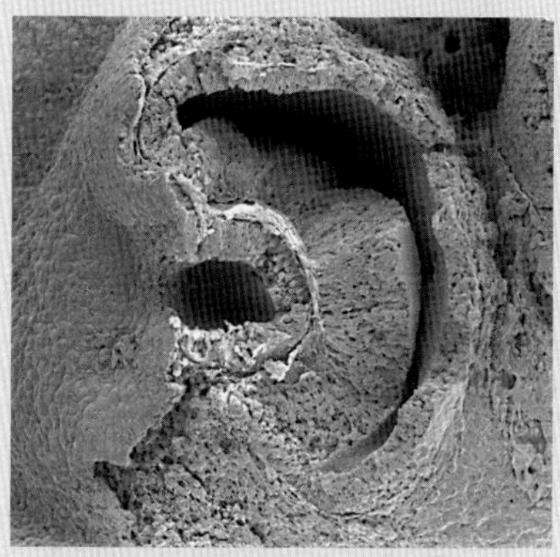

The retina and the pigmented layer have been partly removed.
The lens vesicle has been opened. Embryo at week 6.

6.2.21 / 22

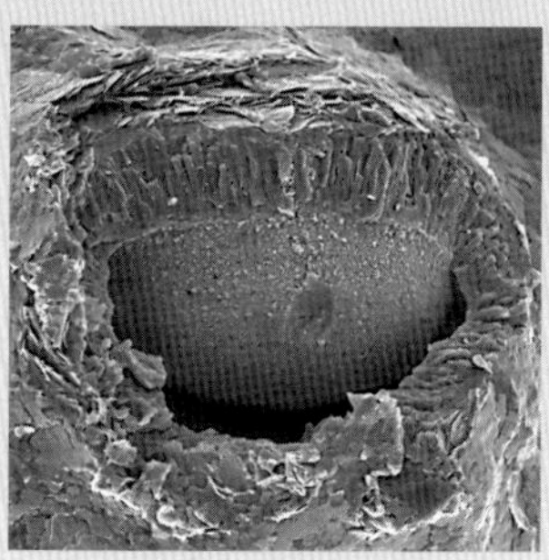

6.2.21
The lens pit is closed and forms the lens vesicle. Cranial view of the opened vesicle. Embryo at week 7.

6.2.22
The lens vesicle has been opened near its cranial pole. The lumen of the lens vesicle is narrowed due to the thickening of the posterior lens epithelium. Embryo at week 8.

6.2.23 / 24

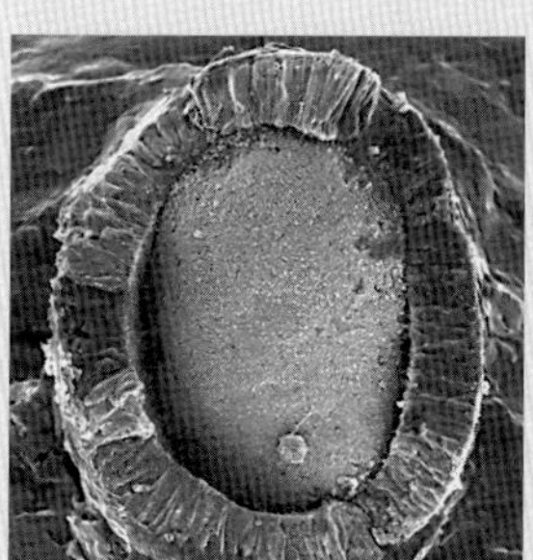

6.2.23
Same embryo as in figure 6.2.22. The cross-fracture of the lens vesicle at its equator shows the primary lens fibres.

6.2.24
The lens vesicle has been opened near its cranial pole. Due to the elongation of the lens fibres the lumen of the lens vesicle has become slit-like. Embryo at the end of week 8.

6.2.01–6.2.33
Developmental stages of the eye.

6.2.15–6.2.30
Development of the lens.

800:1

Week 6 / 8

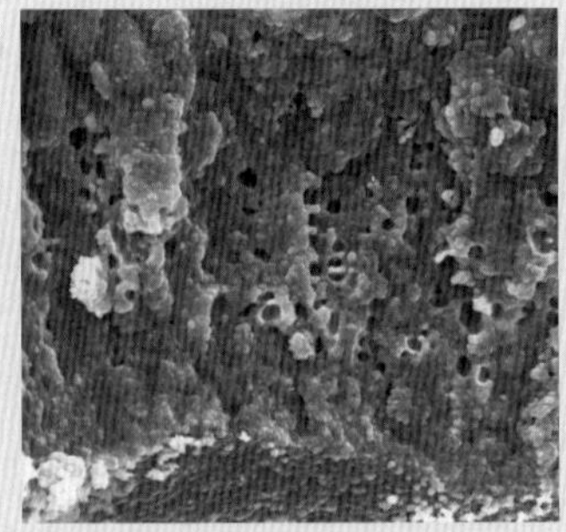

Week 9

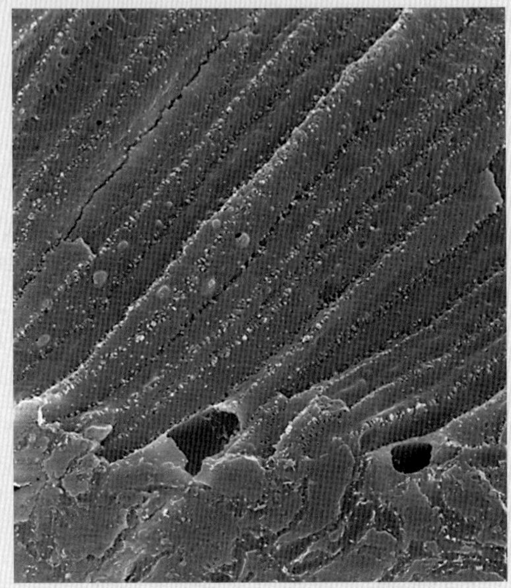

100:1

Week 6

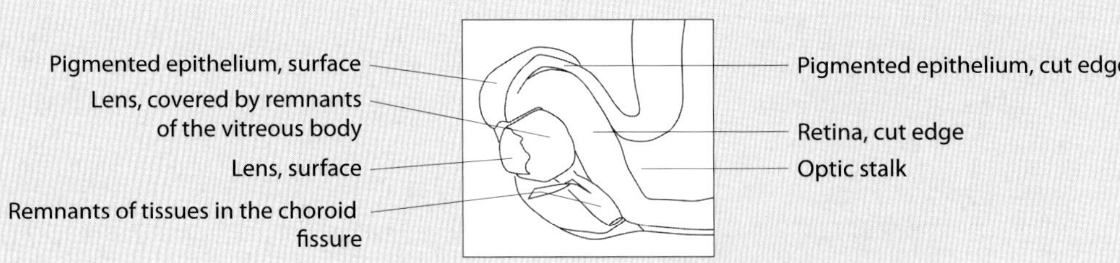

6.2.25 / 26

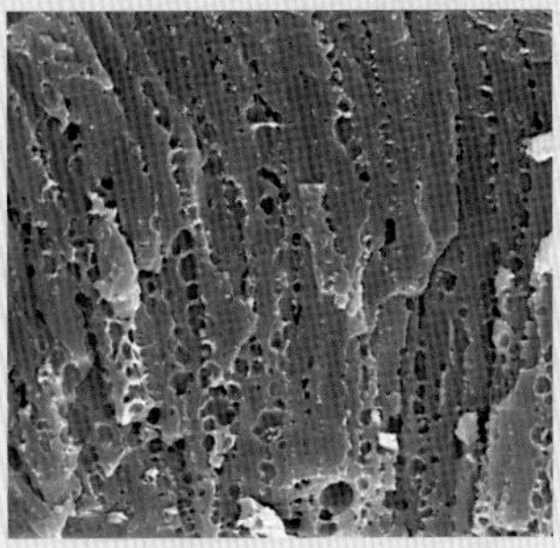

6.2.25
Longitudinal fracture of the primary lens fibres. Embryo at week 6.

6.2.26
Longitudinal fracture of the primary lens fibres. Embryo at week 8.

6.2.27 / 28

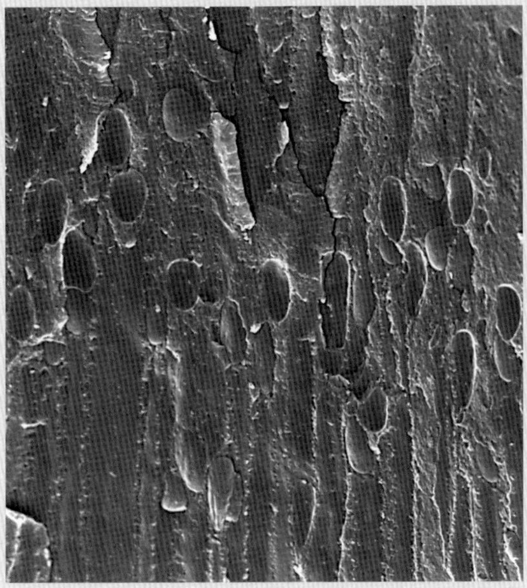

6.2.27
Longitudinal fracture of the primary lens fibres. Embryo at week 9.

6.2.28
The longitudinal fracture of the primary lens fibres at the level of the equator shows the row of nuclei of the lens fibres. Embryo at week 9.

6.2.29

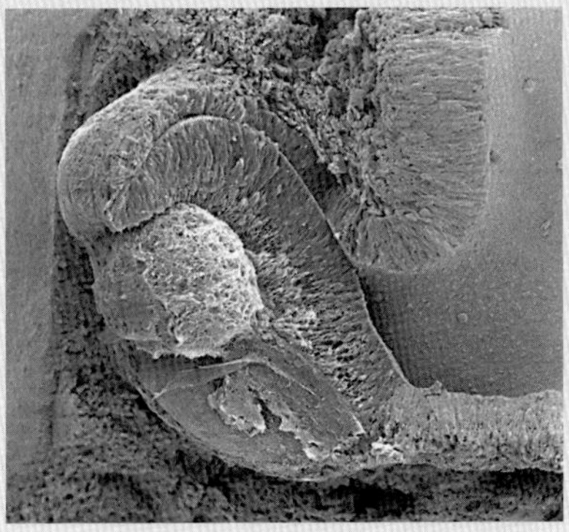

Sagittal fracture of the retina and the pigmented epithelium shows the position of the lens. Lateral-left view. Embryo at week 6.

Week 6

6.2.01–6.2.33
Developmental stages of the eye.

6.2.15–6.2.30
Development of the lens.

Week 6 / 7

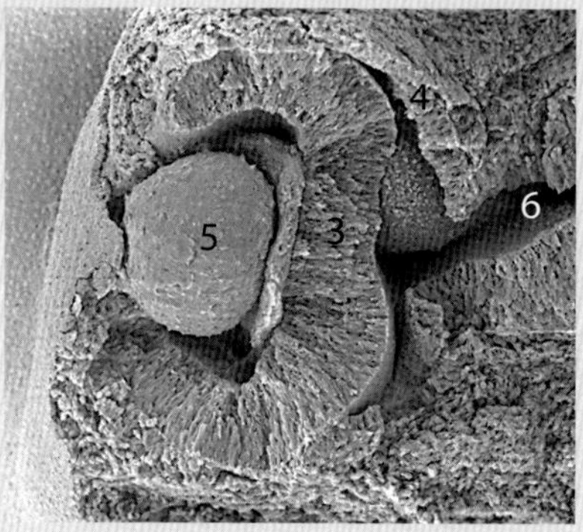

Week 8

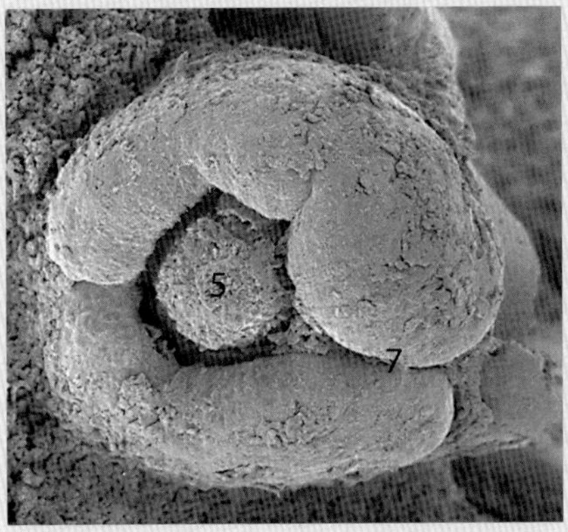

6.2.30 Frontal view of the eye cup and the lens. Embryo at the end of week 6.

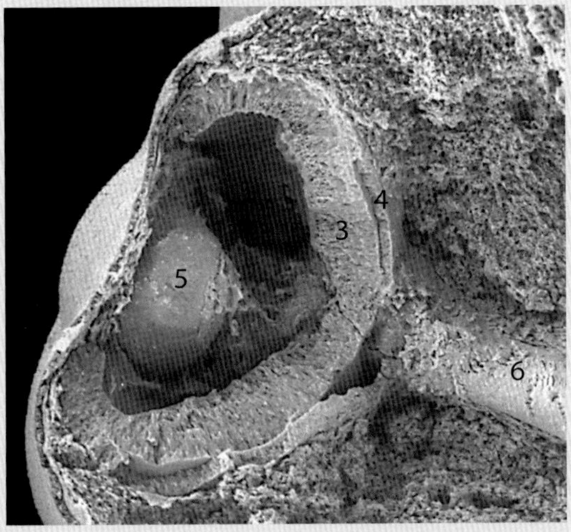

6.2.31 / 32 Sagittal fractures of the eye. Embryos at weeks 6 and 7.

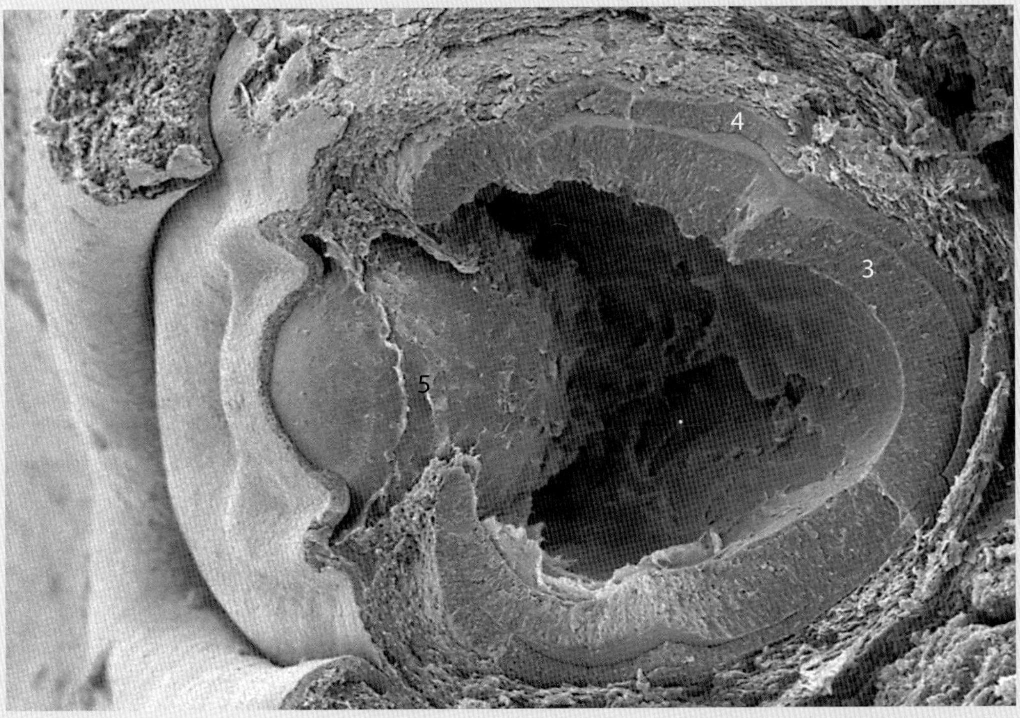

6.2.33 Sagittal fractures of the eye. Embryo at week 8.

6.3
The Middle and the Internal Ear

The internal ear consists of the acoustic organ of hearing, the cochlea, and the organs of balance, the utricle, the saccule and semicircular ducts.

The anlage of the internal ear is the otic pit (fig. 6.3.01, 6.3.02). After its closure, the otic vesicle (fig. 6.3.03) becomes piriform (fig. 6.3.04). The upper (dorsal) portion forms the endolymphatic appendage (fig. 6.3.05) and the lower (ventral) portion forms the utricle, the saccule, the semicircular canals (fig. 6.3.07–6.3.17), and the cochlear duct (fig. 6.3.12–6.3.17). The endolymphatic appendage develops into the endolymphatic duct (fig. 6.3.08–6.3.11).

The semicircular ducts open into the utricle which is connected with the cochlear duct (fig. 6.3.12–6.3.17). In the semicircular ducts, near their opening into the utricle, the ampullae (fig. 6.3.17), in which the thickened epithelium forms the cupula (fig. 6.3.18, 6.3.19), the peripheral ending of the vestibular nerve, arise as swellings.

The middle ear transfers the sound waves from the external ear to the cochlea. It consists of the external acoustic duct, the tympanic cavity, and the auditory tube.

The external acoustic duct develops from the groove of the first pharyngeal arch. The tympanic cavity (fig. 6.3.20–6.3.22) is an outgrowth of the endodermal first pharyngeal pouch and remains connected to the pharynx by the auditory tube (fig. 6.3.23, 6.3.24). The ossicles of the tympanic cavity (fig. 6.3.21, 6.3.22) arise within the mesenchyme of the first (and possibly the second) pharyngeal arch as cartilaginous condensations.

6.3

Abbreviations

1	upper jaw
2	auditory tube
3	pharynx
4	palatal process

6.3.01–6.3.24
Developmental stages of the inner and the middle ear.

6.3.01–6.3.19
The inner ear.

25:1

Week 4

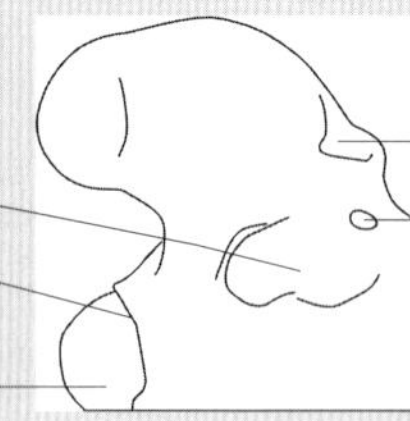

150:1

25:1

Week 5

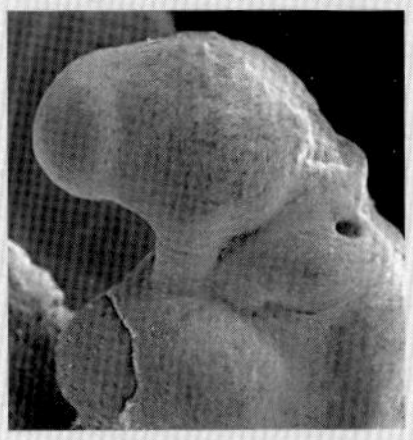

6.3.01 Lateral and dorsal views of the otic pits. Embryo at week 4.

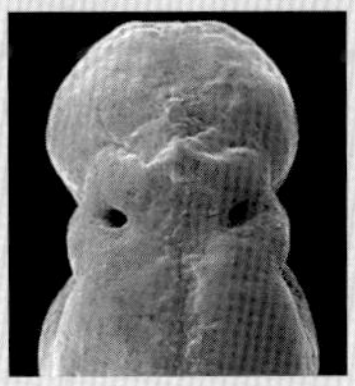

6.3.02

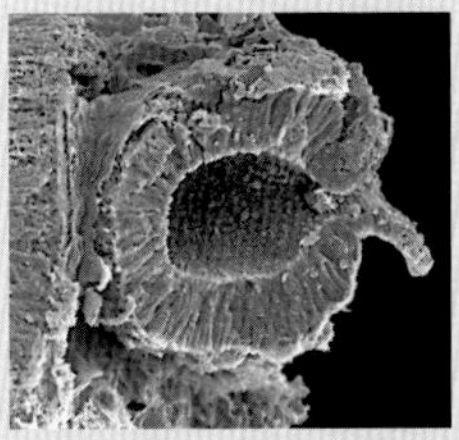

6.3.03 Cross-fraction of a more advanced otic pit closing to form the otic vesicle. Embryo at week 4.

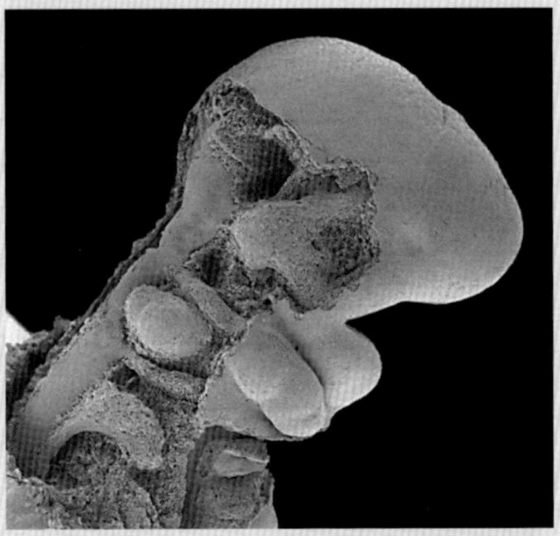

6.3.04 Position of the otic vesicle with an indication of the endolymphatic diverticulum. Lateral-right view. Embryo at week 5.

6.3.05 Lateral-right view of the opened otic vesicle and the endolymphatic diverticulum. Embryo at week 5.

6.
Brain and
Sensory Organs

6.3
The Middle and
the Internal Ear

6.3.01–6.3.24
Developmental stages of the inner and the middle ear.

6.3.01–6.3.19
The inner ear.

6.3.07–6.3.19
Different views of the exposed vestibular apparatus.

25:1

Week 5

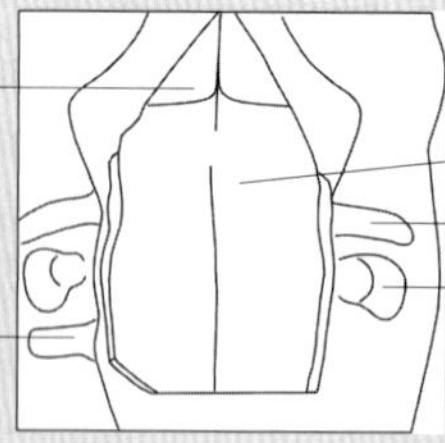

30:1

Week 8

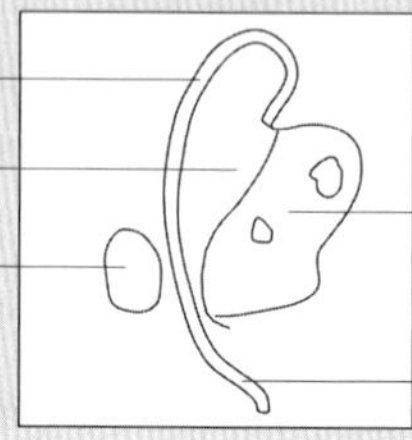

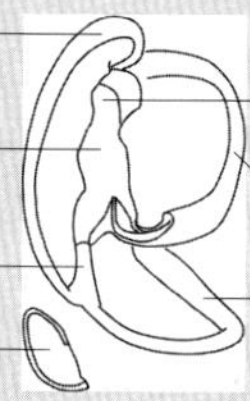

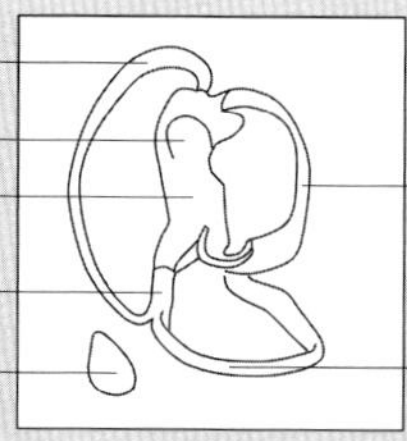

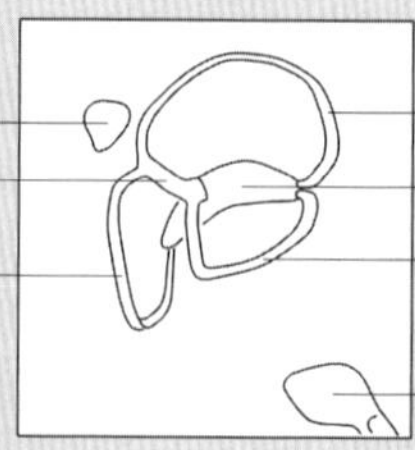

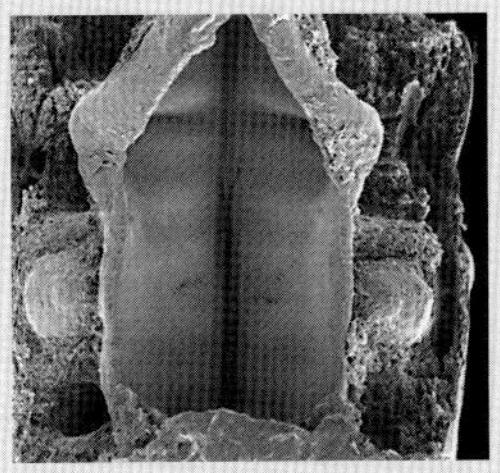

6.3.06 Cranial view of the floor of the rhombencephalon and the otic vesicles. Embryo at week 5.

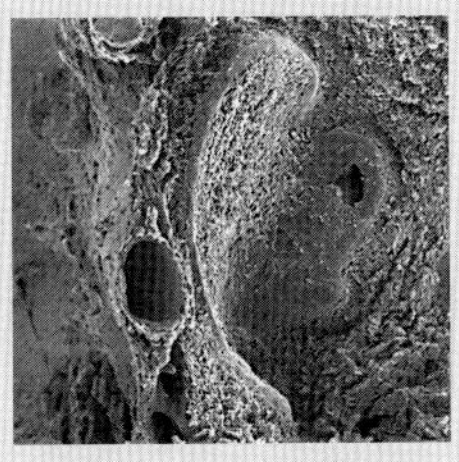

6.3.07 Cranial view of the right vestibular apparatus. Embryo at week 8.

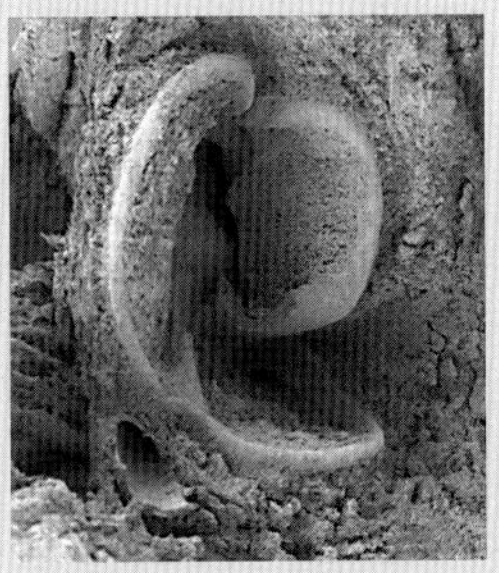

6.3.08 Cranial view of the right vestibular apparatus of a more advanced stage. Embryo at week 8.

6.3.09 Cranial-lateral view of the right vestibular apparatus. Embryo at week 8.

6.3.10 Lateral view of the right vestibular apparatus. Embryo at week 8.

6.3.01–6.3.24
Developmental stages of the inner and the middle ear.

6.3.01–6.3.19
The inner ear.

6.3.07–6.3.19
Different views of the exposed vestibular apparatus.

Week 8

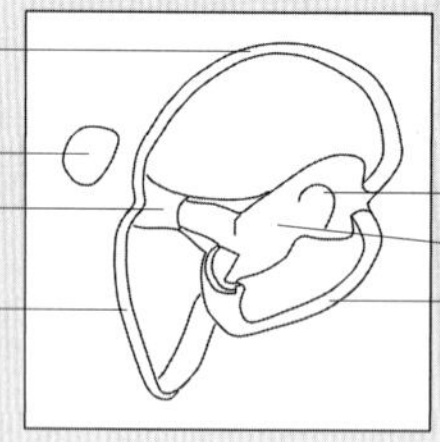

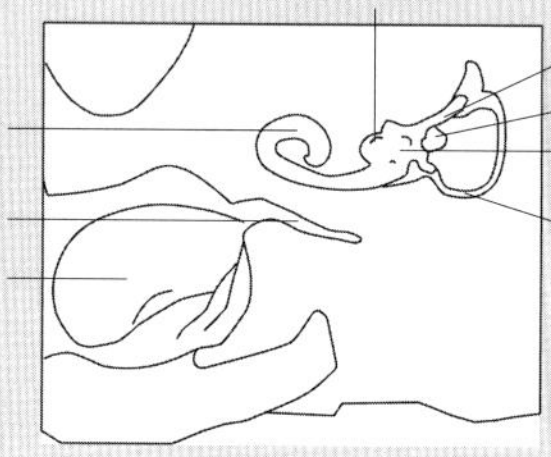

Week 9

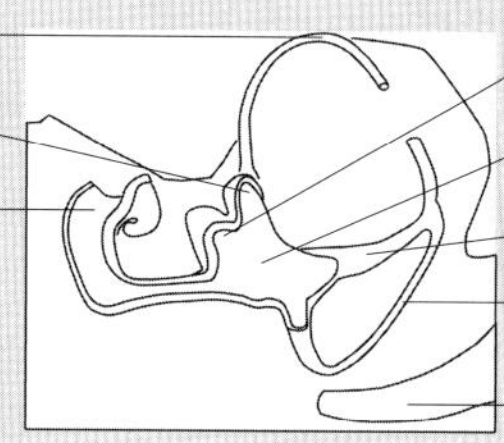

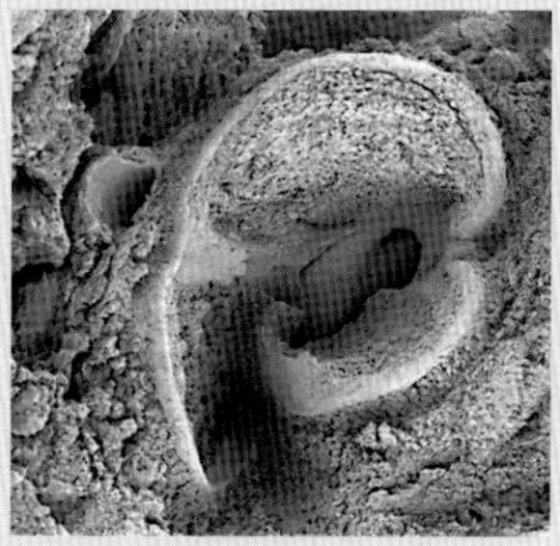

6.3.11 Lateral view. Same embryo as in figure 6.3.10. The utricle has been opened.

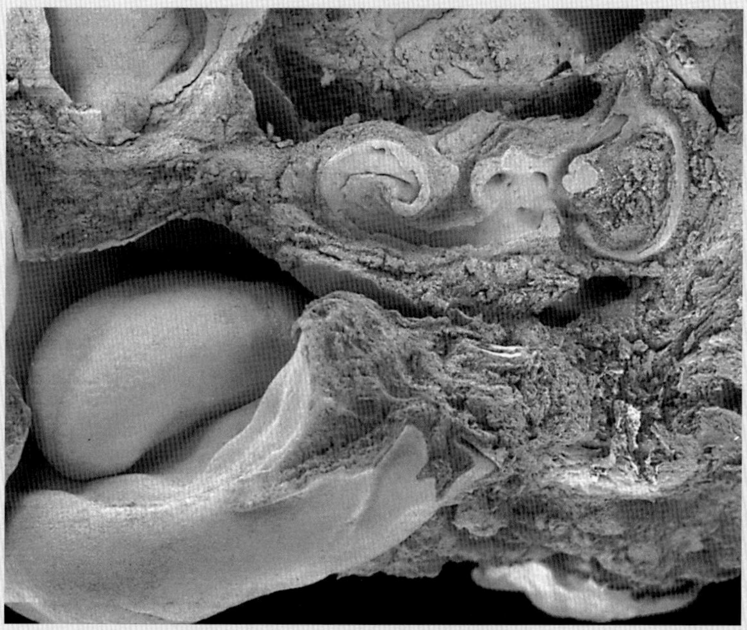

6.3.12 Ventral-left view of the position of the exposed and partly opened left vestibular apparatus and the cochlear duct. Embryo at week 8.

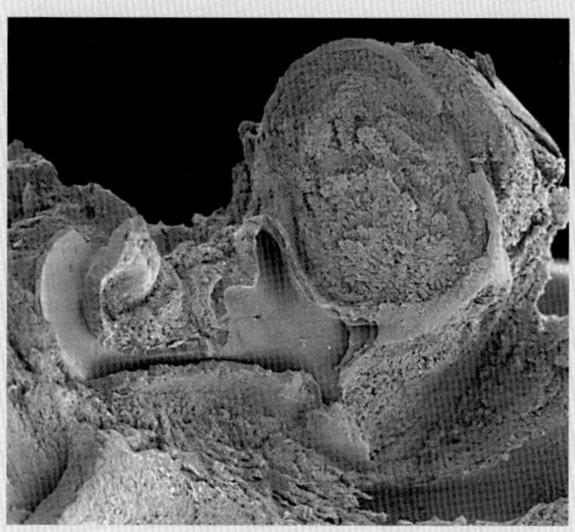

6.3.13 Lateral-left view of the right vestibular apparatus and the cochlear duct. The vestibule, the saccule, and the cochlear duct have been opened. Embryo at week 9.

6.3.01–6.3.24
Developmental stages of the inner and the middle ear.

6.3.01–6.3.19
The inner ear.

6.3.07–6.3.19
Different views of the exposed vestibular apparatus.

Week 9

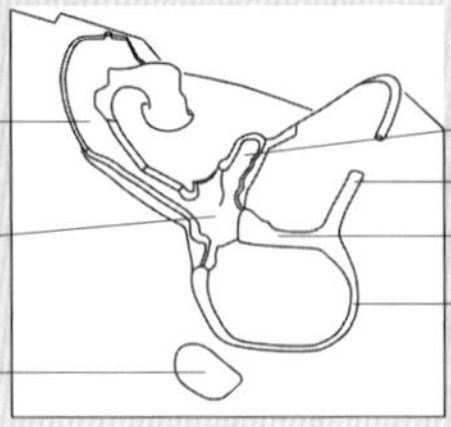

Week 8

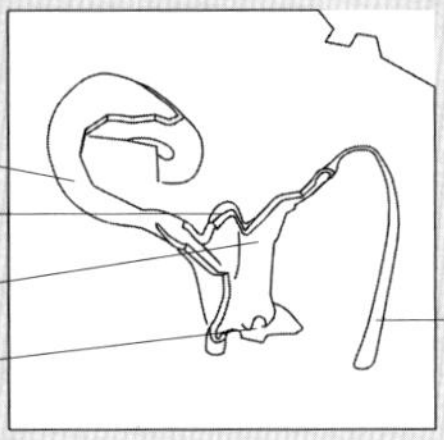

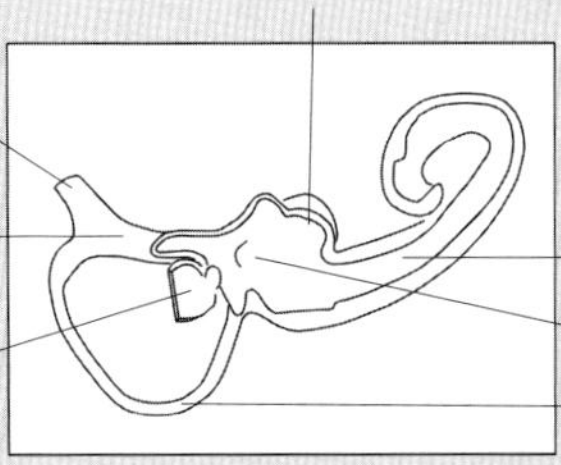

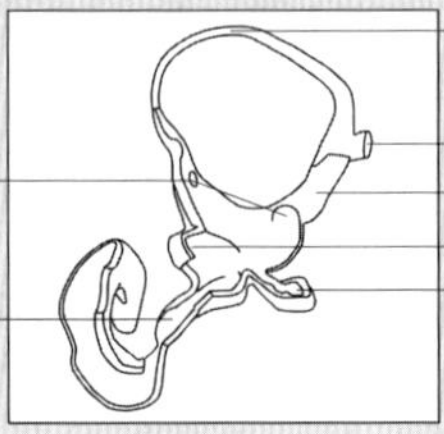

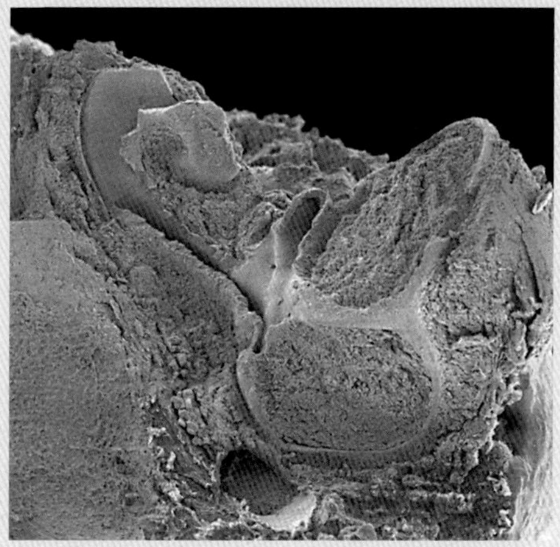

6.3.14 Same embryo as in figure 6.3.13. Dorsal-cranial view of the right vestibular apparatus and the cochlear duct shows the position of the posterior semicircular duct.

6.3.15 Dorsal-left view of the right vestibular apparatus and the cochlear duct. The saccule and the cochlear duct have been opened. Embryo at week 8.

6.3.16 Lateral-left view of the right vestibular apparatus and the cochlear duct. The utricle and the cochlear duct have been opened. Embryo at week 8.

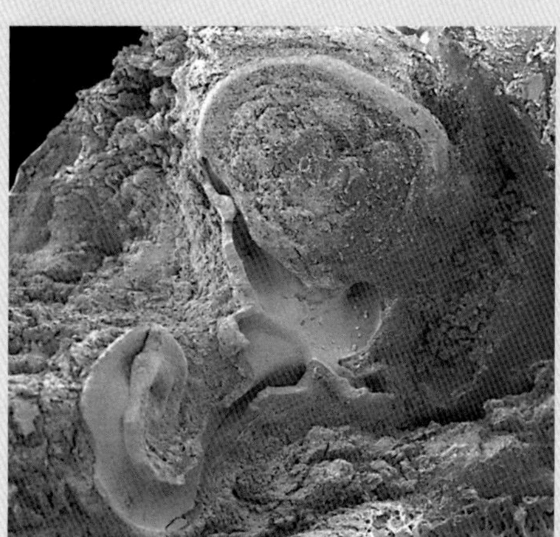

6.3.17 Ventral-left view of the right vestibular apparatus and the cochlear duct. The saccule, the utricle, and the ampullae of the anterior semicircular duct have been opened. The posterior semicircular duct has been partly removed. Embryo at week 8.

6.3.01–6.3.24
Developmental stages of the inner and the middle ear.

6.3.01–6.3.19
The inner ear.

6.3.07–6.3.19
Different views of the exposed vestibular apparatus.

6.3.20–6.3.24
The middle ear.

250:1

Week 8

25:1

Week 6

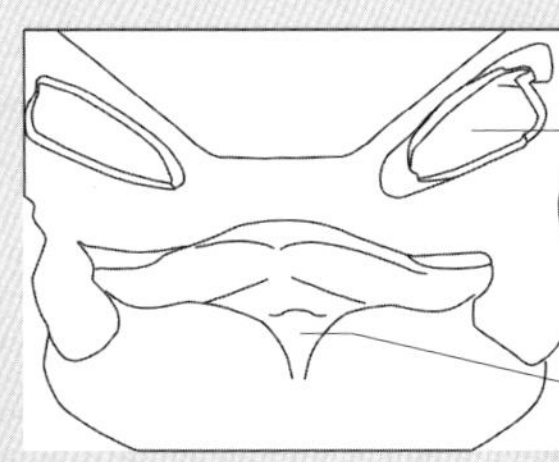

50:1

Week 8

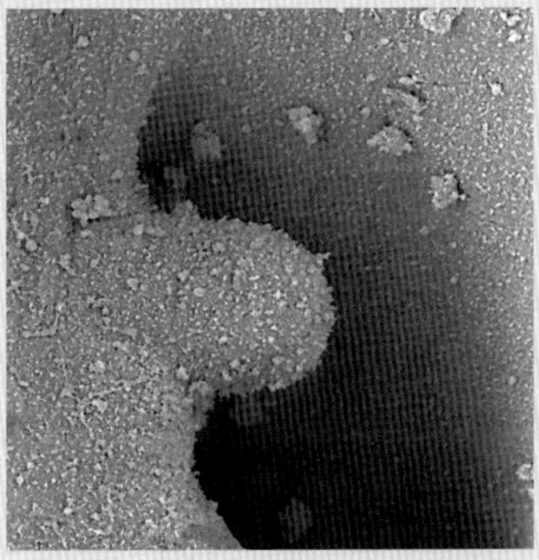

6.3.18 Detailed view of an ampullary crest.

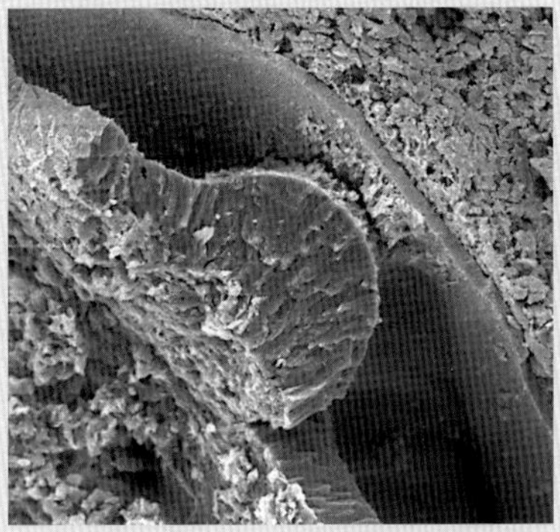

6.3.19 Cross-fracture of the anterior ampullary crest shown in figure 6.3.17.

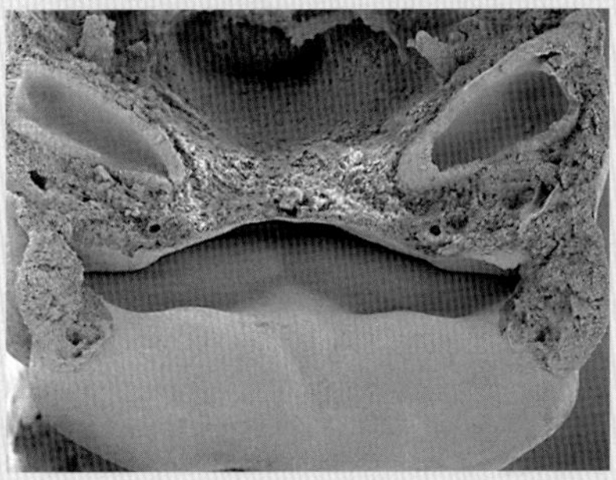

6.3.20 Ventral view of the exposed and opened tympanic cavities. Embryo at week 6.

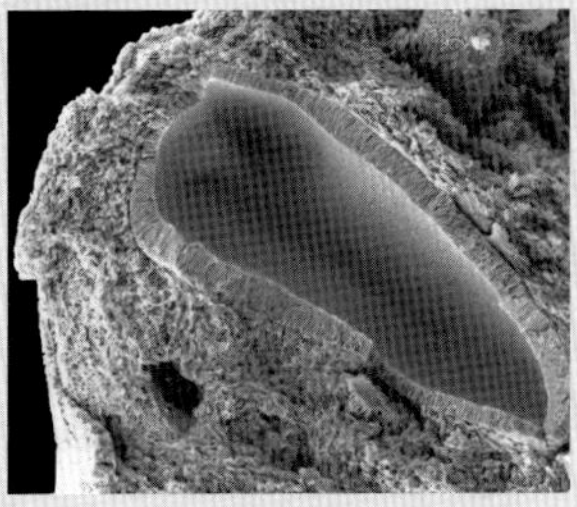

6.3.21 Ventral view of the opened right tympanic cavity. Same embryo as in figure 6.3.07.

6.3.01–6.3.24
Developmental stages of the inner and the middle ear.

6.3.20–6.3.24
The middle ear.

50:1

Week 7

25:1

Week 8

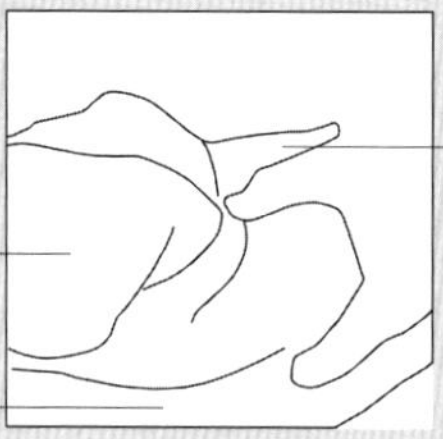

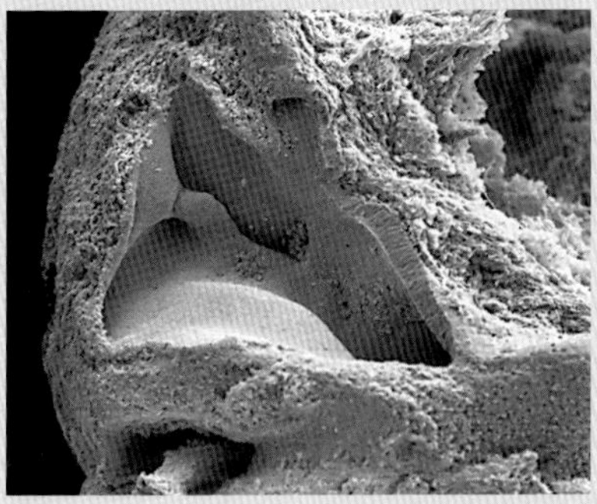

6.3.22 Ventral view of the opened right tympanic cavity with prominence of the malleus. Embryo at week 7.

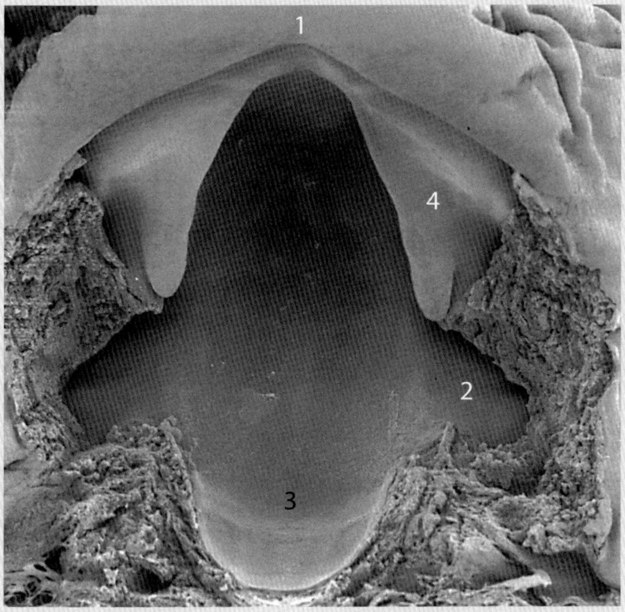

6.3.23 Ventral view of the pharynx showing the origin of the pharyngo-tympanic (auditory) tube. Embryo at week 8.

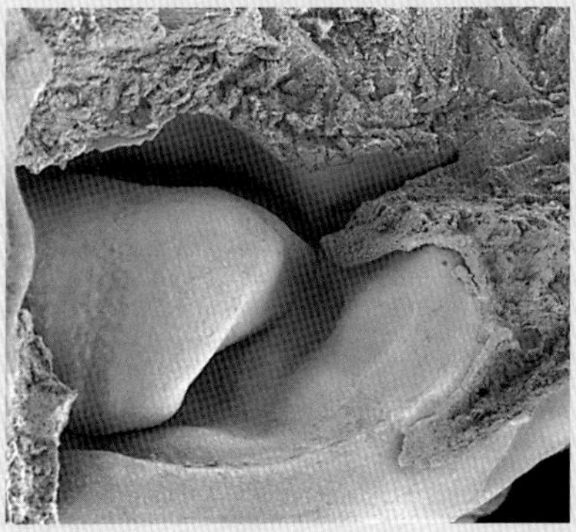

6.3.24 Ventral-left view of the position of the pharyngo-tympanic (auditory) tube. Embryo at week 8.

6.4

The Spinal Cord

The structural development of the spinal cord can best be seen in histological sections and is therefore omitted here.

The closure of the caudal neuropore proceeds from cranial to caudal and is completed by the end of week 4 (fig. 6.4.01–6.4.03).

Some figures have been added to provide an idea of the growth of the spinal cord and the dorsal body wall (fig. 6.4.04, 6.4.05) and to show some details of the spinal ganglia and nerves (fig. 6.4.06, 6.4.07).

6.4.01–6.4.10
The spinal cord.

50:1

Week 4

75:1

Week 5

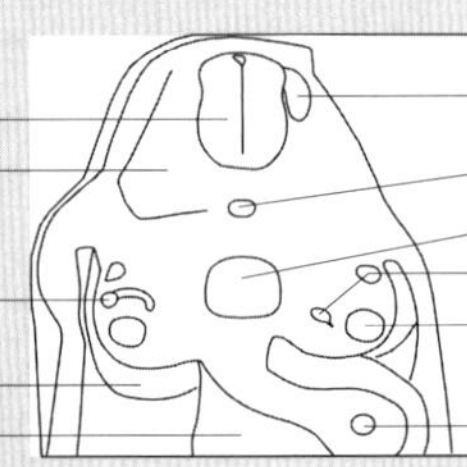

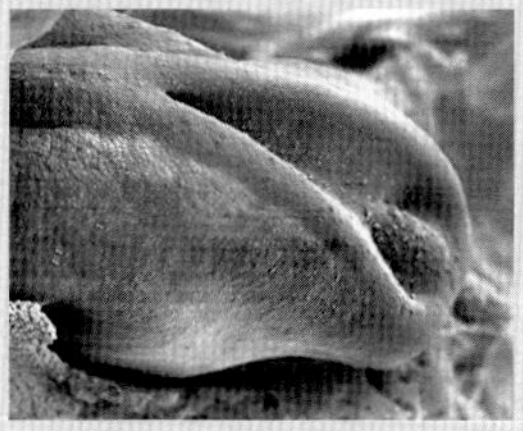

6.4.01

The closure of the caudal neuropore. Lateral and lateral-caudal views of the caudal region of embryos at week 4.

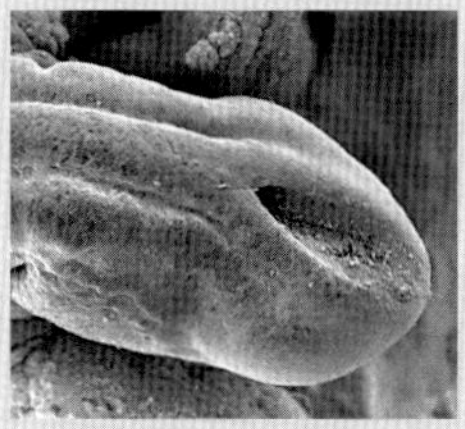

6.4.02

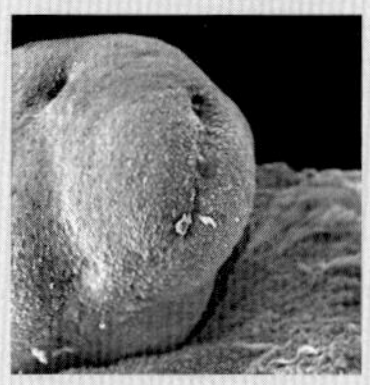

6.4.03

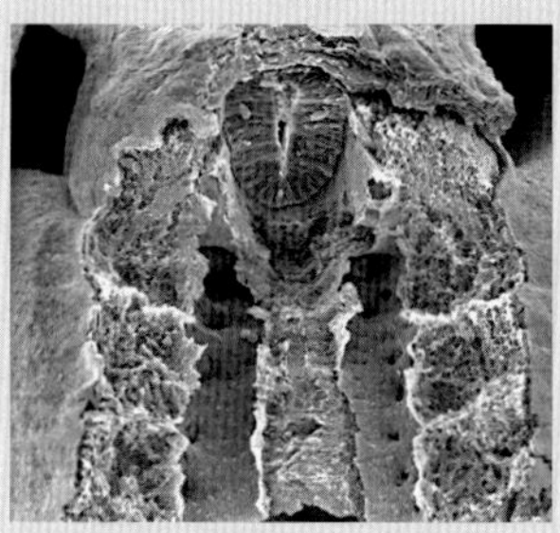

6.4.04

Cross-fractures of embryos at the lumbar level. Embryos at week 4 and 5.

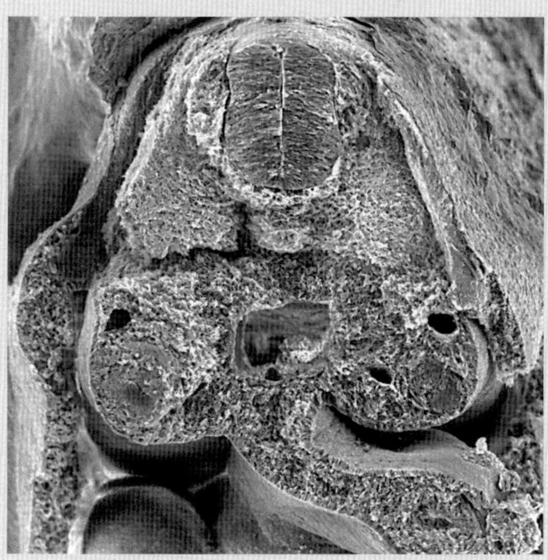

6.4.05

20:1

Week 7

40:1

6.4.01–6.4.10
The spinal cord.

6.4.06

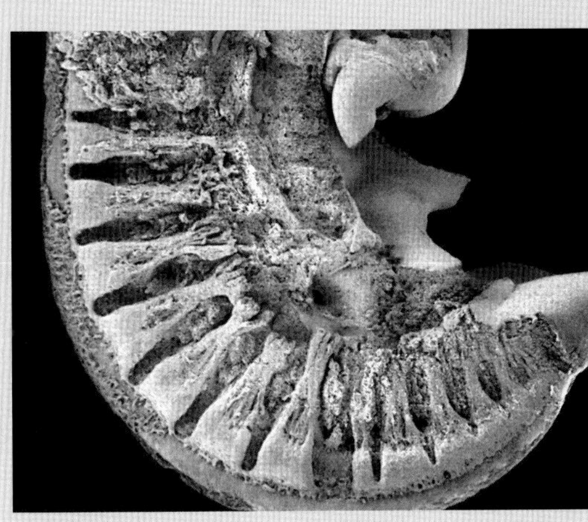

6.4.07

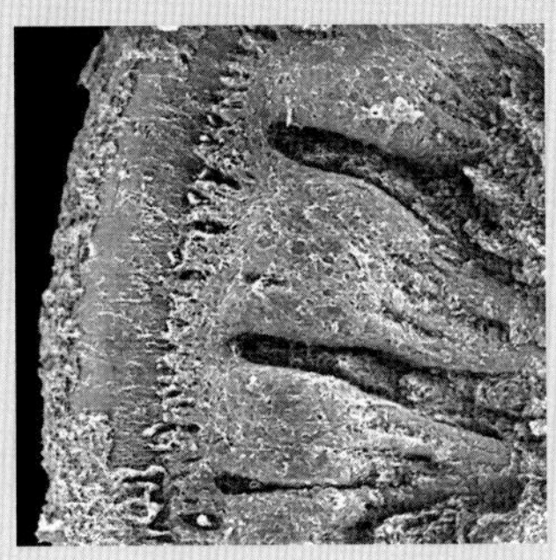

Lateral right views of the exposed spinal ganglia. Embryos at week 7.

Index